Chu-Song Chen · Jiwen Lu
Kai-Kuang Ma (Eds.)

Computer Vision – ACCV 2016 Workshops

ACCV 2016 International Workshops
Taipei, Taiwan, November 20–24, 2016
Revised Selected Papers, Part II

 Springer

Editors
Chu-Song Chen
Institute of Information Science
Academia Sinica
Taipei
Taiwan

Jiwen Lu
Tsinghua University
Beijing
China

Kai-Kuang Ma
School of Electrical and Electronic
  Engineering
Nanyang Technological University
Singapore
Singapore

ISSN 0302-9743   ISSN 1611-3349 (electronic)
Lecture Notes in Computer Science
ISBN 978-3-319-54426-7   ISBN 978-3-319-54427-4 (eBook)
DOI 10.1007/978-3-319-54427-4

Library of Congress Control Number: 2017932642

LNCS Sublibrary: SL6 – Image Processing, Computer Vision, Pattern Recognition, and Graphics

Printed on acid-free paper

This Springer imprint is published by Springer Nature
The registered company is Springer International Publishing AG
The registered company address is: Gewerbestrasse 11, 6330 Cham, Switzerland

Preface

It is our great pleasure to present the workshop proceedings of three LNCS volumes, which contain the papers carefully reviewed and selected from the 17 workshops that were held in conjunction with the 13th Asian Conference on Computer Vision (ACCV), during November 20–24, 2016, in Taipei, Taiwan. There are 134 papers selected from 223 papers submitted to all the 17 workshops as listed below.

1. New Trends in Image Restoration and Enhancement (NTIRE): 14 papers
2. Workshop on Assistive Vision: 6 papers
3. ACCV 2016 Workshop on Hyperspectral Image and Signal Processing: 6 papers
4. Computer Vision Technologies for Smart Vehicle: 7 papers
5. Spontaneous Facial Behavior Analysis: 8 papers
6. 3D Modelling and Applications: 16 papers
7. 4th ACCV Workshop on e-Heritage: 4 papers
8. Multiview Lip-Reading Challenges: 5 papers
9. Workshop on Facial Informatics (WFI): 11 papers
10. Discrete Geometry and Mathematical Morphology for Computer Vision: 4 papers
11. Workshop on Mathematical and Computational Methods in Biomedical Imaging and Image Analysis: 15 papers
12. International Workshop on Driver Drowsiness Detection from Video: 6 papers
13. Workshop on Meeting HCI with CV: 6 papers
14. Workshop on Human Identification for Surveillance (HIS) Methods and Applications: 8 papers
15. Benchmark and Evaluation of Surveillance Task (BEST): 9 papers
16. The Third Workshop on Computer Vision for Affective Computing (CV4AC): 3 papers
17. Workshop on Interpretation and Visualization of Deep Neural Nets: 6 papers

The workshop topics are related to computer vision and its applications, interdisciplinary themes with other application areas, as well as challenges or competitions. Every workshop handles its own paper submission system, and each paper is reviewed by two to three reviewers. We thank all the workshop organizers for their great efforts in holding these successful workshops. We also thank the help of the publication chairs in making this publication possible.

November 2016

Chu-Song Chen
Jiwen Lu
Kai-Kuang Ma

Organization

W01: 3D Modelling and Applications

Chia-Yen Chen	National University of Kaohsiung, Taiwan
Min-Chun Hu	National Cheng Kung University, Taiwan
Li-Wei Kang	National Yunlin University of Science and Technology, Taiwan
Chih-Yang Lin	Asia University, Taiwan
Tang-Kai Yin	National University of Kaohsiung, Taiwan
Guo Shiang Lin	Da Yeh University, Taiwan
Chia-Hung Yeh	National Sun Yat-Sen University, Taiwan

W02: 4th ACCV Workshop on e-Heritage

Katsushi Ikeuchi	Microsoft Research Asia, China
El Mustapha Mouaddib	Université de Picardie Jules Verne, France
Takeshi Masuda	AIST, Japan
Takeshi Oishi	The University of Tokyo, Japan

W03: ACCV 2016 Workshop on Hyperspectral Image and Signal Processing

Keng-Hao Liu	National Sun Yat-sen University, Taiwan
Wei-Min Liu	National Chung Cheng University, Taiwan

W04: Benchmark and Evaluation of Surveillance Task (BEST)

Xiaokang Yang	Shanghai Jiao Tong University, China
Chong-Yang Zhang	Shanghai Jiao Tong University, China
Bingbing Ni	Shanghai Jiao Tong University, China
Lin Mei	The Third Research Institute of the Ministry of Public Security, China

W05: Computer Vision Technologies for Smart Vehicle

Li-Chen Fu	National Taiwan University, Taiwan
Pei-Yung Hsiao	National University of Kaohsiung, Taiwan
Shih-Shinh Huang	National Kaohsiung First University of Science and Technology, Taiwan

W06: Discrete Geometry and Mathematical Morphology for Computer Vision

Jean Cousty	Université Paris-Est, ESIEE Paris, France
Yukiko Kenmochi	Université Paris-Est, CNRS, France
Akihiro Sugimoto	National Institute of Informatics, Japan

W07: International Workshop on Driver Drowsiness Detection from Video

Chen-Kuo Chiang	National Chung Cheng University, Taiwan
Shang-Hong Lai	National Tsing Hua University, Taiwan
Michel Sarkis	Qualcomm Technologies Inc., USA

W08: Large-Scale 3D Human Activity Analysis Challenge in Depth Videos

Gang Wang	Nanyang Technological University, Singapore
Amir Shahroudy	Nanyang Technological University, Singapore
Jun Liu	Nanyang Technological University, Singapore

W09: Multiview Lip-Reading Challenges

Ziheng Zhou	University of Oulu, Finland
Guoying Zhao	University of Oulu, Finland
Takeshi Saitoh	Kyushu Institute of Technology, Japan
Richard Bowden	University of Surrey, UK

W10: New Trends in Image Restoration and Enhancement (NTIRE)

Radu Timofte	ETH Zurich, Switzerland
Luc Van Gool	ETH Zurich, Switzerland
Ming-Hsuan Yang	University of California at Merced, USA

W11: Spontaneous Facial Behavior Analysis

Xiaopeng Hong	University of Oulu, Finland
Guoying Zhao	University of Oulu, Finland
Stefanos Zafeiriou	Imperial College London, UK
Matti Pietikäinen	University of Oulu, Finland
Maja Pantic	Imperial College London, UK

W12: The Third Workshop on Computer Vision for Affective Computing (CV4AC)

Abhinav Dhall	Abhinav Dhall, University of Waterloo, Canada
Roland Goecke	University of Canberra/Australian National University, Australia
O.V. Ramana Murthy	Amrita University, India
Jesse Hoey	University of Waterloo, Canada
Nicu Sebe	University of Trento, Italy

W13: Workshop on Assistive Vision

Chetan Arora	Indraprastha Institute of Information Technology, Delhi, India
Vineeth N. Balasubmanian	Indian Institute of Technology, Hyderabad, India
C.V. Jawahar	International Institute of Information Technology, Hyderabad, India
Vinay P. Namboodiri	Indian Institute of Technology, Kanpur, India
Ramanathan Subramanian	International Institute of Information Technology, Hyderabad, India

W14: Workshop on Facial Informatics (WFI)

Gee-Sern (Jison) Hsu	National Taiwan University of Science and Technology, Taiwan
Moi Hoon Yap	Manchester Metropolitan University, UK
Xiaogang Wang	Chinese University of Hong Kong, Hong Kong, SAR China
Su-Jing Wang	Chinese Academy of Science, China
John See	Multimedia University, Malaysia

W15: Workshop on Meeting HCI with CV

Liwei Chan	National Chiao Tung University, Taiwan and Keio Media Design, Japan
Yi-Ping Hung	National Taiwan University, Taiwan

W16: Workshop on Human Identification for Surveillance (HIS): Methods and Applications

Wei-Shi Zheng	Sun Yat-sen University, China
Ruiping Wang	Institute of Computing Technology, Chinese Academy of Sciences, China

Weihong Deng   Beijing University of Posts and Telecommunications,
                                China
Shenghua Gao   ShanghaiTech University, China

W17: Workshop on Interpretation and Visualization of Deep Neural Nets

Alexander Binder   Singapore University of Technology and Design,
                                Singapore
Wojciech Samek   Fraunhofer Heinrich Hertz Institute, Germany

W18: Workshop on Mathematical and Computational Methods in Biomedical Imaging and Image Analysis

Atsushi Imiya   Chiba University, Japan
Xiaoyi Jiang   Universität Münster, Germany
Hidetaka Hontani   Nagoya Institute of Technology, Japan

Contents – Part II

4th ACCV Workshop on e-Heritage

Multi-view Lip-Reading Challenges

Workshop on Facial Informatics (WFI)

Discrete Geometry and Mathematical Morphology for Computer Vision

Workshop on Mathematical and Computational Methods in Biomedical Imaging and Image Analysis

3D Modelling and Applications

3D Shape Reconstruction in Traffic Scenarios Using Monocular Camera and Lidar

Qing Rao[1](✉), Lars Krüger[1], and Klaus Dietmayer[2]

[1] Daimler AG, Ulm, Germany
{qing.rao,lars.krueger}@daimler.com
[2] Ulm University, Ulm, Germany
klaus.dietmayer@uni-ulm.de

Abstract. In the near future, a self-driving car will be able to perceive and understand its surroundings by composing a 3D environment map at object level. In this map, the 3D shapes of surrounding objects will be precisely reconstructed. The technique to reconstructing 3D object shapes using a monocular camera and a Lidar is presented in this paper. The proposed approach combines deep neural networks with an optimization process called *3D Shaping* in which object pose and shape are jointly optimized. A significant performance improvement by the proposed approach in estimating object 3D orientation and the occupancy bounding box is proven through quantitative evaluation.

1 Introduction

Fast-developing sensor technologies and algorithms have enabled a number of successful autonomous driving pilot projects [1–3]. In the future, a self-driving car shall be able to interpret its entire 3D surroundings at object level. In other words, intelligent cars shall not only be able to reconstruct their surroundings in 3D but, more importantly, *understand* the semantic meaning of the reconstructed scene. This still remains a challenging task for researchers and developers in the automotive industry, mainly due to the following reasons. (1) Traffic scenarios are extremely cluttered and unpredictable, especially in urban areas. (2) The field-of-view and the sensing range of the current generation of automotive sensors are limited. Several important objects in a traffic scenario, such as traffic lights or vulnerable road users, might be out of sight of the sensors. (3) The state-of-the-art classification and reconstruction algorithms are not capable of processing all different traffic scenarios robustly and confidently. Despite of the current success of autonomous prototype vehicles, there is still a long way to go before autonomous driving becomes a regular part of daily life.

The concept of object-level interpretation of the surroundings dates back to [4,5]. The authors of [4,5] attempted to generate indoor semantic 3D object

Electronic supplementary material The online version of this chapter (doi:10. 1007/978-3-319-54427-4_1) contains supplementary material, which is available to authorized users.

C.-S. Chen et al. (Eds.): ACCV 2016 Workshops, Part II, LNCS 10117, pp. 3–18, 2017.
DOI: 10.1007/978-3-319-54427-4_1

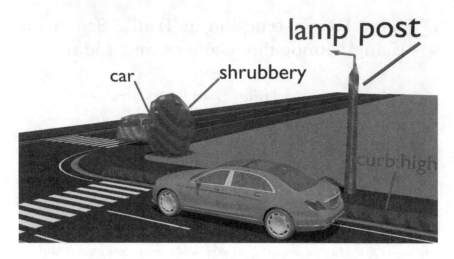

Fig. 1. Example of object-level interpretation of a car's surroundings.

maps for their autonomous home robots from point cloud data. Compared to traditional 3D surface reconstruction, a semantic object map has following advantages: First of all, it is an abstract and compact representation of the environment, which reduces requirements on memory and storage space. Second, the data structure of an object map is preferable for further processing, such as tracking and predicting the motions of other objects. Furthermore, building a 3D object map resembles the way humans perceive their environment. This encourages developers to adapt 3D object mapping to intelligent self-driving cars. Although existing works [6–8] also aimed at reconstructing traffic scenarios using Lidar, monocular/stereo cameras, or combinations thereof, none of them focused intently on object-level interpretation. The authors of this presented work now initiate the discussion of 3D shape reconstruction in traffic scenarios as a first step towards generating a semantic 3D object map for autonomous driving (Fig. 1).

In this paper, a novel approach to reconstructing object 3D shapes in traffic scenarios is presented. The proposed approach fuses 3D point measurements from a Lidar and 2D image cues from a monocular camera. Deep neural networks are used for pixel-level image segmentation and for the initialization of object pose estimation. The authors take advantage of a latent shape space to recover various 3D shapes among the same object class. The extremely low-dimensional representation of 3D shapes using latent variables enables fast convergence of an optimization process in which the 3D shape and the 3D pose of an object are jointly recovered. This optimization process is referred to as *3D Shaping* in the presented work. Quantitative evaluation shows improvement by the proposed approach in both orientation estimation and occupancy bounding box estimation.

The remainder of this paper is structured as follows: Sect. 2 provides a short review of related work. Section 3 briefly explains the energy function used by 3D

Shaping. Sections 4 and 5 explain the training and the recall workflow, respectively. Quantitative evaluations and rendered results of 3D Shaping are presented in Sect. 6. Section 7 concludes this paper with a short summary.

2 Related Work

2.1 Deep Learning

The success in recent years of *deep learning* has been observed by the computer vision community. Algorithms and applications using deep *Convolutional Neural Network* (CNN) continue to set records in object detection and classification [9–11], image segmentation [12–14], viewpoint estimation [15–17], etc. Recent development even extended CNN to solve "nontraditional" deep learning problems such as monocular depth estimation [18,19] and point cloud classification [20]. A more comprehensive survey of deep learning was given in [21].

Specifically for the task of semantic segmentation, Long et al. [13] introduced a novel network architecture named *Fully Convolutional Network* (FCN), which enabled pixel-level segmentation in an image. Since then, a number of improvements [14,22,23] were proposed to resolve the issue of non-sharp boundaries caused by max-pooling layers in FCN. Among them, the method of Lin et al. [23] combining conditional random field with FCN holds the current record for the Cityscapes dataset [24], a dataset for semantic understanding of urban street scenes.

In the presented work, the power of deep learning is exploited to perform semantic image segmentation and to initialize object pose estimation.

2.2 Latent Shape Space

An important issue in composing 3D object maps of surroundings is how to represent 3D shapes efficiently. One way is to encode them implicitly using the Signed Distance Function (SDF) [25,26], also called zero-level embedding function in other publications [27]. Prisacariu et al. [28] further transformed SDF to the frequency domain using *Discrete Cosine Transform* (DCT) in order to increase the "featureness", and they embedded DCT coefficients into an extremely low-dimensional space through *Gaussian Process Latent Variable Model* (GPLVM) [29]. The shape representation in [28] is highly efficient since the search space for the optimal shape is constrained inside the low-dimensional latent space. The latent shape space is adapted in the presented work to represent the 3D shape of the object class *Car*.

Similar works using 3D shape priors to reconstruct surroundings include [30, 31]. Dame et al. [30] used the latent shape of *Cars* to augment the Dense Tracking And Mapping (DTAM) system [32] with object specific identity. Güney and Geiger [31] associated 3D shape priors to resolve stereo matching ambiguities for the object class *Car* as well. Compared to [30,31], the presented work has three major differences. (1) This work focuses on *object-level* reconstruction in traffic scenarios instead of dense reconstruction. (2) Depth measurements directly from a Lidar are available in this work, whereas depth information in [30,31] was first estimated from cameras. (3) The presented work exploits the state-of-the-art deep neural network for image segmentation instead of using hand-crafted features [30] or super-pixel based approaches [31].

The apparently most similar work to the presented paper is [33]. The authors of [33] proposed an approach to 3D shape reconstruction based on a combination of neural networks and a latent shape space as well. However, they only used a monocular camera for reconstruction and only provided evaluations in viewpoint estimation on the PASCAL 3D dataset [34]. In this work, the performance of the proposed reconstruction approach is evaluated on a more challenging dataset – the KITTI vision benchmark suite [35] comprising real traffic scenarios.

3 Energy Function

The 3D Shaping optimization process is performed on each object instance. The energy function in Eq. 1 is minimized during the optimization with respect to the 3D shape and the 3D pose of an object.

$$E(\Phi; \rho) = E_{img}(\Phi; \rho) + \lambda E_{cloud}(\Phi; \rho) \tag{1}$$

In Eq. 1, Φ denotes the SDF that encodes the 3D shape, and ρ denotes the pose parameters. The energy $E(\Phi; \rho)$ combines an image-based energy $E_{img}(\Phi; \rho)$ and a point-cloud-based energy $E_{cloud}(\Phi; \rho)$. The combination is controlled by a weight λ. E_{img} and E_{cloud} are expressed in Eqs. 2 and 3, respectively.

$$E_{img}(\Phi; \rho) = -\ln \sum_{\mathbf{x}^I \in \Omega} P_f(\mathbf{x}^I)\pi(\Phi; \rho) + P_b(\mathbf{x}^I)\left(1 - \pi(\Phi; \rho)\right) \tag{2}$$

$$E_{cloud}(\Phi; \rho) = \sum_{\mathbf{X}^L \in \mathcal{L}} \exp\left\{\frac{\Phi^2(g(\mathbf{X}^L; \rho))}{\Phi^2(g(\mathbf{X}^L; \rho)) + \sigma}\right\}$$
$$= \sum_{\mathbf{X}^O} \exp\left\{\frac{\Phi^2(\mathbf{X}^O)}{\Phi^2(\mathbf{X}^O) + \sigma}\right\} \tag{3}$$

In Eq. 2, $\mathbf{x}^I$ denotes an image pixel in a Region of Interest (ROI) Ω. The superscript I emphasizes the fact that $\mathbf{x}^I$ lies in the **I**mage frame, i.e. the image coordinate system. P_f and P_b denote the functions that estimate the probability of a pixel being foreground (object in focus) and background, respectively. The function π projects the SDF Φ onto the image plane using pose parameters ρ. E_{img} measures the difference between an image-based statistic estimation of foreground/background and a probabilistic derivation from the 3D shape prior. Figure 2 shows an example to help understand the image-based energy and how it is optimized during 3D Shaping.

In Eq. 3, $\mathbf{X}^L$ denotes a 3D point that belongs to a Lidar cluster $\mathcal{L}$. Similarly, the superscript L of $\mathbf{X}^L$ indicates the **L**idar frame. The Lie algebra $g \in SE(3)$

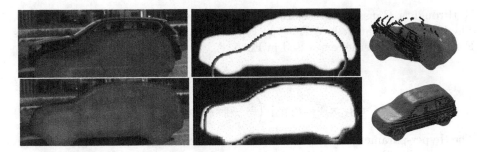

Fig. 2. Initial (first row) and final (second row) 3D Shaping optimization. Image statistics are shown in the middle column, with white pixels being more likely to be foreground pixels. The overlapping area between the projected 3D shape and the foreground is maximized during 3D Shaping. In the right column, 3D object and Lidar measurements are shown. The Lidar points would "stick" to the object surface after an ideally converged 3D Shaping optimization process. Best viewed in color.

transforms a Lidar point $\mathbf{X}^L$ to an object point $\mathbf{X}^O$ lying in the **O**bject frame. The object frame is attached at the geometrical centroid of an object. The exponential term of the Geman-McClure function in Eq. 3 represents the probability of an object point $\mathbf{X}^O$ being on the surface of the object. This probability reaches its minimum if $\mathbf{X}^O$ exactly lies on the zero-level of the SDF Φ, and it increases monotonically with the distance between $\mathbf{X}^O$ and the object surface. The increasing rate is controlled by σ.

In practice, the energy in Eq. 1 can be minimized iteratively through non-linear optimization methods such as gradient descent, Levenberg-Marquadt, etc. Its derivatives with respect to object pose and shape are calculated for each iteration. For more mathematical details of the energy function and its derivatives, please refer to [30].

4 Training

4.1 Latent Shape Space Training

The process of training a latent shape space is briefly explained in this subsection. Given a small number of public available CAD models of the object class *Car*, the objective of the training process is to find a GPLVM mapping between the feature vectors that describe the 3D shapes of the training samples and the so-called latent variables, which span a latent shape space. At first, the scale and the coordinate system of each CAD model in the training set are manually adjusted and normalized. Distance transform and DCT are then applied to each normalized model. These generate a set of feature vectors $\mathbf{Y} = [\mathbf{y}_1, \cdots, \mathbf{y}_n]^T \in \mathbb{R}^{n \times d}$, with n being the number of training samples and d the dimension of a feature vector. Typical values of n and d are $n = 100$ and $d = 8000$, respectively.

Let $\mathbf{X} = [\mathbf{x}_1, \cdots, \mathbf{x}_n]^T \in \mathbb{R}^{n \times q}$, $q \ll d$, denote the set of latent variables being sought. Theoretically, the feature matrix $\mathbf{Y}$ is generated by the latent matrix

X through a Gaussian Processes (GP) controlled by the hyper-parameter Θ. The covariance matrix $\mathbf{K} \in \mathbb{R}^{n \times n}$ in Eq. 4 is kernelized through a Radial Basis Function (RBF) kernel expressed in Eq. 5.

$$P(\mathbf{Y}|\mathbf{X}; \Theta) \sim \mathcal{GP}(\mathbf{0}, \mathbf{K}(\Theta)) \tag{4}$$

$$k_{ij} = \kappa(\mathbf{x}_i, \mathbf{x}_j) = \theta_1 \exp\left(-\frac{\theta_2}{2}\|\mathbf{x}_i - \mathbf{x}_j\|^2\right) + \theta_3 + \theta_4 \delta_{ij} \tag{5}$$

The hyper-parameter Θ and the latent matrix $\mathbf{X}$ are determined through maximum-likelihood estimation. In practice, the negative log-likelihood $L = -\ln P(\mathbf{Y}|\mathbf{X}; \Theta)$ is minimized through a scaled conjugate gradient method. Figure 3 visualizes a trained latent shape space with the latent dimension $q = 2$.

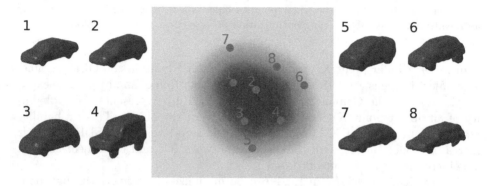

Fig. 3. Visualization of a 2-dimensional latent shape space. Latent variables in the dark area are more likely to generate a *Car*-like 3D shape.

Once a latent shape space is determined, the high-dimensional feature point $\mathbf{y}'$ of an arbitrary latent point $\mathbf{x}'$ can be probabilistically retrieved through Eq. 6.

$$\begin{aligned} E(\mathbf{y}') &= \kappa(\mathbf{x}', \mathbf{X})\mathbf{K}^{-1}\mathbf{Y} \\ Var(\mathbf{y}') &= \kappa(\mathbf{x}', \mathbf{x}') - \kappa(\mathbf{x}', \mathbf{X})\mathbf{K}^{-1}\kappa(\mathbf{x}', \mathbf{X}) \end{aligned} \tag{6}$$

For more details about GPLVM training, please refer to [29].

4.2   Neural Network Training

Neural networks are used for pixel level segmentation and orientation estimation. For the segmentation task, an FCN model trained on the Cityscape dataset [24] is used to distinguish *Car* from the background. The architecture of the FCN is similar to a GoogLeNet [36]. This FCN model is fine-tuned on the KITTI object dataset [35] for the task of orientation estimation.

An overview of the structure of the deep neural network used in this work is depicted in Fig. 4. Three ROI pooling layers are attached to `inception3b`,

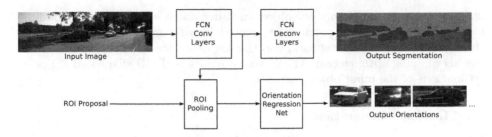

Fig. 4. Structure overview of the deep neural network. Groundtruth bounding boxes and orientations are used to fine-tune the orientation regression network.

`inception4e`, and `inception5b` of the FCN, in order to pool features out of the FCN at different scales. These features are then cascaded and fed into three fully-connected layers in the orientation regression network. The last fully-connected layer outputs an estimated orientation. The orientation is represented by biternion [15]. In other words, the orientation net outputs two values corresponding to $(\cos\hat{\theta}, \sin\hat{\theta})$, with $\hat{\theta}$ being the estimated orientation angle.

During the network training, only the three fully-connected layers of the orientation net are fine-tuned. Detailed parameter settings of the orientation net are provided in the supplemental material. All training processes are carried out through the Caffe Deep Learning Framework [37].

5   Recall

Figure 5 shows the recall workflow of the proposed approach. Object instances are generated in a pre-processing step. The inputs to the system are therefore an

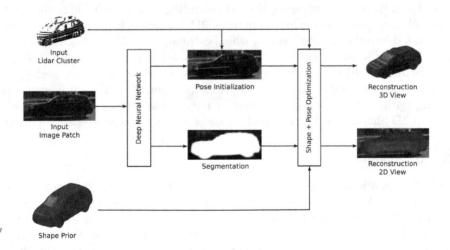

Fig. 5. Recall workflow of the proposed approach.

image patch containing only one object instance and a cluster of corresponding Lidar points. The image patch is segmented into foreground and background through the deep neural network, and the segmentation result is then used for the 3D Shaping optimization. The system outputs both 3D shape and 3D pose estimations of the input object instance.

5.1   Instance Generation

In a pre-processing step, an unorganized Lidar scan of the 3D environment is divided into smaller clusters in order to generate instance proposals. A particular approach to cluster a traffic scene is to remove ground points and group the rest using the nearest-neighbor clustering technique. Ground estimation can be carried out through ransac plane fitting or other more sophisticated methods such as progressive morphological filter. The clustering can be performed based on a kd-tree or octree search. A detailed discussion of Lidar clustering is beyond the scope of this paper. For more details of it, please refer to [38] (Fig. 6).

Fig. 6. Instance generation through Lidar clustering.

5.2   Pose Initialization

After the Lidar scan is clustered, the 3D bounding box of each Lidar cluster is projected onto the image, which generates 2D bounding box proposals. These are then fed into the orientation net, which estimates the 2D orientation of each object instance from the corresponding image patch. The estimates are used to approximate 3D object orientations (yaw angles) at the initialization step.

In fact, the orientation net is not necessarily required by the subsequent 3D Shaping optimization process. However, a reasonable orientation initialization will help 3D Shaping avoid local minima and converge in a shorter time period.

5.3   3D Shaping

In a last step, a joint optimization of 3D pose and 3D shape is performed. The state vector of the 3D Shaping optimization process comprises 3 variables for the translation, 3 variables for the rotation, 1 variable for the scale, and 2 latent variables for the 3D shape. The energy function in Eq. 1 is minimized during the optimization, which terminates if the energy converges below a certain threshold or the number of iterations exceeds a pre-defined maximum.

6   Evaluation

The proposed approach to 3D shape reconstruction using monocular camera and Lidar is evaluated on a subset of the KITTI vision benchmark suite [35]. Thirty-eight raw recording sequences with groundtruth 3D object tracklet labels are used for the evaluation. In total, the evaluation dataset contains 13,151 *Car* objects which are not occluded by other objects. Among them, 3965 objects are closer than 20 meters and thus considered to be at short range. More statistical details of the evaluation dataset are provided in the supplemental material.

Criteria used in the evaluation include the accuracy of orientation estimation and occupancy bounding box estimation. These reflect the qualitative performance of 3D shape reconstruction. Rendered reconstruction results are also presented. A brief discussion of the evaluation is included at the end of this section.

6.1   Orientation Estimation

The 3D Shaping approach using a monocular camera and a Lidar is now compared with the standalone monocular approach [33], which is considered the current state-of-the-art approach to 3D shape reconstruction. In order to evaluate the performance of 3D Shaping independently from the pose initialization step, we manually add a bias with 15° mean and 15° standard deviation to the groundtruth orientation. This initialization bias reflects the performance of the fine-tuned orientation net on short range objects, and it is set to be larger than the average estimation error of the top ranking algorithms listed by KITTI.

As shown in Table 1 and Fig. 10, both Shaping approaches are able to correct orientation estimation within a certain initialization error. At short range, the proposed approach is able to correct an error more than 30° and at full range up to 20°. It clearly outperforms the standalone monocular approach.

Table 1. Evaluation in orientation estimation. $Med(\delta)$ denotes the median of the absolute estimation errors δ in degree. $Acc(< \bullet)$ shows the percentage of estimates that are smaller than a certain threshold.

Range	Approach	$Med(\delta)$	$Acc(<20°)$	$Acc(<10°)$	$Acc(<5°)$
Short ($\leq$ 20m)	Initialization	16.07°	61.23%	31.80%	15.15%
	Orientation net	12.77°	63.00%	42.12%	23.61%
	Shaping cam	13.96°	65.67%	37.95%	21.64%
	Shaping cam+Lidar	**7.02°**	85.45%	67.53%	37.99%
Full	Initialization	16.01°	62.08%	31.87%	15.49%
	Orientation net	20.44°	49.42%	30.61%	16.57%
	Shaping cam	16.08°	60.24%	32.59%	17.91%
	Shaping cam+Lidar	14.14°	61.50%	38.44%	20.22%

6.2   Occupancy Bounding Box

In the presented work, occupancy bounding box is defined as the 2D bounding box of an object from a bird's-eye view. In the context of autonomous driving, it is essential for collision avoidance that an algorithm be able to estimate occupancy bounding boxes of surrounding objects as precisely as possible. Therefore, it is also used as an evaluation criterion for 3D Shaping. The overlap ratio between the estimated bounding box and groundtruth is calculated using Eq. 7, and the results are shown in Table 2 and Fig. 10. According to the results, the proposed approach significantly improves the occupancy estimation performance compared to the standalone monocular approach.

$$Ovl(A, B) = \frac{A \cap B}{A \cup B} \qquad (7)$$

Table 2. Evaluation of occupancy bounding box estimation. $Mean(Ovl)$ denotes the average of the overlap ratios. $Acc(> \bullet)$ indicates the percentage of estimates that are larger than a certain threshold of overlap ratio. The occupancy bounding box of a reconstructed object is considered correct if the overlap ratio is larger than 0.5.

Range	Approach	$Mean(Ovl)$	$Acc(>0.5)$	$Acc(>0.7)$
Short ($\leq$ 20m)	Shaping cam	0.17	8.34%	1.97%
	Shaping cam+Lidar	0.69	**90.57%**	62.77%
Full	Shaping cam	0.11	5.67%	1.21%
	Shaping cam+Lidar	0.53	58.36%	28.87%

Additionally, the median and standard deviation of the absolute translation error is presented in Table 3 and Fig. 10. There is a significant improvement in estimating longitudinal distance for 3D Shaping using Lidar compared to the standalone monocular approach.

Table 3. Absolute translation error. Med and Std denote median and standard deviation of the translation errors, respectively.

Range	Approach	Med (m)		Std (m)	
		Longitudinal	Lateral	Longitudinal	Lateral
Short ($\leq$ 20m)	Shaping cam	2.724	0.522	1.751	0.805
	Shaping cam+Lidar	**0.157**	**0.085**	0.376	0.525
Full	Shaping cam	3.856	0.494	2.540	0.982
	Shaping cam+Lidar	0.359	0.106	0.702	0.655

6.3   3D Shape Reconstruction

Rendered results are shown in this subsection in order to provide readers with an intuitive impression of how 3D Shaping can be used for autonomous driving. Figure 7 shows an example of how it improves occupancy estimation. It is quite common in real traffic scenarios that a Lidar cluster does not cover the entire object surface, due to reflective surface material, occlusion, clustering errors, etc. Estimating occupancy bounding box directly from Lidar clusters could lead to serious autonomous-driving traffic accidents. With help of a 3D shape prior, the occupancy estimation will be more correct even if Lidar measurements are sparse and incomplete.

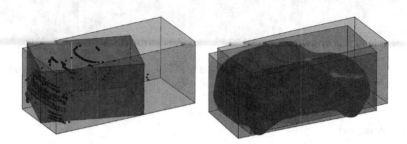

Fig. 7. Improvement in occupancy estimation. The left image shows the bounding box (dark red) estimated directly from a Lidar cluster, whereas the right image shows the estimated bounding box (red) after 3D Shaping. The green bounding box shows the groundtruth. Best viewed in color. (Color figure online)

Fig. 8. Occlusion issue. From left to right: original image patch, FCN segmentation, probability of *Car* pixels. The 3D Shaping optimizer would try to cover the entire white area in the right image with only one object instance.

More rendered results of 3D Shaping are shown in Fig. 9. The proposed approach enables a car to understand and reconstruct its entire surroundings at object level. This is of great importance for autonomous driving, as well as for future automotive technologies such as in-vehicle augmented/virtual reality.

Fig. 9. More reconstruction results of *Cars* with different poses and shapes, at different ranges, and under various lighting conditions. Best viewed in color.

6.4   Discussion

Occlusion and Multiple Instances. Multiple instances and occlusion are not taken into consideration by the energy function proposed in this paper. In other words, the reconstruction of an object that is occluded by other objects in a same object class will be incorrect, as shown in Fig. 8. Instance-aware segmentation [39] would be the key to solving the occlusion problem. Also, temporal tracking of object instances could be introduced in order to increase the consistency of shape reconstruction. The authors will consider defining a global energy function and combine it with instance-aware segmentation and tracking technique in future work.

Dataset. A quantitative evaluation of 3D shape reconstruction algorithms would require groundtruth of the 3D world, which is itself a challenging topic and would cost a great amount of time and effort. The generation of 3D world groundtruth is considered beyond the scope of this paper. Another possible consideration would be to evaluate the proposed approach on a virtual 3D dataset with rendered objects. However, in this case, the neural network for appearance detection must be fine-tuned once again in order to understand rendered images. This is not discussed in the presented paper, since the focus is on real traffic scenarios.

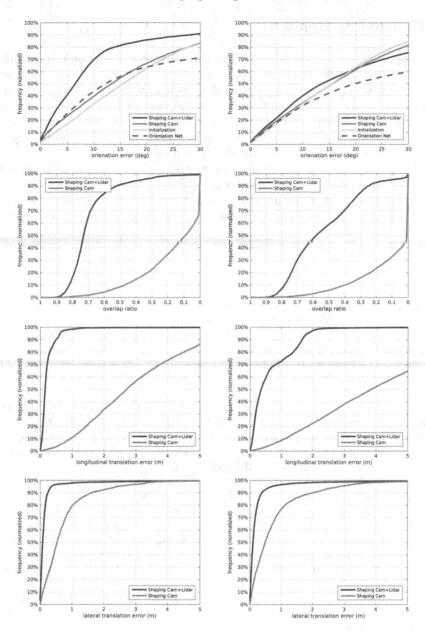

Fig. 10. Evaluation of 3D shape reconstruction. Left: short range objects. Right: all objects. From top to bottom: absolute orientation error, overlap ratio of occupancy bounding box, absolute longitudinal error, and absolute lateral error. The vertical axis shows the percentage of estimates that are smaller or larger than the corresponding threshold on the horizontal axis. Best viewed in color.

7 Concluding Remarks

This paper presented a novel approach to 3D shape reconstruction using monocular camera and Lidar. The focus was on reconstructing *Cars*, the most frequent object in traffic scenarios. Relying on deep neural networks and 3D Shaping optimization, more than 90% of short range objects in the evaluation dataset can be correctly reconstructed. According to the evaluation results, the proposed approach significantly improves the accuracy of pose estimation and occupancy bounding box estimation against the state-of-the-art.

Future work will first focus on resolving the occlusion issue of 3D Shaping by combining it with instance-ware segmentation and temporal tracking. An effort will also be made to realize the proposed reconstruction approach in a test vehicle. Based on this technique, the world is one step closer to the reality of autonomous driving.

References

1. Google X: Google Self-Driving Car Project (2014)
2. Dickmann, J., Appenrodt, N., Klappstein, J., Blöcher, H.L., Muntzinger, M., Sailer, A., Hahn, M., Brenk, C.: Making Bertha see even more: radar contribution. IEEE Access **3**, 1233–1247 (2015)
3. Franke, U., Pfeiffer, D., Rabe, C., Knöppel, C., Enzweiler, M., Stein, F., Herrtwich, R.G.: Making Bertha see. In: ICCV Workshops 2013, pp. 214–221. IEEE (2013)
4. Rusu, R., Blodow, N., Marton, Z., Soos, A., Beetz, M.: Towards 3D object maps for autonomous household robots. In: IROS 2007, pp. 3191–3198. IEEE (2007)
5. Rusu, R., Marton, Z., Blodow, N., Holzbach, A., Beetz, M.: Model-based and learned semantic object labeling in 3D point cloud maps of kitchen environments. In: IROS 2009, pp. 3601–3608. IEEE (2009)
6. Miksik, O., Amar, Y., Vineet, V., Pérez, P., Torr, P.H.S.: Incremental dense multimodal 3D scene reconstruction. In: IROS 2015, pp. 908–915. IEEE (2015)
7. Sengupta, S., Greveson, E., Shahrokni, A., Torr, P.H.S.: Urban 3D semantic modelling using stereo vision. In: ICRA 2013, pp. 580–585. IEEE (2013)
8. Vineet, V., Miksik, O., Lidegaard, M., Niebner, M., Golodetz, S., Prisacariu, V.A., Kahler, O., Murray, D.W., Izadi, S., Pérez, P., Torr, P.H.S.: Incremental dense semantic stereo fusion for large-scale semantic scene reconstruction. In: ICRA 2015, pp. 75–82. IEEE (2015)
9. Girshick, R.B., Donahue, J., Darrell, T., Malik, J.: Rich feature hierarchies for accurate object detection and semantic segmentation. In: CVPR 2014, pp. 580–587. IEEE (2014)
10. Krizhevsky, A., Sutskever, I., Hinton, G.: ImageNet classification with deep convolutional neural networks. In: NIPS 2012, pp. 1097–1105. NIPS Foundation (2012)
11. Redmon, J., Divvala, S., Girshick, R.B., Farhadi, A.: You only look once: unified, real-time object detection. In: CVPR 2016, IEEE (2016, to appear)
12. Hariharan, B., Arbeláez, P., Girshick, R., Malik, J.: Simultaneous detection and segmentation. In: Fleet, D., Pajdla, T., Schiele, B., Tuytelaars, T. (eds.) ECCV 2014. LNCS, vol. 8695, pp. 297–312. Springer, Heidelberg (2014). doi:10.1007/978-3-319-10584-0_20

13. Long, J., Shelhamer, E., Darrell, T.: Fully convolutional networks for semantic segmentation. In: CVPR 2015, pp. 3431–3440. IEEE (2015)
14. Zheng, S., Jayasumana, S., Romera-Paredes, B., Vineet, V., Su, Z., Du, D., Huang, C., Torr, P.H.S.: Conditional random fields as recurrent neural networks. In: ICCV 2015, pp. 1529–1537. IEEE (2015)
15. Beyer, L., Hermans, A., Leibe, B.: Biternion nets: continuous head pose regression from discrete training labels. In: Gall, J., Gehler, P., Leibe, B. (eds.) GCPR 2015. LNCS, vol. 9358, pp. 157–168. Springer, Heidelberg (2015). doi:10.1007/978-3-319-24947-6_13
16. Su, H., Qi, C.R., Li, Y., Guibas, L.: Render for CNN: viewpoint estimation in images using CNNs trained with rendered 3D model views. In: ICCV 2015, pp. 2686–2694. IEEE (2015)
17. Tulsiani, S., Malik, J.: Viewpoints and keypoints. In: CVPR 2015, pp. 1510–1519. IEEE (2015)
18. Eigen, D., Fergus, R.: Predicting depth, surface normals and semantic labels with a common multi-scale convolutional architecture. In: ICCV 2015, pp. 2650–2658. IEEE (2015)
19. Liu, F., Shen, C., Lin, G.: Deep convolutional neural fields for depth estimation from a single image. In: CVPR 2015, pp. 5162–5170. IEEE (2015)
20. Wu, Z., Song, S., Khosla, A., Yu, F., Zhang, L., Tang, X., Xiao, J.: 3D ShapeNets: a deep representation for volumetric shapes. In: CVPR 2015, pp. 1912–1920. IEEE (2015)
21. Schmidhuber, J.: Deep learning in neural networks: an overview. Neural Netw. **61**, 85–117 (2015)
22. Hariharan, B., Arbeláez, P., Girshick, R.B., Malik, J.: Hypercolumns for object segmentation and fine-grained localization. In: CVPR 2015, pp. 447–456. IEEE (2015)
23. Lin, G., Shen, C., van dan Hengel, A., Reid, I.: Efficient piecewise training of deep structured models for semantic segmentation. In: CVPR 2016, IEEE (2016, to appear)
24. Cordts, M., Omran, M., Ramos, S., Rehfeld, T., Enzweiler, M., Benenson, R., Franke, U., Roth, S., Schiele, B.: The cityscapes dataset for semantic urban scene understanding. In: CVPR 2016, IEEE (2016, to appear)
25. Prisacariu, V.A., Reid, I.D.: PWP3D: real-time segmentation and tracking of 3D objects. Int. J. Comput. Vis. **98**, 335–354 (2012)
26. Sandhu, R., Dambreville, S., Yezzi, A., Tannenbaum, A.: Non-rigid 2D–3D pose estimation and 2D image segmentation. In: CVPR 2009, pp. 786–793. IEEE (2009)
27. Ren, C.Y., Reid, I.: A unified energy minimization framework for model fitting in depth. In: Fusiello, A., Murino, V., Cucchiara, R. (eds.) ECCV 2012. LNCS, vol. 7584, pp. 72–82. Springer, Heidelberg (2012). doi:10.1007/978-3-642-33868-7_8
28. Prisacariu, V.A., Segal, A.V., Reid, I.: Simultaneous monocular 2D segmentation, 3D pose recovery and 3D reconstruction. In: Lee, K.M., Matsushita, Y., Rehg, J.M., Hu, Z. (eds.) ACCV 2012. LNCS, vol. 7724, pp. 593–606. Springer, Heidelberg (2013). doi:10.1007/978-3-642-37331-2_45
29. Lawrence, N.: Probabilistic non-linear principal component analysis with Gaussian process latent variable models. J. Mach. Learn. Res. **6**, 1783–1816 (2005)
30. Dame, A., Prisacariu, V.A., Ren, C.Y., Reid, I.D.: Dense reconstruction using 3D object shape priors. In: CVPR 2013, pp. 1288–1295. IEEE (2013)
31. Güney, F., Geiger, A.: Displets: resolving stereo ambiguities using object knowledge. In: CVPR 2015, pp. 4165–4175. IEEE (2015)

32. Newcombe, R.A., Lovegrove, S.J., Davison, A.J.: DTAM: dense tracking and mapping in real-time. In: ICCV 2011, pp. 2320–2327. IEEE (2011)
33. Rao, Q., Krüger, L., Dietmayer, K.: Monocular 3D shape reconstruction using deep neural networks. In: IV 2016, pp. 310–315. IEEE (2016)
34. Xiang, Y., Mottaghi, R., Savarese, S.: Beyond PASCAL: a benchmark for 3D object detection in the wild. In: WACV 2014, pp. 75–82. IEEE (2014)
35. Geiger, A., Lenz, P., Urtasun, R.: Are we ready for autonomous driving? The KITTI vision benchmark suite. In: CVPR 2012, pp. 3354–3361. IEEE (2012)
36. Szegedy, C., Liu, W., Jia, Y., Sermanet, P., Reed, S., Anguelov, D., Erhan, D., Vanhoucke, V., Rabinovich, A.: Going deeper with convolutions. In: CVPR 2015, pp. 1–9. IEEE (2015)
37. Jia, Y., Shelhamer, E., Donahue, J., Karayev, S., Long, J., Girshick, R.B., Guadarrama, S., Darrell, T.: Caffe: convolutional architecture for fast feature embedding. In: MM 2014, pp. 675–678. ACM (2014)
38. Rusu, R.: Semantic 3D object maps for everyday manipulation in human living environments. Ph.D. thesis, Computer Science Department, Technische Universität, München, Germany (2009)
39. Dai, J., Kaiming, H., Sun, J.: Instance-aware semantic segmentation via multi-task network cascades. In: CVPR 2016, IEEE (2016, to appear)

A 3D Recognition System with Local-Global Collaboration

Kai Sheng Cheng, Huei Yung Lin[(⊠)], and Tran Van Luan

Department of Electrical Engineering,
Advanced Institute of Manufacturing with High-Tech Innovation,
National Chung Cheng University, 168 University Road, Chiayi 62102, Taiwan
ram4996@yahoo.com.tw, hylin@ccu.edu.tw, tranvanluan07118@gmail.com

Abstract. To the best of increasing robotic vision in 3D conceptual for recognizing this living world, this paper proposed a 3D recognition system by combining the local feature and global verification technique. To approach this, we modified the state-of-art methods and organized it as a robust hybrid flow. Another contribution to this paper, we release the finest parameters to the Kinect sensor as well as the dataset. In the proposed framework, we expect the pre-process can deal with range filtering, noise reduction, and point cloud refinement. After this, the captured point cloud is more reliable and better to describe the object surface. The Second part is focused on recognition and pose estimation. We here refer two robust methods, SHOT descriptor and Hough Voting, one for the local feature generation and the other contributes to the object alignment. Finally, through the ICP to refine the pose matrix, we remove the false positive while verifying the good instance. Moreover, we design a keypoint selective mechanism after the hypothesis verification stage back into local conception.

1 Introduction

The human population growth is rising, and a huge increase in demand on many products for daily use or consumer electronics is probably inevitable. To face this problem, the industry needs to produce a large amount of standardized products. Mass production uses assembly line to make copies of product quickly, which involves foods, medicine, 3C electronics, apparels and vehicles, etc. Usually a standard factory contains a modern automobile assembly line, but the machinery mass production line is very expensive to ensure its products to be successful output the profit. Due to the high cost of machinery line or partially completed products not well fitting to the robotic arm, some factories employ tremendous manpower to work on each individual step. To increase the product yield rate or produce special material in a factory, employees sometimes need to stay in a hard strict environment and wear anti-dust cloth in a disinfect or high radiation exposure zone.

The high labor cost implies the robotic automation is a key prospect to the growth of manufacturing. Consequently, many research programs focus on

© Springer International Publishing AG 2017
C.-S. Chen et al. (Eds.): ACCV 2016 Workshops, Part II, LNCS 10117, pp. 19–33, 2017.
DOI: 10.1007/978-3-319-54427-4_2

manufacturing applications. The leading research organization Robotics Virtual Organization (Robotics VO) in the US and a large cooperative research center euRobotics AISBL in Europe are currently developing the new technologies for industrial manufacturing. The main research topics include accurate indoor object positioning systems for robotic manipulators (positioning the object), sensor based safety systems, the interaction between human and robot (machine vision), higher levels of realism in filtering system (3D segmentation), reactive planning and controllable in real industrial factory or workshop safety (machine learning).

This work deals with the problem of object recognition and its 3D pose estimation. It is an important issue on visual servoing and provides the information for the robotic manipulator to interact with the target object. In the literature, many 3D recognition techniques have been proposed. The state-of-the-art recognition systems usually adopt two strategies: (1) Use a 2D affine patch dataset with 2D local features to find the correspondences in 2D scenes or 3D point clouds [1–4]. (2) Use a 3D point cloud as a model with local/global feature to find the correspondences in the 3D scene [5–9]. The former is based on 2D invariant local features. It provides the system for recognition from free viewpoints with non-rigid changes. The benefit of these systems is the model can be generalized into multiple 3D viewpoints which link the features between patches and the scene. However, the 2D patch cannot represent all possible 3D conception. The latter is to match the object in a scene by its 3D model. Recent hardware advance allows direct 3D data acquisition from the real scene, and a variety of applications can be developed. These systems use 3D features to group correspondences or generated 3D model to indicate the object (the scene might be 2D). By using the SIFT descriptor in a cluttered environment, Hsiao et al. take different viewpoints of a 3D model and the 2D image of the scene for object recognition [6]. Their approach shows the robustness on finding the pose matrix, but at the cost of losing accuracy. Gomes et al. [5] propose a recognition system for real-time acquisition by extracting the keypoints in different radii for each level (distance). In [7], Drost et al. vote the matched descriptor in an accumulator space, and a point pair scheme is designed for reducing the matching computation requirement. In [8,9], an ideal local pipeline is presented following the steps: keypoint extraction, description, match, correspondence grouping, absolute orientation, ICP, hypothesis verification. All the steps can be roughly partitioned into: pre-processing (for the input data grabbed by the sensor), recognition and pose estimation (including keypoint extraction, description, matching), and post-processing (refining the pose estimation results and evaluating the final outputs).

The approach presented in this paper is similar to the pipeline based on the local feature concept, but with a global verification technique and few pre/post stages. Our system takes a 3D point cloud as input, and assumes the data points are acquired from a single viewpoint. The model datasets of the objects are built under the same environment settings. All keypoints are extracted by normal estimation before SHOT descriptor generation. The local reference frame and support space are computed for each keypoint. For data matching and pose estimation, we use KdTree, FLANN [10] to find the model-scene correspondences of the keypoint pairs. Hough voting is then performed to derive the pose in terms of a rotation matrix and a translation vector. The verification step further refines the pose by ICP, which is able to minimize the mismatch in cluster or occlusion scenes. Finally, a global hypotheses verification is carried out to minimize the false positive and maximize the true positive, followed by filtering the outliers by matched keypoints.

The contribution of this paper contains the formulation and implementation of robust 3D recognition from the real scenes. 3D features are formulated through normal estimation and eigen value decomposition (EVD), and the interpretation is separated into two parts: local feature and global verification. We analyze the techniques for extracting point features, and categorizing the state-of-the-art methods using *signatures* and *histograms*. Due to the way it works for global verification through segmentation and clustering, there are differences compared to the local pipeline. The proposed system architecture in 3D recognition combines the local and global verification to complement the disadvantages. The experiments show the robustness behavior in an environment containing multiple dissimilar objects without suffering from occlusion or clustered scenes.

2   The Unique Signatures of Histograms

The unique signatures of histograms for surfaces and texture description (SHOT) extend the exist works from [11–13], which highlight two major approaches using signatures and histograms. The signature method encodes an invariant by describing the 3D surface into a neighborhood around a given point. It localizes each trait value into coordinate bins, and is highly descriptive due to its individual localized information in the support area. However, small noise can potentially perturb the descriptor. In the histogram concept, the trait value is given according to the specific quantized domain as accumulated count. It is based on local topological entities which map into a histogram. Compared to signatures, histograms gain the robustness, but trading the descriptive accuracy by compressing the trait value into each bin.

For the signature based 3D descriptors, Novatnack and Nishino propose a method based on geometric scale-space to analyze the scale invariant of a range image [14]. The feature normal is encode within the support to ensure the local shape descriptor can be derived and deployed with different global scales. In [15], it indicates the signature is given by the 3D coordinate of each vertex within a support in the local reference frame. Continuing the 2D feature point research, the SIFT descriptor is extended to a hybrid scheme for depth images [16], and the SURF descriptor is adopted for 3D data to compute Haar wavelets as signature trait [17]. For the histogram based 3D descriptors, the spin image computes the 2D image histogram with a volume by measuring a plane spinning around the surface normal [18]. The same concept is used in local surface patches [19] and shape indices [20]. In the 3D shape context, a real full local reference frame that modifies the concept from the spin image and accumulates a 3D histogram to each feature points with a radii around the center is then proposed. Point Feature Histograms (PFH) [21] and Fast Point Feature Histograms (FPTH) [22] accumulate the 3D information into histogram bins that contain three angular values with the normal area overlap among relevant points. More recently, MeshHoG (MH) uses the same hybrid structure as SHOT [23]. It combines the signature and histogram with a unique local reference frame as well as the color information.

SHOT descriptor is generated based on an encoded histogram of normal points, with a local support space. To simulate the inherent signature, a set of local histogram is computed as a 3D sphere with accumulated support. The SHOT signature structure accumulated in the 3D grid is aligned with the axes defined by its local reference frame. Thus, the descriptor performs as a mixture produced by histograms and signatures. In SHOT descriptor, the points are accumulated into several bins from local histograms according to the angle between the point normal and local axis. Several coarser bins can be created by interpolation on normal directions, azimuth planes, elevation planes, and sphere radii. Since each plane contains descriptive information from the local histogram, the sphere grid performs a coarse partition with proper units of descriptor. The sphere grid indicates 32 partitioning volumes from 8 azimuth, 2 elevation, 2 radial divisions. On the other hand, by combining a proper number of bins from internal histograms (11 bins), the total descriptor length is 352. In SHOT descriptor, it is important to avoid boundary effects due to the local histogram.

3 Global Hypotheses Verification

We apply SHOT descriptor to transform feature points to the local reference frame. 3D Hough voting [8] is then performed after point feature registration. In general, Hough voting can be a pose estimation stage for the model (off-line) and the scene. For a reference point C^M in the model coordinates, we can find an exact match C^S in the scene. We give the same EVD process to obtain the local reference frames for the model and the scene, so we can assume the feature points in the model is defined as F_i^M with the centroid C^M. A vector $V_{i,G}^M$ describing the relationship between F_i^M and C^M can be written as

$$V_{i,G}^M = C^M - F_i^M \tag{1}$$

For a global vector $V_{i,G}^M$, we can then find a term $R_{G,L}^M$ representing the rotation invariant to transform to a local vector $V_{i,L}^M$. The relation can be written as

$$V_{i,L}^M = R_{G,L}^M \cdot V_{i,G}^M \tag{2}$$

where $R_{G,L}^M$ is given by

$$R_{G,L}^M = [L_{i,x}^M \quad L_{i,y}^M \quad L_{i,z}^M]^\top \tag{3}$$

Once the rotation matrix from the model and the scene, R_L^{MS}, is derived, $V_{i,L}^S$ can be transformed into the global reference frame, and the equation are given by

$$V_{i,G}^S = R_{LG}^S \cdot V_{i,L}^S + F_j^S \tag{4}$$

$$R_{G,L}^S = [L_{j,x}^S, L_{j,y}^S, L_{j,z}^S] \tag{5}$$

3D Hough voting picks one or more object poses which are higher than a threshold associated with a similar surface in the scene. Global Hypotheses Verification (GHV) is introduced as an additional step to further verify and reject false positives (false detection). First, we consider some notations about GHV after SHOT recognition pipeline. Assume the model set in the library contains m point clouds, $M = \{M_1, \ldots, M_m\}$, and a scene point cloud, S. For a general case, a scene might include several sets of the models. The pose estimation produces the transformation T given by the SHOT pipeline. It relates each model instance to the scene S with 6 DOFs. A pair (M_{h_i}, T_{h_i}), where h_i is a subset from the recognition hypotheses $H = \{h_1, \cdots, h_2\}$, is given by the previous recognition process. In each cue, it tries to determine and minimize the cost function value. The GHV method is designed to maximize the correct recognition items (TPs) belonging to the instance set H, and remove the wrong recognition items (FPs). In addition, a boolean term $X = \{x_0, \cdots, x_n\}$ denotes the ICP converges or not. It considers the case of partial occlusion or rotation in the scene because the model descriptor might be different from the scene or not fit exactly. In the occlusion case, a model might not be visible, i.e., self-occlusion or occluded by the scene parts. We use the binary term X to indicate the corresponding hypothesis is false or valid ($x_i = 0/1$).

Here we introduce the cues in GHV process and adopted in our implementation [24].

Cue (1) Scene Fitting: We assume a model point set $M_{h_i}^v$ has been calculated, and determine the scene fitting points corresponding to the model points. The cue is for examining how the points are explained under a threshold based on the Euclidean distance. For each ICP process, the local fitting measure is given by

$$\omega_{hi}(P) = \delta(p, q) \tag{6}$$

where q represents the model with a pose T and is denoted as $q = N(p, M_{h_i}^v)$, and $\delta(p, q)$ represents the scene point set obtained from 3D Hough voting and is defined by

$$\delta(p, q) = \begin{cases} (-\frac{||p-q||_2}{\rho_e} + 1)(n_p \cdot n_q), & ||p - q||_2 \leq \rho_e \\ 0, & \text{elsewhere} \end{cases} \tag{7}$$

We can take $\delta(p, q)$ as a local alignment of surfaces. Equation (7) checks the normal direction by (n_p, n_q), and it is expected to have two normals in the same direction. If (p, q) distance is smaller than a threshold ρ_e, a weight value (0 to 1) is assigned according to the normal direction to examine where the scene-model fitting is accepted. We conclude the contribution of Cue 1 by

$$\Omega_x(p) = \sum_{i=1}^{n} \omega_{hi}(p) \cdot x_i \tag{8}$$

All $\omega_{hi}(p)$ will be explained if the ICP term $x_i = 1$, and thus $\Omega_x(p) > 0$. If a point $p \in M_{h_i}^v$ but is not fitted in any scene point set according to Eq. (7), we denote it as ϕ_{h_i}.

Cue (2) Multiple Assignment: This cue gives a function for examining the term in Cue 1 by subtraction. The equation is given as follow:

$$\Lambda_X(p) = \begin{cases} \sum_{i=1}^{n} sgn(\omega_{h_i}(p)), & sgn(\omega_{h_i}(p)) > 1 \\ 0, & \text{elsewhere} \end{cases} \tag{9}$$

where

$$sgn(X) = \begin{cases} -1, & X < 0 \\ 0, & X = 0 \\ 1, & X > 0 \end{cases} \tag{10}$$

Cue (3) Cost Function: The cost function concludes Cue 1 and Cue 2 to increase the number of recognized instances as many as possible. It can be simply described as

$$\zeta(X) = f_S(X) + \lambda \cdot f_M(X) \tag{11}$$

where λ is a constant regularization value, and f_S, f_M are

$$f_S(X) = \sum_{p \in S} (\Lambda_X(p) - \Omega_x(p)) \tag{12}$$

$$f_M(X) = \sum_{n=1}^{n} |\phi_{h_i}| \cdot x_i \tag{13}$$

4   System Development and Implementation

In the pre-process stage, we capture 3D point cloud data by Kinect V1 and V2 through Kinect SDK 1.8 and 2.0, respectively. According to the sensor depth range (V1: 1.2–3.5 m, V2: 0.5–4.0 m), we fix a filter range of 1.8 m to remove the background points. For uniform down-sampling of the large point cloud, we set the radius as 0.01 for the model and 0.0125 for the scene. The local features are extracted by normal estimation and SHOT after the pre-processing stage, followed by Hough voting for pose estimation. We set the parameters for the local reference frame radius as 0.08 and the clustering threshold as 10.0. The example model is extracted from the scene exactly, thus the rotation is an identity matrix and the translation vector is zero.

After pose estimation, we fix the pose to minimize the Hough voting errors by ICP. Due to the inherent property of ICP, there are mis-voted cases caused by occlusion or the object placed in a pose but different from the model set. In this cases, we refine the pose from the identity rotation matrix and zero translation vector. The parameters for ICP maximum number of iteration and correspondence distance are 5 and 0.005, respectively. Note that Hough voting gives a rough transformation after 5 ICP iterations no matter the pose converges or not. The global verification process is then carried out after the ICP refinement. It is used to justify the final pose a good or bad instance.

Figure 1 shows an example of hypothesis verification. An offline model dataset is displayed on the right. The red, cyan and violet poses indicate the production of Hough voting, convergence of ICP, and the point correspondences, respectively. The following parameters are used: clutter regularizer, 5.0; inliner threshold, 0.2; clutter radius, 0.015; regularizer value, 3.0; and normal radius: 0.05. The system verifies the actual (true positive) instance as the green pose. In the final step, we pick the highest keypoint matched instance as our result. For example, if there are four instances with matched keypoints $M <30, 20, 75, 70>$, we pick the highest (75) but also giving a threshold K (say, 15). It means that the number of matched points under 75 but greater than 60 is still a good instance. All libraries and codes are built using PCL 1.7.2 [25] on a PC with Intel i7-4790 processor.

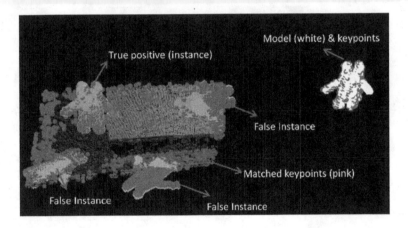

Fig. 1. An example of global hypothesis verification. (Color figure online)

5   Experiments

To evaluate the proposed technique, three datasets are generated for several interested scenarios. Five objects (Alien, Bear, Cbox, Crab, Sulley) are included individually in the occlusion and rotation datasets as shown in Fig. 2, and the cluster dataset contains many objects in the clustered scenes. In the experiments, the objects are placed at about 1 m away from a fixed viewpoint camera.

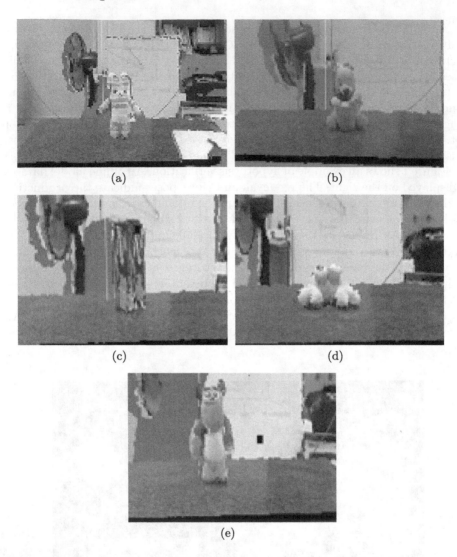

Fig. 2. The test objects used in the experiments. (a) Alien, (b) Bear, (c) Cbox, (d) Crab, and (e) Sulley.

– **Occlusion dataset:** In this dataset, we examine the system limitation by occluding the object with different levels. Five datasets of the scene are collected without any object rotation. Each dataset contains only a single object and can be described by the percentage of occlusion. The non-occluded point cloud is used as the benchmark to calculate the occlusion percentage for the occluded scenes.

- **Rotation dataset:** In this dataset, we rotate the object with different angles but keep the viewpoint still. The scenes only contain a single object without any occlusion event. The non-rotation scene is used as the benchmark dataset.
- **Cluster dataset:** In the cluster dataset, additional objects are placed in the scene. We also investigate the recognition results with different distances between objects and the scene.

Table 1 shows the occlusion experiment results. Due to the sensor frame rate and accuracy (including the point cloud density), Kinect V2 generally works better than Kinect V1. This also illustrates how the system can perform with

Table 1. The occlusion test results with Kinect V1 and V2

Occ. (V1%,V2%)		Kinect V1		Kinect V2	
		<Reg, Match, GHV>	Corr	<Reg, Match, GHV>	Corr
Alien	(6%, 4%)	<1,1,1>	68	<1,1,1>	141
	(10%, 7%)	<4,1,1>	56	<1,1,1>	134
	(22%, 17%)	<4,1,1>	40	<1,1,1>	112
	(41%, 33%)	**<3,0,1>**[a]	31	<1,1,1>	62
	(53%, 51%)	*False*		<1,1,1>	25
Bear	(5%, 12%)	<1,1,1>	107	<1,1,1>	224
	(11%, 21%)	<1,1,1>	109	<1,1,1>	136
	(21%, 30%)	<1,1,1>	76	<1,1,1>	72
	(28%, 47%)	*False*		*False*	
	(48%, 53%)	*False*		*False*	
CBox	(11%, 15%)	<2,1,2>	73	<1,1,1>	116
	(23%, 23%)	<1,1,1>	47	<1,1,1>	94
	(33%, 33%)	<1,1,1>	42	<1,1,1>	53
	(39%, 42%)	*False*		<1,1,1>	34
	(50%, 55%)	*False*		*False*	
Crab	(7%, 11%)	<1,1,1>	94	<1,1,1>	63
	(15%, 19%)	<1,1,1>	56	<1,1,1>	44
	(24%, 30%)	*False*		*False*	
	(34%, 40%)	*False*		*False*	
	(52%, 50%)	*False*		*False*	
Sulley	(7%, 6%)	<1,1,1>	134	<1,1,1>	222
	(14%, 12%)	<1,1,1>	80	<2,1,1>	151
	(24%, 29%)	<1,1,1>	64	<1,1,1>	99
	(34%, 40%)	*False*		<1,1,1>	58
	(50%, 54%)	*False*		*False*	

[a]In this case, <3,0,1>, the system gives 3 Reg instances but without any true positives. It also verifies a wrong instance as a good result.

Fig. 3. (a) Alien fails to match in the occlusion dataset. (b) An example of Crab scene dataset.

a noisy or incomplete point cloud input. We expect the occlusion experiment to have the following specification: (1) The system can keep recognizing an object until it is occluded by a specific percentage. (2) Once the system recognizes an object successfully, the triplet <Reg, Match, GHV> should be at least (Reg ≥ GHV ≥ Match = 1), where <Reg>: recognized instance, <Match>: true matched, <GHV>: instance verified successfully in global hypothesis verification, and Corr: correspondences of the recognized instance. (3) The correspondence keypoint belongs to good instances should decrease till the system reaches its limitation.

Some special cases in the occlusion dataset are given as follows. In Alien V1, the object is occluded by 41% and the system gives <3,0,1>. Three recognized instances are found, with no true positives but a good verification. This is due to V1 sensor can only sense a model without the z-axis information, so the system treats it as a flat surface. On the other hand, Alien V1 dataset outputs more recognized instances than the V2 dataset. In Crab dataset, the object can only be recognized while the occluded region is no more than 20%. This is because the Crab dataset can only show the front view surface less than other datasets. Figure 3 illustrates the special cases in Alien V1 and Crab datasets.

In Bear dataset, the column Corr of Kinect V1 gives similar values for 5% and 11% of occlusion. This is due to the infrared sensor accuracy. In CBox dataset, the system gives the output <2,1,2> for 11% of occlusion. This is caused by two flat areas of the model surface, so that several instances are obtained by the Hough voting process but only one good instance is verified. In Sulley dataset, Kinect V2 gives <2,1,1> output for 12% of occlusion. It indicates that one recognized instance is bad and filtered out by the verification stage. To summarize, the proposed technique is able to recognize the object and filter out bad instances for the occlusion dataset. Moreover, Kinect V2 shows the best results in the point cloud noise reduction and provides almost all <1,1,1> for the match triplet.

In the rotation experiment, we expect the object can be recognized by the system after it is rotated. Let the front view be defined as the 90° direction, and the object is rotated to the left or the right by every 20°. Without the Kinect sensor noise, the correspondences for recognizing the instances should decrease

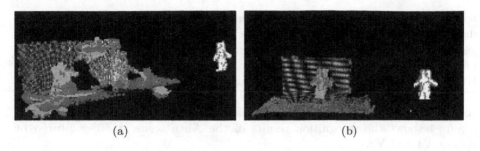

(a)   (b)

Fig. 4. (a) The case of Alien V1 50° fails to recognize. (b) The example of Alien V2 50°.

Table 2. The rotation test results with Kinect V1 and V2

Rotate degree		Kinect V1		Kinect V2	
		<Reg, Match, GHV>	Corr	<Reg, Match, GHV>	Corr
Alien	70°	<5,1,1>	72	<1,1,1>	130
	50°	<5,0,1>	38	<1,1,1>	87
	30°	<1,0,0>	29	<1,1,1>	48
	110°	<6,1,1>	72	<1,1,1>	149
	130°	<8,1,1>	46	<1,1,1>	113
	150°	<2,1,0>	45	<1,1,1>	104
CBox	70°	<1,1,1>	60	<1,1,1>	144
	50°	<1,1,1>	71	<1,1,1>	116
	30°	<2,1,2>	51	<1,1,1>	60
	110°	<1,1,1>	142	<1,1,1>	97
	130°	<1,1,1>	53	<1,1,1>	42
	150°	*False*		*False*	
Crab	70°	<1,1,1>	105	<1,1,1>	112
	50°	*False*		<1,1,1>	51
	30°	*False*		<1,1,1>	26
	110°	*False*		<1,1,1>	141
	130°	*False*		<1,1,1>	54
	150°	*False*		*False*	
Sulley	70°	<1,1,1>	107	<1,1,1>	261
	50°	<1,1,1>	59	<1,1,1>	146
	30°	<1,1,1>	28	<1,1,1>	60
	110°	<1,1,1>	108	<1,1,1>	302
	130°	<1,1,1>	72	<1,1,1>	208
	150°	<1,1,1>	39	<1,1,1>	123

* Bear dataset cannot be recognized by the system in this experiment.

when the rotation angle is increased. Some fail cases are as follows. In Alien V1 dataset, the object surface is incomplete due to the Kinect V1 sensor accuracy issue. Particularly, the system gives an erroneous output <5,0,1> for the rotation angle of 50°. In Bear and Crab datasets, the objects can not be recognized using Kinect V1. Furthermore, Bear dataset fails with Kinect V2 either, and thus the results are not shown in the table. This is mainly due to the self-occlusion of the tall object during rotation which makes the surface more difficult to model. Figure 4 shows the recognition results of the Alien scene at 50° captured by Kinect V1 and V2.

In general, as illustrated in Table 2, Kinect V2 gives better recognition results than V1 due to the noise issue mentioned in the occlusion experiment. Thus, there are more recognition instances shown in Alien V1 dataset. In the rotation experiment, the difficulty is to deal with the vanishing and emerging parts of the object surface. Although some Kinect V1 datasets give good recognition results, the instances are estimated by ICP with a verification process, and more system computation is required.

In the last experiment, we set up three clustered scenes with all objects placed randomly in front of the camera, as shown in Fig. 5. The objective is to recognize Alien, Bear, CBox, Sulley in a scene. The objects in Scene 1 are placed with more occlusion, Scene 2 contains fairly separated objects, and Scene 3 describes an extremely clustered environment. Table 3 shows the results of Cluster dataset, where Alien and Sulley are recognized with good verification. Alien in the clustered scenes is almost not occluded by other objects, but Sulley is placed at different locations with variable revelation. Notice that, for CBox object, Scene 3 shows more surface than Scene 1 but the system gives false

(a)   (b)

(c)

Fig. 5. The dataset with clustered scenes, (a) Scene 1, (b) Scene 2, (c) Scene 3.

Table 3. Cluster dataset test results using Kinect V2 only.

Cluster scene		Kinect V2	
		<Reg, Match, GHV>	Corr
Alien	Scene 1	<1,1,2>	44
	Scene 2	<1,1,1>	78
	Scene 3	<1,1,1>	60
Bear	Scene 1	<1,1,1>	63
	Scene 2	<1,1,1>	112
	Scene 3	*False*	
Cbox	Scene 1	<1,1,1>	43
	Scene 2	<1,1,1>	39
	Scene 3	*False*	
Sulley	Scene 1	<1,1,2>	75
	Scene 2	<1,1,1>	111
	Scene 3	<1,1,1>	55

output. This is due to left part of CBox cannot be estimated for the distance from Kinect V2 sensor and only an incomplete model is obtained.

6 Conclusion

In this paper, we propose a structural hybrid technique for the 3D recognition system. It fully builds using the 3D concept based on local features with global verification of the output instances. Our system takes a 3D point cloud as input, and assumes the data points are acquired from a single viewpoint. The model datasets of the objects are built under the same environment settings. The proposed system architecture in 3D recognition combines the local and global verification to complement the disadvantages. The experiments show the robustness behavior in an environment containing multiple dissimilar objects without suffering from occlusion or clustered scenes. Our system is able to adapt in a general environment and provide better recognition results by verifying good instances in the experiments. The future work will focus on three major issues of the 3D recognition techniques, (1) sensor accuracy: to deal with the resolution of the model and the scene, (2) partial model capability: to increase the recognition rate with partially acquired scenes, and (3) computation requirement: to apply GPU on Hough voting and hypothesis verification, which are not supported by OpenMP.

Acknowledgement. The support of this work in part by the Ministry of Science and Technology of Taiwan under Grant MOST 104-2221-E-194-058-MY2 is gratefully acknowledged.

References

1. Lowe, D.G.: Local feature view clustering for 3D object recognition. In: IEEE Computer Society Conference on Computer Vision and Pattern Recognition (CVPR), Kauai, HI, USA, 8–14 December 2001, pp. 682–688 (2001)
2. Ponce, J., Lazebnik, S., Rothganger, F., Schmid, C.: Towards true 3D object recognition. In: International Conference on Computer Vision and Pattern Recognition (CVPR), Washington, pp. 4034–4041 (2004)
3. Toshev, A., Makadia, A., Daniilidis, K.: Shape-based object recognition in videos using 3D synthetic object models. In: IEEE Computer Society Conference on Computer Vision and Pattern Recognition (CVPR), Miami, Florida, USA, 20–25 June 2009, pp. 288–295 (2009)
4. Hetzel, G., Leibe, B., Levi, P., Schiele, B.: 3D object recognition from range images using local feature histograms. In: IEEE Computer Society Conference on Computer Vision and Pattern Recognition (CVPR), Kauai, HI, USA, 8–14 December 2001, pp. 294–299 (2001)
5. Gomes, R.B., da Silva, B.M.F., de MedeirosRocha, L.K., Aroca, R.V., Velho, L.C.P.R., Gonçalves, L.M.G.: Efficient 3D object recognition using foveated point clouds. Comput. Graph. **37**, 496–508 (2013)
6. Hsiao, E., Collet, A., Hebert, M.: Making specific features less discriminative to improve point-based 3D object recognition. In: IEEE Conference on Computer Vision and Pattern Recognition (CVPR), San Francisco, CA, USA, 13–18 June 2010, pp. 2653–2660 (2010)
7. Drost, B., Ulrich, M., Navab, N., Ilic, S.: Model globally, match locally: efficient and robust 3D object recognition. In: IEEE Conference on Computer Vision and Pattern Recognition (CVPR), San Francisco, CA, USA, 13–18 June 2010, pp. 998–1005 (2010)
8. Tombari, F., di Stefano, L.: Object recognition in 3D scenes with occlusions and clutter by Hough voting. In: Fourth Pacific-Rim Symposium on Image and Video Technology (PSIVT), pp. 349–355 (2010)
9. Aldoma, A., Marton, Z., Tombari, F., Wohlkinger, W., Potthast, C., Zeisl, B., Rusu, R.B., Gedikli, S., Vincze, M.: Tutorial: point cloud library: three-dimensional object recognition and 6 DOF pose estimation. IEEE Robot. Automat. Mag. **19**, 80–91 (2012)
10. Muja, M., Lowe, D.G.: Fast approximate nearest neighbors with automatic algorithm configuration. In: International Conference on Computer Vision Theory and Application (VISSAPP), Lisboa, Portugal, 5–8 February 2009, pp. 331–340 (2009)
11. Hoppe, H., DeRose, T., Duchamp, T., McDonald, J.A., Stuetzle, W.: Surface reconstruction from unorganized points. In: Proceedings of the 19th Annual Conference on Computer Graphics and Interactive Techniques, SIGGRAPH, Chicago, IL, USA, 27–31 July 1992, pp. 71–78 (1992)
12. Mitra, N.J., Nguyen, A., Guibas, L.J.: Estimating surface normals in noisy point cloud data. Int. J. Comput. Geom. Appl. **14**, 261–276 (2004)
13. Tombari, F., Salti, S., Stefano, L.: Unique signatures of histograms for local surface description. In: Daniilidis, K., Maragos, P., Paragios, N. (eds.) ECCV 2010. LNCS, vol. 6313, pp. 356–369. Springer, Heidelberg (2010). doi:10.1007/978-3-642-15558-1_26
14. Novatnack, J., Nishino, K.: Scale-dependent/invariant local 3D shape descriptors for fully automatic registration of multiple sets of range images. In: European Conference on Computer Vision, Marseille, France, 12–18 October 2008, pp. 440–453 (2008)

15. Mian, A.S., Bennamoun, M., Owens, R.A.: On the repeatability and quality of keypoints for local feature-based 3D object retrieval from cluttered scenes. Int. J. Comput. Vis. (IJCV) **89**, 348–361 (2010)
16. Darom, T., Keller, Y.: Scale-invariant features for 3-D mesh models. IEEE Trans. Image Process. **21**, 2758–2769 (2012)
17. Knopp, J., Prasad, M., Willems, G., Timofte, R., Gool, L.J.V.: Hough transform and 3D SURF for robust three dimensional classification. In: 11th European Conference on Computer Vision, Heraklion, Crete, Greece, 5–11 September 2010, pp. 589–602 (2010)
18. Johnson, A.E., Hebert, M.: Using spin images for efficient object recognition in cluttered 3D scenes. IEEE Trans. Pattern Anal. Mach. Intell. (TPAMI) **21**, 433–449 (1999)
19. Dorai, C., Jain, A.K.: COSMOS - a representation scheme for 3D free-form objects. IEEE Trans. Pattern Anal. Mach. Intell. (TPAMI) **19**, 1115–1130 (1997)
20. Chen, H., Bhanu, B.: 3D free-form object recognition in range images using local surface patches. Pattern Recogn. Lett. **28**, 1252–1262 (2007)
21. Rusu, R.B., Blodow, N., Marton, Z.C., Beetz, M.: Aligning point cloud views using persistent feature histograms. In: International Conference on Intelligent Robots and Systems, 22–26 September 2008, pp. 3384–3391 (2008)
22. Rusu, R.B., Blodow, N., Beetz, M.: Fast point feature histograms (FPFH) for 3D registration. In: IEEE International Conference on Robotics and Automation (ICRA), Kobe, Japan, 12–17 May 2009, pp. 3212–3217 (2009)
23. Zaharescu, A., Boyer, E., Horaud, R.: Keypoints and local descriptors of scalar functions on 2D manifolds. Int. J. Comput. Vis. (IJCV) **100**, 78–98 (2012)
24. Aldoma, A., Tombari, F., Stefano, L., Vincze, M.: A global hypotheses verification method for 3D object recognition. In: Fitzgibbon, A., Lazebnik, S., Perona, P., Sato, Y., Schmid, C. (eds.) ECCV 2012. LNCS, vol. 7574, pp. 511–524. Springer, Heidelberg (2012). doi:10.1007/978-3-642-33712-3_37
25. PointClouds.org: Point cloud library (2014). http://pointclouds.org/

Comparison of Kinect V1 and V2 Depth Images in Terms of Accuracy and Precision

Oliver Wasenmüller[(✉)] and Didier Stricker

German Research Center for Artificial Intelligence (DFKI),
Kaiserslautern, Germany
{oliver.wasenmueller,didier.stricker}@dfki.de

Abstract. RGB-D cameras like the Microsoft Kinect had a huge impact on recent research in Computer Vision as well as Robotics. With the release of the Kinect v2 a new promising device is available, which will – most probably – be used in many future research. In this paper, we present a systematic comparison of the Kinect v1 and Kinect v2. We investigate the accuracy and precision of the devices for their usage in the context of 3D reconstruction, SLAM or visual odometry. For each device we rigorously figure out and quantify influencing factors on the depth images like temperature, the distance of the camera or the scene color. Furthermore, we demonstrate errors like *flying pixels* and *multipath interference*. Our insights build the basis for incorporating or modeling the errors of the devices in follow-up algorithms for diverse applications.

1 Introduction

Since a couple of years RGB-D cameras have a huge impact on the research in the Computer Vision community as well as on related fields like Robotics and Image Processing. These cameras provide dense depth estimations together with color images at a high frame rate. This considerably pushed forward several research fields such as: 3D reconstruction [1,2], camera localization and mapping (SLAM) [3,4], gesture and object recognition [5,6], bilateral filtering [7,8], and many more. Recently, several algorithms have been developed using the Microsoft Kinect v1, since it is one of the most common RGB-D devices. With the release of the Microsoft Kinect v2 a new promising device is available, which uses a new Time-of-Flight (ToF) camera and will – most probably – be the basis for the development and evaluation in many future research.

Our contribution in this paper is a rigorous evaluation and comparison of the depth images of Kinect v1 and Kinect v2. We concentrate on the depth images of the two devices, since they are the core input for many algorithms. The gained results on accuracy and precision can be incorporated or modeled in numerous follow-up algorithms [9]. This includes especially RGB-D 3D reconstruction, SLAM or visual odometry, since their accuracy is directly related to

Electronic supplementary material The online version of this chapter (doi:10.1007/978-3-319-54427-4_3) contains supplementary material, which is available to authorized users.

C.-S. Chen et al. (Eds.): ACCV 2016 Workshops, Part II, LNCS 10117, pp. 34–45, 2017.
DOI: 10.1007/978-3-319-54427-4_3

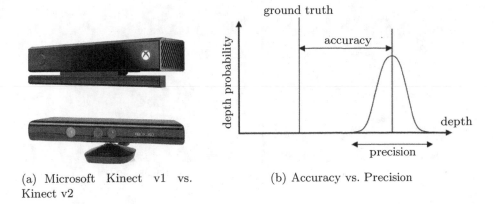

(a) Microsoft Kinect v1 vs. Kinect v2

(b) Accuracy vs. Precision

Fig. 1. We present our systematic comparison and evaluation of the Microsoft Kinect v1 and Kinect v2. More precisely, we investigate the accuracy and precision of the captured depth images.

the inaccuracies and noise in the used depth images. We analyze in this paper the influence of temperature, camera distance and scene color on the depth values of both devices. Furthermore, we analyze errors like *flying pixels* and *multipath interference*. We hope to provide a fruitful basis for future research and development with the devices. We also summarize and illustrate all our results in the supplementary video.

Because of its recent release, only little work has been published on the Kinect v2 [10]. The precision of the depth images of the single sensors (Kinect v1 or Kinect v2) was already assessed in some publications [10–13] by analyzing the noise properties addressing special applications. Other publications comparing the two devices target towards special application fields of the Kinect like motion tracking [14], face tracking [15] or multimedia [16]. We compare the two sensor in identical environments and in identical experiments in order to draw repeatable conclusions on precision and accuracy of the captured depth images. To the best of our knowledge the accuracy in terms of a metrically correct depth estimation was not assessed so far. State-of-the-art papers measure the distance from the camera case to a seen object with a tape [10] or a laser [11]. But, depth is defined from the camera center to an object, which is hard to measure with their approaches. In our approach we determine ground truth depth estimation for planar surfaces with a checkerboard. This delivers accurate results and enables easy repetition for other researcher using their own Kinect sensors or even other cameras. Our experiments enable us to directly compare the results for the two devices.

2   Preliminaries

We evaluate and characterize in this paper the Microsoft Kinect v1 and Kinect v2, which are RGB-D cameras consisting of one depth and one color camera.

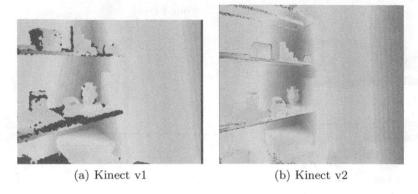

(a) Kinect v1   (b) Kinect v2

Fig. 2. Captured depth images of the same scene for the Kinect v1 and Kinect v2.

The depth image records in each pixel the distance from the camera to a seen object (Fig. 2). The Kinect v1 measures the depth with the Pattern Projection principle, where a known infrared pattern is projected into the scene and out of its distortion the depth is computed. The Kinect v2 contains a Time-of-Flight (ToF) camera and determines the depth by measuring the time emitted light takes from the camera to the object and back. Therefore, it constantly emits infrared light with modulated waves and detects the shifted phase of the returning light [17,18]. In the following, we refer to both cameras (Pattern Projection and ToF) as depth camera. We recorded all images in the raw output conditions using the *OpenNI* driver [19] for Kinect v1 and the unofficial *libfreenect2* driver [20] for Kinect v2. This means we recorded the images with the resolutions and frame rates of Table 1. We performed all recordings in an air-conditioned room with constant temperature and without direct sunlight illumination to assure reliable results.

Table 1. Resolution and frame rate of the images captured by a Microsoft Kinect v1 and Kinect v2.

	Kinect v1		Kinect v2	
	Resolution	Frame rate	Resolution	Frame rate
	[Pixel × Pixel]	[Hz]	[Pixel × Pixel]	[Hz]
Color	640 × 480	30	1920 × 1080	30
Depth	640 × 480	30	512 × 424	30
Infrared	640 × 480	30	512 × 424	30

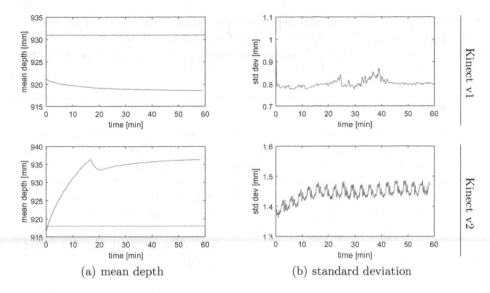

(a) mean depth   (b) standard deviation

Fig. 3. Evaluation of depth values over time, while the camera heats up and captures a flat wall. For the Kinect v1 (top) the depth values are slightly deceasing but almost constant over time. For the Kinect v2 (bottom) the depth values strongly correlate to the device temperature. The red line depicts the ground truth distance. (Color figure online)

3   Evaluation

In this section, we analyze and describe the properties of Kinect v1 and Kinect v2 depth images. The goal is to investigate the accuracy and precision of the two devices, because this information is required for algorithms like 3D reconstruction, SLAM or visual odometry. Accuracy is defined as the difference/offset of a measured depth value compared to the ground truth distance. Precision is defined as the repeatability of subsequent depth measurements under unchanged conditions. The two definitions are also illustrated in Fig. 1b. For determining the accuracy we need to know the ground truth depth. Therefore, we estimate the pose of planar surfaces relative to the camera center and compute the ground truth depth from that relation. Other than state-of-the-art papers [10,11] we generate the ground truth precisely with the help of a 12×10 checkerboard as visible in Fig. 1b. The corners of the board can be easily detected in the captured infrared images within subpixel precision [21]. Since the dimensions of the checkerboard are known, we can apply the PnP algorithm [22] to estimate the relative pose of the board. With this information we can describe the wall as a plane and compute a ground truth depth value for each pixel individually. While capturing depth images for the evaluation, the checkerboard was not visible in the scene to avoid its influence (cp. Sect. 3.3).

Since the images of both cameras exhibit a relative high level of noise, we want to be robust against it and on the other hand describe it. Therefore, we always capture a set of 300 depth images – unless otherwise mentioned – while the camera stands on a stable tripod. For the evaluation of absolute depth values we use the mean depth of the image set in each pixel. The standard deviation is computed based on the deviation in an image set.

3.1 Influence of Temperature

First, we investigate the influence of temperature on the captured depth images. Especially the Time-of-Flight (ToF) camera – or more precisely the infrared emitter – is getting warm while capturing. Therefore, the Kinect v2 has an integrated fan with a non-influenceable control. Nevertheless, the temperature of the device varies. We mounted cold and recently unused Kinect v1 and Kinect v2 on a stable tripod facing a flat white wall. Then we captured all depth images for a period of one hour and analyzed them. The results of processing these 108,000 images are depicted in Fig. 3 showing the mean measured distance and the mean standard deviation over time for both cameras.

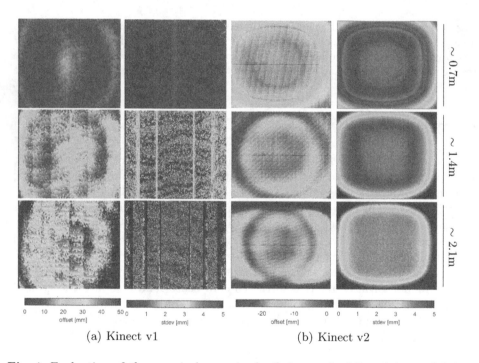

(a) Kinect v1   (b) Kinect v2

Fig. 4. Evaluation of the per-pixel error in depth images in 0.7 m, 1.4 m and 2.1 m distance for (a) Kinect v1 and (b) Kinect v2.

The Kinect v1 shows a weak correlation to the temperature. While capturing the measured depth values are decreasing for less than 2 mm. The standard deviation is on an almost constant level of 0.8 mm. In contrast, for the Kinect v2 the distance measurements exhibit a strong correlation to the temperature. In the first 16 min the distance increases constantly for around 20 mm. Then, the fan starts to rotate leading to a distance decrease for 4 min of around 3 mm. Afterwards, the distance increases again in a converging manner of around 3 mm. The standard deviation correlates only weakly to the temperature. It slightly increases until the fan starts to rotate and stays on an almost constant level afterwards. Concluding, we recommend to run the Kinect v2 for at least 25 min before capturing in order to avoid temperature influences. Kinect v1 can already be used after a short warm-up and constant depth values are delivered. The measured distances will be compared to the ground truth distance in Sect. 3.2 in order to draw conclusions on the absolute accuracy and precision. For the remaining experiments of this paper we let both cameras warm up for at least one hour.

3.2   Influence of Camera Distance

In this section, we investigate the influence of the camera distance to the scene. Therefore, we again capture a flat wall in several distances with a warm Kinect v1 and Kinect v2 standing on a stable tripod.

The left column of images in Fig. 4a and b show the offset of depth pixels to the ground truth for three different distances. For the Kinect v1 we detected

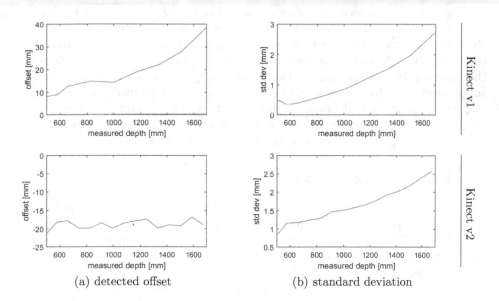

(a) detected offset   (b) standard deviation

Fig. 5. Comparison of the captured depth with the ground truth distance for Kinect v1 (top) and Kinect v2 (bottom). (a) While the Kinect v2 has a (almost constant) offset of −18 mm, the Kinect v1 has an exponentially increasing offset of up to 40 mm in the analyzed distances. (b) The standard deviation is exponentially increasing for both cameras. Please not the different scales on the y-axis.

a stripe pattern in the depth images. The number of stripes increases with the distance to the wall. The stripes lead to an irregular and difficult to model offset in the depth images. In addition, pixels in the image corners have a huge offset. For the Kinect v2 we detected a variable per-pixel offset, which mainly depends on the distance of the pixel to the image center. The corner pixels have a much higher offset than the inner. The reason for it is the infrared light cone, which does not illuminate the scene homogeneously. The infrared light cone and the offset pattern coincide.

Furthermore, we detected for both cameras a mean offset as more detailed in Fig. 5a. In this figure only the central pixels are considered, since outer pixels are too unreliable. We define the central pixels as a circle with a radius of 300 pixels around the camera center. For the Kinect v1 we detected an exponentially increasing offset for increasing distances. While the offset for 0.5 m is below 10 mm, the offset increases more than 40 mm for 1.8 m distance. In contrast, Kinect v2 we detected a offset of on average −18 mm. This means the measured depth values of the Kinect v2 are too deep respectively long. The slight variation is negligible compared to other influence factors.

Next we have a look on the standard deviation, which is shown in the right column of Fig. 4a and b. For Kinect v1 the standard deviation contains again the stripe pattern and increases with the distance. In contrast, for Kinect v2 the standard deviation in the central pixels is almost constant and increases considerably for the outer pixels. As shown in Fig. 5b the standard deviation correlates to the distance for both cameras. However, the standard deviation is lower for Kinect v1 than for Kinect v2 in given distances. Summarized, the precision and accuracy of Kinect v1 decreases with increasing distances. In contrast, the accuracy of Kinect v2 is almost constant over different distances, whereas the precision is also decreasing. Another property, which is visible best in the supplementary video, is the noise behavior. The Kinect v2 incorporates a per-pixel noise, meaning that in case of imprecise measurements the depth values of neighboring pixels strongly differ. The Kinect v1 shows in contrast a per-patch noise, meaning that neighboring pixels have similar values and errors.

3.3 Influence of Object Color

In this section, we evaluate how the color of a scene influences the depth estimation of the two devices. Therefore, we capture a planar x-rite ColorChecker [23] with 24 different colors in around one meter distance. Figure 6 shows the offset from the ground truth and the standard deviation of the depth images for both cameras. It can be clearly seen that the depth estimation of Kinect v2 depends on the scene color, whereas Kinect v1 does not. Dark colors have an up to 10 mm higher depth value than lighter colors for Kinect v2. The scene color has an even more obvious influence on the standard deviation of the depth values (see Fig. 6b). The black surface has a standard deviation of up to 4 mm, whereas light colors have a deviation of around 1 mm. By trend, for scene parts with less reflective colors it is less reliable to estimate the depth with the Kinect

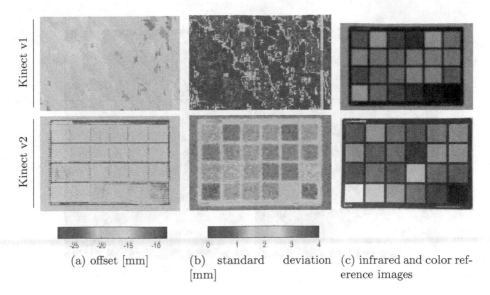

Kinect v1

Kinect v2

(a) offset [mm]   (b) standard deviation   (c) infrared and color ref-
                         [mm]   erence images

Fig. 6. Evaluation of the influence of the scene color on the depth values. Whereas Kinect v1 (top) is not influenced by the color, the Kinect v2 (bottom) is affected in terms of (a) offset and (b) standard deviation. As a reference (c) shows the infrared and color image captured by Kinect v2.

v2. Thus, for the evaluation of depth offsets and variations it is required to use only colors with similar reflectivity.

3.4   Flying Pixel

In this section, we analyze the so-called *flying pixel*; a well-known artifact for Time-of-Flight (ToF) cameras [24], since all ToF cameras suffer from this problem. *Flying pixels* are erroneous depth estimates which occur close to depth discontinuities as visible in Fig. 7b and also on the image boundaries. In this experiment we placed two boards upon each other with a distance of around 200 mm. We captured the scene with a Kinect v1 and a Kinect v2 perpendicular to the boards. Although there should be no geometry in between, there are several 3D points captured with Kinect v2. The effect is also noticeable in Fig. 1 and even more in our supplementary video. In contrast, Kinect v1 does not contain any flying pixels, which makes it much more precise close to depth discontinuities.

3.5   Multipath Interference

In this section we analyze the so-called *multipath interference* effect. Since rays of light are being sent out from the cameras, light can reflect off surfaces in numerous ways and a particular pixel may receive light originally sent out for

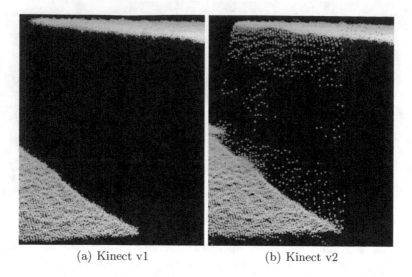

(a) Kinect v1   (b) Kinect v2

Fig. 7. To show the *flying pixel* effect we recorded two boards lying upon each other. The erroneous pixels in between for the Kinect v2 are *flying pixel*. The effect is even more noticeable in our supplementary video.

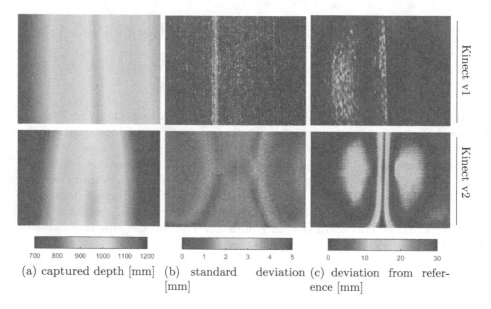

(a) captured depth [mm]   (b) standard   deviation (c) deviation from reference [mm]

Fig. 8. We captured two perpendicular walls in order to analyze the *multipath interference* effect. For Kinect v1 (top) this effect is not visible; only the usual noise is visible (cp. Fig. 4). In contrast, for Kinect v2 (bottom) on each wall a clear bulge with up to 19 mm deviation is visible, caused by reflected light of the other wall.

other pixels as well [25]. This appears often in concave geometries, even without highly reflective surfaces. Both cameras might be affected by this interference, since both cameras send out light in order to measure the depth. To analyze this effect we captured images of two perpendicular walls leading to the results of Fig. 8, where we considered only central pixels in order to eliminate other effects. In Fig. 8c we compare the depth values with the reference values. For Kinect v1 this effect is not visible at all. Only the usual noise is visible that we already investigated in Sect. 3.2. In contrast, for Kinect v2 on both walls a clear bulge is visible with up to 19 mm deviation. This effect is even more noticeable in our supplementary video. The standard deviation of both cameras is negligibly influenced by this artifact as shown in Fig. 8b. For Kinect v2 we detected a huge offset in the intersection of the two walls, which is caused by smoothing algorithms inside the camera hardware that can not be influenced. Kinect v1 does not show this smoothing effect and captures a sharp edge.

4   Conclusion

In this paper, we systemically evaluated and analyzed the two versions of Microsoft Kinect. We concentrated on the depth images, since they are the core input for algorithms like 3D reconstruction, SLAM or visual odometry. The goal was to investigate the accuracy and precision of the depth images of both devices in order to give suggestions for research in follow-up algorithms.

First of all we figured out a strong correlation of depth accuracy and temperature for the Kinect v2, resulting in the recommendation to pre-heat the device for at least 25 min in order to achieve reliable results. The Kinect v1 captures reliable images already after a few initial images. With our precise ground truth generation using a checkerboard we are able to proof the accuracy and precision of the captured depth images in different distances. While the accuracy decreases exponentially with increasing distance for Kinect v1, Kinect v2 has a constant accuracy in form of an offset of −18 mm. This is a very important fact for the above mentioned applications, since a constant offset can be easily modeled. In addition, Kinect v1 incorporates the stripe pattern in the depth images, which is difficult to compensate. For Kinect v2 all central pixels show a similar accuracy; only the image corners deviate. On the other hand, the precision of the depth images is higher for Kinect v1. This holds for flat surfaces, but especially for depth discontinuities, where *flying pixels* occur for Kinect v2. In follow-up algorithms these imprecisions must be modeled or compensated. The respective literature presents several approaches for that such as bilateral filtering [7,8] or fusion of subsequent depth images [2,26]. Furthermore, the depth estimation of Kinect v2 is influenced by the scene color, whereas Kinect v1 is unaffected. This has to be considered in bilateral filtering approaches, since they rely on coinciding color and depth changes [7,8]. In contrast to Kinect v1, Kinect v2 depth images are influenced by the *multipath interference* effect, meaning that concave geometry is captured with bulges. The respective literature proposes approaches to compensate for this effect [25].

Summarized, we recommend to use Kinect v2 in the context of 3D reconstruction, SLAM or visual odometry. Kinect v2 has the higher accuracy, which is difficult to enhance in an algorithmic way. However, due to the lower precision we recommend to apply many pre-processings on the depth images before using them. This includes compensations of random noise, *flying pixels* and *multipath interference*. These pre-processings are not necessary (in that extend) for Kinect v1, which makes it suitable for fast prototypes. The results of our evaluation can be used to incorporate or model the errors of the respective device in follow-up algorithms. We hope to provide a fruitful basis that considerably pushes forward further research with the devices.

References

1. Newcombe, R.A., Izadi, S., Hilliges, O., Molyneaux, D., Kim, D., Davison, A.J., Kohi, P., Shotton, J., Hodges, S., Fitzgibbon, A.: Kinectfusion: real-time dense surface mapping and tracking. In: IEEE International Symposium on Mixed and Augmented Reality (ISMAR) (2011)
2. Wasenmüller, O., Meyer, M., Stricker, D.: Augmented reality 3d discrepancy check in industrial applications. In: IEEE International Symposium on Mixed and Augmented Reality (ISMAR), pp. 125–134. IEEE (2016)
3. Kerl, C., Sturm, J., Cremers, D.: Robust odometry estimation for RGB-D cameras. In: IEEE International Conference on Robotics and Automation (ICRA), pp. 3748–3754. IEEE (2013)
4. Wasenmüller, O., Meyer, M., Stricker, D.: CoRBS: comprehensive RGB-D benchmark for SLAM using kinect v2. In: IEEE Winter Conference on Applications of Computer Vision (WACV), IEEE (2016)
5. Chen, C., Jafari, R., Kehtarnavaz, N.: UTD-MHAD: a multimodal dataset for human action recognition utilizing a depth camera and a wearable inertial sensor. In: IEEE International Conference on Image Processing (ICIP), 168–172. IEEE (2015)
6. Chandra, S., Chrysos, G.G., Kokkinos, I.: Surface based object detection in RGBD images. In: Proceedings of the British Machine Vision Conference (BMVC), pp. 187.1–187.13. BMVA Press (2015)
7. Vianello, A., Michielin, F., Calvagno, G., Sartor, P., Erdler, O.: Depth images super-resolution: an iterative approach. In: IEEE International Conference on Image Processing (ICIP), pp. 3778–3782 (2014)
8. Wasenmüller, O., Bleser, G., Stricker, D.: Combined bilateral filter for enhanced real-time upsampling of depth images. In: International Conference on Computer Vision Theory and Applications (2015)
9. Wasenmüller, O., Ansari, M.D., Stricker, D.: DNA-SLAM: dense noise aware SLAM for ToF RGB-D cameras. In: Asian Conference on Computer Vision Workshop (ACCV Workshop), Springer (2016)
10. Fankhauser, P., Bloesch, M., Rodriguez, D., Kaestner, R., Hutter, M., Siegwart, R.: Kinect v2 for mobile robot navigation: Evaluation and modeling. In: International Conference on Advanced Robotics (ICAR) (2015)
11. Lachat, E., Macher, H., Mittet, M., Landes, T., Grussenmeyer, P.: First experiences with kinect v2 sensor for close range 3d modelling. In: International Archives of the Photogrammetry, Remote Sensing and Spatial Information Sciences (ISPRS) (2015)

12. Butkiewicz, T.: Low-cost coastal mapping using kinect v2 time-of-flight cameras. In: Oceans-St. John's, pp. 1–9. IEEE (2014)
13. Fürsattel, P., Placht, S., Balda, M., Schaller, C., Hofmann, H., Maier, A., Riess, C.: A comparative error analysis of current time-of-flight sensors. IEEE Trans. Comput. Imaging **2**, 27–41 (2016)
14. Samir, M., Golkar, E., Rahni, A.A.A.: Comparison between the kinect v1 and kinect v2 for respiratory motion tracking. In: IEEE International Conference on Signal and Image Processing Applications (ICSIPA), pp. 150–155. IEEE (2015)
15. Amon, C., Fuhrmann, F., Graf, F.: Evaluation of the spatial resolution accuracy of the face tracking system for kinect for windows v1 and v2. In: Proceedings of the 6th Congress of the Alps Adria Acoustics Association (2014)
16. Zennaro, S., Munaro, M., Milani, S., Zanuttigh, P., Bernardi, A., Ghidoni, S., Menegatti, E.: Performance evaluation of the 1st and 2nd generation kinect for multimedia applications. In: 2015 IEEE International Conference on Multimedia and Expo (ICME), pp. 1–6. IEEE (2015)
17. Bell, J., O'Connor, P.: The xbox one system on a chip and kinect sensor. In: IEEE Micro, pp. 44–53 (2014)
18. Lefloch, D., Nair, R., Lenzen, F., Schäfer, H., Streeter, L., Cree, M.J., Koch, R., Kolb, A.: Technical foundation and calibration methods for time-of-flight cameras. In: Grzegorzek, M., Theobalt, C., Koch, R., Kolb, A. (eds.) Time-of-Flight and Depth Imaging. Sensors, Algorithms, and Applications. LNCS, vol. 8200, pp. 3–24. Springer, Heidelberg (2013). doi:10.1007/978-3-642-44964-2_1
19. (OpenNI) http://www.openni.org
20. Blake, J., Echtler, F., Kerl, C.: (libfreenect2). https://github.com/OpenKinect/libfreenect2
21. Rufli, M., Scaramuzza, D., Siegwart, R.: Automatic detection of checkerboards on blurred and distorted images. In: IEEE/RSJ International Conference on Intelligent Robots and Systems (IROS), pp. 3121–3126. IEEE(2008)
22. Lepetit, V., Moreno-Noguer, F., Fua, P.: EPnP: an accurate O(n) solution to the PnP problem. Int. J. Comput. Vis. (IJCV) **81**, 155–166 (2009)
23. x rite: ColorChecker Classic. (http://xritephoto.com/ph_product_overview.aspx ?ID=1192). Accessed 2016
24. Gottfried, J., Nair, R., Meister, S., Garbe, C., Kondermann, D.: Time of flight motion compensation revisited. In: IEEE International Conference on Image Processing (ICIP), pp. 5861–5865. IEEE (2014)
25. Freedman, D., Smolin, Y., Krupka, E., Leichter, I., Schmidt, M.: SRA: fast removal of general multipath for ToF sensors. In: Fleet, D., Pajdla, T., Schiele, B., Tuytelaars, T. (eds.) ECCV 2014. LNCS, vol. 8689, pp. 234–249. Springer, Heidelberg (2014). doi:10.1007/978-3-319-10590-1_16
26. Cui, Y., Schuon, S., Thrun, S., Stricker, D., Theobalt, C.: Algorithms for 3d shape scanning with a depth camera. IEEE Trans. Pattern Anal. Mach. Intell. **35**, 1039–1050 (2013)

3D Line Segment Reconstruction in Structured Scenes via Coplanar Line Segment Clustering

Kai Li, Jian Yao[✉], Li Li, and Yahui Liu

School of Remote Sensing and Information Engineering, Wuhan University,
Wuhan, Hubei, People's Republic of China
jian.yao@whu.edu.cn
http://cvrs.whu.edu.cn/

Abstract. This paper presents a new algorithm aiming for 3D Line Segment (LS) reconstruction in structured scenes that are comprised of a set of planes. Due to location imprecision of image LSs, it often produces many erroneous reconstructions when reconstructing 3D LSs by triangulating corresponding LSs from two images. We propose to solve this problem by first recovering space planes and then back-projecting image LSs onto the recovered space planes to get reliable 3D LSs. Given LS matches identified from two images, we estimate a set of planar homographies and use them to cluster the LS matches into groups such that LS matches in each group are related by the same homography induced by a space plane. In each LS match group, the corresponding space plane can be recovered from the 3D LSs obtained by triangulating all the LS correspondences. To reduce the incidence of incorrect LS match grouping, we formulate to solve the LS match grouping problem into solving a multi-label optimization problem. The advantages of the proposed algorithm over others in this area are that it can generate more complete and detailed 3D models of scenes using much fewer images and can recover the space planes where the reconstructed 3D LSs lie, which is beneficial for upper level applications, like scene understanding and building facade extraction.

1 Introduction

3D reconstruction from images has been a widely studied field of research and some remarkable works have been done through exploiting feature points extracted from images [1,9,30,37]. However, objects in man-made scenes are often structured and can be outlined by a bunch of LSs. It is therefore advantageous to get the 3D wireframe model of a scene by exploiting LSs on the images. For example, for the house shown in Fig. 1(d), our proposed 3D LS reconstruction method to be introduced generates the 3D model shown in Fig. 1(e) using only two images. It is easy to recognize the house from this 3D model, but hardly possible to achieve this from the extremely sparse point clouds obtained by some point based 3D reconstruction methods. Some works [12,28] also proved that 3D modeling by exploiting both feature points and line segments on images resulted in more accurate and complete results.

© Springer International Publishing AG 2017
C.-S. Chen et al. (Eds.): ACCV 2016 Workshops, Part II, LNCS 10117, pp. 46–61, 2017.
DOI: 10.1007/978-3-319-54427-4_4

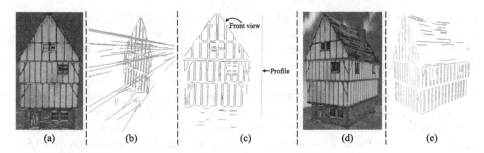

Fig. 1. An example showing problems in 3D LS reconstruction and the results obtained by our proposed solution. (a) An image of a roughly planar scene and the extracted LSs. (b) The 3D LS reconstruction result for the scene shown in (a) by triangulating LS correspondences from two images. (c) The 3D LS reconstruction result obtained by the proposed method for the scene shown in (a) by using two images. The front view and profile of the obtained 3D model are shown. (d) An image of a scene comprised of multiple planes, and the extracted LSs. (e) The 3D LS reconstruction result for the scene shown in (d) obtained by the proposed method using two images. Different colors are used to differentiate 3D LSs lying on different space planes. (Color figure online)

Despite of the above benefits of exploiting LSs on images for 3D reconstruction, it is often hard to reliably reconstruct 3D LSs because of the unstableness and low location accuracy of image LSs. Image LSs are the straight fittings of curve edges detected on images so that sometimes a 3D edge results in two straight fittings that are not precisely corresponding on two images. This fact makes it difficult to reliably reconstruct 3D LSs through triangulating corresponding LSs from two images. For example, to reconstruct 3D LSs in the scene shown in Fig. 1(a), all of which can approximately be regarded to lie on a single space plane, triangulating LS correspondences from two images capturing the scene produced 3D LSs shown in Fig. 1(b). Obviously, there are many mistakes. To solve this problem, some methods [12–14,17] resorted to exploit multiple (three or more) images photographing a scene to eliminate the mistakes. These methods involved establishing LS correspondences among multiple images, or some sophisticated hypothesizing-and-testing procedures. In this paper, we propose a simple, yet effective, solution to this problem. Our solution requires only two images and can be easily extended to more images, if available.

We observed from Fig. 1(b) that despite of many false reconstructions, the 3D model contains a big fraction of 3D LSs approaching to one space plane. This fact makes it possible to recover the space plane from the 3D LSs using RANSAC. Once the space plane being recovered, reliable 3D LSs can be obtained easily by back-projecting image LSs onto the space plane. With this idea, errors shown in Fig. 1(b) can be completely eliminated, as shown Fig. 1(c). The scene shown in Fig. 1(a) comprises of only one main space plane, which enables us to use all 3D LSs obtained by image LS triangulation to recover the space plane. But for scenes comprised of multiple planes, such as the one shown in Fig. 1(d), it is not clear which part of the obtained 3D LSs come from one plane, while some others come from another one. We need to cluster the 3D LSs according to their coplanarity. Instead of directly analyzing the 3D LSs, we propose to achieve this by exploiting LS matches obtained from the images.

Our solution is based on the fact that the projections of 3D LSs from a space plane onto two images shall be related by the planar homography induced by the space plane. Given LS matches identified from two images, we first estimate a set of planar homographies and use them to cluster the LS matches into groups such that LS matches in each group are related by a homography induced by a space plane. Then, in each LS match group, the corresponding 3D LSs are supposed to come from the same space plane, so that the final reliable 3D LSs can be obtained as the single plane case shown in Fig. 1(a). To reliably cluster LS matches, we formulate to solve the LS match clustering problem by solving a multi-label optimization problem. With our solution, for the multi-plane scene shown in Fig. 1(d), the 3D model shown in Fig. 1(e) are obtained. We can observe that a big fraction of the scene LSs are correctly reconstructed and categorized according to the space planes they lie.

In summary, the major contributions of this paper are twofold: First, we propose a new solution for solving the ambiguities in 3D LS reconstruction through LS match grouping, space plane estimation and image LS back-projection. Second, we formulate to solve the LS match grouping problem by solving a multi-label optimization problem.

2   Related Works

We divide existing 3D LS reconstruction methods into two categories: methods that require LS matching before 3D reconstruction and those do not. Many methods in the former category focus on the exploitation of different mathematical representations for a 3D line to establish the projective relationship between a 2D line and its 3D correspondence, which is the foundation of 3D LS reconstruction and camera calibration based on lines. A series of representations for a line in 3D space have been proposed. They are plücker coordinates [3,20,25], pair of points [2,10,11,22,29,36], pair of planes [11], a unitary direction vector and a point on a line [34], the intersections a line with two orthogonal planes [32], and a more recent one, Cayley representation [38]. With these representations, researchers proposed various methods for reconstructing 3D lines and/or estimating camera parameters. Some methods in the first category aim to reconstruct 3D LSs in certain types of scenes, like scenes meeting Manhattan World assumption [16,27], piecewise planar scenes [28] and poorly textured scenes [4]. The prior knowledge of these scenes decreases reconstruction uncertainties and often benefits for remarkable results.

Some recent algorithms in this area attempt to free the reconstruction procedure from the heavy dependence on the LS matching procedure because it is hard to get reliable LS correspondences in some kinds of scenes, such as poorly textured indoor environments [21] and scenes containing wiry structures (e.g., power pylons [13]). Most of these methods adopted the strategy of firstly generating a set of 3D hypotheses for each extracted LSs, either by sampling the depths of the endpoints of 3D LSs to camera centers [17], or triangulating putative LS correspondences after enforcing some soft constraints on the extract LSs [12,13], and then validating the hypotheses by projecting them back to images. In [26], a novel algorithm is proposed to obtain 3D LSs with an unknown global scale from a single image capturing a Manhattan World scene. It is possible to do so because 3D LSs in this special type of scenes can only distribute in three orthogonal directions. This fact tremendously deceases the degrees of freedom when to reconstruct the scene LSs.

Our method belongs to the first category and we focus only on 3D LS reconstruction. The camera parameters are obtained by some external camera calibration methods, or some existing SFM pipelines. The most similar method to ours is the one proposed by Kim and Manduchi [16], which also focuses on recovering planar structures of a scene from LSs. But their method is confined to be only applicable for structured scenes which meet Manhattan World assumption. Our method is a more general one and do not underlie this pretty strong assumption. Besides, their method exploits parallel LSs to determine their spatial coplanarity, while our method instead uses planar homographies.

3   Algorithm

This section first presents our method for 3D LS reconstruction from two views (images), and then introduces how we extend the two-view based method into multiple views. To be clear, in this paper, when we say *multiple views*, we mean three or more views, as a differentiation with two views.

3.1   Two-View Based 3D Line Segment Reconstruction

Given two images $\mathbf{I}$ and $\mathbf{I}'$, suppose their corresponding camera poses are $\mathbf{C}$ and $\mathbf{C}'$, which can be obtained by some existing SFM pipelines, such as the famous *Bundler* [30,31], or some camera calibration methods [25,39]. Suppose LS matches obtained from $\mathbf{I}$ and $\mathbf{I}'$ by a LS matcher is $\mathcal{M} = \{(\mathbf{l}_m, \mathbf{l}'_m)\}_{m=1}^M$. The 3D LS reconstruction procedures begin with estimating a set of planar homographies.

Homography from Line Segment Matches. A planar homography $\mathbf{H}$ is determined by eight degrees of freedom, necessitating 8 independent constraints to find a unique solution. However, when the fundamental matrix $\mathbf{F}$ between the two images is known, $\mathbf{H}^\top \mathbf{F}$ is skew-symmetric [19],

$$\mathbf{H}^\top \mathbf{F} + \mathbf{F}^\top \mathbf{H} = 0. \tag{1}$$

The above equation gives five independent constraints on $\mathbf{H}$, and the other three are required to fully describe a homography. The fundamental matrix $\mathbf{F}$ can be obtained easily by using some point matching methods, or computing from the projection matrices of the two images [11], as they are known in our case.

The homography induced by a 3D plane π can be represented as

$$\mathbf{H} = \mathbf{A} - \mathbf{e}'\mathbf{v}^\top, \tag{2}$$

where the 3D plane is represented by $\pi = (\mathbf{v}^\top, 1)$ in the projective reconstruction with camera matrices $\mathbf{C} = [\mathbf{I}|\mathbf{0}]$ and $\mathbf{C}' = [\mathbf{A}|\mathbf{e}']$. A homography maps a point from one 2D plane to another 2D plane. For a line segment match $(\mathbf{l}, \mathbf{l}')$, suppose $\mathbf{x}$ is an endpoint of $\mathbf{l}$, $\mathbf{H}$ maps it to its correspondence $\mathbf{x}'$ as: $\mathbf{x}' = \mathbf{Hx}$. Since $\mathbf{l}$ and $\mathbf{l}'$ correspond with each other, $\mathbf{x}'$ must be a point lying on $\mathbf{l}'$, that is, $\mathbf{l}'^\top \mathbf{x}' = 0$. Therefore, we obtain

$$\mathbf{l}'^\top (\mathbf{A} - \mathbf{e}'\mathbf{v}^\top)\mathbf{x} = 0. \tag{3}$$

Arranging the above equations, we get

$$\mathbf{x}^\top \mathbf{v} = \frac{\mathbf{x}^\top \mathbf{A}^\top \mathbf{l}'}{\mathbf{e}'^\top \mathbf{l}'}, \tag{4}$$

which is linear in $\mathbf{v}$. Each endpoint of a LS in a LS match provides one such equation. Two line segment matches, which totally provide four such equations, are sufficient to compute $\mathbf{v}$, and accordingly $\mathbf{H}$ from Eq. (2). If more such LS matches are available, as long as they are induced by 3D LSs from space plane π, additional constraints can be used to help more robust homography estimation.

Homography from Point Matches (Optional). If point matches from the two images which are induced by 3D points also coming from space plane π are available, they can be incorporated into the above LS match based local homography estimation method. Note that point matches are optional for the proposed method, and they are used only to provide additional constraints for homography estimation. Suppose $(\mathbf{p}, \mathbf{p}')$ is a such point match, there exists $\mathbf{p}' = \mathbf{Hp}$. Replacing $\mathbf{H}$ using Eq. (2), we get

$$\mathbf{p}' = \mathbf{Ap} - \mathbf{e}'(\mathbf{v}^\top \mathbf{p}). \tag{5}$$

From this equation, we know vectors $\mathbf{p}'$ and $\mathbf{Ap} - \mathbf{e}'(\mathbf{v}^\top \mathbf{p})$ are parallel, and their vector product is supposed to be zero:

$$\mathbf{p}' \times (\mathbf{Ap} - \mathbf{e}'(\mathbf{v}^\top \mathbf{p})) = (\mathbf{p}' \times \mathbf{Ap}) - (\mathbf{p}' \times \mathbf{e}')(\mathbf{v}^\top \mathbf{p}) = \mathbf{0}. \tag{6}$$

It holds when using Eq. (6) to form the scalar product with the vector $\mathbf{p}' \times \mathbf{e}'$

$$\mathbf{p}^\top \mathbf{v} = \frac{(\mathbf{p}' \times (\mathbf{Ap}))^\top (\mathbf{p}' \times \mathbf{e}')}{(\mathbf{p}' \times \mathbf{e}')^\top (\mathbf{p}' \times \mathbf{e}')}. \tag{7}$$

This equation is also linear in $\mathbf{v}$ and provides one constraint.

Line Segment Match Grouping. The last section presents how to estimate a local homography from (at least two) LS matches (and optional point matches when available) under the condition that they are induced by coplanar 3D LSs (or points). It is yet hard to determine which LS matches meet this condition only from images. However, due to spatial adjacency, the projections of coplanar 3D LSs onto image planes are likely to be adjacent. Therefore, it is alternative to use spatially adjacent LS matches to estimate local homographies.

For every LS match $(\mathbf{l}_m, \mathbf{l}'_m) \in \mathcal{M}$, we search its spatial neighbors in $\mathcal{M}$ by finding matched LSs from $\mathbf{I}$ which are adjacent to $\mathbf{l}_m$. A matched LS from $\mathbf{I}$ which has at least one of its two endpoints dropping in the rectangle centered around $\mathbf{l}_m$ is regarded as a neighbor of $\mathbf{l}_m$, and the corresponding LS match is regarded as a neighbor of $(\mathbf{l}_m, \mathbf{l}'_m)$. For example, if matched LS $\mathbf{l}_n$ is found to be adjacent with $\mathbf{l}_m$, then LS match $(\mathbf{l}_n, \mathbf{l}'_n)$ is treated as a neighbor of $(\mathbf{l}_m, \mathbf{l}'_m)$. The rectangle around $\mathbf{l}_m$ has the width equaling to the length of $\mathbf{l}_m$ and the height of 20 pixels (10 pixels in both sides of $\mathbf{l}_m$) in this paper. When point matches are available, we can also find point match neighbors for $(\mathbf{l}_m, \mathbf{l}'_m)$ using the same strategy. Having found the neighbors for $(\mathbf{l}_m, \mathbf{l}'_m)$, we estimate the corresponding local homography using the method presented in the last section.

A set of homographies, $\mathcal{H} = \{\mathbf{H}_i\}_{i=1}^{H}$, can be obtained after processing all LS matches in $\mathcal{M}$. H denotes the number of homographies obtained and it is often smaller than the number of elements of $\mathcal{M}$ because we sometimes cannot find for a LS match even one neighbor, and a LS match alone is insufficient to define a unique homography.

The projections of 3D LSs from a space plane onto two images would be related by the homogrpahy induced by the space plane. Based on this fact, we cluster LS matches in $\mathcal{M}$ using homographies in $\mathcal{H}$. For a LS match $(\mathbf{l}, \mathbf{l}') \in \mathcal{M}$, we find its most consistent homography matrix $\mathbf{H} \in \mathcal{H}$ which minimizes the distance of a pair of LSs according to a homography:

$$ d = \frac{\mathbf{l}'^{\top}\mathbf{H}\mathbf{x}_1 + \mathbf{l}'^{\top}\mathbf{H}\mathbf{x}_2 + \mathbf{l}^{\top}\mathbf{H}^{-1}\mathbf{x}'_1 + \mathbf{l}^{\top}\mathbf{H}^{-1}\mathbf{x}'_2}{4}, \tag{8} $$

where $\mathbf{x}_{i=1,2}$ and $\mathbf{x}'_{j=1,2}$ denote the two endpoints of $\mathbf{l}$ and $\mathbf{l}'$, respectively. Note that each of the four components of the right side of the above equation measures the distance from an endpoint of one LS to the other LS according to the given homography. For example, $\mathbf{l}'^{\top}\mathbf{H}\mathbf{x}_1$ measures the distance from $\mathbf{x}_1$ to $\mathbf{l}'$ according to $\mathbf{H}$. In other words, it is the distance between point $\mathbf{x}_1^h = \mathbf{H}\mathbf{x}_1$ and $\mathbf{l}'$: $\mathbf{l}'^{\top}\mathbf{x}_1^h = \mathbf{l}'^{\top}\mathbf{H}\mathbf{x}_1$, where $\mathbf{x}_1^h$ is the mapping of $\mathbf{x}_1$ under $\mathbf{H}$ from $\mathbf{I}$ to $\mathbf{I}'$.

After finding for each LS match in $\mathcal{M}$ a most consistent homography, some homographies in $\mathcal{H}$ are assigned with some LS matches from $\mathcal{M}$, forming a LS match group set $\mathcal{S} = \{\mathcal{G}_i\}_{i=1}^{N_s}$, where $\mathcal{G}_i$ denotes the i-th LS match group whose elements are from $\mathcal{M}$. Each LS match group in $\mathcal{S}$ is formed based on a homography, induced by a space plane. Next, we merge some groups in $\mathcal{S}$ to ensure LS matches induced by 3D LSs coming from the same space plane are clustered into only one group.

For two LS match groups, $\mathcal{G}_i$ and $\mathcal{G}_j$, suppose they are formed based on homographies, $\mathbf{H}_i$ and $\mathbf{H}_j$, respectively. If LS matches in $\mathcal{G}_i$ are *consistent* with $\mathbf{H}_j$, and the same goes for $\mathcal{G}_j$ and $\mathbf{H}_i$, we merge the two groups into one. Here, a group of LS matches are "consistent" with a homography means the average of their distances according to the homography (the distance measure is defined in

(a)   (b)   (c)   (d)

Fig. 2. An example used to illustrate some important steps of the proposed two-view based 3D LS reconstruction method. (a) The LS match grouping result before the refinement procedure. The grouping result of the matched LSs in the first used images is shown. LSs drawn in the same color are regarded to belong to the same group. (b) The Delaunay triangles constructed using the middle points of matched LSs in the first image to define the adjacent relationship among the LS matches. (c) The LS match grouping result after applying the refinement procedure. (d) The final 3D LS reconstruction result for the scene. (Color figure online)

Eq. (8)) is a small value (2 pixels in this paper). After this, we obtain an updated LS match group set $\mathcal{S}$, in which the elements decrease significantly.

Line Segment Match Grouping Result Refinement. We found that it often brought in mistakes when we grouped LS matches only based on the distance of two LS correspondences according to estimated homographies, such that some LS matches which should be assigned into one group but were clustered into another group mistakenly. This kinds of mistakes frequently occur when there are several similar space planes in the scene and the estimated homographies are not so accurate. For instance, Fig. 2(a) shows an example of the LS match grouping result using the strategy presented above. We draw in different colors the matched LSs in one of the two used images to differentiate the groups they belong. LSs drawn in the same color are supposed to appear on the same scene plane if they have been correctly grouped. But, as we can see, a considerable number of them are mistakenly clustered.

We propose to refine the LS match grouping result by enforcing spatial smoothness constraint that requires LS matches induced by coplanar 3D LSs are more likely to be adjacent with each other. We formulate to solve the LS match grouping problem by solving a multi-label optimization problem and minimizing

$$E = \sum_p D_p(l_p) + \sum_{p,q} V_{p,q}(l_p, l_q), \tag{9}$$

where the data term, D_p measures the cost of an object p being assigned with the label l_p, and the smoothness term, $V_{p,q}$ encourages a piecewise smoothness labeling by assigning a cost whenever neighboring objects p and q are assigned with labels l_p and l_q, respectively. Specifically to our problem, the data term D_p is the cost of a LS match $p = (\mathbf{l}_p, \mathbf{l}'_p)$ being labeled to belong to a group l_p. Suppose the homography relating LS matches in l_p is $\mathbf{H}_{l_p}$, D_p can then be calculated from Eq. (8). The smoothness term $V_{p,q}$ measures the cost of two neighboring LS matches p and q being labeled to belong to groups l_p and l_q, respectively. To define it, an adjacency graph among the LS matches needs to be constructed. Inspired by [8, 24], which constructed Delaunay triangles for feature points to define their adjacency, we construct Delaunay triangles using the midpoints of matched LSs in the first image to define the adjacent relationship among the LS matches, as shown in Fig. 2(b). Under this adjacency graph, we set the smoothness term as

$$V_{p,q}(l_p, l_q) = \begin{cases} sw_{pq} & l_p \neq l_q \\ 0 & l_p = l_q, \end{cases}$$

where w_{pq} is the weight for the edge linking vertexes p and q in the adjacency graph. It is assigned by Gaussian function according to the distance between the two vertexes to encourage vertexes with smaller distances being assigned with a same label in a higher possibility. s is a constant to amplify the differences of weights. It is empirically set as 4 pixels in this paper. Having defined all the terms, we resort to graph cuts [5] to minimize the objective function. The refined LS match grouping result corresponding to the minimum of the objective function is shown in Fig. 2(c). Comparing Figs. 2(a) and (c), we can observe that almost all mistakes have been corrected.

Space Plane Estimation and Trimming. For a LS match group $\mathcal{G}_i \in \mathcal{S}$, triangulating all the pairs of corresponding LSs obtains a group of 3D LSs, $\mathcal{L}_i$. All 3D LSs in $\mathcal{L}_i$ are supposed to lie on a space plane $\mathbf{P}_i$. We estimate $\mathbf{P}_i$ from the endpoints of 3D LSs in $\mathcal{L}_i$ using RANSAC. Next, we recompute the homography induced by $\mathbf{P}_i$ and use it to check if LS matches in $\mathcal{G}_i$ are consistent with it or not. We accept $\mathbf{P}_i$ as a correct plane only when the majority (0.8 in this paper) of LS matches in $\mathcal{G}_i$ are consistent with it. This step can ensure only robust space planes are kept for further processing because an accidentally formed LS match group would not result in a robust space plane such that the majority of the LS matches are consistent with its induced homography. If $\mathbf{P}_i$ is accepted, the final reliable 3D LSs corresponding to LS matches in $\mathcal{G}_i$ can be obtained simply by back-projecting matched LSs from one image onto $\mathbf{P}_i$, producing an updated $\mathcal{L}_i$. After processing all LS match groups in $\mathcal{S}$, we obtain a space plane set $\mathcal{P} = \{\mathbf{P}_i\}_{i=1}^K$, and the corresponding 3D LS set $\hat{\mathcal{L}} = \{\mathcal{L}_i\}_{i=1}^K$.

To remove some falsely reconstructed 3D LSs brought by a few falsely grouped matches that exist even after enforcing the smoothness constraint, we intersect adjacent 3D planes, trim each plane at the intersection and keep the half plane on which there are more 3D LSs than those on the other half plane. It is reasonable to do so because only a minor (if any) fraction of 3D LSs on a plane are falsely reconstructed and they are sure to lie on the opposite side (according to the intersection) of the correctly reconstructed majority. Illustration of this plane trimming strategy is shown in Fig. 3(a).

The way we determine the adjacency of space planes is as follows: We project all groups of 3D LSs in $\hat{\mathcal{L}}$ onto the first image, generating 2D LS set $\hat{\mathcal{L}}^{2d} = \{\mathcal{L}_i^{2d}\}_{i=1}^K$. Refer to Fig. 3(b), for two space planes $\mathbf{P}_i, \mathbf{P}_j \in \mathcal{P}$, suppose their corresponding 2D LS sets are $\mathcal{L}_i^{2d}$ and $\mathcal{L}_j^{2d}$. Let the convex hulls determined by $\mathcal{L}_i^{2d}$ and $\mathcal{L}_j^{2d}$ be CH_i and CH_j, respectively, and the convex hull determined by

Fig. 3. Illustration of the strategy of removing falsely reconstructed 3D LSs. (a) Adjacent space plane intersection and trimming. (b) Finding adjacent space planes. (Color figure online)

both $\mathcal{L}_i^{2d}$ and $\mathcal{L}_j^{2d}$ be CH_w (the region outlined by dashed red line in Fig. 3(b)). If there exists a third 2D LS set $\mathcal{L}_m^{2d} \in \hat{\mathcal{L}}^{2d}$, which determines a convex hull CH_m that has a big overlapping ratio (0.6 in this paper) with CH_w, we deem there is a third space plane lying between $\mathbf{P}_i$ and $\mathbf{P}_j$, and do not regard $\mathbf{P}_i$ and $\mathbf{P}_j$ to be adjacent. Otherwise, we treat $\mathbf{P}_i$ and $\mathbf{P}_j$ as adjacent planes. This strategy makes sense because it is very likely to be true in structured scenes that two space planes are adjacent if there is not a third space plane between them.

In Fig. 2, we show the final 3D LSs for the scene in sub-figure (d). We can see that the three main planes in this scene are recovered and all 3D LSs are correctly reconstructed and clustered w.r.t. the space planes they lie.

3.2   Multi-view Based 3D Line Segment Reconstruction

If more than two images are available, it is easy to extend the above two-view based 3D LS reconstruction method to deal with multiple views. We just need to combine the results obtained from every adjacent pair of images. Specifically, we begin to use the first two images to generate a set of space planes $\mathcal{P}_1$, and the corresponding set of 3D LSs $\hat{\mathcal{L}}_1$. The two sets are used to initialize the global space plane set $\mathcal{P}^g = \mathcal{P}$, and the global 3D LS set $\hat{\mathcal{L}}^g = \hat{\mathcal{L}}_1$, for the whole scene. The subsequent images are used to refine the two global sets. Each subsequent image is used to reconstruct 3D LSs with its previous image (we assume the input images are aligned), and generate a new space plane set $\mathcal{P}_i$, and a new 3D LS set $\hat{\mathcal{L}}_i$. For each space plane $\mathbf{P}_{ij} \in \mathcal{P}_i$, suppose its corresponding 3D LS set is $\mathcal{L}_{ij} \in \hat{\mathcal{L}}_i$, if $\mathcal{L}_{ij}$ is consistent with a space plane $\mathbf{P}_m \in \mathcal{P}^g$, whose corresponding 3D LS set is $\mathcal{L}_m$, we merge $\mathbf{P}_{ij}$ and $\mathbf{P}_m$ into a new space plane using 3D LSs in $\mathcal{L}_{ij}$ and $\mathcal{L}_m$. Next, we project 3D LSs in $\mathcal{L}_{ij}$ and $\mathcal{L}_m$ onto the new space plane. Otherwise, we regard $\mathbf{P}_{ij}$ as a new plane and insert it into $\mathcal{P}^g$, and meanwhile insert $\mathcal{L}_{ij}$ into and $\hat{\mathcal{L}}^g$.

After processing all images, there would exist a considerable number of duplications in $\hat{\mathcal{L}}^g$ because a same 3D LS can be visible in more than two views and be reconstructed in multiple times. We need to remove these duplications. Since 3D LSs in our case are organized w.r.t. space planes, the duplications of a 3D LS must lie on the same space plane. We can therefore conduct duplication removal plane by plane in 2D space. For each space plane $\mathbf{P}_i \in \mathcal{P}^g$, we project 3D LSs on it to a 2D plane $\mathbf{P}_i^{2d}$. For a LS $\mathbf{l}_m$ on $\mathbf{P}_i^{2d}$, we search its neighbors in a band around it. The band has the width equaling to the length of $\mathbf{l}_m$ and the height

---

Algorithm 1. 3D Line Segment Reconstruction

Input: Images $\mathcal{I} = \{I_i\}_{i=1}^N (N \geqslant 2)$, line segment matches $\hat{\mathcal{M}} = \{\mathcal{M}_i\}_{i=1}^{N-1}$
Output: 3D line segments $\hat{\mathcal{L}}^g$, space planes $\mathcal{P}^g$
 1: Initialize $\hat{\mathcal{L}}^g = \emptyset$, $\mathcal{P}^g = \emptyset$
 2: **for each** $\mathcal{M}_i \in \hat{\mathcal{M}}$ **do**
 3:     Estimate local homographies $\mathcal{H}_i$ using $\mathcal{M}_i$.
 4:     Group line segment matches in $\mathcal{M}_i$ using $\mathcal{H}_i$ into clusters as $\mathcal{S}_i = \{\mathcal{G}_j\}_{j=1}^M$.
 5:     Refine $\mathcal{S}_i$ through multi-label optimization.
 6:     **for each** $\mathcal{G}_j \in \mathcal{S}_i$ **do**
 7:         Estimate the corresponding space plane $\mathbf{P}_j$.
 8:         Project line segments in $\mathcal{G}_j$ onto $\mathbf{P}_j$ and obtain 3D line segment set $\mathcal{L}_j$.
 9:         **if** $\mathbf{P}_j$ can be merged with a space plane $\mathbf{P}_m \in \mathcal{P}$ **then**
10:             Merge $\mathbf{P}_j$ and $\mathbf{P}_m$, update $\mathcal{P}^g$ and $\hat{\mathcal{L}}^g$.
11:         **else**
12:             Insert $\mathcal{L}_j$ into $\hat{\mathcal{L}}^g$, and $\mathbf{P}_j$ into $\mathcal{P}^g$.
13:         **end if**
14:     **end for**
15: **end for**
16: Remove duplications in $\hat{\mathcal{L}}^g$.

---

of 6 pixels (3 pixels in both sides of $\mathbf{l}_m$) in this paper. A LS $\mathbf{l}_n$ is regarded as a neighbor of $\mathbf{l}_m$ if it meets the two condition: First, both its two endpoints drop in the band around $\mathbf{l}_m$. Second, the direction difference between $\mathbf{l}_m$ and $\mathbf{l}_n$ is less than 5°. In this way, we obtain a set of neighbors for $\mathbf{l}_m$. All neighbors of $\mathbf{l}_m$ and $\mathbf{l}_m$ itself are merged into a single LS. After that, we project the merged new LSs from $\mathbf{P}_i^{2d}$ back to $\mathbf{P}_i$.

The above duplication removal strategy has advantages over existing methods because it is easier and more reliable for us to define which LSs are adjacent enough to be merged into one. We only need to search in the band around a LS to find its possible duplications in a 2D plane, rather than in a cylinder in 3D space as that done in [12,17]. Therefore, the cases that the 3D reconstructions of multiple scene LSs being falsely regarded as the duplications of one scene LS, and one scene LS being reconstructed with multiple 3D representations are rare in our method. This contributes to the benefit of our method on delivering more details of scenes.

Algorithm 1 outlines the main steps of the proposed method.

4 Experiments

This section presents the experimental results of the proposed method. All images employed for experiments come from public datasets [17,18,33]. We used the method presented in [36] for LS extraction and the method presented [15] for LS matching.

4.1 Two Views

We presented in Figs. 1 and 2 two sample 3D LS reconstruction results based on two views, and Fig. 4 shows four additional such results. We can observe from

Fig. 4. Two-view based 3D LS reconstruction results. The top row shows the first ones of two images used for 3D LS reconstruction and the extracted LSs; the bottom row shows the obtained 3D LSs.

these results that the proposed method successfully reconstructed a large part of space LSs lying on main planes of the scenes, and correctly clustered them according to the space planes they lie. The main structures of the scenes are outlined by the reconstructed 3D LSs. This proves the feasibility of the proposed two-view based 3D LS reconstruction method.

4.2   Multiple Views

In this part, we present the experiments of our method on two image datasets, a synthetic image dataset and a real image dataset.

Synthetic Images. The synthetic image dataset has $80 \times 3 = 240$ images photographing around a CAD model from the upper, middle and bottom viewpoints. Each round consists of 80 images separated by a constant angle interval. An example image from the dataset is shown in Fig. 5(a). We employed for experiments only the 80 images for the middle round because we found in our initial experiments that the reconstruction result generated by our method based on the 80 images is negligibly different from that based on all 240 images, but the running time dropped significantly. The result model, O_{80} is shown in Fig. 5(b). We can observe from O_{80} that the main planes in this scene are correctly recovered, and LSs in the scene are precisely presented and correctly clustered w.r.t. the planes they lie. We overlapped O_{80} with the ground truth CAD model to qualitatively evaluate the reconstruction accuracy, as shown in Fig. 5(c). As we can see, the vast majority of the reconstructed LSs (in black) cling to or closely approach the ground truth model, which indicates the high reconstruction accuracy. To test the robustness of the proposed method for 3D LS reconstruction from a small number of images, we sampled from the 80 used images by taking one from every three images, producing a new image sequence containing 27 images. Taking as input this new image sequence, our method generated the 3D model, O_{27} shown in Fig. 5(d). Comparing O_{27} with O_{80}, we can see that there is no significant difference between them, except some missing LSs on the roof and bottom of the captured house in O_{27}; LSs on the walls of the house are identically and completely reconstructed in both models; LSs in O_{27} are also correctly clustered w.r.t. their respective planes.

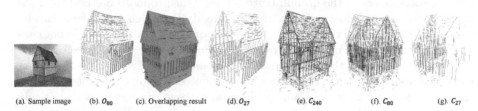

(a). Sample image   (b). O_{80} (c). Overlapping result   (d). O_{27} (e). C_{240} (f). C_{80} (g). C_{27}

Fig. 5. 3D LS reconstruction results on a synthetic image dataset. (a) One of the used images. (b) The 3D model (referred later as O_{80}) obtained by the proposed method using 80 images. (c) The overlapping result of O_{80} with the ground truth model of the scene. (d) The 3D model (O_{27}) obtained by the proposed method using 27 images. (e)–(g) The 3D models generated by Line3D++ [14] using 240, 80, and 27 images, respectively. The three models will orderly be referred later as C_{240}, C_{80} and C_{27}.

For comparison, we show in Fig. 5(e)–(g) the reconstruction models of a recent algorithm, Line3D++ [14], using the whole 240 images of the dataset, our used 80 images and 27 images, respectively. We can see that the reconstruction result of Line3D++ degenerates dramatically as the number of used images decreases. Line3D++ is able to generate good result when plentiful images are available, but cannot guarantee good performance with a small number of images. Our method, on other hand, is much less dependable on the availability of abundant images. Comparing Line3D++'s best model C_{240} with our model O_{80}, we can see that although C_{240} presents more details at the bottom of the house, our model is much neater and contains less short LSs that are arbitrarily distributed, which, to some extent, indicates our model is a better wireframe model for the scene. Besides, through carefully inspection, we can observe that for some scene LSs, C_{240} presents several duplications, while these cases are rare in our model. This proves the benefit of our duplication removal strategy.

To quantitatively evaluate the reconstruction accuracy, following [12,13,17], we calculated the Hausdorff distances between densely sampled points along the 3D LSs in our models and the ground truth CAD model, and computed the Mean Error (ME) and Root Mean Square Error (RMSE). We do not directly compare our measure data with that of Line3D++ because Line3D++ is based on the point clouds and camera parameters generated by some existing SFM systems, whose outputs are under arbitrary coordinates. 3D models generated by Line3D++ are hence inherently under the input arbitrary coordinates. This fact hinders the quantitative evaluations of models generated by Line3D++ because the underlying coordinates are inconsistent with that of the ground truth model. ICP [6] is a powerful way to align point clouds from different coordinates, which makes it possible to quantitatively evaluate the 3D models generated by Line3D++. However, ICP itself shall introduce alignment errors, and these errors would be counted into the errors between models generated by Line3D++ and the ground truth model. Therefore, the error data calculated in this situation cannot reflect the true accuracy of the models. On the other hand, the camera matrices corresponding to images in the synthetic dataset are provided and they are consistent with the coordinates of the ground truth model. Our proposed method took the provided camera matrices as input and produced 3D models that are naturally aligned with the ground truth model. So, the error data of our models do not contain alignment errors. For this reason, it is an unfair comparison if we compare our error data (without alignment errors) with that of Line3D++ (with alignment errors). Alternatively, since Line3D++ is a promoted version of the methods presented in [13] and [12] by the same authors, and these two methods do not rely on SFM results, a comparison between our measure data with the report data in [13] and [12] is also some kind of meaningful[1]. Meanwhile, we will show that this indirect comparison does not affect us to reach a conclusion about the accuracy performances of our method and Line3D++.

Table 1. The Mean Error (ME) and Root Mean Square Error (RMSE) data of the reconstruction results obtained by our method and several other ones on a synthetic dataset. "–" denotes the corresponding measure datum was not reported in the paper.

	$\rho = 1.0$					$\rho = 0.6$				
	O_{27}	O_{80}	[17]	[13]	[12]	O_{27}	O_{80}	[17]	[13]	[12]
ME	0.077	0.89	0.162	0.065	–	0.075	0.082	0.137	0.044	0.029
RMSE	0.114	0.135	0.291	0.196	–	0.104	0.109	0.189	0.080	0.046

Table 1 shows the measure data. We can see that when we set the cutoff distance threshold (distance values greater than this threshold are treated as gross errors and excluded for ME and RMSE calculations) $\rho = 1.0$, as that applied in [13], the RMSEs of our two models O_{27} and O_{80}, are much better than the others, while the MEs are slightly inferior to that of [13]. When we set $\rho = 0.6$ as that used in [12], our two models are better than [17], but worse than both [12,13]. Since Line3D++ is promoted from [12,13], its generated model is supposed to be of even higher accuracy. It is thus reasonable to infer that the reconstruction accuracy of C_{240} is better than our models. But as can be obviously seen from Fig. 5, it is unlikely that the reconstruction accuracy of C_{80} and C_{27} is better than our two models, O_{80} and O_{27}, when the same numbers of images are used. Therefore, we can reach the conclusion that Line3D++ can produce 3D models with higher accuracy than our method, when plentiful images are available, but in the case that there are only a small number of images, our method produces more accurate 3D models.

Real Images. The real image dataset contains 30 images. Figure 6 shows the result models of our method and Line3D++ generated from these images. As we can see, in our model, the 3D LSs lying on the main planes of the scene are well reconstructed; the details of the scene are precisely presented (see the bricks and

---

[1] The authors of Line3D++ made the source code of Line3D++ publicly available, but did not do so for its preliminary versions. So, we can only compare our measure data with the reported data in the papers.

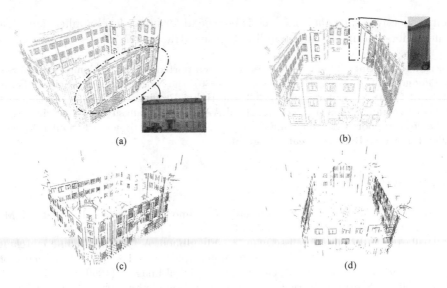

Fig. 6. The 3D LS reconstruction results of the proposed method and Line3D++ on a real image dataset. The top row shows our result model from two different viewpoints, while the bottom row shows that of Line3D++.

windows of the selected dashed elliptical region shown in Fig. 6(a)). Our method failed to reconstruct 3D LSs on the main planes of this scene shown in the selected rectangle region in Fig. 6(b). This is because only several LSs were extracted on these two planes and even fewer LS matches were obtained. Our method is unable to reliably estimate a space plane when LS matches induced by 3D LSs on it are too few, and hence incapable to reconstruct the 3D LSs on it. Comparing our model with that of Line3D++, it is obvious that our model is much more complete and detailed.

Running Time. The algorithm is currently implemented based on MATLAB. The unrefined codes took 631 s on the 80 synthetic images and 1021 s on the real image dataset on a 3.4 GHz Inter(R) Core(TM) processor with 12 GB of RAM. It is expected that the code can be substantially accelerated after refinements and being reimplemented in C++.

5 Conclusions

We have presented in this paper a new method about 3D LS reconstruction in structured scenes. A new solution is proposed to solve the uncertainties in 3D LS reconstruction by estimating space planes from clustered LS matches and back-projecting image LSs onto the space planes. We introduce a multi-label optimization framework to improve LS match grouping results. Experiments show the superiority of the proposed method to others in this area for its better performance in using small numbers of images and its ability of clustering 3D LSs w.r.t.

their respective space planes, which is beneficial for upper level applications, like scene understanding [23] and building facade extraction [7, 35].

Acknowledgment. This work was partially supported by the National Natural Science Foundation of China (Project No. 41571436), the National Natural Science Foundation of China under Grant 91438203, the Hubei Province Science and Technology Support Program, China (Project No. 2015BAA027), the Jiangsu Province Science and Technology Support Program, China (Project No. BE2014866), and the South Wisdom Valley Innovative Research Team Program.

References

1. Agarwal, S., Furukawa, Y., Snavely, N., Simon, I., Curless, B., Seitz, S.M., Szeliski, R.: Building Rome in a day. Commun. ACM **54**, 105–112 (2011)
2. Baillard, C., Schmid, C., Zisserman, A., Fitzgibbon, A.: Automatic line matching and 3D reconstruction of buildings from multiple views. In: ISPRS Conference on Automatic Extraction of GIS Objects from Digital Imagery (1999)
3. Bartoli, A., Sturm, P.: Structure-from-motion using lines: representation, triangulation, and bundle adjustment. Comput. Vis. Image Underst. **100**, 416–441 (2005)
4. Bay, H., Ess, A., Neubeck, A., Van Gool, L.: 3D from line segments in two poorly-textured, uncalibrated images. In: 3DPVT (2006)
5. Boykov, Y., Veksler, O., Zabih, R.: Efficient approximate energy minimization via graph cuts. IEEE Trans. Pattern Anal. Mach. Intell. **36**, 1222–1239 (2001)
6. Chetverikov, D., Svirko, D., Stepanov, D., Krsek, P.: The trimmed iterative closest point algorithm. In: ICPR (2002)
7. Delmerico, J.A., David, P., Corso, J.J.: Building facade detection, segmentation, and parameter estimation for mobile robot stereo vision. Image Vis. Comput. **31**, 841–852 (2013)
8. Delong, A., Osokin, A., Isack, H.N., Boykov, Y.: Fast approximate energy minimization with label costs. Int. J. Comput. Vis. **96**, 1–27 (2012)
9. Furukawa, Y., Ponce, J.: Accurate, dense, and robust multiview stereopsis. IEEE Trans. Pattern Anal. Mach. Intell. **32**, 1362–1376 (2010)
10. Habib, A.F., Morgan, M., Lee, Y.R.: Bundle adjustment with selfcalibration using straight lines. Photogram. Rec. **17**, 635–650 (2002)
11. Hartley, R., Zisserman, A.: Multiple View Geometry in Computer Vision. Cambridge University Press, Cambridge (2003)
12. Hofer, M., Maurer, M., Bischof, H.: Improving sparse 3D models for man-made environments using line-based 3D reconstruction. In: 3DV (2014)
13. Hofer, M., Wendel, A., Bischof, H.: Incremental line-based 3D reconstruction using geometric constraints. In: BMVC (2013)
14. Hofer, M., Maurer, M., Bischof, H.: Efficient 3D scene abstraction using line segments. Comput. Vis. Image Underst. (2016). doi:10.1016/j.cviu.2016.03.017
15. Li, K., Yao, J., Lu, X., Xia, M., Li, L.: Joint point and line segment matching on wide-baseline stereo images. In: WACV (2016)
16. Kim, C., Manduchi, R.: Planar structures from line correspondences in a Manhattan World. In: Cremers, D., Reid, I., Saito, H., Yang, M.-H. (eds.) ACCV 2014. LNCS, vol. 9003, pp. 509–524. Springer, Heidelberg (2015). doi:10.1007/978-3-319-16865-4_33

17. Jain, A., Kurz, C., Thormahlen, T., Seidel, H.P.: Exploiting global connectivity constraints for reconstruction of 3D line segments from images. In: CVPR (2010)
18. Jensen, R., Dahl, A., Vogiatzis, G., Tola, E.: Large scale multi-view stereopsis evaluation. In: CVPR (2014)
19. Luong, Q.-T., Viéville, T.: Canonical representations for the geometries of multiple projective views. Comput. Vis. Image Underst. 64, 193–229 (1996)
20. Matinec, D., Pajdla, T.: Line reconstruction from many perspective images by factorization. In: CVPR (2003)
21. Micusik, B., Wildenauer, H.: Structure from motion with line segments under relaxed endpoint constraints. In: 3DV (2014)
22. Micusik, B., Wildenauer, H.: Descriptor free visual indoor localization with line segments. In: 3DV (2015)
23. Pan, J.: Coherent scene understanding with 3D geometric reasoning. Ph.D. thesis, Carnegie Mellon University (2014)
24. Pham, T.T., Chin, T.J., Yu, J., Suter, D.: The random cluster model for robust geometric fitting. IEEE Trans. Pattern Anal. Mach. Intell. 36, 1658–1671 (2014)
25. Přibyl, B., Zemčík, P., Čadík, M.: Camera pose estimation from lines using Plücker coordinates. In: BMVC (2015)
26. Ramalingam, S., Brand, M.: Lifting 3D Manhattan lines from a single image. In: ICCV (2013)
27. Schindler, G., Krishnamurthy, P., Dellaert, F.: Line-based structure from motion for urban environments. In: 3DPVT (2006)
28. Sinha, S.N., Steedly, D., Szeliski, R.: Piecewise planar stereo for image-based rendering. In: ICCV (2009)
29. Smith, P., Reid, I.D., Davison, A.J.: Real-time monocular SLAM with straight lines. In: BMVC (2006)
30. Snavely, N., Seitz, S.M., Szeliski, R.: Photo tourism: exploring photo collections in 3D. ACM Trans. Graph. 25, 835–846 (2006)
31. Snavely, N., Seitz, S.M., Szeliski, R.: Modeling the world from internet photo collections. Int. J. Comput. Vis. 80, 189–210 (2008)
32. Spetsakis, M.E., Aloimonos, J.Y.: Structure from motion using line correspondences. Int. J. Comput. Vis. 4, 171–183 (1990)
33. Strecha, C., Hansen, W.V., Gool, L.V., Fua, P., Thoennessen, U.: On benchmarking camera calibration and multi-view stereo for high resolution imagery. In: CVPR (2008)
34. Taylor, C.J., Kriegman, D.J.: Structure and motion from line segments in multiple images. IEEE Trans. Pattern Anal. Mach. Intell. 17, 1021–1032 (1995)
35. Teboul, O., Simon, L., Koutsourakis, P., Paragios, N.: Segmentation of building facades using procedural shape priors. In: CVPR (2010)
36. Werner, T., Zisserman, A.: New techniques for automated architectural reconstruction from photographs. In: Heyden, A., Sparr, G., Nielsen, M., Johansen, P. (eds.) ECCV 2002. LNCS, vol. 2351, pp. 541–555. Springer, Heidelberg (2002). doi:10.1007/3-540-47967-8_36
37. Wu, C.: Towards linear-time incremental structure from motion. In: 3DV (2013)
38. Zhang, L., Koch, R.: Structure and motion from line correspondences: representation, projection, initialization and sparse bundle adjustment. J. Vis. Commun. Image Represent. 25, 904–915 (2014)
39. Zhang, Z.: Flexible camera calibration by viewing a plane from unknown orientations. In: ICCV (1999)

Bio-Inspired Architecture for Deriving 3D Models from Video Sequences

Julius Schöning$^{(\boxtimes)}$ and Gunther Heidemann

Institute of Cognitive Science, Osnabrück University, Osnabrück, Germany
juschoening@uos.de

Abstract. In an everyday context, automatic or interactive 3D reconstruction of objects from one or several videos is not yet possible. Humans, on the contrary, are capable of recognizing the 3D shape of objects even in complex video sequences. To enable machines for doing the same, we propose a bio-inspired processing architecture, which is motivated by the human visual system and converts video data into 3D representations. Similar to the hierarchy of the ventral stream, our process reduces the influence of the position information in the video sequences by object recognition and represents the object of interest as multiple pictorial representations. These multiple pictorial representations are showing 2D projections of the object of interest from different perspectives. Thus, a 3D point cloud can be obtained by multiple view geometry algorithms. In the course of a detailed presentation of this architecture, we additionally highlight existing analogies to the view-combination scheme. The potency of our architecture is shown by reconstructing a car out of two video sequences. In case the automatic processing cannot complete the task, the user is put in the loop to solve the problem interactively. This human-machine interaction facilitates a prototype implementation of the architecture, which can reconstruct 3D objects out of one or several videos. In conclusion, the strengths and limitations of our approach are discussed, followed by an outlook to future work to improve the architecture.

1 Introduction

Nowadays the amount of video data on online video-sharing platforms like *YouTube*, *Vimeo*, and *Flickr* is increasing exponentially by the minute. However, even with state of the art technologies, it is not yet possible to reconstruct 3D objects out of video sequences, if the videos are showing the objects in their natural context. For humans, it is quite easy to recognize the 3D shape of an object of interest (OOI) within video sequences, even if the videos do not show the OOI from every side, or if the OOI is partially occluded.

For enabling machines to reconstruct a 3D model of any OOI, which is shown from 360° in their natural environment in one or several video sequences,

Electronic supplementary material The online version of this chapter (doi:10.1007/978-3-319-54427-4_5) contains supplementary material, which is available to authorized users.

C.-S. Chen et al. (Eds.): ACCV 2016 Workshops, Part II, LNCS 10117, pp. 62–76, 2017.
DOI: 10.1007/978-3-319-54427-4_5

we developed and implemented a bio-inspired processing architecture. With our general architecture of deriving 3D models out of video sequences, we are moving computer vision from controlled environments with specific data sets, conditions, setups, etc. into real world scenarios. The real world environment causes challenges, which are not yet completely solvable with fully-automatic approaches. Anyhow, to facilitate a prototypical implementation of our processing architecture, some modules put the user in the loop in order to perform the task with close cooperation between human and machine.

As input data, the proposed bio-inspired processing architecture requires one or several videos showing the same OOI from different views. The system first recognizes the OOI in all frames to eliminate the background as well as the various positions of the OOI. Then multiple 2D projections of the OOI from different views are extracted. Finally, a structure from motion (SfM) algorithm calculates the corresponding 3D model out of these multiple views.

The detailed architecture of our bio-inspired process for deriving 3D models out of videos sequences is illustrated and described in Sect. 2. Regarding this architectural overview, related work associated with the processing modules is introduced in Sect. 3. The following Sect. 4 describes the prototypical implementation of the proposed architecture and discusses its analogies to the view-combination scheme by Ullman [43]. Based on two video sequences, showing the same OOI in its natural environment, we show in Sect. 5 that our architecture can reconstruct a 3D model of an OOI. We discuss the first experimental results in Sect. 6 and give an outlook how the process can be further automatized as well as optimized in Sect. 7.

2   Bio-Inspired Architecture for Deriving 3D Models

The architecture for reconstructing 3D models of an OOI from any video sequence, recorded by a regular monocular camera consists of six processing modules, as shown in Fig. 1. The first four processing modules recognize the 2D projection of the OOI on every frame of each video sequence. Similar to the main processing direction of the ventral stream pathway of the visual cortex [44, 45]—starting with the retina and passing through $V1$, $V2$, $V4$, and the inferior temporal (IT) cortex—these processing modules are responsible for object recognition. Thus, these four modules are sequential and organized bottom-up in our architecture. The following two processing modules reconstruct the 3D model of the OOI by using the recognized 2D projection of the OOI as multiple views of the 3D object. These last two processing modules are primarily technically inspired by common image based 3D reconstruction approaches [1, 3, 18, 26, 47].

There are two major branches in biological image processing on which we could draw for motivation of the technical system: Firstly, the ventral visual stream [44, 45] of the visual cortex, which is associated with object recognition, in other words, the identification of "what" objects one is seeing. Secondly, the dorsal stream [44, 45], which responds to the spatial aspect "where" the object is located. The architecture presented here is motivated by the ventral visual

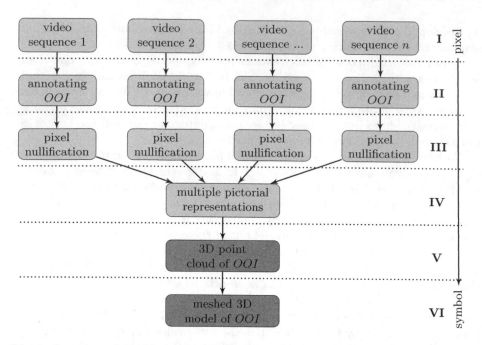

Fig. 1. Overview of the bio-inspired architecture for deriving 3D models out of video sequences powered by viewpoint-specific object recognition. It consists of six processing modules: **I** reading input videos as pixel cuboids, **II** annotating pixels of cuboids—viewpoint-specific pixel-accurate recognition, **III** annotation based pixel nullification, **IV** creating multiple pictorial representations, **V** building a point cloud, and **VI** meshing the point cloud. Yellow marked modules show analogies to human vision; blue marked modules are technically inspired and transform the virtual model into viewpoint-invariant 3D models for technical use. (Color figure online)

stream since for any reconstruction of a single object, the only relevant information is "what" the object looks like, not "where" the object is located. "Where" would be relevant only in case trajectory tracking becomes necessary for object identification, but here we assume that objects can always uniquely be recognized by their appearance. So the context around an object is irrelevant for recognizing its 3D shape. Nevertheless, we keep in mind that for general object recognition in the real world, context is relevant [25].

I. **Reading input videos as pixel cuboids:** Translate the video sequences into a processable data format, like the retina of human eyes perceives the environment and translates it into a brain-processable input signal.

In our case, video sequences compressed by codecs are translated into 3D cuboids of pixels $C_1, C_2, \ldots C_n$, where n is the number of available video sequences $S_1, S_2, \ldots S_n$ showing the OOI. For each $i = 1, \ldots n$, the dimensions of cuboid C_i are defined by the horizontal resolution h_i, by the vertical resolution v_i and the number of frames f_i of each video sequence, thus C_i has $h_i \times v_i \times f_i$ pixels. Furthermore, every pixel $p_i(x, y, t)$ is given as a color triple $\langle R, G, B \rangle$ representing the colors red R, green G, and blue B. An individual frame $F_{i,t}$ in cuboid i at time t consists of the pixels $p_i(\cdot, \cdot, t)$.

II. **Annotating pixels of cuboids** states, whether a pixel $p_i(x, y, t)$ belongs to the OOI. This essential processing step is only possible by a pixel-accurate object recognition in natural scenes. In addition to this binary pixel-wise classification, global information is provided, specifying whether the OOI is occluded by another object or if parts of the OOI are outside a certain frame $F_{i,t}$. Both information boost the result of the following processing modules.

III. **Annotation based pixel nullification** sets the pixels mark as not belonging to the OOI to null. Thus, after this module the cuboids $C_1, C_2, \ldots C_n$ represent the 2D projections of the OOI, their horizontal and vertical position within each frame as well as the temporal ordering.

IV. **Creating multiple pictorial representations** extracts and crops all non-empty frames from the cuboids. A frame is considered empty if it only contains nulls. Cropping means extracting the minimum bounding rectangle of all non-null pixels of a frame. This results in a sequence of views $V_1, V_2, \ldots, V_z$ showing 2D projections of the OOI from different perspectives. The different projections are caused by rotation, translation, and scaling of the 3D OOI. In addition, the temporal ordering of views is exported.

V. **Building a point cloud** of the OOI using the set of views $V_1, V_2, \ldots, V_z$. Therefore, feature points of each view are extracted. If more than two views have three feature points in common, one can compute their 3D position $\langle u, v, w \rangle$ in $\mathbb{R}^3$ using multiple view geometry. Following this principle, image feature based SfM algorithms calculate a dense point cloud of the OOI.

VI. **Meshing the point cloud** creates the surface of the OOI based on a point cloud. From the several different approaches that exist, we suggest the ball-pivoting algorithm by Bernardini et al. [4] because it does not interpolate points of the point cloud thus it preserves their position during meshing. However, unlike for example the Poisson meshing [16], the resulting mesh is in most instances not watertight.

3   Related Work

For an overview of related work, we describe only the three major topics: (i) image processing software and frameworks, (ii) image segmentation and annotation tools, and (iii) 3D reconstruction techniques.

3.1   Image Processing Software and Frameworks

Image processing software and frameworks are the essential keys for fast and successful implementation of architecture prototypes. They offer various functionalities for standard images and video handling and processing. For example, *MATLAB* [42], a tool that is often used in the field of academic and industrial development, provides ready-to-use classes, e.g., *VideoReader* for reading video files compressed by common codecs like *MPEG-1*, *MPEG-4*, *H.264*, *3GPP*, *AVCHD* and *AVI*. Furthermore, *MATLAB* as a mathematical software tool efficiently implements mathematical operations such as matrix multiplications, sorting algorithms as well as standard process flow elements like loops and branches.

Similar to the proprietary software *MATLAB*, the Open Source Computer Vision library (*OpenCV*) [15] offers a wide variety of implemented algorithms for computer vision and image processing. Due to its implementation in C and C++ and its quite active community, it is maintained quite well and constantly introduces new state of the art algorithms. Its implementation in C/C++ enables the combination of *OpenCV* with other frameworks like *QT* for creating user interfaces (UI) and software libraries like the *EIGEN* library for mathematical operations or the *OpenNI* library for the integration of natural UIs like the *Kinect* camera.

Like *MATLAB*, the python computer vision library (*PCV*) [39] is interpreted at runtime, so it is executable without any compilation. Although the book *Programming Computer Vision with Python* promotes the *PCV* library, it has only a small community. Rather than *PCV*, the python-bindings of *OpenCV* are more popular.

3.2   Image Segmentation and Annotation Tools

Fully automatic segmentation methods for pixel-wise object detection from images, especially from image sequences, seems far from being a solved problem. Further, it can be observed that the performance scores [5] of segmentation algorithms drop significantly from easier data sets, i.e., showing the complete *OOI*, to natural data sets, i.e., showing a partially occluded *OOI*. Even the run time of segmentation methods [5] is, in our opinion, not applicable for high resolution images. To overcome inaccurate segmentation, interactive segmentation algorithms like *GrabCut* [30], interactive graph cut [6], or interactive 3D labeling with SemanticPaint [46] are promising approaches.

Annotation tools for polygon-shaped labeling of *OOI* can be seen as an extension of interactive segmentation algorithms. Thus, existing video annotation tools could be used for the implementation of processing modules **I** and **II** of our proposed architecture. According to a review of seven video annotation tools by Dasiopoulou et al. [8] only one tool, the *Video and Image Annotation tool* (*VIA*) [22] provides pixel-wise *OOI* annotations. Unfortunately, *VIA* seems to be incapable of processing video sequences in high resolution and only supports *AVI* as input format.

The Video Performance Evaluation Resource (*ViPER*) [10] is still quite popular for video annotation due to its properly defined and specified *XML* output format. The object labeling process with this tool can be sped up by an automatic linear 2D propagation of the annotations.

Recently, we developed and released an open source video annotation tool named *iSeg* [32, 33]. It came with a novel interactive and semi-automatic segmentation process for creating pixel-wise labels of several objects in video sequences. The latest version of *iSeg*, implemented in *OpenCV* and *QT*, supports multicore processing and provides a semantic timeline for defining if an object is visible, partially occluded, or completely occluded in a certain frame. Its output format is inspired by *ViPER*'s *XML* format.

3.3   3D Reconstruction Techniques

In the wide research area of 3D reconstruction, one can observe two main strategies: *fully automatic reconstruction*, e.g., for robot navigation [14,24,27] and *interactive reconstruction*, e.g., for object replication [7,17,41,49]. For our bio-inspired architecture, only 3D reconstruction approaches are relevant that work with images or videos captured by monocular cameras. Techniques using stereo input data or even special equipment [37] are neglected here.

Fully automatic reconstruction methods based on large image collections have been shown to produce sufficiently accurate 3D models [18,36]. In 2006, Snavely et al. [38] published the *photo explorer*, which uses an unstructured image collection of a scene, e.g., acquired from the Internet, and converts the images to a 3D model. Using a monocular hand-held camera instead of a set of images, Pollefeys et al. [29] build visual models of scenes, used for one of the first augmented reality approaches. Filling the gap between the fully automated and the interactive approaches, Tanskanen et al. [41] ask a user to move around with a consumer smart phone to record *OOI*. Then the 3D object is reconstructed and visualized on the smart phone. While the scene is static and the camera is moving in [41], Pan et al. solve the opposite problem: In near real-time, the probabilistic feature-based online rapid model acquisition tool reconstructs objects rotated freely in front of a static video camera [26]. This system guides the user actively to acquire the views necessary for reconstruction (e.g., "please rotate the object clockwise").

Major challenges for such automatic reconstruction techniques are the handling of delicate structures, textureless surfaces, hidden boundaries, illumination, specularity, or even dynamic or moving objects, etc. [36]. All of these challenges are natural to real world videos. Motivated by these issues, Kowdle et al. [18] came up with a semi-automatic approach, which embeds the user in the process of reconstruction. Their interactive approach yields high quality reconstructions of an *OOI* or a complete scene from an image collection. Other interactive approaches are based on the presence of common geometric primitives, such as man-made architecture [2,9,17,23] for reconstruction. Usually, such approaches consist of automatic pre-processing for edge extraction or feature point detection for the computation of an initial polygon based 3D model. In the following interactive processing step, the user iteratively refines the model by editing or expanding the polygon model. Hengel et al. [12,13] expanded these approaches and purpose sketch-like 3D drawing for 3D model reconstruction in videos.

All of the mentioned techniques have in common that the reconstructed 3D objects are monolithic, that is, the models cannot be split into semantically meaningful entities such as parts and subparts. Today, monolithic 3D models are sufficient for e.g., indoor navigation [14,27] and reconstruction of urban environments [28]. For advanced tasks, however, the detailed reconstruction of objects using, e.g., a 3D printer, part-based information might be necessary. As a consequence, further research in 3D reconstruction should also be focusing on semantic part detection both in 3D models [31] and, as a starting point, in the input data.

4   Implementation of the Architecture

We give a proof of concept of the proposed bio-inspired architecture using a first implementation (Fig. 1). In addition, we have developed an object label exchange and visualization format. Further, we will point out analogies between the proposed system and the view-combination scheme by Ullman [43].

4.1   Implementations

For a bottom-up implementation of processing module **I**, the first task is to review programming languages and programming tools. Of major importance is the handling of compressed video sequences and efficient performance on matrix operations for image analysis. So we started the prototype implementation in *MATLAB*, which facilitates working independently on different platforms. After re-programming an automatic segmentation algorithm [11], we noticed that the recognized OOI often do not match the OOI we want to reconstruct. Further, we noticed that the accuracy of the segmentation differs strongly from man-made segmentation, as pointed out by Borji et al. [5].

So, for a proof of concept, the automatic segmentation algorithm is substituted by our interactive and more accurate tool *iSeg* [32,33]. *iSeg*'s current implementation completely covers the processing modules **I** and **II**. For creating a well-defined interface—here a general exchange format—from processing module **II** to module **III**, we have defined a multimedia container format containing the video sequence, the polygon-shaped pixel-wise label of the OOI and additional information (e.g., about occlusion) [34,35]. Using this exchange format, it will be possible, e.g., to replace the interactive OOI segmentation with other automatic approaches easily.

For designing a general exchange format, we chose the open *Matroška* container format [20] (*mkv*) as encapsulation format. The major advantage of *mkv* is allowing different types of payload besides video files. The pixel-wise labels of the OOIs, as well as additional information, are encoded by the extended version V4.00+ [40] of the *Sub Station Alpha* (*SSA*) video subtitle file format— also known as *Advanced SSA* (*ASS*). A side effect of using the existing subtitle format *ASS* is that our exchange format provides an immediate visualization of pixels which have been recognized to be part of an OOI (it is provided that up-to-date media players are used). As shown in Fig. 2, the additional information is visualized, as well. Based on the source code of *iSeg*, the export functionality is altered such that it directly exports the labeled OOIs as *mkv* files.

The processing modules **III** and **IV**, implemented in *MATLAB*, read video sequences containing the annotated OOI, create the cuboids $C_1, C_2, \ldots C_n$ and perform the pixel nullification based on the OOI annotation. The resulting views

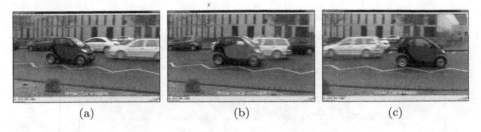

(a)   (b)   (c)

Fig. 2. Example views of the object label exchange and visualization format, which is used between the processing modules **II** and **III**; generated by *iSeg*. Using an up-to-date media player, the proposed *mvk*-based format visualizes the annotated *OOI* (here: the white car) together with additional information: (a) white car is visible—(b) next time white car is partially occluded—(c) next time white car is visible again.

$V_1, V_2, \ldots V_z$, where the background is completely masked out, are exported as multiple *jpg* images. These multiple images are a pictorial representation of the *OOI*. As argued in Sect. 2, getting the pictorial representations, is the changing part of the process because the *OOI* must be recognized on every frame. Using these pictorial representations, out of the box SfM algorithms can create the point cloud of the *OOI* out of the views $V_1, V_2, \ldots, V_z$. In this implementation, *VisualSfM* [47] is used as a tool for creating the point cloud out of the pictorial representation $V_1, V_2, \ldots, V_z$ of the *OOI*. Based on this point cloud, the ball-pivoting algorithm [4] using *Meshlab* [21] is applied to create the surface of the *OOI*. An overview which software or tool is used for realizing a certain processing module can be obtained from row *tools & comments* in Fig. 3.

4.2   Analogies to the View-Combination Scheme

In the view-combination scheme, Ullman stated that "recognition by multiple pictorial representations and their combinations constitutes a major component of 3D object recognition" [43, p. 154] of humans and "unlike some of the prevailing approaches ... object recognition ... is simpler, more direct, and pictorial in nature." [43, p. 154]. Identified analogies between the view-combination and our architecture are illustrated in Fig. 3.

The first four processing modules **I–IV** of our bio-inspired architecture reduces the complexity of the input information, from unclassified pixels over segmented pixels to multiple pictorial representations without any contextual information. Further, in this process translational invariance is achieved, which parallels biological processing: "Cells along the hierarchy from *V1* to *V4* also show an increasing degree of tolerance for the position and size of their preferred stimuli" [43, p. 152]. In the next step, "an object appears to be represented in *IT* by multiple units, tuned to different views of the object" [43, p. 152], which describes, on a general level, the idea of the multi view geometry—processing module **V**.

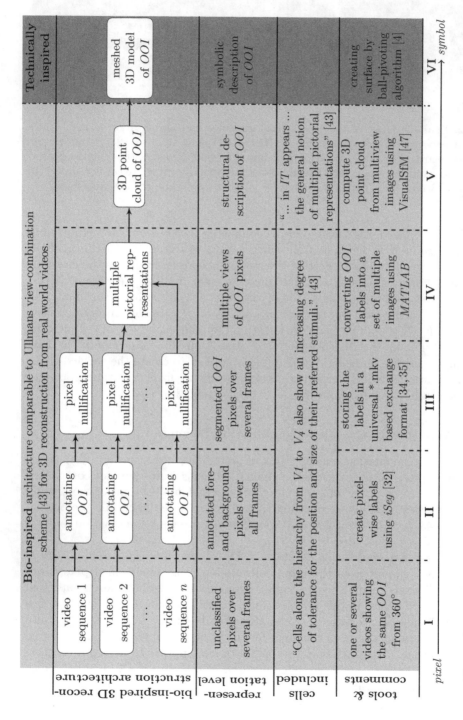

Fig. 3. Comparison of the bio-inspired 3D reconstruction architecture to the view-combination scheme by Ullman [43].

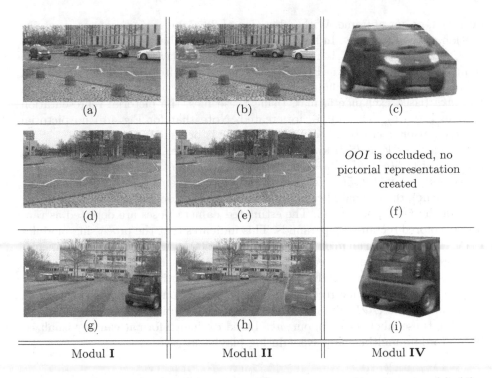

(a)	(b)	(c)
(d)	(e)	*OOI* is occluded, no pictorial representation created (f)
(g)	(h)	(i)
Modul **I**	Modul **II**	Modul **IV**

Fig. 4. Exemplary result of applying our architecture to two video sequence; (a), (d), (g) unprocessed video; (b), (e), (h) *OOI* is pixel-wised labeled; (c), (i) multiple pictorial representation; (f) no pictorial representation due to occlusion of *OOI*.

Up to processing module **V**, one can see a strong analogy between the bottom-up idea of the view-combination scheme and the architecture proposed in this paper. Processing module **VI** might be a part of the top-down process, representing an internal object model. However, in our opinion, a full 3D representation of an object is mainly technically motivated and cannot be mapped to Ullman's scheme.

5   Experimental Application of the Architecture

We use the first two video sequences of the benchmark data set for evaluating visualization and analysis techniques for eye tracking for video stimuli of Kurzhals et al. [19]. They show the same rigid *OOI*, a red car, driving along roads. Both video sequences provide a resolution of 1920×1080 pixels at a frame rate of $25 fps$, exemplary frames are shown in Fig. 4(a), (d), and (g). The first sequence, with a duration of 25 s, contains both object motion and camera movement. It includes frames, where the *OOI* is partially occluded, completely occluded, or out of frame. With a duration of 28 s, the second sequence contains motion of the *OOI* only. In both sequences, the number plates are blurred.

Using *iSeg* the *OOI* is recognized in an interactive process in both video sequences, as illustrated in Fig. 4(b), (e), and (h). After processing module **II**, both video sequences including the pixel-wise labels of the *OOI* are stored in the *mkv* based exchange format. Using these files, the pictorial representation seen in Fig. 4(c), (f), and (i) are created. Note that for Fig. 4(e) no pictorial representation is created, because the red car is occluded.

Using Wu's *VisualSfM* system, a point cloud of the *OOI* with 22 548 vertices, shown in Fig. 5(c) and (d), is calculate out of the multiple pictorial representations. As seen in Fig. 5(a) (without bundle adjustment) and (b) (with bundle adjustment), the estimated camera poses are mainly located around 3D rotating axes of the *OOI* point cloud. The estimated camera poses are depicted as randomly colored rectangular pyramids. This indicates that the processing modules **I–IV** eliminate both all position changes of the *OOI* and the camera movements. The resulting surface, created by ball-pivoting is presented in Fig. 5(e) and (f).

As supplementary material, a video demonstrates the application of our bio-inspired architecture for reconstructing 3D models out of video sequences. In addition, all the *OOI* labels used for this reconstruction can be downloaded[1]. Besides, these labels saved in our *mkv* based exchange format can be visualized by up-to-date multimedia players during playback.

6   Results

If the resulting meshed model (Fig. 5(e) and (f)) is regarded only in a quantitative manner, one can assume that the proposed bio-inspired architecture has failed to reconstruct the *OOI* from natural, non-laboratory made video sequences which show an *OOI* from 360°. However, taking into account the quantitative quality of the resulting dense point cloud, our architecture successfully reconstructed the *OOI*. Caused by the missing top and bottom views in the videos the number of points in these regions is very limited. From the side, frontal and back view of the car the point cloud provides a sufficient amount of points.

This result does not contradict our prospects because only the processing modules **I–V** exhibit similarities to the ventral visual stream [44, 45] and to the view-combination scheme [43]. As a consequence, the technically inspired processing module **VI** must be replaced by another module. Focusing on Ullman's work, the module **VI** might be replaced with a top-down approach. For example, a wire-frame [50] or a CAD [48] model might be used to fit surfaces onto the created point cloud. So the final model including surfaces would result out of the internal intermediate 3D representation and out of a knowledge data base containing generic 3D models.

---

[1] https://ikw.uos.de/%7Ecv/publications/3DMA16.

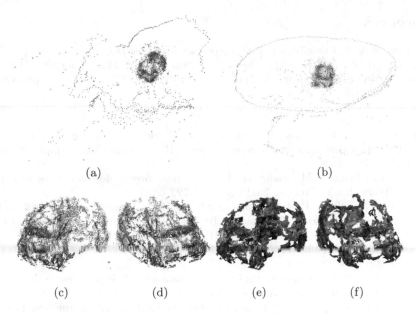

Fig. 5. Exemplary result of processing module **V**: (c), (d) computed point cloud by *VisualSfM* with 22 548 vertices, estimated camera poses without (a) and with (b) bundle adjustment; and processing module **VI**: (e) and (f) resulting mesh by ball pivoting.

7 Conclusion and Further Work

In conclusion, it can be said that our bio-inspired architecture for deriving 3D models from video sequences can reconstruct a 3D point cloud of an *OOI* if the video sequences contain different views of the *OOI*. The creation of fully meshed models is not yet possible, due to the quite sparse point cloud. Further, we conclude that the proposed architecture can work autonomously as long as the problem of semantic pixel-wise object annotation and recognition is solved. So far, accurate pixel-wise object annotation and recognition are possible with a little user interaction. Thus, this architecture can already be implemented and used for deriving 3D models from videos.

For the future, one should focus on three main topics: (i) creating a benchmark data set with video sequences showing the same *OOI* including the corresponding CAD models of the *OOI* as ground truth, (ii) improving our architecture by replacing processing module **VI** with a top-down approach based on geometric primitives like wire-frames or generic 3D models, and (iii) integrating feedback loops between the processing modules, which is common between the cell areas in the ventral visual stream.

References

1. Agisoft: Agisoft PhotoScan (2016), http://www.agisoft.ru/
2. Arikan, M., Schwärzler, M., Flöry, S., Wimmer, M., Maierhofer, S.: O-Snap: optimization-based snapping for modeling architecture. ACM Trans. Graph. **32**(1), 6:1–6:15 (2013)
3. Autodesk Inc.: Autodesk 123D Catch|3D model from photos (2016). http://www.123dapp.com/catch
4. Bernardini, F., Mittleman, J., Rushmeier, H., Silva, C., Taubin, G.: The ball-pivoting algorithm for surface reconstruction. IEEE Trans. Vis. Comput. Graph. **5**(4), 349–359 (1999)
5. Borji, A., Cheng, M.M., Jiang, H., Li, J.: Salient object deection: a benchmark. IEEE Trans. Image Process. **24**(12), 5706–5722 (2015)
6. Boykov, Y., Jolly, M.P.: Interactive graph cuts for optimal boundary and region segmentation of objects in N-D images. In: IEEE International Conference on Computer Vision (ICCV), pp. 105–112 (2001)
7. Chen, T., Zhu, Z., Shamir, A., Hu, S.M., Cohen-Or, D.: 3-Sweep. ACM Trans. Graph. **32**(6), 1–10 (2013)
8. Dasiopoulou, S., Giannakidou, E., Litos, G., Malasioti, P., Kompatsiaris, Y.: A survey of semantic image and video annotation tools. In: Paliouras, G., Spyropoulos, C.D., Tsatsaronis, G. (eds.) Knowledge-Driven Multimedia Information Extraction and Ontology Evolution. LNCS (LNAI), vol. 6050, pp. 196–239. Springer, Heidelberg (2011). doi:10.1007/978-3-642-20795-2_8
9. Debevec, P.E., Taylor, C.J., Malik, J.: Modeling and rendering architecture from photographs: a hybrid geometry-and image-based approach. In: Computer Graphics and Interactive Techniques - SIGGRAPH, pp. 11–20 (1996)
10. Doermann, D., Mihalcik, D.: Tools and techniques for video performance evaluation. Int. Conf. Recogn. (ICPR) **4**, 167–170 (2000)
11. Grundmann, M., Kwatra, V., Han, M., Essa, I.: Efficient hierarchical graph-based video segmentation. In: IEEE Computer Vision and Pattern Recognition (CVPR), pp. 2141–2148 (2010)
12. van den Hengel, A., Dick, A., Thormählen, T., Ward, B., Torr, P.H.S.: VideoTrace: rapid interactive scene modelling from video. ACM Trans. Graph. **26**(3), 86:1–86:6 (2007)
13. van den Hengel, A., Hill, R., Ward, B., Dick, A.: In situ image-based modeling. In: IEEE International Symposium on Mixed and Augmented Reality (ISMAR), pp. 107–110 (2009)
14. Henry, P., Krainin, M., Herbst, E., Ren, X., Fox, D.: RGB-D mapping: using depth cameras for dense 3D modeling of indoor environments. In: Khatib, O., Kumar, V., Sukhatme, G. (eds.) Experimental Robotics, pp. 477–491. Springer, Heidelberg (2014)
15. Itseez: OpenCV — OpenCV (2016). http://opencv.org/
16. Kazhdan, M., Hoppe, H.: Screened poisson surface reconstruction. ACM Trans. Graph. **32**(3), 1–13 (2013)
17. Kholgade, N., Simon, T., Efros, A., Sheikh, Y.: 3D object manipulation in a single photograph using stock 3D models. ACM Trans. Graph. **33**(4), 127:1–127:13 (2014)
18. Kowdle, A., Chang, Y.J., Gallagher, A., Batra, D., Chen, T.: Putting the user in the loop for image-based modeling. Int. J. Comput. Vis. **108**(1), 30–48 (2014)
19. Kurzhals, K., Bopp, C.F., Bässler, J., Ebinger, F., Weiskopf, D.: Benchmark data for evaluating visualization and analysis techniques for eye tracking for video stimuli. In: Workshop on BELIV, pp. 54–60 (2014)

20. Matroska: Matroska media container (2016). https://www.matroska.org/
21. MeshLab: Meshlab (2016). http://meshlab.sourceforge.net/
22. Multimedia Knowledge and Social Media Analytics Laboratory: Video image annotation tool (2015). http://mklab.iti.gr/project/via
23. Musialski, P., Wonka, P., Aliaga, D.G., Wimmer, M., Gool, L., Purgathofer, W.: A survey of urban reconstruction. Comput. Graph. Forum. **32**, 146–177 (2013)
24. Newcombe, R.A., Izadi, S., Hilliges, O., Molyneaux, D., Kim, D., Davison, A.J., Kohi, P., Shotton, J., Hodges, S., Fitzgibbon, A.: KinectFusion: real-time dense surface mapping and tracking. In: IEEE International Symposium on Mixed and Augmented Reality (ISMAR), pp. 127–136 (2011)
25. Oliva, A., Torralba, A.: The role of context in object recognition. Trends Cogn. Sci. **11**(12), 520–527 (2007)
26. Pan, Q., Reitmayr, G., Drummond, T.: ProFORMA: probabilistic feature-based on-line rapid model acquisition, pp. 112:1–112:11. British Machine Vision Conference (BMVC) (2009)
27. Pintore, G., Gobbetti, E.: Effective mobile mapping of multi-room indoor structures. Vis. Comput. **30**(6), 707–716 (2014)
28. Pollefeys, M., Nistér, D., Frahm, J.M., Akbarzadeh, A., Mordohai, P., Clipp, B., Engels, C., Gallup, D., Kim, S.J., Merrell, P., Salmi, C., Sinha, S., Talton, B., Wang, L., Yang, Q., Stewénius, H., Yang, R., Welch, G., Towles, H.: Detailed real-time urban 3D reconstruction from video. Int. J. Comput. Vis. **78**(2), 143–167 (2008)
29. Pollefeys, M., Van Gool, L., Vergauwen, M., Verbiest, F., Cornelis, K., Tops, J., Koch, R.: Visual modeling with a hand-held camera. Int. J. Comput. Vis. **59**(3), 207–232 (2004)
30. Rother, C., Kolmogorov, V., Blake, A.: GrabCut. ACM Trans. Graph. **23**(3), 309–314 (2004)
31. Schöning, J.: Interactive 3D reconstruction: new opportunities for getting CAD-ready models. In: Imperial College Computing Student Workshop (ICCSW). OpenAccess Series in Informatics (OASIcs), vol. 49, pp. 54–61. Schloss Dagstuhl-Leibniz-Zentrum fuer Informatik, Dagstuhl, Germany (2015)
32. Schöning, J., Faion, P., Heidemann, G.: Semi-automatic ground truth annotation in videos: an interactive tool for polygon-based object annotation and segmentation. In: International Conference on Knowledge Capture (K-CAP), pp. 17:1–17:4. ACM, New York (2015)
33. Schöning, J., Faion, P., Heidemann, G.: Pixel-wise ground truth annotation in videos - an semi-automatic approach for pixel-wise and semantic object annotation. In: International Conference on Pattern Recognition Applications and Methods (ICPRAM), pp. 690–697. SCITEPRESS (2016)
34. Schöning, J., Faion, P., Heidemann, G., Krumnack, U.: Eye tracking data in multimedia containers for instantaneous visualizations. In: IEEE VIS Workshop on Eye Tracking and Visualization (ETVIS), IEEE (2016)
35. Schöning, J., Faion, P., Heidemann, G., Krumnack, U.: Providing video annotations in multimedia containers for visualization and research. In: IEEE Winter Conference on Applications of Computer Vision (WACV) (2017)
36. Schöning, J., Heidemann, G.: Evaluation of multi-view 3D reconstruction software. In: Azzopardi, G., Petkov, N. (eds.) CAIP 2015. LNCS, vol. 9257, pp. 450–461. Springer, Heidelberg (2015). doi:10.1007/978-3-319-23117-4_39

37. Schöning, J., Heidemann, G.: Taxonomy of 3D sensors - a survey of state-of-the-art consumer 3D-reconstruction sensors and their field of applications. In: Joint Conference on Computer Vision, Imaging and Computer Graphics Theory and Applications (VISAPP), vol. 3, pp. 194–199. SCITEPRESS (2016)
38. Snavely, N., Seitz, S.M., Szeliski, R.: Photo tourism: exploring photo collections in 3D. ACM Trans. Graph. **25**(3), 835–846 (2006)
39. Solem, J.E.: Programming Computer Vision with Python: Tools and Algorithms for Analyzing Images. O'Reilly Media Inc., Sebastopol (2012)
40. Sub Station Alpha: Sub station alpha v4.00+ script format (2016). http://moodub.free.fr/video/ass-specs.doc
41. Tanskanen, P., Kolev, K., Meier, L., Camposeco, F., Saurer, O., Pollefeys, M.: Live metric 3D reconstruction on mobile phones. In: IEEE International Conference on Computer Vision (ICCV), pp. 65–72. IEEE (2013)
42. The MathWorks Inc: MATLAB - MathWorks (2016). http://mathworks.com/products/matlab
43. Ullman, S.: High-level Vision: Object Recognition and Visual Cognition, 2nd edn. MIT Press, Cambridge (1997)
44. Ungerleider, L.: What and where in the human brain. Curr. Opin. Neurobiol. **4**(2), 157165 (1994)
45. Ungerleider, L., Mishkin, M.: Two cortical visual systems. In: Ingle, D., Goodale, M., Mansfield, R. (eds.) Analysis Visual Behavior, pp. 549–586. MIT Press, Boston (1982)
46. Valentin, J., Torr, P., Vineet, V., Cheng, M.M., Kim, D., Shotton, J., Kohli, P., Niener, M., Criminisi, A., Izadi, S.: Semanticpaint. ACM Trans. Graph. **34**(5), 1–17 (2015)
47. Wu, C.: VisualSfM: a visual structure from motion system (2016). http://ccwu.me/vsfm/
48. Xiang, Y., Mottaghi, R., Savarese, S.: Beyond PASCAL: a benchmark for 3D object detection in the wild. In: IEEE Winter Conference on Applications of Computer Vision (WACV), pp. 75–82 (2014)
49. Zhang, Y., Gibson, G.M., Hay, R., Bowman, R.W., Padgett, M.J., Edgar, M.P.: A fast 3D reconstruction system with a low-cost camera accessory. Sci. Rep. **5**, 10909:1–10909:7 (2015)
50. Zhang, Z., Tan, T., Huang, K., Wang, Y.: Three-dimensional deformable-model-based localization and recognition of road vehicles. IEEE Trans. Image Process. **21**(1), 113 (2012)

DSLIC: A Superpixel Based Segmentation Algorithm for Depth Image

Ali Suryaperdana Agoes$^{(\boxtimes)}$, Zhencheng Hu, and Nobutomo Matsunaga

Department of Information Technology on Human and Environmental Science,
Kumamoto University, Kumamoto, Japan
ali@st.cs.kumamoto-u.ac.jp, huzc@wissenstar.com,
matunaga@cs.kumamoto-u.ac.jp

Abstract. Limited illumination outdoor and indoor environment leads to the lack of object's color information. Faced with this situation, it is not always possible to generate superpixel by using RGB or LaB features. To tackle this scenario, we propose a superpixel generation algorithm solely on depth image. We aim the semantically-incoherent superpixel problem on depth image, caused by identical depth value in the vicinity of the border. Our algorithm is an adaptation of Simple Linear Iterative Clustering (SLIC) with a novel utilization of depth and gradient direction as an alternate of LaB color space features. Our novel approach is demonstrated perform favorably to over-segment large planar area in an unlit environment.

1 Introduction

Superpixel term introduced by Ren and Malik [9] is a group of pixels with similar perceptual properties. A Group of pixels that form a superpixel may derived from an RGB image, a depth image or a RGBD image accordingly. In an RGB image, the light intensity is the only information gathered regarding the scene. RGB image considered prone to illumination change, partially illuminated object, and background-object's color separation. Encouraged by those, utilization of depth image as addition of color information then begin to widely used in computer vision research field.

In conjunction with that, more researcher adopting RGBD image based superpixel into their framework in the computer vision field. For example scene understanding [2], image segmentation [15], indoor segmentation and support inference [12]. They have similarities, which are utilization of RGBD image in oversegmentation steps. Nathan's [12] superpixel initialization set use watershed algorithm on g-Pb boundary, in which the faint intensity pixel was overcome with the aid of 3D plane region generation. Based on the similar superpixel algorithm, [15] used a combination of depth and intensity image in color difference [5] to overcome indistinct plane boundaries. Gupta's [2] works to give scene understanding built upon a watershed algorithm generated superpixel, their superpixel boundary approach used features from monocular cues and depth image.

© Springer International Publishing AG 2017
C.-S. Chen et al. (Eds.): ACCV 2016 Workshops, Part II, LNCS 10117, pp. 77–87, 2017.
DOI: 10.1007/978-3-319-54427-4_6

However, in outdoor and indoor environment which has limited illumination, it is not always possible to realize superpixel using RGB or LaB features. Figure 1(a) captured from NYU depth dataset V2 scenes, shows the typical unlit indoor situation. In this scenario, subtle different of object-background color will inadequate to produce semantically-coherent superpixel. On the other hand, as a consequence the limitation of color information, whereas depth is the alternative scenes information. Adjacent pixel which contains identical depth amplitude will be grouped into 1 superpixel with respect to its size. In some scene, pixels located in the vicinity of surface border tends to cross clustered. Or in other word depth semantically-incoherent superpixel generated in these circumstances, to illustrate see Fig. 1(c).

Several superpixel algorithms introduced in recent years, one of the them that draw researcher attention is SLIC. Achanta et al. [1] reported SLIC, a superpixel generator algorithm which has contribution of region shape compactness and efficient computation time [7]. We modify SLIC features components to work solely on depth image. We name this modification as Depth Simple Linear Iterative Clustering (DSLIC). DSLIC propose the utilization of depth and gradient direction as an alternate of LaB color space feature, and incorporate $x - y$ pixel spatial coordinate feature. First, we employed median filter to filtered hole and noise in depth image. Then, we reduced depth amplitude local variation by the means of L0 gradient minimization [16]. Finally, we iteratively clustered pixel based on proposed features.

The main contribution of this research paper is a novel approach to generate superpixel solely on depth image. We demonstrate that the employment of depth and gradient direction features can cope depth image semantically-incoherent superpixel problem. A depth image scene taken from NYU depth dataset V2, preprocessed then over-segmented into superpixel, SLIC is shown in Fig. 1(c). And, DSLIC our approach result is shown in Fig. 1(d).

This paper presented as follows. Section 2 overview of SLIC related superpixel generation algorithm. Section 3 describes our depth image preprocessing. Section 4 shows our proposed DSLIC method. Experiment result presented in Sect. 5, and conclusions are given in Sect. 6.

2   Related Work

In recent years, many studies that have been done in the scope of SLIC superpixel. For example, Schick et al. [11] has reported better object boundary adherence by applying a distance measure solely to the pixel belonging superpixel boundary. Ren's worked in [8], improve SLIC computation time performance by implementing parallel computation in GPU and the NVIDIA CUDA framework. The most recent SLIC modification called nSLIC done by Jia et al. [3], they resolve the compactness parameter issue by the original SLIC with adjustable distance measure developed in five dimensional space. However, none of this work takes into account the limited color information to implement SLIC. The Superpixel generation method presented in this paper differs from the references

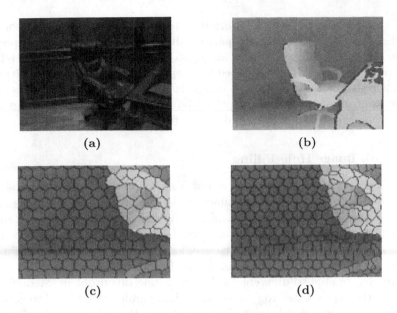

Fig. 1. (a) and (b) Show RGB and raw depth image from NYU depth dataset V2. (c) SLIC algorithm, and (d) DSLIC, method proposed in this paper to solve depth image semantically-incoherent superpixel problem. (Color figure online)

above in our approach to perceive the environment. We show that utilization of depth image solely is adequate to generate semantically-coherent superpixel.

The most related to our work was presented by Yang et al. [17]. Their work, 3-D geometrically enhanced SLIC to distinguish adjacent similar color object. A 3-D coordinate of the pixel was procured by incorporating depth and pixel location $(x - y)$. Furthermore, they employed normalized 8 dimensional $[Lab, XYZ, xy]$ features space. In which Lab denotes the CIELAB color space, $[XYZ]$ denotes 3-D geometry coordinate, and $[xy]$ denotes pixel coordinates in the image plane. Next, pixels clustered into superpixel iteratively based on these. Our work differs from them, in the way we express features space. Since our work incorporates only depth image, our features space expressed in $[d_{ddg}, d_{dgrad}, xy]$. In which $[d_{ddg}]$ stands for depth amplitude, $[d_{dgrad}]$ stands for depth gradient, and $[xy]$ stands for pixel spatial coordinate. Other than that, our distance metric separates the depth amplitude from pixel spatial coordinates. Thus, the depth amplitudes itself contribute independently into the distance metric.

3   Depth Image Preprocessing

Raw disparity image provided by NYU depth dataset V2 comes along with a set of toolbox to process with. However, the toolbox they provide devoted to work on the RGBD image. We did a minor adjustment in some functions that we used. Furthermore, our way to process raw disparity image similar to what was on the [12]. Firstly, we undistorted it to remove Kinect's IR sensor lens defect. Secondly, we convert the disparity value into depth amplitude expressed in meter. Then, we aligned depth image to the world coordinates and RGB camera of Kinect. Finally, we cropped the depth image. Once the raw disparity image has been processed into depth image. It undergoes into next process which are holes filling and reducing the depth amplitude local variation by L0 gradient minimization. We describe in more detail in following.

3.1 Depth Image Hole Filling

In some scenario, Kinect's IR part is not able to perceive objects shape adequately. This observed as holes contained in depth image, as shown in Fig. 2(a). The hole formed by reflection of the emitted ray from Kinect out of sensor range. Another possibility is that, the surface itself has a bad reflection characteristic. To solve this problem, NYU depth dataset V2 toolbox provides functions to fill the hole. However, this function employs the RGB image as reference.

Therefore, we applied different method to tackle this problem. Median filter took solely the depth image to process. In which, parallel to our goal to generate superpixel based on a depth image features. Other than that, the median filter has widely known as an edge preserving filter. In addition to the interesting nature of a median filter, the hole in the sliding window was excluded from the median calculation. In this way, the object geometrical shape preserved. Consequently, this would replace the missing value while object's shape not deteriorate. Furthermore, to make the filtering computation more efficient. An image mask has been applied to drive the sliding window peculiarly at holes part. The median value inside window has been calculated to alter the missing depth amplitude. Holes filled image results through the utilization masked median filter (MMF) shown in Fig. 2(b).

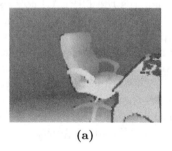   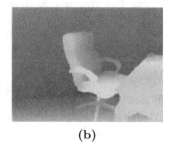

(a)   (b)

Fig. 2. (a) Holes (displayed as dark blue spot or dark gray as viewed in BW prints, located at the table and chair edge) contained in undistorted and aligned depth image, (b) after MMF applied. (Color figure online)

3.2   Depth Image Smoothing via L0 Gradient Minimization

The depth image composed of insignificant local variation depth amplitude. It confines the direct computation of the gradient direction into DSLIC features. Our way to reduce the depth amplitude local variation, by utilization of L0 gradient minimization introduced by Xu et al. [16]. It imposes non zero gradient number which lessens the depth amplitude local variation in image while object boundary globally preserved.

Implementation of L0 gradient minimization is straight forward. Depth image input treated as a 2D array. Then, 2 fold depth gradient for each neighboring pixel computed along x and y direction. While the constriction of depth gradient achieved through alternating optimization of L0 norm. The authors introduced variable κ to weight the L0 regularization, smaller values of κ require longer running time. Another variable is λ, where structure coarseness controlled by this variable. In our experiment we used value of $\kappa = 8$ and $\lambda = 0.01$ to reduce depth amplitude local variation. We found these combinations provide the adequate smoothness of depth amplitude while prominent object boundary preserved. Figures 3(a)–(h), Illustrate original and smoothen depth image through various method compared to L0 gradient minimization.

4   DSLIC Superpixel

Our proposed superpixel generation method employs 4 features $[d_{ddg}, d_{dgrad}, xy]$ which are depth amplitude $[d_{ddg}]$, depth gradient direction $[d_{dgrad}]$, and pixel spatial location $[xy]$ for x and y direction. Those features are modified from original SLIC, in which the spatial proximity term for depth features included in

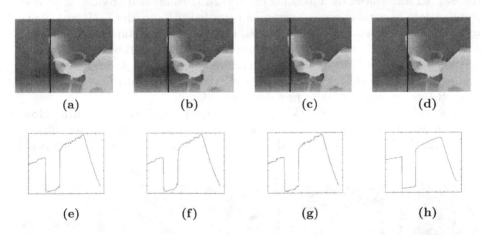

(a)   (b)   (c)   (d)

(e)   (f)   (g)   (h)

Fig. 3. Upper and lower, pairs of smoothed depth image using various method and its cross section accordingly, (a) and (e) Original depth image, (b) and (f) Colorization, (c) and (g) CBF, (d) and (h) L0 gradient minimization. (Color figure online)

the depth image extension. Our distance metric measures D_s in Eq. (1) separate the contribution factor of d_{ddg} which is the distance of depth amplitude d as in Eq. (2), d_{dgrad} which is the distance of depth gradient direction $ddgrad$ as in Eq. (3), and pixel spatial coordinates d_{xy} as in Eq. (4). Similar to the previous, we adopted variable m to control superpixel compactness in use with an image grid interval S.

$$D_S = d_{ddg} + d_{dgrad} + \frac{m}{S}d_{xy} \tag{1}$$

$$d_{ddg} = \sqrt{(d_k - d_i)^2} \tag{2}$$

$$d_{dgrad} = \sqrt{(ddgrad_k - ddgrad_i)^2} \tag{3}$$

$$d_{xy} = \sqrt{(x_k - x_i)^2 + (y_k - y_i)^2} \tag{4}$$

The DSLIC superpixel algorithm was commenced from the image division, some initial amount of superpixel corresponding to user input. The center of each area calculated as a starting point reference later in the clustering process. Next, the depth amplitude gradient calculated by the means of central difference filter on both x and y direction. Subsequently, gradient direction calculation step performed. Later on, iteratively correlate pixel features with respect to the center of each superpixel on each step. Lastly, spurious region merged into larger correspondence labeled segment.

5   Experimental Result

In this section, we present our experimental result evaluated on the NYU depth dataset V2 introduced by Silberman et al. [12]. It comprised of video sequences from variety of actual indoor scenes, as recorded in both the RGB and depth images. The captured indoor environment contains various cluttered scenes with a broad range of illumination level. However, in this experiment we take into account scene's low light condition. Therefore, we highlight several scenes that accommodates our consideration, shown in Fig. 4(a)–(e). Further on, we solely process the raw disparity data provided to work with.

The NYU depth dataset V2 provides the labeled ground truth. However, in some label images contain small unlabeled region between object, shown in Fig. 5(a). We adopted the technique from [10] to fill the unlabeled

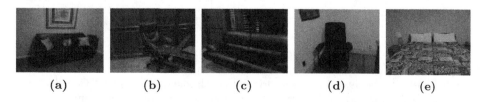

| (a) | (b) | (c) | (d) | (e) |

Fig. 4. NYU depth dataset V2 scenes that we utilized for experiments. (a) Scene 359, (b) Scene 385, (c) Scene 386, (d) Scene 536, and (e) Scene 1133.

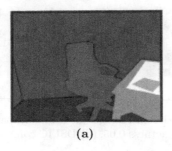 

(a)   (b)

Fig. 5. An example of label image from scene 385, (a) Unlabeled region displayed as black region, compared to (b) Filled unlabeled region.

region. And to fit to our experiment, filled label image cropped to match the depth image size. An example of filled region in label ground truth shown in Fig. 5(b).

Comparable result is imperative for superpixel generator algorithm method. For that purpose, we test our algorithm's performance against SLIC to work with depth image. We choose dim condition room with different exposing wall-border portion. Furthermore, we also evaluate filter addition in conjunction to our algorithm performance. Finally, we split superpixel performance evaluation into qualitative and quantitative assessments.

5.1 DSLIC Performance on Unfiltered Depth Image

The significance of depth preprocessing in our algorithm is not principal. As our main concern to generate semantically-coherent superpixel in the vicinity of wall-floor border. Its characteristic depends on the depth gradient, that we calculated on depth image. Which then, to be included in our proposed features. Moreover, the role of L0 gradient minimization filter is more likely to reduce spurious region on the generated superpixel without relocating edge. Number of spurious regions will later contribute to superpixel compactness. Additionally, MMF role to fill the hole in depth image also does not directly related to superpixel separation. We demonstrate the influence of the absence of the additional filter both on SLIC: Fig. 6(a) and (b), and DSLIC: Fig. 6(c) and (d).

5.2 Qualitative Assessment

In this subsection we present the qualitative assessment of superpixel, which determined visually by observing each result. Preprocessed depth images over-segmentation result are shown in Fig. 7. The results show that DSLIC outperform the SLIC in scenes which contain steep depth amplitude transition. For examples, in the scenes 359 and 385 pixels located on the horizontal vicinity wall-floor border. And also, pixels located around the vertical wall border in the scenes 536 and 1133. Superpixel generated by SLIC grouped pixels in that area into 1,

| (a) | (b) | (c) | (d) |

Fig. 6. Evaluating SLIC and DSLIC on unfiltered depth image. (a) SLIC compactness 0.05, (b) SLIC compactness 0.1, (c) DSLIC compactness 0.05, (d) DSLIC compactness 0.1.

because lack of gradient direction feature. On the other hand, DSLIC is able to produce semantically-coherent superpixel as a result of incorporating depth amplitude direction changes.

5.3 Quantitative Assessment

In this subsection we provide the quantitative assessment. We employ the error metric as suggested in [13]. We realize 2 from 6 of them which are boundary recall introduced in [6] and undersegmentation error defined in [4,14]. Boundary recall is a fraction of ground truth boundaries that are correctly detected by superpixel. Here, superpixel number contribute to the high boundary recall value. The more superpixel number, chance to hit the ground truth boundaries rise. Undersegmentation error is a number of superpixel that occupying outside the specific ground truth. In which, smaller size of superpixel will lead to reduced region leakage. By the means of these two variables, superpixel performance assessment will be given in Fig. 8.

Figure 8 show that our method performs consistently toward Fig. 7. Our method outperforms SLIC significantly in boundary recall for a scene containing a large portion of steep depth amplitude, such in scene 1133. Although DSLIC

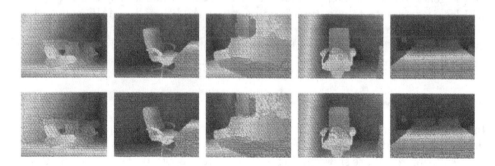

Fig. 7. Visual comparison for scene 359, 385, 386, 536, and 1133 from left to right. Top and bottom images are obtained by SLIC and DSLIC accordingly.

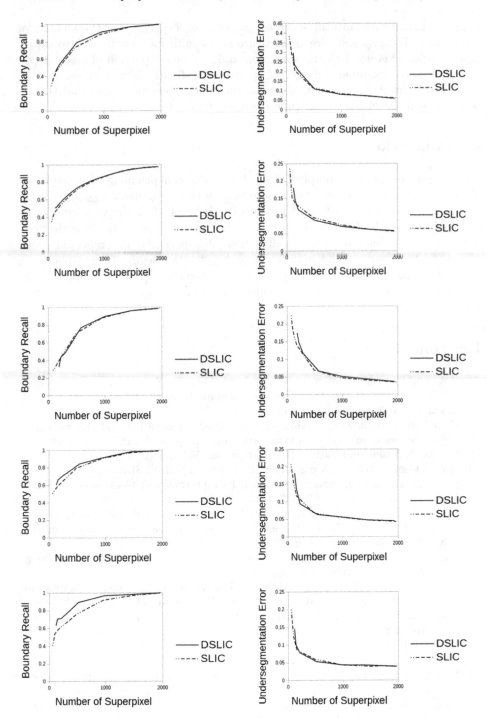

Fig. 8. Quantitative assessment for SLIC and DSLIC performance. Left side for boundary recall, and right side for undersegmentation error. From top and bottom, scene 359, 385, 386, 536, and 1133.

able to generate semantically-coherent superpixel, especially in the vicinity of the border. The gradient direction feature has significant contribution to super-pixel shape. Therefore, at the scene containing a small portion of steep depth amplitude. We examine that DSLIC performed slightly increased toward the SLIC counterpart. Both quantitative and qualitative results suggest that DSLIC produce semantically-coherent superpixel on the depth image.

6   Conclusion

We present a dedicated superpixel algorithm based on depth image segmentation. We cluster iteratively pixels into superpixel, by the introduced $[d_{ddg}, d_{dgrad}, xy]$ features. By the means of these features, merged pixel in the vicinity of bor-der problem is solved. Both visual and quantitative comparisons show that our method able to generate semantically-coherent superpixel. The reported result also state that our method perform favorably to over segment large planar in dim light condition. The result is preliminary since the experiment conducted in a still image. As our future work, we need to reveal the superpixel stability over an image sequence.

References

1. Achanta, R., Shaji, A., Smith, K., Lucchi, A., Fua, P., Susstrunk, S.: SLIC super-pixels compared to state-of-the-art superpixel methods. IEEE Trans. Pattern Anal. Mach. Intell. **34**, 2274–2282 (2012)
2. Gupta, S., Arbelaez, P., Malik, J.: Perceptual organization and recognition of indoor scenes from RGB-D images. In: Proceedings of the IEEE Conference on Computer Vision and Pattern Recognition, pp. 564–571 (2013)
3. Jia, S., Geng, S., Gu, Y., Yang, J., Shi, P., Qiao, Y.: NSLIC: SLIC superpixels based on nonstationarity measure. In: 2015 IEEE International Conference on Image Processing (ICIP), pp. 4738–4742. IEEE (2015)
4. Liu, M.Y., Tuzel, O., Ramalingam, S., Chellappa, R.: Entropy rate superpixel seg-mentation. In: 2011 IEEE Conference on Computer Vision and Pattern Recognition (CVPR), pp. 2097–2104. IEEE (2011)
5. Meyer, F.: Color image segmentation. In: International Conference on Image Processing and Its Applications, pp. 303–306. IET (1992)
6. Mori, G., Ren, X., Efros, A.A., Malik, J.: Recovering human body configurations: combining segmentation and recognition. In: Proceedings of the 2004 IEEE Com-puter Society Conference on Computer Vision and Pattern Recognition, CVPR 2004, vol. 2, pp. II-326. IEEE (2004)
7. Neubert, P., Protzel, P.: Superpixel benchmark and comparison. Proc. Forum Bild-verarb. 1–12 (2012)
8. Ren, C.Y., Reid, I.: gSLIC: a real-time implementation of SLIC superpixel segmen-tation. Technical report, Department of Engineering, University of Oxford (2011)
9. Ren, X., Malik, J.: Learning a classification model for segmentation. In: Proceed-ings of the Ninth IEEE International Conference on Computer Vision, 2003, pp. 10–17. IEEE (2003)

10. Ren, X., Bo, L., Fox, D.: RGB-(D) scene labeling: features and algorithms. In: 2012 IEEE Conference on Computer Vision and Pattern Recognition (CVPR), pp. 2759–2766. IEEE (2012)
11. Schick, A., Fischer, M., Stiefelhagen, R.: Measuring and evaluating the compactness of superpixels. In: 2012 21st International Conference on Pattern Recognition (ICPR), pp. 930–934. IEEE (2012)
12. Silberman, N., Hoiem, D., Kohli, P., Fergus, R.: Indoor segmentation and support inference from RGBD images. In: Fitzgibbon, A., Lazebnik, S., Perona, P., Sato, Y., Schmid, C. (eds.) ECCV 2012. LNCS, vol. 7576, pp. 746–760. Springer, Heidelberg (2012). doi:10.1007/978-3-642-33715-4_54
13. Stutz, D., Hermans, A., Leibe, B.: Superpixel segmentation using depth information. RWTH Aachen University, Aachen, Germany (2014)
14. Veksler, O., Boykov, Y., Mehrani, P.: Superpixels and supervoxels in an energy optimization framework. In: Daniilidis, K., Maragos, P., Paragios, N. (eds.) ECCV 2010. LNCS, vol. 6315, pp. 211–224. Springer, Heidelberg (2010). doi:10.1007/978-3-642-15555-0_16
15. Wan, J., Xia, T., Tang, S., Li, J.: Robust range image segmentation based on coplanarity of superpixels. In: International Conference on Pattern Recognition (ICPR), pp. 3618–3621 (2012)
16. Xu, L., Lu, C., Xu, Y., Jia, J.: Image smoothing via L0 gradient minimization. ACM Trans. Graph. (TOG) **30**, 174 (2011). ACM
17. Yang, J., Gan, Z., Gui, X., Li, K., Hou, C.: 3-D geometry enhanced superpixels for RGB-D data. In: Huet, B., Ngo, C.-W., Tang, J., Zhou, Z.-H., Hauptmann, A.G., Yan, S. (eds.) PCM 2013. LNCS, vol. 8294, pp. 35–46. Springer, Cham (2013). doi:10.1007/978-3-319-03731-8_4

Monocular Depth Estimation of Outdoor Scenes Using RGB-D Datasets

Tianteng Bi, Yue Liu$^{(\boxtimes)}$, Dongdong Weng, and Yongtian Wang

School of Optoelectronics, Beijing Institute of Technology, Beijing, China
liuyue@bit.edu.cn

Abstract. Depth estimation is a classical topic in computer vision, however, inferring the depth of a scene from a single image remains an extremely difficult problem. In this paper, a non-parametric method is adopted to obtain the depth of a single image. To this end, RGB-D datasets are exploited as the inference basis. Given a query image, a global scene-level retrieval is performed against the dataset, followed by a superpixel-level matching. The superpixels-based scene representation is introduced to model the depth jointly in terms of superpixel centroid. The depth estimation is formulated as contextual inference and the depth propagation. The contextual inference is expressed as a Markov random field (MRF) energy function defined on a sparse depth map obtained by the matching process and implemented in a graphical model whose edges encode the interactions between the superpixel centroids. Then the depth propagation generates the final dense depth map from the inferred result. The benefits of the proposed method is demonstrated on the standard dataset.

1 Introduction

Depth estimation from images is a classical topic in computer vision with extensive applications such as 3D reconstruction, robotics and virtual reality. In recent years, many researchers propose various approaches that can be divided into two categories based on the number of the scene images. The majority of existing methods requires multiple images of a scene and expensive capture devices. More specifically, stereo vision based approaches [3,9] utilize the computed disparities between a pair of images of the same scene taken from two different viewpoints to recover the depth. Shape from motion (SFM) [8,27] uses the correspondences between the images to obtain the 2D motion field to recover the 3D motion and the depth. Depth from focus (DFF) [2,20] captures a set of images using different focus settings and measures the sharpness of image for each pixel, in which the depth of the pixel depends on the image that the pixel is selected from. Such methods not only rely on correspondences between images, but also suffer from the occlusion problem. More importantly, they cannot work for a single image scenario.

Humans are good at judging depth from a single image by combining such monocular cues as texture and defocus, while it is still an extremely difficult

© Springer International Publishing AG 2017
C.-S. Chen et al. (Eds.): ACCV 2016 Workshops, Part II, LNCS 10117, pp. 88–99, 2017.
DOI: 10.1007/978-3-319-54427-4_7

problem for computer. For example, the texture of an object is different at different location. In recent years, researchers propose single image-based depth estimation methods according to such cues and the process of human to obtain the knowledge about depth. Among the single view depth cues, defocus is one of the strongest that allows humans to understand the order of the objects in a scene. Such depth cue has been extensively investigated in depth estimation from a single viewpoint [11]. Single-image based depth from defocus (DFD) approaches only need one image to compute the depth of the scene, which simplifies the capture procedure. Levin et al. proposed an algorithm using a coded aperture which is more sensitive to the depth variation [14]. The depth can be obtained by a set of calibrated blur kernels after a deconvolution process. Chen et al. represented the defocus blur amount by the energy spectra of the point spread function and detected the defocused step edge to recover the depth with camera settings [7]. Zhuos approach employed edge-detection methods to first estimate the defocus blur of the step edge based on Gaussian gradient ratio, then generated the dense defocus map by using interpolation [28]. This approach can be divided into step edge detection, defocus blur amount estimation and defocus map interpolation, during which the author uses a parameterized model to formulate the edge blurred by the point spread function and recover the depth by estimating the parameters.

Defocus-based methods rely on the blurring in the image and cannot obtain an accurate result. A new trend is learning a model to predict the local depth similar to the human's cognitive process over the years. For example, Saxena et al. proposed a supervised strategy to directly predict the depth of image pixels or superpixels by capturing the relationship between depths, which actually solved the parameters in their assumed models [23–25]. Liu et al. achieved better depth estimation accuracy by combining semantic labels in their MRF model [16]. Similarly, Russell et al. exploited user annotation [22] as the additional information to improve their result. Besides, more complexed neural networks are also adopted to predict each pixels depth in the image. Eigen et al. used two deep network stacks to generate dense depth map [10]. One network makes a coarse depth map based on the entire image and the other one modifies the result generated by the first one locally. Liu et al. combines the deep convolutional neural networks (DNN) with the conditional random field (CRF) into a unified framework for estimating the depth from a single image [17].

Recently, non-parametric methods are introduced to handle this problem under the circumstance that the above-mentioned approaches need many unavailable information. Karsch et al. obtained the depth map by warping the candidate images and depths to a query image with an optimization procedure to smooth the warped depth [12]. Liu's method introduced a CRF inference including discrete and continuing variables that encoded the depths of superpixels and relationships between superpixels respectively [18]. Zhuo constructed a multilayer graph that jointly leveraged the mid-level information and global structures of scene to infer local depth [29]. Given a query image, these approaches firstly retrieved certain similar images as the candidates in a RGB-D dataset in which the depth of each image is known. The depths of the candidates are used for estimating a depth map with smooth constraints.

Inspired by the above-mentioned methods, in this paper we introduce a simple non-parametric depth estimation method from a single image instead of complexed models and restrictive assumptions, which addresses this issue by concatenating the contextual inference and the depth propagation. In particular, the contextual inference only operates on a graph defined by the superpixel centroids and the edges between centroids, while the depth propagation performs the edge-aware depth interpolation from the sparse depth map composed of the depths of superpixel centroids.

In addition to the similar process of image retrieval, the superpixel-level matching is also adopted to determine the depth of each superpixel centroid in the input image, followed by the inference using MRF, with the depth propagation generating the final dense depth map similar to [28]. More specifically, given a query image, we exploit a non-parametric approach to retrieve similar images in a RGB-D dataset. The depth of these similar images will be used for original data terms in our system. Furthermore, we adopt feature-based superpixel matching in the superpixel pool composed of superpixels in the retrieved similar images to find the depth of each superpixel centroid in the query image. Then the centroid depths form a sparse depth map and are considered as our original inference terms. The feature-based superpixel matching only finds similar partners for each isolated superpixel in the query image, while the contextual inference can take the relationships between superpixels into consideration. After the inference, the edge-aware depth propagation can fill the sparse depth map according to the centroid depth.

Different from the existing methods, the proposed approach avoids such complex procedure as the warping in [12] and adopts simple graphical model rather than the complex one in [18,29]. More importantly, the depth propagation keeps the discontinuities at the edge location in the image, endowing our method more detailed information. Experimental results show the benefits of the proposed method that yields qualitatively accuracies on the Make3D outdoor scenes dataset.

The rest of this paper is organized as follows. Section 2 presents the system configuration. Section 3 describes the experimental results. Finally the conclusion is presented in Sect. 4.

2   System Description

This section presents the details of all the components of the proposed system which is based on non-parametric strategy, meaning that there is almost no offline training except certain pre-processing procedures; given a query image, our system dynamically selects the relevant samples and transfers the depth to the query. Our system can be divided into following four parts:

(1) Image retrieval (Sect. 2.1). Given an input image, finding a set of similar images to the query based on some kinds of image features using nearest neighbor search in a RGB-D dataset.

(2) Sparse depth map acquisition (Sect. 2.2). Segment the input image into superpixels and extract feature vectors from each superpixel in it. For each superpixel in the query image, we concatenate all types of feature vectors and find the nearest neighbor superpixels in the superpixel pool constructed by the superpixels in the similar images to the query. After finding the similar superpixels for each superpixel in the query image, we set the averaged centroid depth of the similar superpixels as the centroid depth of the corresponding superpixel.

(3) Contextual inference (Sect. 2.3). The sparse depth map is determined by obtaining each superpixel's centroid depth without considering the relationship between superpixels. We adopt the sparse depth map together with pairwise co-occurrence energies in a Markov Random Field (MRF) framework to put a modification on the sparse depth map.

(4) Depth propagation (Sect. 2.4). To generate a full depth map, an edge-aware interpolation method is adopted to solve this case based on the inferred sparse depth map.

2.1 Image Retrieval

Our first step to estimate the depth of a single image is to retrieve a small set of similar images as the source of further superpixel-level matching from the dataset. A good retrieval result will include the images with similar scene structure, objects and color distribution to the query image. To this end, three common features (gist [21], PHOG [4], color histogram) are adopted to find the image set. Given an input image, we compute the three types of feature vector of the image and compare each type of feature vector with the corresponding one of the images in the RGB-D dataset by Euclidean distance. According to the Euclidean distance to the query, each feature type can show an increasing rank, meaning that each image in the dataset has three ranks corresponding to three feature types. We choose the minimum rank as the final rank of each image and take the top seven images as our similar image set, which improves the retrieval result as reported in [26]. The images in the set will provide the computational efficiency for the superpixel matching and avoid some local disruptive factors.

2.2 Sparse Depth Map Acquisition

In the attempt of obtaining the sparse depth map, we perform superpixel-level matching to determine the depth of superpixel centroid. We use SLIC algorithm [1] to segment the image into superpixels that can not only help us choose depth sampling points, but also provide solution for getting the depth of these points, meaning that the superpixels representing a small region in the image let us take the superpixel centroid as the nodes of graphical model normally and exploit the depth of the candidate images reasonably by superpixel-level matching. The 14 different features [19] are used for describing superpixel in our system, including different aspects of shape, texture, color and location for a superpixel. All the features are computed for the superpixels in the query image and similar images. We directly concatenate all feature types into a vector as $\mathbf{v} = \{shape, texture, color, location\}$. Then each superpixel is represented by a 1632 dimensional vector and a nearest neighbor search is performed on the superpixel pool composed of all the superpixels belonging to the similar images for finding 80 most similar superpixels to the corresponding one in the query image. The depths of 80 superpixel centroids are averaged as the original depth of the query superpixel.

2.3 Contextual Inference

After extracting the features of the superpixels in the query image, a sparse depth map constructed by the depths at the superpixel centroid is obtained. Next, we would like to enforce contextual constraints on the sparse depth map. For example, a superpixel with a little depth surrounded by large ones is not very plausible. In keeping with our nonparametric strategy, we formulate the problem as the minimization of a MRF energy function defined on our sparse depth map:

$$E(L) = \sum_{p \in C} E_p(l_p) + \sum_{(p,q) \in A} E_{p,q}(l_p, l_q) \tag{1}$$

where C is the set of superpixel centroids and A is the set of adjacent superpixel pairs. E_p is a unary term encoding the penalty of the assigned label l_p to the superpixel centroid p. $E_{p,q}$ is a pairwise term encouraging the coherence between the superpixel centroids p and q.

More specifically, let $L = \{ l_1, l_2, ..., l_N \}$ represent N superpixel centroids in the input image and l_i can take V discrete depth values defined by quantizing the depth range of the dataset. The unary term relies on the sparse depth map measuring how close the inferred superpixel centroid depth is to the sparse case and can be defined as:

$$E_p(l_p) = (d_p(l_p) - d_p^s)^2 \tag{2}$$

where d_p is the depth of the superpixel centroid p for a particular label l_p and d_p^s is the depth of the superpixel centroid p in the sparse depth map.

Since there is some obvious relationship between image appearance and depth, we assume that superpixels in the image with similar color are likely to have similar depth and color differences in the image are likely to correspond to the break in the depth. Therefore, the pairwise term can be expressed as:

$$E_{p,q}(l_p, l_q) = \exp(-dist(c_p, c_q))(d_p(l_p) - d_q(l_q))^2 \tag{3}$$

where the $dist(c_p, c_q)$ is the color distance between the superpixels at which the centroid p and q are located respectively. Here, the color distance is defined by the chi square distance between the color histograms of two superpixels corresponding to the centroid p and q. If $h^{(1)}$ and $h^{(2)}$ are the two color histograms, the chi square distance is:

$$d(h^{(1)}, h^{(2)}) = \sum_{i=1}^{n} \frac{(h_i^{(1)} - h_i^{(2)})^2}{h_i^{(1)} + h_i^{(2)}} \tag{4}$$

In our experiment, the Graph Cuts is used for the optimization of the MRF model. The gco-v 3.0 toolbox is adopted to implement the graph cut operation [5,6,13].

2.4 Depth Propagation

After minimizing the energy function, the depth estimation method described in the previous sections generates a smoothed sparse depth map. In this section, we propagate the depth from the superpixel centroids to the entire image and obtain a full depth map similar to [28]. This method is proposed by Levin in [15] for natural image matting and can be used as an edge-aware interpolation that keep the discontinuities in the depth with the image edges. Here, we use this method for estimating the depth by combining the image appearance and readers can refer to [15,28] for more details.

3 Experiments

We evaluate the performance of the proposed approach on a RGB-D dataset: Make3D depth dataset [23,25]. The depth dataset is composed of pairs of RGB images and depth maps in a variety of outdoor scenes. For quantitative evaluation, three commonly used metrics - averaged relative error (**rel**), root mean squared error (**rms**) and averaged log_{10} error (**log10**) are used for reporting the errors of our algorithm:

$$\mathbf{rel} = \frac{1}{N} \sum_{i=1}^{N} \frac{|D_i - D_i^*|}{D_i^*} \tag{5}$$

$$\mathbf{rms} = \sqrt{\frac{1}{N} \sum_{i=1}^{N} (D_i - D_i^*)^2} \tag{6}$$

$$\mathbf{log10} = \frac{1}{N} \sum_{i=1}^{N} |log_{10} D_i - log_{10} D_i^*| \tag{7}$$

where N is the number of total pixels in all images, D_i is the depth of the ith pixel in the estimated depth map and D_i^* is the groundtruth of the ith pixel in the corresponding map.

In all the experiments, each image is segmented into approximate 500 super-pixels. The discrete depth includes 16 values from 5 to 80 by steps of 5. In the depth propagation step, we set a scalar parameter $\lambda = 0.01$ to balance between fidelity to the sparse depth map and smoothness of interpolation.

The Make3D dataset contains 534 images covering a variety of outdoor scenes, partitioned into 134 test images and 400 training images. All the images

Table 1. Depth estimation errors on the Make3D dataset of the proposed method.

Method	rel	rms	log10
Our method	0.8981	18.8052	0.2693

Image   Ground-truth   Ours

Fig. 1. Qualitative comparison of the depths estimated with our method on the Make3D dataset. Color indicates depth (red is far, blue is close). (Color figure online)

were resized to 460 × 345 pixels to keep the aspect ratio of the original images. The errors of our method are reported in Table 1. Compared with the state of art single image estimation methods results as reported in [12], our method have a loss of accuracy, but more detailed information is provided. Analysis results show that direct concatenation of all the feature vectors of the query superpixel to find the nearest neighbor ones ignores the characteristic of each feature. Furthermore, that the averaging the similar superpixel centroid depths

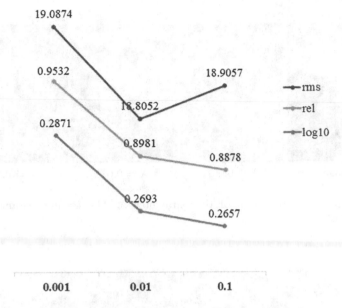

Fig. 2. Effect of the balance parameter λ. Errors are computed on the Make3D dataset with varying values of λ.

is assigned to the corresponding one cannot take the depth distribution of the nearest neighbor superpixels into consideration. Besides, the depth propagation algorithm generates more details in the final depth map, leading to more errors to the groundtruth that looks smoother.

As shown in Fig. 1, a qualitative comparison between the proposed method and the groundtruth is shown. Similar to the quantitative result, our method can correctly estimate the scene depth with many details corresponding to the image appearance. However, these details cannot appear in the groundtruth that is obtained by laser scanner, making the detailed pixels in our depth map similar to the noises against the groundtruth.

We also evaluate how the parameter λ affects the performance of the proposed approach. Such metrics (same as the above-mentioned **rel, rms, log10**) are computed with different values of the parameter as shown in Fig. 2. The parameter is used for balancing between fidelity to the sparse depth map and smoothness of interpolation in the depth propagation. It can be seen from Fig. 2 that the parameter can reduce the certain errors while the other is increasing, meaning that the parameter leads the error fluctuation. A possible trend is that the balance parameter cannot improve the accuracy obviously.

The qualitative comparison of the different parameter values is shown in Fig. 3. We can see that the depth map is smoother when the parameter is greater, meaning that some holes generated by the trees around the sky become blurring in the right depth image, while they are clearer in the left one.

Image   λ = 0.001   λ = 0.01   λ = 0.1

Fig. 3. Qualitative comparison of the parameter λ. The greater parameter make smoother depth map.

Image   Depth map

Fig. 4. Depth estimation of Dashuifa in Yuanmingyuan Garden (the former Summer Palace). The maximum range is set to about 80 m by the dataset.

Image   Depth map

Fig. 5. Computing result of the Internet image. The depth of sky is a bit closer, but the relative depth between the hill and the sky is correct.

Image   Layered model

Fig. 6. The reconstructed result of Dashuifa. The 2D image can be converted into a stereoscopic layered model according to the depth.

Furthermore, the proposed method is performed on the images captured by ourselves and downloaded from the Internet. Figure 4 shows the correct estimation result of a single image captured at the Dashuifa in Yuanmingyuan Garden (the former Summer Palace). Note that the depth of sky is provided by the Make3D dataset in which 80 m is the maximum range.

It can be seen from Fig. 5 that the proposed method give a closer estimation on the sky region, but the relationship between the tower and the sky is accurate.

The proposed method can be used for a layered reconstruction of outdoor scenes. After depth estimation, a pixel is assigned to foreground if its depth is smaller than a threshold. As shown in Fig. 6, the Dashuifa is reconstructed from its 2D image according to its depth.

4   Conclusion

We have introduced a single image based depth estimation approach by concatenating the contextual inference and depth interpolation. Experimental results have demonstrated such benefits as an image appearance-aware approach over local depth prediction. In particular, the experimental results show that the proposed method can provide more detailed depth in coherence with the image appearance, although the accuracy is less in some image regions. In the future, we intend to investigate this phenomenon in more details, and study whether the acquisition of the sparse depth map and the depth propagation can be incorporated in a unified optimization framework. Furthermore, we plan to make use of multi-layer MRF/CRF models or booming DNN in our depth estimation framework.

Acknowledgement. This work has been supported by the National Key Technology R&D Program (2015BAK01B05) and Natural Science Foundation of China (61370134). The authors would like to thank Dr. Nai for his suggestions and helps for the realization of the proposed framework.

References

1. Achanta, R., Shaji, A., Smith, K., Lucchi, A., Fua, P., Susstrunk, S.: SLIC superpixels compared to state-of-the-art superpixel methods. IEEE Trans. Pattern Anal. Mach. Intell. **34**, 2274–2282 (2012)
2. Asada, N., Fujiwara, H., Matsuyama, T.: Edge and depth from focus. Int. J. Comput. Vis. **26**, 153–163 (1998)
3. Barnard, S.T., Fischler, M.A.: Computational stereo. ACM Comput. Surv. (CSUR) **14**, 553–572 (1982)
4. Bosch, A., Zisserman, A., Munoz, X.: Representing shape with a spatial pyramid kernel. In: ACM International Conference on Image and Video Retrieval, CIVR 2007, Amsterdam, The Netherlands, pp. 401–408, July 2007
5. Boykov, Y., Kolmogorov, V.: An experimental comparison of min-cut/max-flow algorithms for energy minimization in vision. IEEE Trans. Pattern Anal. Mach. Intell. **11**, 1124–1137 (2001)
6. Boykov, Y., Veksler, O., Zabih, R.: Efficient approximate energy minimization via graph cuts. IEEE Trans. Pattern Anal. Mach. Intell. **20**, 1222–1239 (2001)
7. Chen, C.-W., Chen, Y.-Y.: Recovering depth from a single image using spectral energy of the defocused step edge gradient. In: 2011 18th IEEE International Conference on Image Processing (ICIP), pp. 1981–1984. IEEE (2011)
8. Dellaert, F., Seitz, S.M., Thorpe, C.E., Thrun, S.: Structure from motion without correspondence. In: IEEE Conference on Computer Vision and Pattern Recognition, 2000. Proceedings, pp. 557–564. IEEE (2000)
9. Dhond, U.R., Aggarwal, J.K.: Structure from stereo-a review. IEEE Trans. Syst. Man Cybern. **19**, 1489–1510 (1989)
10. Eigen, D., Puhrsch, C., Fergus, R.: Depth map prediction from a single image using a multi-scale deep network. In: Advances in Neural Information Processing Systems, pp. 2366–2374 (2014)
11. Favaro, P., Soatto, S.: 3-D Shape Estimation and Image Restoration: Exploiting Defocus and Motion-Blur. Springer Science and Business Media, Heidelberg (2007)
12. Karsch, K., Liu, C., Kang, S.B.: Depth extraction from video using non-parametric sampling. IEEE Trans. Pattern Anal. Mach. Intell. **36**, 775–788 (2012)
13. Kolmogorov, V., Zabih, R.: What energy functions can be minimized via graph cuts? IEEE Trans. Pattern Anal. Mach. Intell. **26**, 147–159 (2004)
14. Levin, A., Fergus, R., Durand, F., Freeman, W.T.: Image and depth from a conventional camera with a coded aperture. ACM Trans. Graph. (TOG) **26**(3), 70 (2007). ACM
15. Levin, A., Lischinski, D., Weiss, Y.: A closed-form solution to natural image matting. IEEE Trans. Pattern Anal. Mach. Intell. **30**, 228–242 (2008)
16. Liu, B., Gould, S., Koller, D.: Single image depth estimation from predicted semantic labels. In: IEEE Conference on Computer Vision and Pattern Recognition, pp. 1253–1260 (2010)
17. Liu, F., Shen, C., Lin, G.: Deep convolutional neural fields for depth estimation from a single image. In: Proceedings of the IEEE Conference on Computer Vision and Pattern Recognition, pp. 5162–5170 (2015)
18. Liu, M., Salzmann, M., He, X.: Discrete-continuous depth estimation from a single image. In: IEEE Conference on Computer Vision and Pattern Recognition, pp. 716–723 (2014)
19. Malisiewicz, T., Efros, A.A.: Recognition by association via learning per-exemplar distances, pp. 1–8 (2008)

20. Nayar, S.K., Nakagawa, Y.: Shape from focus. IEEE Trans. Pattern Anal. Mach. Intell. **16**, 824–831 (1994)
21. Oliva, A., Torralba, A.: Modeling the shape of the scene: a holistic representation of the spatial envelope. Int. J. Comput. Vis. **42**, 145–175 (2001)
22. Russell, B.C., Torralba, A.: Building a database of 3D scenes from user annotations. In: IEEE Conference on Computer Vision and Pattern Recognition, pp. 2711–2718 (2009)
23. Saxena, A., Chung, S.H., Ng, A.Y.: Learning depth from single monocular images. In: Advances in Neural Information Processing Systems, pp. 1161–1168 (2005)
24. Saxena, A., Chung, S.H., Ng, A.Y.: 3-D depth reconstruction from a single still image. Int. J. Comput. Vis. **76**, 53–69 (2008)
25. Saxena, A., Sun, M., Ng, A.Y.: Make3D: learning 3D scene structure from a single still image. IEEE Trans. Pattern Anal. Mach. Intell. **31**, 824–840 (2009)
26. Tighe, J., Lazebnik, S.: SuperParsing: scalable nonparametric image parsing with superpixels. In: European Conference on Computer Vision, pp. 352–365 (2010)
27. Tomasi, C., Kanade, T.: Shape and motion from image streams under orthography: a factorization method. Int. J. Comput. Vision **9**, 137–154 (1992)
28. Zhuo, S., Sim, T.: Defocus map estimation from a single image. Pattern Recogn. **44**, 1852–1858 (2011)
29. Zhuo, W., Salzmann, M., He, X., Liu, M.: Indoor scene structure analysis for single image depth estimation. In: Computer Vision and Pattern Recognition, pp. 614–622 (2015)

Reconstruction of 3D Models Consisting of Line Segments

Naoto Ienaga[✉] and Hideo Saito

Department of Information and Computer Science, Keio University, Yokohama, Japan
{ienaga,saito}@hvrl.ics.keio.ac.jp

Abstract. Reconstruction of three-dimensional (3D) models from images is currently one of the most important areas of research in computer vision. In this paper, we propose a method to recover 3D models using the minimum number of line segments. By using structure-from-motion, the proposed method first recovers a 3D model of line segments detected from an input image sequence. We then detect overlapping 3D line segments that redundantly represent a single line structure so that the number of 3D line segments representing the target scene can be reduced without losing the detailed geometry of the structure. We apply matching and depth information to remove redundant line segments from the model while keeping the necessary segments. In experiments, we confirm that the proposed method can greatly reduce the number of line segments. We also demonstrate that the accuracy and computational time for camera pose estimation can be significantly improved with the 3D line segment model recovered by the proposed method. Moreover, we have applied the proposed method to see through occluded areas.

1 Introduction

Reconstruction of three-dimensional (3D) models is one of the most studied topics in computer vision. For 3D model reconstruction from images, image features that provide the input data of structure-from-motion need to be matched between different viewpoints [1,2]. To achieve the matching of feature points, techniques, such as scale-invariant feature transform [3] or speeded up robust features [4], are often used. However, only a few feature points are detected in artificial situations consisting mainly of texture-less objects. For adapting such artificial environments, line segments can be used as image features.

For 3D reconstruction, line segments should also be matched between images from different viewpoints. In comparison with feature points, a line segment feature has a less distinctive appearance, which makes it difficult to correctly match line segments. Many existing methods [5–7], therefore, do not use appearance-based matching of line segments. However, some studies about descriptors for line segment features have been reported. The mean standard-deviation line descriptor [8] is a representative example and a line-based eight-directional histogram feature (LEHF) has also been introduced [9]. To make the LEHF more robust, directed LEHF has been proposed [10] and 3D reconstruction has been done with

© Springer International Publishing AG 2017
C.-S. Chen et al. (Eds.): ACCV 2016 Workshops, Part II, LNCS 10117, pp. 100–113, 2017.
DOI: 10.1007/978-3-319-54427-4_8

this method [11]. Reconstructed 3D models are used for various purposes, including estimating camera pose for augmented reality [12]. In [12], the accuracy of camera poses has been improved by optimization through image sequences and plane segmentation. The reconstructed 3D models have a problem.

The method used in [12] is problematic because it retains all detected line segments of all frames. The final 3D model, therefore, has many redundant line segments. As shown in Fig. 2(a), many redundant line segments overlap at each position, although they should be represented by only one line segment. This redundancy leads to longer computation times and lower accuracy when we use the model to estimate camera poses or leads to worse visual information on the 3D model.

Line segment-based 3D models are superior to point cloud-based models in terms of visual information because a line segment indicates outlines. Some studies [13, 14] have used line segments to enhance models, while other studies have suggested that using all of the available depth information (e.g., [15]) is better than using only line segments as visual information. In terms of data size, a line segment is superior [16]. We therefore conclude that line segment-based models are very useful, especially for improving the visual appearance of a model.

In this paper, we propose a method to reconstruct 3D models represented by line segments that can be adapted to texture-less situations. In this method, we first recover a 3D model based on the method in [12]. We then remove redundant line segments that overlap the same 3D line segments. Our contributions are as follow:

- We remove many redundant line segments of a 3D model that was reconstructed using the method in [12] without losing the detailed geometry of the structure.
- We experimentally demonstrate that the 3D model reconstructed by the proposed method provides improved accuracy of camera poses and requires less computation time.
- We improve the visual appearance of the model by removing line segments that do not actually exist.

Numerous studies investigating line segments detect them with a line segment detector (LSD) [17], which was also used in [12]. Line segments can also be considered as edges of a plane. In some studies (e.g., [18]), planes rather than line segments were used to create a 3D model. When we consider line segments to be edges *of* a plane, we cannot detect the line segments *on* a plane, such as textures, small depressions, and edges of thin objects like paper. For example, in Fig. 1(c), papers are attached on the whiteboard, and we can detect their edges, as shown in Fig. 1(f). They cannot be detected if we consider line segments to be edges of a plane.

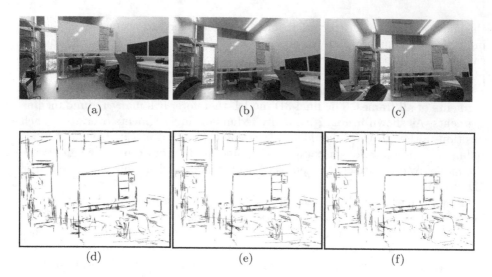

Fig. 1. Input images and 3D models of a whiteboard scene. (a) Input image 1 of 5. (b) Input image 3. (c) Input image 5. (d) 3D model reconstructed by [12]. (e) Reconstructed model with matching results. (f) Our final reconstructed model with matching results and depth information. Model (d) has 2266 line segments; model (f) has 1266 line segments.

2   Removing Redundant Line Segments

Our proposed method has two ways of removing redundant line segments from a 3D model; one utilizes matching information and the other utilizes depth information. Figure 1 shows three of five input images and the reconstructed models. Figure 1(d) is reconstructed by [12], and Fig. 1(f) is our final output.

2.1   Removing Redundant Line Segments with a Matching Result

When we consider the removal of redundant line segments, the information obtained by matching the line segments is very useful. Zhang and Faugeras [19] proposed a method to merge matched line segments to make a new line segment. In our case, the longest line segment of each matching line segment group is the most appropriate to keep because the method we are using [12] carefully removes outliers many times; this results in good line segment matching. We therefore remove redundant line segments by keeping only the longest ones.

Figure 2(b) is part of the model shown in Fig. 1(e). We removed all line segments except for the longest one of each group. As can be seen, we successfully removed redundant line segments while keeping the outlines.

2.2   Removing Redundant Line Segments with Depth Information

The model reconstructed by the previous process is not yet satisfactory as some line segments do not actually exist. Figure 3(a) shows such line segments.

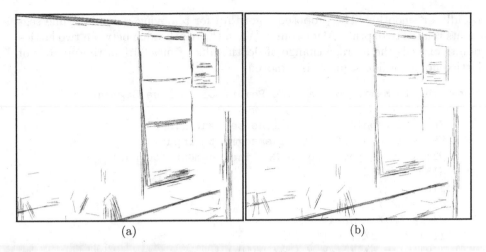

(a)   (b)

Fig. 2. Result of removing redundant line segments by matching. (a) Part of Fig. 1(d). Many redundant line segments are at each actual line structure position. (b) Part of Fig. 1(e). Compared with (a), many redundant line segments have clearly been removed.

(a)   (b)

Fig. 3. Result of removing redundant line segments using the depth information. (a) Part of Fig. 1(e). Some line segments that are unnaturally projected do not actually exist. (b) Part of Fig. 1(f). Some line segments that do not actually exist in Fig. 3(a) are completely removed.

We utilize LSD [17] to detect the line segments from 2D images and use the method in [12] to make lattice points of line segments in 3D space from the start and end points detected by LSD in 2D space. We then remove redundant line segments. This method is displayed in Algorithm 1.

Figure 4 illustrates why nonexisting line segments are drawn. The depth values on a line segment detected on the boundary of an object in a 2D image are sometimes mixed depth values of the foreground and the background because of depth noise. Some line segments in Fig. 3(a) are detected on the boundary of objects. To remove these line segments, we check the depth values of the lattice points of the line segment and check whether the depth values change rapidly or smoothly. The proposed algorithm for removing redundant line segments related to depth is Algorithm 2. When the depth value between two lattice points exceeds the average change of depth values times 90% of the number of lattice points, a line segment is removed.

```
Algorithm 1. Making 3D Lattice Points of 2D Line Segment
   var
      SP: start point of a 2D line segment (input)
      EP: end point of a 2D line segment (input)
      P3: lattice points of a 3D line segment (output)
      STEP, dY, dX
   begin
      dY := EP.Y - SP.Y
      dX := EP.X - SP.X
      if | dY / dX | < 1
         STEP := dX / (dY^2 + dX^2)^0.5
         while (STEP > 0 and SP.X < EP.X) or
                                    (STEP < 0 and SP.X > EP.X)
            convert SP to a 3D point and add it to P3
            SP.X := SP.X + STEP
            SP.Y := Y value of a 2D line segment when X == SP.X
         endwhile
      else
         STEP := dY / (dY^2 + dX^2)^0.5
         while (STEP > 0 and SP.Y < EP.Y) or
                                    (STEP < 0 and SP.Y > EP.Y)
            convert SP to a 3D point and add it to P3
            SP.Y := SP.Y + STEP
            SP.X := X value of a 2D line segment when Y == SP.Y
         endwhile
      endif
end.
```

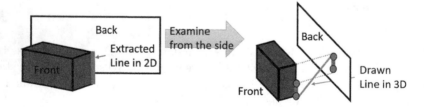

Fig. 4. Sometimes the depth values of the front of an object and its background on a line segment extracted from the boundary of an object in 2D are mixed. We draw a line segment in 3D space using the depth of both end points of the line segment extracted in 2D space. If one of the end points is located in front of the object and the other is located in the background, the line segment drawn in 3D becomes an incorrect line segment. To remove such a line segment, we check the depth values of the lattice points of each line segment.

```
Algorithm 2. Removing Redundant Line Segments with Depth
   output is TRUE or FALSE. FALSE means removing a line segment.
   var
     Z: Z values of lattice points of a 3D line segment (input)
     N: length of ARR
     MAX, MARGIN, ITER
   begin
     MAX := floor( N * 0.9 + 0.5 ) * | Z[0] - Z[N-1] | / N
     for ITER := 1 to N-1
       MARGIN := | Z[ITER-1] - Z[ITER] |
       if MARGIN > MAX
         return FALSE
       endif
     endfor
     return TRUE
end.
```

Figure 3(b) shows the result of Algorithm 2. Some line segments that do not actually exist are completely removed.

2.3 Summary of Algorithm for Removing Redundant Line Segments

We have explained the method for removing redundant line segments from a 3D model in two ways: first, with a matching result and second, with depth information. In practice, we first remove line segments with depth and then we keep the longest line segments of each matching group. The result of [12] is shown in Fig. 1(d) and that of our proposed method is shown in Figs. 1(f) and 5. The model in (d) [12] has 2266 line segments. In contrast, the model in (f) of the proposed method has 1266 line segments and successfully retains the outlines of the model. Moreover, we improve the visual appearance of the model by removing line segments that do not actually exist.

3 Experiments

We conducted experiments to demonstrate that we can remove only redundant line segments while keeping the necessary segments. In the experiments, we estimated camera poses by using both the 3D models reconstructed by [12] and those reconstructed by the proposed method.

We used input images of two scenes, shown in Fig. 6. The desk scene and the door scene consist of 12 and 60 frames, respectively. Table 1 shows the number of line segments of the 3D models reconstructed by [12] and by the proposed method. The proposed method greatly reduced the number of segments used.

Table 2 shows that the proposed method leaves only the necessary line segments. We used 11 images to estimate the camera pose. The desk scene consisted of five images, and the door scene consisted of six images. In the desk

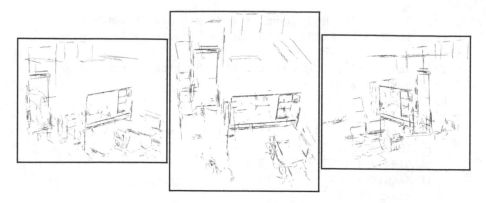

Fig. 5. Result of 3D reconstruction of the whiteboard scene. Three views for the model shown in Fig. 1(f) are presented.

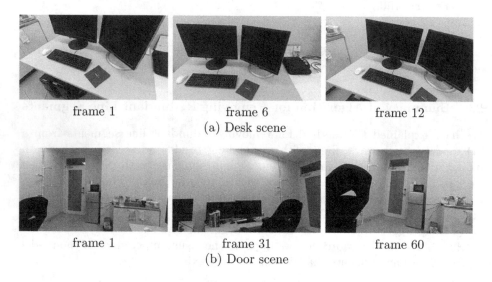

Fig. 6. Input images of the desk scene (a) and the door scene (b). The desk scene and the door scene consist of 12 and 60 frames, respectively.

Table 1. Number of line segments by the method in [12] and the proposed methods.

Scene	[12]	Proposed
Desk	3083	**1301**
Door	14339	**4775**

scene (Fig. 7), we tracked the purple book on the right side of the desk, and in the door scene (Fig. 8), we tracked the inside frame of the door on the left side. Each blue rectangle in both figures is the result of reprojecting four 3D points

Table 2. Average of reprojection errors and the computational times to estimate camera poses. The bold numbers indicate the lowest number of errors (pixels) and lowest times (s).

Image	Comparison	[12]	Proposed
Desk, image 1	Error	**4.70**	4.72
	Time	4.47	**3.93**
Desk, image 2	Error	1300.71	**294.01**
	Time	4.97	**4.44**
Desk, image 3	Error	37.13	**8.87**
	Time	4.74	**3.94**
Desk, image 4	Error	76.97	**56.46**
	Time	5.42	**4.68**
Desk, image 5	Error	**2.79**	2.96
	Time	4.48	**4.03**
Door, image 1	Error	49.51	**39.89**
	Time	6.47	**4.42**
Door, image 2	Error	**32.17**	35.31
	Time	9.04	**5.69**
Door, image 3	Error	38.01	**36.88**
	Time	6.64	**4.73**
Door, image 4	Error	**26.09**	37.21
	Time	6.28	**4.37**
Door, image 5	Error	32.57	**24.10**
	Time	7.97	**5.22**
Door, image 6	Error	**17.78**	24.26
	Time	6.74	**4.50**

for tracking, which were calculated by using the estimated camera poses. The average reprojection errors of the four points (in pixels) and the computational times (in seconds) to estimate the camera poses are shown in Table 2. The computational times might seem long compared with the ones reported in [12], but it should be noted that the experimental environments, such as the machine and the scenes, are different. We can clearly say that our models provided better accuracy of camera poses than those of the method in [12] in less computational time. When the camera positions of reconstructing a model and of tracking differ significantly, the reprojection error is large. Models reconstructed by the proposed method removed the incorrect line segments with depth, and the reprojection errors were therefore constrained considerably, as shown in images 2, 3, and 4 in the desk scene in Fig. 7. Although the proposed method sometimes has a large error, the difference is not so large compared with the difference when the proposed method has a smaller error.

Fig. 7. Tracking results of the desk scene. (a) The method in [12]. (b) The proposed method. The reprojection errors and computational times are listed in Table 2. (Color figure online)

4   Application to Diminished Reality

As an example application of camera pose estimation using a reconstructed 3D model by the proposed method, we implemented a system for visualizing an area occluded by some obstacles, which is called diminished reality (DR). In this section, we show some results of the DR system based on the proposed method.

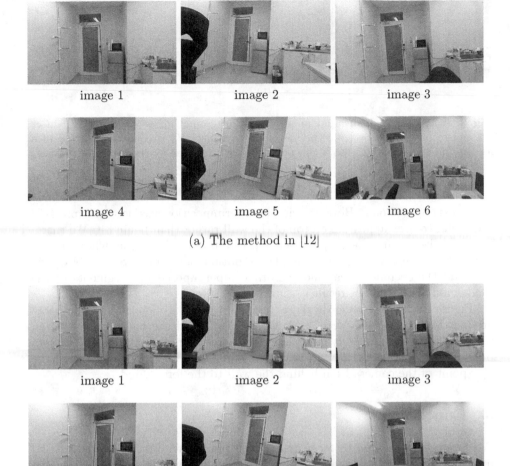

image 1   image 2   image 3

image 4   image 5   image 6

(a) The method in [12]

image 1   image 2   image 3

image 4   image 5   image 6

(b) The proposed method

Fig. 8. Tracking results of the door scene. (a) The method in [12]. (b) The proposed method. The reprojection errors and computational times are listed in Table 2. (Color figure online)

Using this DR system, we visualized a scene behind a wall of the room. Here, we reconstructed a line segment-based 3D model of the room by using the proposed method. The occluded area by the wall of the room was captured by an RGB-D camera, Kinect V2 and reconstructed using the method in [20]. In the performance of the DR stage, camera pose was estimated by the proposed method for each frame. The predefined area was tracked, and the occluded area

(a)   (b)   (c)

Fig. 9. (a) An example of input images. (b) Background image captured by Kinect V2 and reconstructed by [20]. The viewpoint of Kinect V2 is converted to that of a foreground camera, which has a viewpoint that is estimated by the proposed method. (c) DR result.

was superimposed on it. However, simple superimposition was poor at expressing the occluded area appearing on top of the wall rather than behind it. We therefore used the two approaches proposed in [21] to provide depth cues when viewing the occluded area. One approach used a gradation along the border between the wall and the occluded area, and the other superimposed line segments to the predefined area. The DR result is shown in Fig. 9(c).

Figure 10 shows a person walking outside a room. To capture a wide background, two Kinect V2s were used. We reconstructed the background in each frame in order to handle dynamic scenes. Because we can estimate the camera pose even in a scene with white walls, we could achieve DR by converting the viewpoint of the camera from behind the wall to the foreground.

5   Conclusion

We proposed a method to reconstruct a 3D model. We first recovered the 3D model using the method in [12] and removed the redundant line segments. In the method proposed in this paper, we used matching information to keep the minimum number of necessary line segments, and we removed line segments that were unnaturally projected due to depth noise. By removing such line segments, we improved the visual appearance.

In the experiments, we demonstrated that the proposed method could greatly reduce the number of line segments while keeping the detailed geometry of the structure. By using the proposed method to reconstruct a 3D model, the accuracy of estimated camera poses was better than that with the method in [12], and there were fewer line segments. By reducing the number of line segments, we can estimate camera poses with less computational time compared to [12]. Moreover, our proposed method enables seeing through walls.

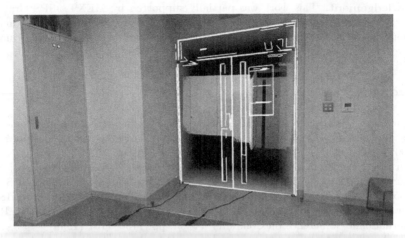

Fig. 10. DR results with a walking human.

Acknowledgement. This work was partially supported by MEXT/JSPS Grant-in-Aid for Scientific Research(S) 24220004, and JST CREST "Intelligent Information Processing Systems Creating Co-Experience Knowledge and Wisdom with Human-Machine Harmonious Collaboration".

References

1. Agarwal, S., Snavely, N., Simon, I., Seitz, S.M., Szeliski, R.: Building Rome in a day. In: 2009 IEEE 12th International Conference on Computer Vision, pp. 72–79 (2009)
2. Sturm, P., Triggs, B.: A factorization based algorithm for multi-image projective structure and motion. In: Buxton, B., Cipolla, R. (eds.) ECCV 1996. LNCS, vol. 1065, pp. 709–720. Springer, Heidelberg (1996). doi:10.1007/3-540-61123-1_183
3. Lowe, D.G.: Object recognition from local scale-invariant features. In: 7th IEEE International Conference on Computer Vision, pp. 1150–1157 (1999)
4. Bay, H., Tuytelaars, T., Gool, L.: SURF: speeded up robust features. In: Leonardis, A., Bischof, H., Pinz, A. (eds.) ECCV 2006. LNCS, vol. 3951, pp. 404–417. Springer, Heidelberg (2006). doi:10.1007/11744023_32
5. Hofer, M., Wendel, A., Bischof, H.: Line-based 3D reconstruction of wiry objects. In: 18th Computer Vision Winter Workshop (2013)
6. Hofer, M., Wendel, A., Bischof, H.: Incremental line-based 3D reconstruction using geometric constraints. In: 24th British Machine Vision Conference, pp. 92.1–92.11 (2013)
7. Jain, A., Kurz, C., Thormählen, T., Seidel, H.P.: Exploiting global connectivity constraints for reconstruction of 3D line segments from images. In: 2010 IEEE Computer Vision and Pattern Recognition, pp. 1586–1593 (2010)
8. Wang, Z., Wu, F., Hu, Z.: MSLD: a robust descriptor for line matching. Pattern Recogn. **42**, 941–953 (2009)
9. Hirose, K., Saito, H.: Fast line description for line-based slam. In: 23th British Machine Vision Conference, pp. 83.1–83.11 (2012)
10. Nakayama, Y., Honda, T., Saito, H., Shimizu, M., Yamaguchi, N.: Accurate camera pose estimation for KinectFusion based on line segment matching by LEHF. In: 22nd International Conference on Pattern Recognition. 2149–2154. (2014)
11. Nakayama, Y., Saito, H., Shimizu, M., Yamaguchi, N.: 3D line segment based model generation by RGB-D camera for camera pose estimation. In: Jawahar, C.V., Shan, S. (eds.) ACCV 2014. LNCS, vol. 9010, pp. 459–472. Springer, Heidelberg (2015). doi:10.1007/978-3-319-16634-6_34
12. Nakayama, Y., Saito, H., Shimizu, M., Yamaguchi, N.: Marker-less augmented reality framework using on-site 3D line-segment-based model generation. J. Imaging Sci. Technol. **60**, 20401-1–20401-24 (2016)
13. Hofer, M., Maurer, M., Bischof, H.: Improving sparse 3D models for man-made environments using line-based 3D reconstruction. In: 2014 2nd International Conference on 3D Vision, pp. 535–542 (2014)
14. Sugiura, T., Torii, A., Okutomi, M.: 3D surface reconstruction from point-and-line cloud. In: 3rd International Conference on 3D Vision, pp. 264–272 (2015)
15. Newcombe, R.A., Izadi, S., Hilliges, O., Molyneaux, D., Kim, D., Davison, A.J., Kohi, P., Shotton, J., Hodges, S., Fitzgibbon, A.: KinectFusion: real-time dense surface mapping and tracking. In: 2011 10th IEEE International Symposium on Mixed and Augmented Reality, pp. 127–136 (2011)

16. Yonezawa, T., Ueda, K.: Real-time 3D data reduction and reproduction of spatial model using line detection in RGB image. In: Joint 7th International Conference on Soft Computing and Intelligent Systems and 15th International Symposium on Advanced Intelligent Systems, pp. 727–730 (2014)
17. von Gioi, R.G., Jakubowicz, J., Morel, J.M., Randall, G.: LSD: a line segment detector. Image Process. On Line **2**, 35–55 (2012)
18. Nguyen, T., Reitmayr, G., Schmalstieg, D.: Structural modeling from depth images. IEEE Trans. Vis. Comput. Graph. **21**, 1230–1240 (2015)
19. Zhang, Z., Faugeras, O.: Building a 3D world model with a mobile robot: 3D line segment representation and integration. In: 10th International Conference on Pattern Recognition, pp. 38–42 (1990)
20. Holz, D., Behnke, S.: Fast range image segmentation and smoothing using approximate surface reconstruction and region growing. In: Lee, S., Cho, H., Yoon, K.-J., Lee, J. (eds.) Intelligent Autonomous Systems 12, vol. 194, pp. 61–73. Springer, Heidelberg (2013)
21. Avery, B., Sandor, C., Thomas, B.H.: Improving spatial perception for augmented reality X-Ray vision. In: 2009 IEEE Virtual Reality Conference, pp. 79–82 (2009)

3D Estimation of Extensible Surfaces Through a Local Monocular Reconstruction Technique

S. Jafar Hosseini[✉] and Helder Araujo

Institute of Systems and Robotics,
Department of Electrical and Computer Engineering,
University of Coimbra, Coimbra, Portugal
jafar@isr.uc.pt

Abstract. This paper deals with the monocular reconstruction of an extensible surface by proposing a novel approach for the determination of the 3D positions of a set of points on images of the deformed surface. Given a 3D template, this approach is applied to each image independently of the others. To proceed with the reconstruction, the surface is divided into small patches that overlap in chain-like form. We model these surface patches as being uniformly extensible. Using a linear mapping from the template onto a patch, the variation of the patch shape is split into a rigid body transformation and a pure deformation. To estimate the pure deformation, we use an optimization procedure that minimizes the reprojection error along with the error over a constraint associated with uniform expansion. Having estimated the pure deformation, the rigid body transformation can be determined by decomposing the essential matrix between the current image and the virtual image that results from projecting the 3D positions that correspond to pure deformation of the template. This enables complete estimation of the linear mapping, thereby obtaining the 3D positions of the surface patch up to scale. To define a common scale, the surface smoothness is enforced by considering that the overlapping points of the patches are the same. The experimental results show the feasibility of the approach and that the accuracy of the reconstruction is good.

1 Introduction

Monocular reconstruction of deformable surfaces has been extensively studied in the last few years [1–7] and is important for many applications. The use of prior knowledge on the shape and motion has been a popular approach in order to constrain the solution. The priors can be divided in two main categories: the statistical and physical priors. For instance, the methods relying on the low-rank factorization paradigm [1,2] can be classified as statistical approaches. Learning approaches such as [7,8] also belong to the statistical approaches. Physical constraints include spatial and temporal priors on the surface to be reconstructed [9,10]. A physical prior of particular interest in this case is the hypothesis of having an inextensible (i.e. isometric) surface [11–13]. Isometric reconstruction from perspective camera views has been the subject of significant research efforts.

© Springer International Publishing AG 2017
C.-S. Chen et al. (Eds.): ACCV 2016 Workshops, Part II, LNCS 10117, pp. 114–123, 2017.
DOI: 10.1007/978-3-319-54427-4_9

This hypothesis means that the length of the geodesics between every two points on the surface should not change over time. Early approaches relax the non-convex isometric constraints to inextensibility with the so-called maximum depth heuristic [12], better known as upper-bound approach. The idea is to maximize point depths so that the Euclidean distance between every pair of points is upper bounded by its geodesic distance, computed in the template [7,13]. In these papers, a convex cost function combining the depth of the reconstructed points and the negative of the reprojection error is maximized while enforcing the inequality constraints arising from the surface inextensibility. The resulting formulation can be easily turned into a SOCP problem. This problem is convex and gives accurate reconstructions. A similar approach is explored in [11]. The approach of [12] is a point-wise method. The approaches of [7,11,13] use a triangular mesh as a surface model, and the inextensibility constraints are applied to the vertices of the mesh. In [14], the authors propose the mapping between a template image and the 3D surface to obtain smooth reconstructions of the surface using differential geometry. The reconstruction of deformable surfaces has been recently extended to non-developable surfaces undergoing conformal deformation [15]. An analytical approach was applied to isometric and conformal deformations using Partial Differential Equations [16], where the approach was developed for weak-perspective projection. A SLAM method for elastic surfaces was tried with fixed boundary conditions [17]. In [18], the authors formulate the reconstruction problem of a generic surface in terms of the minimization of the stretching energy and impose a set of fixed boundary 3D points to constrain the solution. This approach deals with a general group of elastic surfaces without applying any constraints explicitly associated with conformal deformation.

1.1   Problem Statement

In this work, we tackle the problem of reconstructing extensible surfaces, specially those that may extend or be enlarged in volume and take the form of volumetric shapes similar to the sphere, cylinder and the ellipsoid. Examples are plastic balloons, hearts, balls - see Fig. 1. Such objects undergo a continuous, relatively uniform expansion over any small patch of the surface. This expansion usually occurs along a certain axis. Given a 3D template consisting of known 3D points and its image, the objective is to determine the 3D positions of a set of points in any image, from a video sequence acquired with a calibrated camera. Correspondences are established with their positions in the template image. In our formulation of the reconstruction problem, we assume that the whole surface deformation is locally homothetic i.e. homothetic deformation implies uniform expansion. As a result, the surface is treated as being built from patches connected together, with the deformation of each patch being considered homothetic. This kind of deformation can be represented by a linear function that transforms the 3D template into a deformed patch. This function is decoupled into a pure deformation and a rigid body transformation. Our approach is developed in such a way that we start by estimating the pure deformation first. Therefore an optimization procedure is defined so that only the pure deformation is estimated. The rigid body transformation can be then estimated. This is done by decomposing the essential matrix between the current image and the virtual image that corresponds to the projection of the deformed points following the application of the pure deformation. Finally, as the result of the estimation of the linear mapping, we can compute the 3D positions of the points for each surface patch, separately, up to a scaling factor. Each patch is assumed to overlap only up to two neighboring patches. That assumption allows a smooth reconstruction up to a global scale, by enforcing the smoothness of the surface.

Fig. 1. Test objects that can be expanded by inflation.

In Sect. 2 we describe the decomposition of the linear mapping. The optimization itself is explained in detail in Sect. 3. In Sect. 4, the experimental results are presented. Section 5 contains conclusions and future work.

2   Proposed Approach: Deformation Model

A deformation can be modeled and specified by the point-to-point mapping function $\mathbf{p}_i = M\left(\mathbf{p}_i^\circ\right)$ from the undeformed shape S° onto the deformed shape S, as shown in Fig. 2. The deformation gradient is the fundamental measure of deformation in continuum mechanics. It is the second order tensor which maps line elements on S° into line elements (consisting of the same material particles) on S. Let a point i on S° be defined as $\mathbf{p}_i^\circ = \left[x_i^\circ \ y_i^\circ \ z_i^\circ\right]^T$ and the corresponding point on S as $\mathbf{p}_i = \left[x_i \ y_i \ z_i\right]^T$, the deformation gradient $\mathbf{F}$ is given by:

$$\mathbf{F} = \begin{bmatrix} \frac{\delta x_i}{\delta x_i^\circ} & \frac{\delta x_i}{\delta y_i^\circ} & \frac{\delta x_i}{\delta z_i^\circ} \\ \frac{\delta y_i}{\delta x_i^\circ} & \frac{\delta y_i}{\delta y_i^\circ} & \frac{\delta y_i}{\delta z_i^\circ} \\ \frac{\delta z_i}{\delta x_i^\circ} & \frac{\delta z_i}{\delta y_i^\circ} & \frac{\delta z_i}{\delta z_i^\circ} \end{bmatrix} \tag{1}$$

$\mathbf{F}$ can be written as either $\mathbf{R.U}$ or $\mathbf{V.R}$ through a polar decomposition. In each case, $\mathbf{R}$ is the rotation matrix, and $\mathbf{U}$ and $\mathbf{V}$ are symmetric matrices describing stress and strain, which contribute to deformations.

Uniformly extensible deformations are homothetic. These deformations are characterized by having the same gradient at every point. As a result, the mapping from the undeformed shape, which is chosen as the template, onto the deformed surface can be formulated as:

$$\mathbf{p}_i = \mathbf{F}.\mathbf{p}_i^\circ + \mathbf{c} \tag{2}$$

where $\mathbf{c}$ is a 3×1 vector that denotes the rigid body translation. We refer to this linear mapping as deformation model. Every surface patch is then represented separately by means of this mapping. Let us denote $\mathbf{F}$ as $\mathbf{R}.\mathbf{U}$. We propose an algorithm for calculating the rigid body transformation depending on the deformation matrix $\mathbf{U}$ to be given in advance. The algorithm below shows how to perform the calculation.

For a surface patch, the 3D positions that result from the equation $\hat{\mathbf{p}}_i = \mathbf{U}.\mathbf{p}_i^\circ$ are on a virtual patch that can be viewed as intermediate between the template and the real deformed patch. The corresponding points on both the virtual and the real patch are related as follows:

$$\mathbf{p}_i = \mathbf{R}.\hat{\mathbf{p}}_i + c \tag{3}$$

This equation basically defines a transformation between 2 coordinate systems. However, in our case, $\mathbf{R}$ and c specifies a transformation between $\hat{\mathbf{p}}_i$ and $\mathbf{p}_i$ in the same coordinate system (i.e. camera coordinates). To deal with this transformation, we consider it differently, in the sense that, with no loss of generality, we can assume that $\mathbf{R}$ and c indicate the transformation between the current image and a virtual image - see Fig. 2. This is a dual configuration of the real problem under consideration. The virtual image, in fact, represents only 2D image points obtained by projecting the intermediate patch onto the image plane with the calibration matrix. As a result, we now have one rigid surface patch projected onto two separate images that are related by $\mathbf{R}$ and c. By decomposing the essential matrix, $\mathbf{R}$ and c are then estimated. To do so, the following procedure can be followed:

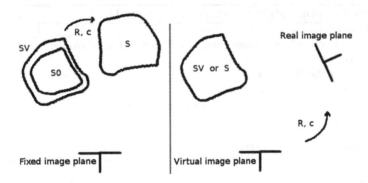

Fig. 2. Left-side image: the real problem. Right-side image: the dual problem. **S0**, **SV** and **S** indicate the template, the intermediate patch and the real deformed patch respectively.

1. Estimate the essential matrix between the two images by using the corresponding image points for the patch.

2. Decompose the essential matrix into rotation and translation.
3. This decomposition yields 4 possible solutions [19]. To find the one that is feasible in our problem, the positive depth constraint can be applied.

Having estimated $\mathbf{R}$ and c, the deformed surface can be reconstructed up to scale. However, the algorithm just described depends on our knowledge of the deformation matrix U. Since this matrix is unknown, we have to determine it. For this purpose, we formulate an optimization procedure in which $\mathbf{U}$ is the unknown. Such optimization requires an initial estimate of $\mathbf{U}$. The 3-step algorithm above forms the core of our optimization procedure. For simplicity's sake, we call it *pose estimation algorithm*.

3   Proposed Approach: Optimization

Consider that $\mathbf{K}$ is a known calibration matrix with focal length f and that $\mathbf{q} = (u, v)$ denotes the image points - corresponding to $\mathbf{p}$ - projected in the image of the deformed surface. For point i, the projection equations are defined by $f.p_{x,i} - u_i.p_{z,i} = 0$ and $f.p_{y,i} - v_i.p_{z,i} = 0$. These equations are used to define the reprojection error to be minimized. We also define a second constraint, intended to ensure that the deformation is approximately homothetic. This is equivalent to saying that the off-diagonal entries of $\mathbf{U}$ should be approximately zero. This results from the fact that the diagonal entries of the deformation matrix $\mathbf{U}$ are dominant for a homothethic surface, compared with the off-diagonal ones. The optimization procedure includes the pose estimation algorithm, reprojection error and the constraint for homothetic deformation. Algorithm 1 details the different steps of the optimization procedure.

Algorithm 1. Optimization procedure to estimate $\mathbf{U}_0$.

Initialization of $\mathbf{U}_0$.
while the total error is not lower than a specified threshold **do**
    Update $\mathbf{U}_0$.
    $[\mathbf{R}_0, \mathbf{U}] = \text{PolarDecom}(\mathbf{U}_0)$
    Pose estimation with $\mathbf{U}$ as input.
    Minimization of the total error including reprojection error and the other constraint error.
end while
The iteration is stopped and the current value for $\mathbf{U}_0$ is used.

Next we describe this algorithm in detail:

Step 1: $\mathbf{U}_0$ is a 3×3 matrix that is to be estimated as the result of the optimization. This matrix is initialized with random values.

Step 2: $\mathbf{U}_0$ is factorized by a polar decomposition into a deformation matrix $\mathbf{U}$, which is symmetric, positive semi-definite and a rotation matrix $\mathbf{R}_0$ which is not used.

Step 3: $\mathbf{U}$, is used in the pose estimation algorithm. At this point, $\mathbf{R}$ and c are determined.

Step 4: After having estimated all the parameters of the deformation model, approximate 3D positions can be computed using Eq. 2. Then, the following error is minimized with respect to $\mathbf{U}_0$:

$$\min_{\mathbf{U}_0} \sum_{i=1}^{n} \left[(f.p_{x,i} - u_i.p_{z,i})^2 + (f.p_{y,i} - v_i.p_{z,i})^2 \right] + w. \sum \text{OffDiagonalEntries}\,(\mathbf{U})^2 \tag{4}$$

where the latter term indicates the sum of squares of off-diagonal entries of $\mathbf{U}$ and w is a weighting factor which is set empirically. n is the number of points.

The above optimization is a non-linear sum of squares problem that can be solved using the Levenberg-Marquardt algorithm.

Step 5: If the step error (that is, the size of the change in the location where the objective function was evaluated) is not lower than a specified threshold, the optimization is iterated starting with a new random value for $\mathbf{U}_0$.

Step 6: Once the value for matrix $\mathbf{U}_0$ is estimated, we can solve for all the parameters of the deformation model by using steps 2 and 3. As a result, the 3D positions of the points are determined up to a scaling factor. This factor is due to the fact that translation t is estimated only up to a scale factor. The whole approach proposed so far is applied separately to all surface patches. Consequently, each will be reconstructed on a possibly different scale. In order to obtain a common scale, surface smoothness is enforced by considering the points where the patches overlap. Assume that the patches are labeled and that there are a total of n_a patches. Every 2 successive patches overlap in chain-like form,

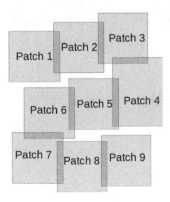

Fig. 3. The chain-like form of patches

as shown in Fig. 3. The smoothness enforcement is accomplished with this formula:

$$scale = \frac{1}{n_c} \sum_{i=1}^{n_c} \left(\frac{1}{3} \sum_{j=1}^{3} p_{i,j,l}/p_{i,j,k} \right) \tag{5}$$

where n_c is the number of common points belonging to patches l and k. Subscript j denotes the x, y and z coordinates. Patch 1 is selected to be the scaling reference and all the patches will be rescaled with respect to it. Equation 5 is employed for every 2 overlapping patches, sequentially, up to patch n_a.

4   Experimental Results

In this section, we validate the proposed approach by conducting a set of experiments on data that conforms to the type of deformation we are addressing. For that purpose we consider simple examples, namely the cylinder, ellipsoid and the sphere- see Fig. 4. Their mathematical models can be used to generate the set of n point correspondences required. For example, the sphere is expanded/contracted uniformly by changing its radius. Although these volumetric shapes are simple surfaces, they constitute adequate models for the tests with synthetic data, so that the efficiency of the approach can be evaluated quantitatively. Points **p** are assumed to be visible from the virtual camera, as illustrated in Fig. 4. There are a total of 185 points which are divided into 10 patches, each of which has the

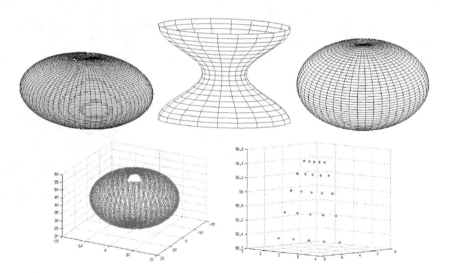

Fig. 4. Synthetic extensible surfaces. Top: from left to right: ellipsoid, deformed cylinder, sphere. Bottom: left image shows a patch of the sphere, while the right image shows the points on the surface. The estimated and the ground-truth points are marked with '+' and 'o', respectively.

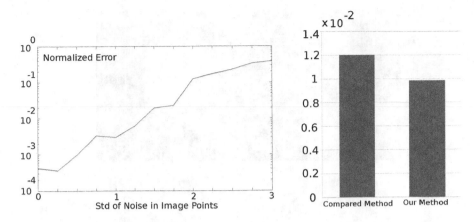

Fig. 5. Left: reconstruction error with respect to noise added to image points coordinates. Right: a comparison of reconstruction error between our approach and the approach of [18].

same number of points. This is done by considering 5 common points between any two overlapping patches and so we have 25 points per patch. The accuracy of the approach is evaluated using the 3D reconstruction error which is defined by:

$$RE = \frac{1}{n} \sum_{i=1}^{n} \left[\|\mathbf{p}_i - \hat{\mathbf{p}}_i\|^2 / \|\hat{\mathbf{p}}_i\|^2 \right] \tag{6}$$

This error corresponds to the normalized Euclidean distance between the observed ($\hat{\mathbf{p}}_i$) and the estimated ($\mathbf{p}_i$) world points where n is the total number of points. Taking into consideration that the reconstruction estimates the 3D coordinates up to scale with respect to the ground-truth data, an estimate of the scaling factor is used in the calculation of the reconstruction error. In addition to the case of synthetic data without noise, we also estimate the reconstruction error by adding noise to the coordinates of the 2D image points (Fig. 5). The results shown in the left-side plot correspond to average values obtained by repeating the estimation 100 times for each one of 15 different random deformations used. Five different random deformations were considered for each synthetic surface. As expected, the number of iterations required for the optimization to converge increases with noise. In the absence of noise the average number of iterations required for convergence is 150.

The experiments with real images were performed using a plastic baloon. The desired deformations are obtained by inflating this object. 3D ground-truth data corresponding to a template shape is also computed- see Fig. 6. The image of the deformed surface is acquired by a calibrated camera. The surface includes a total of 80 points which are grouped into 7 patches with 3 points on each overlapping area. Each patch has 14 points. The input data contains 2D point correspondences matched to the template image.

Fig. 6. Top: real extensible deformations using plastic baloons with the reconstructed points indicated by dots. Bottom: the surface fit to the points.

We compare the performance of the proposed approach with the approach proposed in [18], where the authors formulate the reconstruction of a generic surface in terms of the minimization of stretching energy while imposing a set of fixed boundary 3D points to constrain the solution - see Fig. 5. The results given on the right-side plot show the averages obtained by repeating the estimation 100 times for each of 10 deformations of the plastic baloon.

5   Conclusions and Future Work

In this paper, we have proposed a reconstruction method for elastic surfaces that can expand freely. This approach is based on the assumption that the deformation is locally homothetic (i.e. uniformly extensible), and reconstructs the surface by splitting it into overlapping patches. Each local deformation was then modeled by a linear mapping between the image and a known 3D template. The mapping is defined by a combination of a pure deformation and a rigid-body transformation. The deformation and rigid transformation were estimated by means of an optimization procedure. As a result, estimates of the 3D coordinates (up to scale) of the points on each patch are obtained. A global scale factor for all the patches is obtained by enforcing the smoothness of the reconstruction. The results demonstrated the efficiency of the approach. As ongoing project, we are working on expanding our reconstruction method to medical imaging.

References

1. Aans, H., Kahl, F.: Estimation of deformable structure and motion. In: Workshop on Vision and Modelling of Dynamic Scenes, ECCV, Denmark (2002)
2. Del-Bue, A., Llad, X., Agapito, L.: Non-rigid metric shape and motion recovery from uncalibrated images using priors. In: IEEE Conference on Computer Vision and Pattern Recognition, New York (2006)
3. Bregler, C., Hertzmann, A., Biermann, H.: Recovering non-rigid 3D shape from image streams. In: IEEE Conference on Computer Vision and Pattern Recognition, pp. 2690–2696 (2000)
4. Bartoli, A., Gay-Bellile, V., Castellani, U., Peyras, J., Olsen, S., Sayd, P.: Coarse-to-fine low-rank structure-from-motion. In: IEEE Conference on Computer Vision and Pattern Recognition (2008)
5. Brand, M.: Morphable 3D models from video. In: CVPR (2001)
6. Bookstein, F.: Principal warps: thin-plate splines and the decomposition of deformations. IEEE Trans. Pattern Anal. Mach. Intell. 11(6), 567–585 (1989)
7. Salzmann, M., Fua, P.: Reconstructing sharply folding surfaces: a convex formulation. In: IEEE Conference on Computer Vision and Pattern Recognition (2007)
8. Gay-Bellile, V., Perriollat, M., Bartoli, A., Sayd, P.: Image registration by combining thin-plate splines with a 3D morphable model. In: International Conference on Image Processing (2006)
9. Gumerov, N., Zandifar, A., Duraiswami, R., Davis, L.S.: Structure of applicable surfaces from single views. In: Pajdla, T., Matas, J. (eds.) ECCV 2004. LNCS, vol. 3023, pp. 482–496. Springer, Heidelberg (2004). doi:10.1007/978-3-540-24672-5_38
10. Prasad, M., Zisserman, A., Fitzgibbon, A.: Single view reconstruction of curved surfaces. In: IEEE Conference on Computer Vision and Pattern Recognition, pp. 1345–1354 (2006)
11. Shen, S., Shi, W., Liu, Y.: Monocular 3-D tracking of inextensible deformable surfaces under L2-norm. IEEE Trans. Image Process. 19, 512–521 (2010)
12. Perriollat, M., Hartley, R., Bartoli, A.: Monocular template-based reconstruction of inextensible surfaces. Int. J. Comput. Vis. 88(1), 85–110 (2010)
13. Salzmann, M., Moreno-Noguer, F., Lepetit, V., Fua, P.: Closed-form solution to non-rigid 3D surface registration. In: Forsyth, D., Torr, P., Zisserman, A. (eds.) ECCV 2008. LNCS, vol. 5305, pp. 581–594. Springer, Heidelberg (2008). doi:10.1007/978-3-540-88693-8_43
14. Brunet, F., Bartoli, A., Hartley, R.: Monocular template-based 3D surface reconstruction: convex inextensible and nonconvex isometric methods, pp. 157–186, April 2014 (accepted)
15. Malti, A., Bartoli, A., Collins, T.: Template-based conformal shape-from-motion-and-shading for laparoscopy. In: Abolmaesumi, P., Joskowicz, L., Navab, N., Jannin, P. (eds.) IPCAI 2012. LNCS, vol. 7330, pp. 1–10. Springer, Heidelberg (2012). doi:10.1007/978-3-642-30618-1_1
16. Pizarro, D., Bartoli, A., Collins, T.: Isowarp and conwarp: warps that exactly comply with weak-perspective projection of deforming objects. In: BMVC (2013)
17. Agudo, A., Calvo, B., Montiel, J.M.M.: FEM models to code non-rigid EKF monocular SLAM. In: ICCVW (2011)
18. Malti, A., Hartley, R., Bartoli, A., Collins, T.: Monocular template-based 3D reconstruction of extensible surfaces with local linear elasticity (2013)
19. Hartley, R., Zisserman, A.: Multiple View Geometry in Computer Vision. Cambridge University Press, Cambridge (2004)

Disparity Estimation by Simultaneous Edge Drawing

Dexmont Peña$^{(\boxtimes)}$ and Alistair Sutherland

Dublin City University, Glasnevin, Dublin, Ireland
dexmont.penacarrillo2@mail.dcu.ie, alistair.sutherland@dcu.ie

Abstract. This work presents a new low-level real-time algorithm for simultaneous edge drawing and disparity calculation in stereo image pairs. It works by extending the principles from the ED algorithm, a fast and robust edge detector able to produce one pixel-wide chains of pixels for the edges in the image. In this paper the ED algorithm is extended to run simultaneously on both images in a stereo-pair. The disparity information is obtained by matching only a few anchor points and then propagating those disparities along the image edges. This allows the reduction of computational costs compared to other edge-based algorithms, as only a few pixels require to be matched, and avoids the problems present in other edge-point based approaches. The experiments show that this new approach is able to obtain accuracies similar to other state-of-the-art approaches but with a reduced number of computations.

1 Introduction

The use of 3D information has many applications from navigation in mobile robots to object reconstruction and augmented reality. The computer vision approach for extracting 3D information has been widely studied, creating a variety of algorithms which could be broadly classified into dense and sparse. Dense approaches try to obtain depth information for every pixel in the input images. These approaches take advantage of the smoothness constraints in surfaces in order to increase the accuracy of the obtained depths but this comes at a high computational cost. In contrast sparse approaches obtain depth information only for important features in the images. Image features could correspond to pixels, blobs, edges or ridges. By using only those important features, the amount of computation required is decreased as only a few pixels are processed for the calculation of depth.

Point-based approaches have been successfully used to reconstruct highly textured areas but they fail in areas with low texture [1]. In contrast, image edges are present at object boundaries (borders, colour or texture changes). Based on this easy availability of edges, in this work a new algorithm is proposed. The proposed approach extracts disparity information by matching only a few anchor points and then propagates the disparities while the edges are simultaneously detected in a stereo-pair.

C.-S. Chen et al. (Eds.): ACCV 2016 Workshops, Part II, LNCS 10117, pp. 124–135, 2017.
DOI: 10.1007/978-3-319-54427-4_10

1.1   Related Work

Due to the reduced cost involved in the computation of depth maps using image edges, different approaches have been proposed to take advantage of this. The approaches found in the literature can be divided into line-based and curve-based. Line-based approaches assume a straight line may be fitted to the image edges. This set of approaches have been successfully used for wide-baseline stereo [2], Visual Odometry [3,4] and Structure from Motion [5]. Although effective in some cases, if few or no straight lines are present in the image, those approaches would fail, as is the case in outdoors scenarios where curved objects like trees and humans are present.

Only a few approaches have been proposed for using curves as image edges. Rao et al. [6] used Bezier curves to solve the SLAM problem. But only the endpoints of the curves are used to perform the tracking. To our knowledge the approaches presented by Witt and Weltin [7,8] and Ohta and Kanade [9] are the only ones able to obtain disparity maps in real time for edge pixels without fitting any model to the edges.

In [7] Witt and Weltin use Canny edge detection [10] to extract the image edges. Then every edge pixel is matched across the stereo pair by using different support regions in an area based approach. A confidence measure is obtained for the match. Then interpolation is performed to increase the confidence of poor or not-found matches. In [8] Witt and Weltin use Canny edge detection for finding the edge pixels. Then they match the edge pixels by using a strip as support region in an area based approach. Only highly confident matches are kept and the remaining disparities are recovered by dynamic programming. In [9] Ohta and Kanade detect edges by using image gradients. Then the edges on the same scanline are matched by using dynamic programming. After this, dynamic programming is used intra-scanline to identify and recover wrong matches. Although fast, these approaches require the computation of descriptors, matching cost and optimization for every edge pixel.

1.2   Contributions

This paper introduces an approach able to extract chains of connected 3D points which represent the 3D edges of the objects in the scene captured by the stereo-pair. In summary the main advantages of this approach over the state-of-the-art edge-based approaches are:

- Edge matching robust to partial occlusions.
- Only a small fraction of the edge-points require to be matched, reducing the number of descriptors and cost computations required.
- The extraction of chains of connected 3D points which represent the edges of the objects in 3D, comes as a by-product of the simultaneous smart routing.

The approach for obtaining this is detailed in the following section.

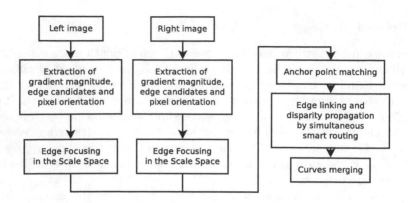

Fig. 1. Pipeline of simultaneous edge drawing.

2 Disparity Estimation by Simultaneous Edge Drawing

The proposed algorithm is able to obtain the disparity of edge pixels by simultaneously detecting the edges in a stereo pair. This is achieved by extending the Edge Drawing (ED) algorithm proposed by Topal and Akinlar [11–13], so that it runs simultaneously on both images in the stereo pair. By doing this, it reduces the required amount of computation as only a few anchor points are matched across the image-pair. The obtained disparities are propagated along the edges as they are detected by linking the pixels along the image gradient. The pipeline of the proposed algorithm is shown in Fig. 1.

The algorithm for sparse disparity calculation by Simultaneous Edge Drawing (SED) consists of the following stages:

1. Computation of the gradient magnitude, edge candidates and orientations.
2. Extraction of anchor points in the Scale Space by Edge Focusing [14].
3. Anchor matching by an area based approach.
4. Edge linking and disparity propagation by simultaneous smart routing.
5. Curve merging.

The algorithm for computing of the gradient magnitude, edge candidates and orientations is inherited from ED. The extraction of the anchor points is performed across the scale space in order to reduce the number of anchor points which might be created by noise or texture. The anchor matching is common to other pixel-based stereo-matching approaches. The simultaneous smart routing is extended from ED to run on a stereo-pair. The curve merging uses a simple approach for recovering fragmented edges. Each of these steps is detailed in the following sections.

2.1 Computation of the Gradient Magnitude, Edge Candidates and Orientations

The first step in ED is to compute the gradient magnitude $G(p)$, the edge candidates $E(p)$ and orientations $H(p)$ for every pixel p in the stereo-pair. This step is performed in a smoothed version of the input images. As in ED, the smoothing is performed by convolving the input images with a Gaussian kernel with $\sigma = 1$. The calculation of the gradient magnitude is performed as in the parameter-free version of ED (EDPF [13]). The horizontal and vertical gradients, $g_x(p)$ and $g_y(p)$ respectively, are calculated by using the Prewitt operator [15]. Then the gradient magnitude is calculated using the formula

$$G(p) = \sqrt{(g_x(p))^2 + (g_y(p))^2} \tag{1}$$

The edge orientation at every pixel p is calculated by comparing the horizontal and vertical gradients. If $|g_x(p)| < |g_y(p)|$ it is assumed that the pixel is traversed by an horizontal edge, $H(p) = 1$. Otherwise, the edge is assumed to be vertical, $H(p) = 0$ following the same criteria as in ED.

The edge candidates are the pixels which may contain an edge, they are identified by gradient thresholding. Pixels with a gradient magnitude larger than a threshold t_g are taken as edge candidates. The threshold t_g is used as in the parameter free version of ED where only gradient values produced by a quantization error are discarded. For the Prewitt operator this occurs at the value of $t_g = 8.48$ as in ED [13].

2.2   Extraction of the Anchor Points

The anchor points in ED are pixels with a high chance of being edge pixels, in this work we took only those pixels with a high chance of being edge pixels across the scale space. This is done in order to reduce the number of anchors which might correspond to edges created by low textures or noise. In order to perform this, the Edge Focusing algorithm [14] is applied. The Edge Focusing algorithm applies non-maxima suppression at a coarse level and then the detected pixels are tracked through the scale space until a fine level is reached.

As in [14], the maximum sigma for the scale space is set to $\sigma_{MAX} = 4$ and the size of the step between scale space levels is set to $\delta\sigma = 0.5$ in order to guarantee that the detected edge pixels do not move more than one pixel between scales.

Additionally, as in ED, the scan interval s is used to reduce the number of processed anchor points. The scan interval is used to process only one out of s rows, i.e. if $s = 2$ only one of every two rows is processed, if $s = 3$ only one of every three rows is processed.

2.3   Anchor Matching

The anchor points are matched by using an area-based stereo-matching approach. The approach is selected as in [8] but using the Complete Rank Transform [16] as pixel descriptor. The Complete Rank Transform uses the morphological order of the pixels in a window to create a descriptor invariant to changes in illumination. An example of this image transform is shown in Fig. 2. As in [8], the descriptors are computed at each side of the anchor points. The Sum of Absolute Differences (SAD) is taken as match cost and a WTA stategy is applied

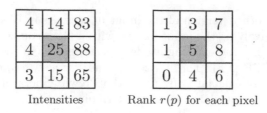

Intensities   Rank $r(p)$ for each pixel

Fig. 2. Example of the complete rank transform, CRT(p) = 1, 3, 7, 1, 5, 6, 0, 4, 6.

for selecting the best match as in [8]. Left-right confidence check is used in order to identify occluded anchors and discard them [17].

Additionally, the Peak-Ratio Naive Inverse (PKRNI) and Max Cost Ratio (MCR) confidence metrics are applied in order to remove possible wrong matches. The PKRNI is based on the PKR confidence metric [18] but bounded to the interval $[0, 1]$. This is:

$$PKRNI = 1 - \frac{c_1 + 1}{c_{2m} + 1} \tag{2}$$

where c_1 is the minimum found cost and c_{2m} corresponds to the second minimum cost. In the case where there is only one candidate, c_{2m} is taken as the maximum cost for the descriptor c_{MAX}. For the CRT, c_{MAX} is computed as:

$$c_{MAX} = (mn - 1) * (mn) \tag{3}$$

where m and n are the number of rows and columns used for the CRT. The MCR confidence metric is simply a ratio between the minimum cost and the maximum cost bounded to the interval $[0,1]$, this is expressed as:

$$MCR = 1 - \frac{c_1}{c_{MAX}} \tag{4}$$

Only high confident matches are kept, to determine this thresholds t_{PKRI} and t_{MCR} are applied to the PKRI and MCR confidence metrics.

It is important to note that only non-linked anchors are matched one at the time. This means that for every matched anchor a simultaneous smart routing is performed and if an anchor is already linked it would not be processed. By doing this the number of required descriptors and cost computations is decreased as only a small portion of the anchors is processed.

2.4 Disparity Propagation by Simultaneous Smart Routing

The matched anchor points are used as seeds for the simultaneous smart routing. The simultaneous smart routing procedure extends the smart routing from ED to run simultaneously on each image in the stereo pair and produces a chain of 3D points. Every point in this chain has the form (x, y, d) where x and y are the coordinates on the left image and d is the disparity of the pixel.

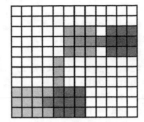

Fig. 3. Location of windows for a 3×3 CRT used for creating the anchor point descriptors. The red pixels are the anchor points, the yellow are the pixels used for the left descriptor and the blue ones are the pixels used for the right descriptor. Gray pixels are edge pixels. (Color figure online)

The simultaneous smart routing takes as input a matched anchor pair and then by using the same principle from ED it moves one pixel in each image of the image-pair. This displacement across the edge candidates is performed until any of the following stopping criteria is met:

- The next pixel to be linked in any of the images in the stereo pair is not in the edge map E for the left and right images, i.e., the thresholded gradient of the next pixel is zero (as in ED).
- A previously linked pixel is found (as in ED).
- The difference in the y location of the pixels is larger than an epipolar threshold t_e.

In order to add tolerance to errors in the image rectification the threshold t_e is introduced. This threshold is computed as the difference in the y location of the current linked pixels:

$$t_e = |y_c^l - y_c^r| \tag{5}$$

where y_c^l and y_c^r are the y location in the left and right image of the current pixel being linked.

In order to determine the next pixel to be linked an approach similar to that used in ED is applied: starting at a pair of matched anchor points, for the left image if a vertical edge passes through the matched pair the linking is started up and down, if the edge is horizontal the linking is started to the left and right. The same is applied to the right image.

As in [11,13,19] it is not clear how to handle the changes in the orientation of the edges in the smart routing procedure, Fig. 4 describes the criteria applied in this work. For an horizontally oriented pixel if the last displacement was to the right then the pair of points is moved to the right, otherwise the pair of points is moved to the left. A similar approach is applied for vertical edges, if the last displacement was up, then the pair is moved up, otherwise the pair is moved down. The direction of movement is calculated for each image independently. After obtaining the direction of move, the three immediate neighbours in that direction are considered and the one having the maximum gradient is taken as the next pixel.

```
Symbols used:
(x_c, y_c): Current location.
(x_l, y_l): Last pixel linked.

get_direction( (x_c, y_c), (x_l,y_l), h(x_c,y_c) ){
    if (Pixel is horizontal)
        if ( x_c > x_l )
            direction = RIGHT
        else
            direction = LEFT
    else
        if ( y_c > y_l )
            direction = DOWN
        else
            direction = UP
}
```

Fig. 4. Procedure used to get the direction of the next pixel to be linked in SED.

Figure 5 illustrates the linking procedure in the left and right images. Starting at the anchor point (red) the linking is started to the left. By comparing the three immediate neighbours on the left $\{50, 55, 47\}$ and $\{49, 53, 46\}$ for the left and right images respectively, the pixel with the maximum gradient is selected as the next for each image, i.e. 55 and 53 respectively. Then the compared elements are $\{47, 50, 45\}$ and $\{43, 49, 43\}$ selecting 50 and 49 for the left and right images respectively. For the next pixels $\{30, 35, 47\}$ and $\{29, 34, 48\}$ are compared selecting 47 and 48 as the next for each image. At this stage a change in orientation is found and the direction of linking is determined by the algorithm shown in Fig. 4. The new direction for linking is $DOWN$ for each image and the compared elements are $\{50, 48, 32\}$ and $\{51, 46, 32\}$ selecting 50 and 51 for the left and right image respectively. This is repeated until one of the termination conditions is reached.

If required, in order to obtain sub-pixel accurate disparities efficiently, the edges are located at the sub-pixel level by following the approach in [20], then the difference in the horizontal location is taken as disparity.

2.5   Curve Merging

After the extraction of the edge segments, they are merged in order to have longer curves. Two endpoints are merged if their distance is smaller than a radius r and the difference in their disparity values is smaller than 1. This value for disparity is selected in order to only merge pixels which are in the same disparity region. After the edges are merged, only those with a minimum length of 1% of the image diagonal are kept as in [21].

After linking and merging, the output of the algorithm is a set of chains of connected, well-localized and one-voxel wide 3D points in the form (x, y, d)

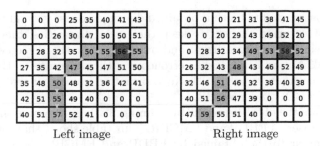

Left image   Right image

Fig. 5. Linking procedure. (Color figure online)

where x and y are the location on the left of right image and d is the disparity at the pixel. As the resulting chains of 3D points are obtained from the edges in the image, they could be used to identify the 3D objects by template matching or reconstruction.

3 Evaluation

In this section the performance of the Simultaneous Edge Drawing (SED) is tested on the Middlebury [22] and the KITTI [23] datasets, in order to compare the performance with other existing methods. All of the experiments were run on a Laptop with a processor Intel Core i7 2675QM at 2.1 GHz and 8 GB of RAM using only one thread.

To our knowledge the only methods found able to obtain disparity values in real time for edge pixels without fitting a line or a curve to the edge pixels prior to stereo matching are: EMCBR [7], EBDP [8]. As only a subset of the Middlebury is used in [7,8], the same subset is used for comparisonn.

The parameters used for SED are kept fixed for all the experiments except for the computation of sub-pixel values, when comparing against EBDP and EMCBR sub-pixel are computed meanwhile for the KITTI dataset they are not used as the KITTI dataset do not provide sub-pixel values. The scan interval is set to $s = 2$ in order to reduce the number of processed anchors, larger values could decrease further the number of anchors at the cost of avoiding the detection of small structures. As pixel descriptor a CRT of size 9×9 is used, larger windows do not produce a significant increase in accuracy but increase the required number of computations. No aggregation is used as it did not show any improvement in accuracy for transformation windows larger than 9×9. Additionally, in order to keep only confident matches, the confidence thresholds are set to $t_{PKRI} = 0.44$ and $t_{MCR} = 0.88$, the values were selected as they increased the accuracy without significantly decreasing the number of obtained matches. For the linking stage the epipolar threshold is set to $t_e = 1$ to allow a small offset on the epipolar lines. Finally, for the merging stage a radius of 5 is used, this value was found experimentally.

The edge-based ground truth is obtained dilating a dense ground truth image with a 3×3 structuring element to keep the larger disparities, then edges are detected by using Canny algorithm, and only values at the detected edges are kept as in [7,8]. A pixel is marked as erroneous if the difference on the obtained disparity and the ground truth is > 1. No information from other error thresholds is provided by EBDP and EMCBR.

Table 1 shows the obtained result by running SED on the same subset of the Middlebury dataset as EBDP and EMCBR. This table shows that although the accuracy is similar to the obtained by EBDP and EMCBR the running time is twice as fast considering only one core is used on our experiments whereas two cores are used on EBDP and EMCBR. This reduction in the number of computations is important in low-cost embedded systems where multiple cores and GPUs may not be available. It is important to note that no information is provided for EBDP and EMCBR but for SED an error threshold of 3 obtained accuracies of around 99%. The running times on the Middlebury dataset include 30–50ms for the Edge Focusing algorithm, this being the most computational expensive part of SED; 9–22ms for anchor matching and 1–3ms for simultaneous smart routing. The reduction in the computing times is explained as typically less than 50% of the anchor points are matched avoiding the computation of many descriptors and cost values. Figure 6 shows the obtained images by SED and EBDP.

Table 1. Comparisson between SED, EBDP and EMCBR.

	Tsukuba		Teddy		Cones		Venus		Sawtooth		Running
	Matches	Errors	Matches	Errors	Matches	Errors	Matches	Errors	Matches	Errors	Time (ms)
SED	6178	10.9 %	9396	8.7 %	14188	7.7 %	8527	3.4 %	11694	7.2%	45–83, 1 core
EBDP	9920	7.6 %	11755	11.9 %	15155	5.4 %	11610	1.8 %	14614	2.7 %	60–85, 2 core
EMCBR	8550	8.8 %	10514	5.3 %	16147	5.3 %	11816	2.0 %	14202	2.4 %	20–45, 2 core

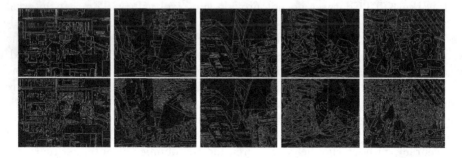

Fig. 6. Disparity results for SED (top) and EBDP (bottom). Green pixels have an error threshold $\epsilon \leq 1$. Red pixels have an error threshold $\epsilon > 1$. White pixels are unmatched pixels. The images for EBDP were taken from [8]. (Color figure online)

Table 2. Results from SED on the KITTI stereo evaluation 2015 using only the estimated pixels. D1-bg = outliers on background regions, D1-fg = outliers on foreground regions, D1-all = outliers over all ground truth pixels. This would place our approach in the top 15 at the time of writing, being the only one with a running time smaller than 1 s. without the use of multiple cores or GPUs.

Algorithm	D1-bg	D1-fg	D1-all	Density	Time	Environment
SED (ours)	4.9%	8.29%	5.74%	4.02%	0.68 s	1 core @ 2.0 GHz (C/C++)

Table 3. Results from SED on the Middlebury v3 stereo evaluation using only the estimated pixels. The selected metric is bad 2.0 and the mask is for non-occluded pixels only. These values are the default for the ranking.

Name	Avg	Austr	AustrP	Bicyc2	Class	ClassE	Compu	Crusa	CrusaP
SED (ours)	0.26	0.38	0.31	0.28	0.22	0.2	0.36	0.13	0.15
contd.	Djemb	DjembL	Hoops	Livgrm	Nkuba	Plants	Stairs		
	0.28	0.72	0.22	0.21	0.19	0.2	0.2		

Table 2 shows the obtained results on the KITTI stereo dataset 2015. The obtained results would place our algorithm among the top-15 of the ranking at the time of writing being the only one with a running time smaller than 1 s without the use of multiple cores or GPUs.

Table 3 shows the obtained results on the Middlebury v3 stereo dataset. The obtained results would place our algorithm among the top performing on the ranking at the time of writing.

4   Conclusions

This paper presented a new approach to use edge information in order to reduce the number of matched pixels in a local approach and to produce sparse disparity maps which are able to represent the image semantics. The proposed approach extends a robust edge detector to run simultaneously on a stereo-pair producing chains of ordered 3D points. The key speed up is in the reduction of the number of descriptors and the cost computed as only a few pixels require to be matched. Further research would include the implementation of the approach in a low-cost embedded system and the use of the obtained chains of 3D points for obstacle detection.

Potential improvements can be identified in terms of speed as the Edge Focussing is the main time consuming step in the algorithm, followed by the matching. Improvements in accuracy could be obtained by using a different pixel descriptor which allows to better describe horizontal edges and by using an alternate approach for identifying occlusion. The connectivity information could additionally be used in a post-processing in order to recover missing edges but it is out of the scope of this work.

Acknowledgement. This research has been funded by the Irish Research Council under the EMBARK initiative, application No. RS/2012/489.

References

1. Fabbri, R., Kimia, B.: 3D curve sketch: flexible curve-based stereo reconstruction and calibration. In: 2010 IEEE Computer Society Conference on Computer Vision and Pattern Recognition, pp. 1538–1545. IEEE (2010)
2. Al-Shahri, M., Yilmaz, A.: Line matching in wide-baseline stereo: a top-down approach. IEEE Trans. Image Process. **23**, 1–1 (2014)
3. Witt, J., Weltin, U.: Robust stereo visual odometry using iterative closest multiple lines. In: 2013 IEEE/RSJ International Conference on Intelligent Robots and Systems, pp. 4164–4171 (2013)
4. Koletschka, T., Puig, L., Daniilidis, K.: MEVO: multi-environment stereo visual odometry. In: 2014 IEEE/RSJ International Conference on Intelligent Robots and Systems, pp. 4981–4988. Number IROS, IEEE (2014)
5. Hofer, M., Donoser, M., Bischof, H.: Semi-global 3D line modeling for incremental structure-from-motion. In: Valstar, M., French, A., Pridmore, T. (eds.) Proceedings of the British Machine Vision Conference. BMVA Press (2014)
6. Rao, D., Chung, S.J., Hutchinson, S.: CurveSLAM: an approach for vision-based navigation without point features. In: 2012 IEEE/RSJ International Conference on Intelligent Robots and Systems, pp. 4198–4204. IEEE (2012)
7. Witt, J., Weltin, U.: Robust real-time stereo edge matching by confidence-based refinement. In: Su, C.-Y., Rakheja, S., Liu, H. (eds.) ICIRA 2012. LNCS (LNAI), vol. 7508, pp. 512–522. Springer, Heidelberg (2012). doi:10.1007/978-3-642-33503-7_50
8. Witt, J., Weltin, U.: Sparse stereo by edge-based search using dynamic programming. In: 2012 21st International Conference on Pattern Recognition (ICPR), pp. 3631–3635. Number ICPR (2012)
9. Ohta, Y., Kanade, T.: Stereo by intra- and inter-scanline search using dynamic programming. IEEE Trans. Pattern Anal. Mach. Intell. **PAMI-7**, 139–154 (1985)
10. Canny, J.: A computational approach to edge detection. IEEE Trans. Pattern Anal. Mach. Intell. **PAMI-8**, 679–698 (1986)
11. Topal, C., Akinlar, C.: Edge drawing: a combined real-time edge and segment detector. J. Vis. Commun. Image Represent. **23**, 862–872 (2012)
12. Akinlar, C., Topal, C.: EDLines: a real-time line segment detector with a false detection control. Pattern Recogn. Lett. **32**, 1633–1642 (2011)
13. Akinlar, C., Topal, C.: EDPF: a real-time parameter-free edge segment detector with a false detection control. Int. J. Pattern Recogn. Artif. Intell. **26**, 1255002 (2012)
14. Bergholm, F.: Edge focusing. IEEE Trans. Pattern Anal. Mach. Intell. **PAMI-9**, 726–741 (1987)
15. Prewitt, J.M.S.: Object enhancement and extraction. In: Lipkin, B., Rosenfeld A. (eds.) Picture Processing and Psychopictorics, pp. 75–149. Academic Press, New York (1970)
16. Demetz, O., Hafner, D., Weickert, J.: The complete rank transform: a tool for accurate and morphologically invariant matching of structures. In: Proceedings of the 2013 British Machine Vision Conference, Bristol. BMVA Press (2013)
17. Xiaoyan, H., Mordohai, P.: A quantitative evaluation of confidence measures for stereo vision. IEEE Trans. Pattern Anal. Mach. Intell. **34**, 2121–2133 (2012)

18. Haeusler, R., Klette, R.: Disparity confidence measures on engineered and out-door data. In: Alvarez, L., Mejail, M., Gomez, L., Jacobo, J. (eds.) CIARP 2012. LNCS, vol. 7441, pp. 624–631. Springer, Heidelberg (2012). doi:10.1007/978-3-642-33275-3_77
19. Topal, C., Akinlar, C., Genc, Y.: Edge drawing: a heuristic approach to robust real-time edge detection. In: 2010 20th International Conference on Pattern Recognition, pp. 2424–2427 (2010)
20. Devernay, F.: A non-maxima suppression method for edge detection with sub-pixel accuracy. Technical report, INRIA (1995)
21. Hofer, M., Wendel, A., Bischof, H.: Incremental line-based 3D reconstruction using geometric constraints. In: Procedings of the British Machine Vision Conference 2013, British Machine Vision Association, pp. 92.1–92.11 (2013)
22. Scharstein, D., Szeliski, R.: High-accuracy stereo depth maps using structured light. In: Proceedings of the 2003 IEEE Computer Society Conference on Computer Vision and Pattern Recognition, vol. 1, pp. I-195–I-202 (2003)
23. Menze, M., Geiger, A.: Object scene flow for autonomous vehicles. In: 2015 IEEE Conference on Computer Vision and Pattern Recognition (CVPR), pp. 3061–3070. IEEE (2015)

Image-Based Camera Localization for Large and Outdoor Environments

Chin-Hung Teng[1(✉)], Yu-Liang Chen[2], and Xuejie Zhang[2]

[1] Department of Information Communication and Innovation Center for Big Data and Digital Convergence, Yuan Ze University, Chung-Li, Taiwan
chteng@saturn.yzu.edu.tw
[2] School of Information Science and Engineering, Yunnan University, Kunming, China

Abstract. Locating camera position and orientation is an important step for many augmented reality (AR) applications. In this paper, we develop a system for estimating camera pose for large and outdoor environments. A large set of images for outdoor environments are collected and 3D structure of the scenes are recovered using a structure from motion technique. To improve image indexing accuracy and efficiency, a convolutional neural network (CNN) is employed to extract image features and a set of locality sensitive hashing (LSH) functions are used to classify CNN features. With these techniques, camera localization is achieved by first indexing the nearest images by CNN and LSH and then a set of 2D-3D correspondences are established from the indexed images and the recovered 3D structure. A perspective-n-point (PnP) algorithm is then applied on the 2D-3D correspondences to estimate camera pose. A series of experiments are conducted and the results confirm the effectiveness of proposed system. The nearest neighbors to query image can be accurately and efficiently extracted and the camera pose can be accurately estimated.

1 Introduction

Estimating camera position and orientation from images is a fundamental task for many computer vision related applications such as augmented reality (AR). AR is a technique for superimposing virtual objects on real scene images, producing the illusion that the virtual objects are part of the real scene. To correctly register a virtual object on a real scene, camera pose estimation is a necessary step since we need to know the transformation between the camera and the scene so that we can translate and rotate the virtual object to the desired position and orientation.

Over the past decades, many real-time camera pose estimation techniques had been proposed such as SLAM (simultaneous localization and mapping) [1,2] and PTAM (parallel tracking and mapping) [3]. SLAM is a technique to simultaneously establishing the scene map and locating the position of a device. SLAM was originally developed in robot area but now it was also used in computer vision community and some vision-based algorithms such as monoSLAM [4] were

© Springer International Publishing AG 2017
C.-S. Chen et al. (Eds.): ACCV 2016 Workshops, Part II, LNCS 10117, pp. 136–147, 2017.
DOI: 10.1007/978-3-319-54427-4_11

developed. PTAM is a technique similar to SLAM. It performs camera tracking and scene mapping in different threads and thus achieve quite efficient and stable augmented reality application. Although SLAM and PTAM perform quite well in AR applications, both of them estimate camera pose with respect to a local coordinate frame. That is, they work in their own workspace and hence can be seen as local AR systems. Recently, an AR system which can achieve global localization is proposed [5]. This system runs a local keyframe-based SLAM in a mobile phone and maintains a globally-registered map in a remote server. When the local map expands, the local keyframes are registered with the global map to provide a global and stable tracking of the mobile phone. Because of the global localization, the system can work in very large and outdoor environments.

For large and outdoor environments, there may have several tens of thousands of keyframes in the system. Hence, indexing nearest keyframes and performing feature matching is a critical issue for implementing the system. In [5], GPS information is employed to quickly find the neighbor keyframes. However, for GPS-denied environments this approach cannot work well. Hence, in this paper we incorporate an accurate image retrieval technique, the convolutional neural network (CNN) [6], into a camera localization system to quickly index neighbor images. CNN is a deep neural network and it has been proved that after large scale image training, CNN can extract effective image visual features and produce very accurate image retrieval results [7,8]. Moreover, to make our system more efficient, a locality sensitive hashing (LSH) is also employed in our system to effectively group similar images, greatly reducing the number of images to be matched. With these techniques, we achieve an accurate and efficient camera localization system which is suitable for large and outdoor environments. The system cam work well when the keyframe set contains tens of thousands images. The remainder of this paper is organized as follows. In Sect. 2, a related work is given. Subsequently, the proposed camera localization system is introduced. The experimental results are presented in Sect. 4. Finally, we conclude in Sect. 5.

2   Related Work

Camera pose estimation is a key step in AR applications and in AR domain the technologies for tracking camera can be roughly classified as marker- and markerless-based tracking. Marker-based approaches estimate camera pose by detecting a fiducial marker in the images and then compute the relative transformation of the camera with respect to the marker using some geometric entities such as parallel lines or homographies [9, 10]. ARToolKit is a representative tool for marker-based AR.

On the other hand, markerless AR tracks camera by some features in the images such as edges and points. Recently, point-based tracking technologies receives more attentions because of the advent of some robust or efficient feature detection and matching technologies such as SIFT [11], SURF [12], and ORB [13]. In additional to vision community, the notable SLAM systems were developed in robotic community. SLAM achieves device localization by observing a number of landmarks in space and then update device position and orientation by these observations. The systems of SLAM can simply be classified into filter-based and vision-based approaches. Filter-based approaches estimate device position via some well-known filters such as Kalman filter or particle filter. On the other hand, vision-based SLAM achieves camera pose estimation by the so-called bundle adjustment [14]. One can refer to the two tutorial papers [1,2] to deep understand SLAM and its development. In literature, the process of estimating the egomotion of an agent using camera has another terminology called visual odometry (VO). Vision-based SLAM can be seen as one type of VO. One can refer to the introductory papers of VO [15,16] to further understand vision based camera pose estimation.

As discussed in Sect. 1, SLAM estimates camera pose with respect to a local coordinate frame. Ventura et al. proposed a global localization method that can register a local map with respect to a global frame and can work well on very large and outdoor environments [5]. Another AR system that works on wide area environment is proposed by [17]. In this system, keyframe recognition is achieved by a technique called random ferns. In this paper, we employ a state of the art deep neural network, the CNN, to achieve accurate image indexing. To reduce the number of searched images, a set of locality sensitive hashing (LSH) functions is used to classify CNN image features. By the powerful recognition capability of CNN, we believe that this approach can produce accurate and efficient camera localization for large and outdoor environments.

3   Proposed Camera Localization

3.1   System Architecture

Figure 1 depicts the system diagram of proposed system. Our system can be decomposed into two parts: offline feature extraction and 3D reconstruction and online camera pose estimation. The images from the same scene are grouped in the dataset and the SURF features of each image in the same group are extracted. These SURF features are matched and a structure from motion algorithm is applied to reconstruct a 3D point cloud for the scene. The 2D SURF features as well as their 3D correspondences are served as the feature map of the scene. Meanwhile, the CNN feature of each image in the database is extracted and several sets of LSH functions are used to generate a number of binary codes for CNN feature. The images with the same binary code are put into the same bucket as shown in Fig. 2 where the system flow for CNN and LSH based nearest neighbor search is illustrated.

For the online process, when a query image is received, its CNN feature is extracted and the same sets of LSH functions are used to transform the feature into binary codes to quickly index a number of candidate images. Following this, the distances of CNN features between the query image and the candidate images are computed and finally some nearest neighbor images are retrieved (see Fig. 2). After identify these nearest images, the SURF features of the query image

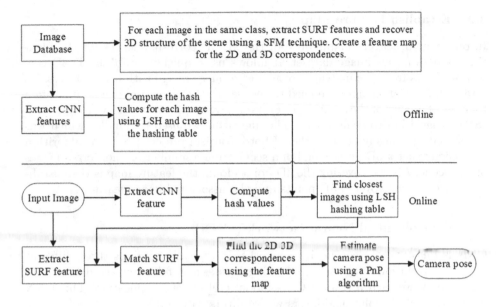

Fig. 1. System diagram

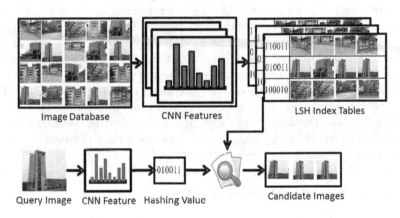

Fig. 2. Extract nearest neighbor images using CNN and LSH

are extracted and matched with the features in the indexed neighbor images. Since we have recovered the 3D structure for these neighbor images, by feature matching results we can quickly identify a set of 2D-3D correspondences for the query image. Finally, a perspective-n-point algorithm is applied to estimate camera pose of the query image.

In the following subsections, the technical details of feature map creation, CNN feature extraction, and LSH generation are discussed.

3.2   Establish Feature Map

In our system, feature map stores all the 2D-3D correspondences for the identified scenes in our image dataset. It allows us to quickly establish 2D-3D correspondences from the matching results of a query image with its neighbors. To establish the feature map, we need to recover the 3D structure of a scene from a number of images of the scene. In our system, we employ a structure from motion (SFM) algorithm to achieve this. The algorithm we used is Bundler which was developed by Snavely et al. in their Photo Tourism project [18]. This algorithm can reconstruct a 3D point cloud of a scene from a number of unordered images of the scene. After recovering the 3D point cloud, the feature map is then established from the recovered 3D points and the associated image features.

3.3   Convolutional Neural Network (CNN)

CNN is one type of deep neural networks. It typically contains three different layers: convolutional layers, max-pooling layers, and fully-connected layers. In our system, we used a trained CNN model [19] for our application. This CNN has a stack of convolutional layers where filters with a 3×3 receptive field are used. Spatial pooling is performed by five max-pooling layers. Max-pooling is carried out over a 2×2 pixel window with stride 2. A stack of convolutional filters is followed by three fully-connected layers. The output of the network is a feature with 4096 dimensions. The model is pre-trained on a dataset from ImageNet Large-Scale Visual Recognition Challenge (ILSVRC). When performing on image retrieval, we can directly estimate the similarity of two images by comparing their CNN features, i.e., the Euclidean distance between two CNN features. In our system, we use a set of LSH functions to generate binary codes from CNN feature. The final image indexing is achieved by ranking the CNN feature distances from a query image to a set of candidates produced from LSH hashing tables and select the top-k nearest neighbors.

3.4   Locality Sensitive Hashing (LSH)

LSH is a hashing based approximate nearest neighbors (ANN) search scheme. It states that two data points with distance smaller than a value R will have the probability greater than P_1 to get the same hashing value. On the other hand, if the distance is greater than a value cR, the probability of producing the same hashing value is smaller than P_2. Therefore, the hashing value generated by LSH can in some senses be used to evaluate the similarity of two data points. Typically, the hashing value generated by LSH is a binary number and by fast XOR operation the processing of LSH hashing value is very efficient. Moreover, the generated binary code can also be used to classify data points and thus achieve ANN search.

In our system, we use the inner product similarity to define the hashing function [20]. Given a random vector $\mathbf{r}$ from a d-dimensional normal distribution

$N(0,1)$, a hashing function $h_{\mathbf{r}}$ can be defined as:

$$h_{\mathbf{r}}(\mathbf{q}) = \begin{cases} 1 & \text{if } \mathbf{r} \cdot \mathbf{q} \geq 0 \\ 0 & \text{if } \mathbf{r} \cdot \mathbf{q} < 0 \end{cases} \tag{1}$$

where $\mathbf{q}$ is a CNN feature of an image. By this way, a number of sets of hashing functions can be defined and a set of binary codes for a CNN feature can be generated. The images with the same binary code are grouped to form the hashing tables for subsequent nearest neighbor image indexing. In our system, we used 10 sets of LSH with each set contains 8 functions, thus we have 10 hashing tables and the length of generated binary code is 8 bits.

3.5   Camera Pose Estimation

When neighbor images are identified, the SURF features of the query image are extracted and compared with the features of indexed neighbor images. If the matched features have corresponding 3D points in the feature map, we can then establish a set of 2D-3D correspondences for the query image. With this information as well as the camera focal length extracted from the EXIF tag of the query image, a perspective-n-point (PnP) algorithm is applied to estimate camera pose. In our system, we employ the robust version of PnP in OpenCV, the **solvePnPRansac**, to determine camera pose.

4   Experiments and Results

We conducted a series of experiments to evaluate the performance of proposed system. The image dataset we used contains 12,000 images. Among these images, 600 images were taken from 10 scenes of our campus with each scene has 60 images. The other images were collected from the Internet. The image resolution for these 600 images is 2592×1936. For the remained images, resolution is ranged from 640×480 to 3264×2448. The images for the 10 scenes are fed to a structure from motion algorithm, the Bundler, to recover a set of 3D point clouds. The reprojection errors of the recovered 3D point clouds for the 10 scenes are ranged from 1.5 to 2.6 pixels and for such image resolution (2592×1936) we think that these are acceptable. Figure 3 shows images from three scenes and the recovered 3D points. From the figure, we can observe that, except some outliers, the reconstructed 3D points as well as the recovered camera poses (yellow pyramids in Fig. 3) are reasonable.

To understand the effectiveness of CNN and LSH, we used images in the 600-image set as query images to index neighbor images in the 12,000-image set. Since the 600 images had been manually classified, we identified the correctness by checking whether the retrieval images belong to the same class as query image. Because the nearest image is the query image itself, the image with zero CNN distance is discarded. We treated the top-5 retrieval images as the indexing

Fig. 3. Three scenes in our image database and the reconstructed 3D point clouds. (Color figure online)

results. In our experiment, CNN + LSH approach can return the correct neighbors for all the query images using previous definition. Table 1 shows the average match number of CNN features for CNN only and CNN + LSH approaches in the image indexing experiment. For CNN only approach, we need to compare the CNN feature of the query image with those in the whole database, thus the match number is always 12,000. For CNN + LSH, the images had been categorized according to the generated LSH binary codes, thus the average match number differs for the 10 scenes. In our experiment, we have 10 hashing tables and the length of binary code is 8. Images with the same binary code to the query image in the 10 tables are the resulting candidates that pass the LSH fil-

Table 1. Average matching number of CNN features

Scene	1	2	3	4	5	6	7	8	9	10
CNN	12,000	12,000	12,000	12,000	12,000	12,000	12,000	12,000	12,000	12,000
CNN + LSH	1144.6	1318.2	1376.2	1534.0	1150.9	1097.6	1526.6	1593.3	1104.5	1457.7

tering. From Table 1, we can see that CNN + LSH can filter out a large portion of undesired images and greatly reduce the average match number. Our system was run on a PC with Intel i7-4790K 4 GHz CPU and 8 GB RAM. The average matching time is 0.5671 s for CNN and 0.1319 s for CNN + LSH. This again confirms the efficiency of CNN + LSH over CNN only approach.

To further check the correctness of CNN and CNN + LSH image indexing, we used the results of camera localization to evaluate the performance. Since we do not know the true camera poses of the query images, we employed the estimated camera pose from SFM algorithm as the ground truth. This is not the best choice, but in some sense it can still reflect the performance of proposed system. Moreover, since the recovered 3D structure differs the true structure in an ambiguous scale, we cannot evaluate absolute camera position error. Thus, only the camera orientation error is evaluated in our experiment. Nevertheless, from the resulting images in Fig. 4, we still can observe the high accuracy of estimated camera position.

Similar to previous experiment, for a query image from 600-image set, we used the approaches of CNN and CNN + LSH to index neighbor images. The query image is not considered in the neighbor set. Subsequently, camera pose of the query image is estimated from the 2D-3D correspondences, created from the indexed neighbor images and the established feature map. Camera orientation error is defined as follows:

$$E_{ori} = \cos^{-1} \left(\frac{\mathbf{v}_{est} \cdot \mathbf{v}_{SFM}}{\|\mathbf{v}_{est}\| \, \|\mathbf{v}_{SFM}\|} \right) \cdot \frac{180}{\pi} \tag{2}$$

where $\mathbf{v}_{est}$ and $\mathbf{v}_{SFM}$ are the camera orientation vectors obtained by proposed system and SFM algorithm, respectively.

Table 2 shows the camera orientation errors for the three scenes in Fig. 3 using CNN only and CNN + LSH approach. In this table, we also show the results using different numbers of neighbor images. That is, top-k indicates we used the nearest k images in camera pose estimation. From the table, we can see that the estimation accuracy for CNN only and CNN + LSH approach is comparable. It even produces the same results for top-1 and top-2 cases, which indicates that if we only retrieve 1 or 2 nearest images, CNN only and CNN + LSH approach return the same images. In additional to this, this table also tells us the high accuracy of proposed system in camera orientation estimation. The estimated camera orientation errors in these three cases are all less than 1.5°.

Figure 4 provides us a visual way to evaluate the performance of estimated camera pose. In this figure, the green pyramids indicate the recovered camera poses by SFM algorithm. The red pyramids, which closely overlapped with the

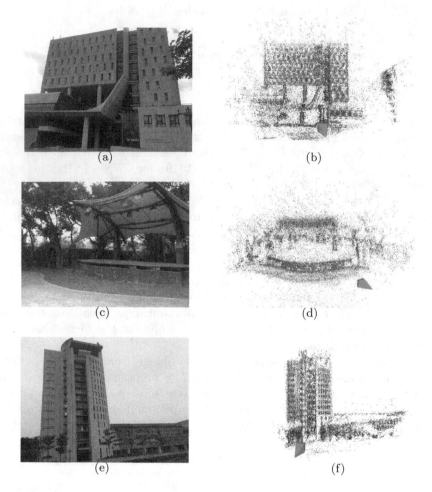

Fig. 4. Three query images and the estimated camera poses. The green pyramids indicate the camera poses from SFM reconstruction and the red ones are the camera poses produced by proposed system. (Color figure online)

Table 2. Camera orientation errors (degrees) for the three scenes in Fig. 3

		Top-1	Top-2	Top-3	Top-4	Top-5
Figure 3 top	CNN	0.4263	0.6087	0.8254	0.5116	0.7200
	CNN + LSH	0.4263	0.6087	0.8254	0.5220	0.6941
Figure 3 middle	CNN	0.9054	0.3952	0.6717	0.5210	0.4932
	CNN + LSH	0.9054	0.3952	0.6640	0.5297	0.5226
Figure 3 bottom	CNN	0.8849	0.7442	1.2832	1.2886	1.4517
	CNN + LSH	0.8849	0.7442	1.2756	1.2900	1.4433

Fig. 5. Two query images not from the image dataset and the estimated camera poses (red pyramids) with respect to the recovered point cloud. (Color figure online)

green ones, are the estimated camera poses by proposed camera localization system. This figure displays the high consistency of proposed system with SFM algorithm, both in position and orientation.

In Fig. 4, the query image is one of the image in SFM 3D reconstruction although we did not employ it in camera pose estimation. To further check the results when the query images are not part of the set of SFM reconstruction, two other images were fed to our system to check its performance on a more physical situation. Figure 5 shows some results for these cases. However, in these cases we do not know the exact camera poses, thus we can only show the estimated camera poses with respect to the recovered 3D structure. Nevertheless, the results are visually consistent with our expectation. By visual inspection, the estimated camera is about at the expected position and orientation.

5   Conclusion

In this paper, we presented a camera localization system suitable for large and outdoor environments. A SFM algorithm is first employed to recover 3D structure of scenes. A CNN + LSH scheme is used to accurately and efficiently index nearest neighbor images. By combining these techniques, camera pose can be accurately estimated as illustrated by our experiments. Our future work is to extend the image database to accommodate more scenes to make it applicable to more environments.

Acknowledgement. This work was supported in part by the Ministry of Science and Technology, Taiwan, under Grant Nos. MOST 104-2221-E-155-032 and MOST 104-3115-E-155-002.

References

1. Durrant-Whyte, H., Bailey, T.: Simultaneous localization and mapping (SLAM): Part I the essential algorithms. IEEE Robot. Autom. Mag. **13**, 99–110 (2006)
2. Bailey, T., Durrant-Whyte, H.: Simultaneous localization and mapping (SLAM): Part II state of the art. IEEE Robot. Autom. Mag. **13**, 108–117 (2006)
3. Klein, G., Murray, D.: Parallel tracking and mapping for small AR workspaces. In: 6th IEEE and ACM International Symposium on Mixed and Augmented Reality (ISMAR 2007), pp. 225–234 (2007)
4. Davison, A.J., Reid, I.D., Molton, N.D., Stasse, O.: MonoSLAM: real-time single camera SLAM. IEEE Trans. Pattern Anal. Mach. Intell. **29**, 1052–1067 (2007)
5. Ventura, J., Arth, C., Reitmayr, G., Schmalstieg, D.: Global localization from monocular SLAM on a mobile phone. IEEE Trans. Vis. Comput. Graph. **20**, 531–539 (2014)
6. Krizhevsky, A., Sutskever, I., Hinton, G.E.: ImageNet classification with deep convolutional neural networks. In: Advances in Neural Information Processing Systems, vol. 25, pp. 1106–1114 (2012)
7. Razavian, A.S., Azizpour, H., Sullivan, J., Carlsson, S.: CNN features off-the-shelf: an astounding baseline for recognition. In: IEEE Conference on Computer Vision and Pattern Recognition Workshops, pp. 512–519 (2014)
8. Xie, L., Hong, R., Zhang, B., Tian, Q.: Image classification and retrieval are ONE. In: International Conference on Multimedia Retrieval (2015)
9. Kato, H., Billinghurst, M.: Marker tracking and HMD calibration for a video-based augmented reality conferencing system. In: International Workshop on Augmented Reality (IWAR 1999) (1999)
10. Lepetit, V., Fua, P.: Monocular model-based 3D tracking of rigid objects: a survey. Found. Trends Comput. Graph. Vis. **1**, 1–89 (2005)
11. Lowe, D.G.: Distinctive image features from scale-invariant keypoints. Int. J. Comput. Vis. **60**, 91–110 (2004)
12. Bay, H., Ess, A., Tuytelaars, T., van Gool, L.: Speeded-up robust features (SURF). Comput. Vis. Image Underst. **110**, 346–359 (2008)
13. Rublee, E., Rabaud, V., Konolige, K., Bradski, G.: ORB: an efficient alternative to SIFT or SURF. In: International Conference on Computer Vision, pp. 2564–2571 (2011)
14. Hartley, R.I., Zisserman, A.: Multiple View Geometry in Computer Vision, 2nd edn. Cambridge University Press, Cambridge (2004). ISBN 0521540518
15. Scaramuzza, D., Fraundorfer, F.: Visual odometry: Part I the first 30 years and fundamentals. IEEE Robot. Autom. Mag. **18**, 80–92 (2011)
16. Scaramuzza, D., Fraundorfer, F.: Visual odometry: Part II matching, robustness, optimization, and applications. IEEE Robot. Autom. Mag. **19**, 78–90 (2012)
17. Guan, T., Duan, L., Yu, J., Chen, Y., Zhang, X.: Real-time camera pose estimation for wide-area augmented reality applications. IEEE Comput. Graph. Appl. **31**, 56–68 (2011)

18. Snavely, N., Seitz, S.M., Szeliski, R.: Photo tourism: exploring photo collections in 3D. In: ACM Transactions on Graphic (SIGGRAPH 2006), vol. 25, pp. 835–846 (2006)
19. Simonyan, K., Zisserman, A.: Very deep convolutional networks for large-scale image recognition. Technical report (2014). arXiv:1409.1556
20. Charikar, M.: Similarity estimation techniques from rounding algorithm. In: ACM Symposium on Theory of Computing, pp. 380–388 (2002)

An Efficient Meta-Algorithm for Triangulation

Qianggong Zhang$^{(\boxtimes)}$ and Tat-Jun Chin

The University of Adelaide, Adelaide, SA 5000, Australia
qianggong.zhang@adelaide.edu.au

Abstract. Triangulation by ℓ_∞ minimisation has become established in computer vision. State-of-the-art ℓ_∞ triangulation algorithms exploit the quasiconvexity of the cost function to derive iterative update rules that deliver the global minimum. Such algorithms, however, can be computationally costly for large problem instances that contain many image measurements. In this paper, we exploit the fact that ℓ_∞ triangulation is an instance of generalised linear programs (GLP) to speed up the optimisation. Specifically, the solution of GLPs can be obtained as the solution on a small subset of the data called the *support set*. A meta-algorithm is then constructed to efficiently find the support set of a set of image measurements for triangulation. We demonstrate that, on practical datasets, using the meta-algorithm in conjunction with all existing ℓ_∞ triangulation solvers provides faster convergence than directly executing the triangulation routines on the full set of measurements.

1 Introduction

Triangulation refers to the task of estimating the 3D coordinates of a scene point from multiple 2D image observations of the point, given that the pose of the cameras are known. Triangulation is fundamentally important to 3D vision, since it enables the recovery of the 3D structure of a scene. Whilst in theory structure and motion must be obtained simultaneously, there are many settings, such as large scale reconstruction [10,24] and SLAM [17], where the camera poses are first estimated with a sparse set of 3D points, before a denser scene structure is produced by triangulating other points using the estimated camera poses.

An established approach for triangulation is by ℓ_∞ minimisation [12]. Specifically, we seek the 3D coordinates that minimise the maximum reprojection error across all views. Unlike the sum of squared error function which contains multiple local minima, the maximum reprojection error function is quasiconvex and thus contains a single global minimum. Algorithms that take advantage of this property have been developed to solve such quasiconvex problems exactly [2,5,7,9,13,14,19]. In particular, Agarwal et al. [2] showed that some of the most effective algorithms belong to the class of generalised fractional programming methods [6,11].

Although algorithms for ℓ_∞ triangulation have steadily improved, there is still room for improvement. In particular, on large scale reconstruction problems or SLAM where there are usually a significant number of views per point (recall

© Springer International Publishing AG 2017
C.-S. Chen et al. (Eds.): ACCV 2016 Workshops, Part II, LNCS 10117, pp. 148–161, 2017.
DOI: 10.1007/978-3-319-54427-4_12

that the size of a triangulation problem is the number of 2D observations of a scene point), the computational cost of many of the algorithms [2] can be considerable; we will demonstrate this in Sect. 4. A major reason is that the algorithms need to repeatedly solve convex programs (LPs or SOCPs) constructed using *all the measurements* to determine the update direction. Although theoretically convex programs can be solved efficiently, the runtime of many existing solvers are still non-trivial given large input sizes. It is thus of significant practical interest to develop faster algorithms.

Contributions. We propose the usage of a meta-algorithm that works in conjunction with existing ℓ_∞ solvers to achieve fast triangulation. The meta-algorithm exploits the fact that ℓ_∞ triangulation is a GLP. Specifically, the solution of a GLP given a set of input data can be obtained as the solution of the same problem on a small subset of the data called the *support set*. The purpose of the meta-algorithm is to efficiently find the support set. In contrast to existing ℓ_∞ solvers, the meta-algorithm does not require to solve convex problems on the full data simultaneously. This property is desirable on large scale problems. We demonstrate the benefits of the meta-algorithm on publicly available large scale 3D reconstruction datasets.

2 Background

Let $\{\mathbf{P}_i, \mathbf{u}_i\}_{i=1}^N$ be a set of data for triangulation, consisting of camera matrices $\mathbf{P}_i \in \mathbb{R}^{3\times4}$ and observed image positions $\mathbf{u}_i \in \mathbb{R}^2$ of the same scene point $\mathbf{x} \in \mathbb{R}^3$. In this paper, by a "datum" we mean a specific camera and image point $\{\mathbf{P}_i, \mathbf{u}_i\}$, and let $\mathcal{X} = \{1, \ldots, N\}$ index the set of data. The ℓ_∞ technique estimates $\mathbf{x}$ by minimising the maximum reprojection error

$$\min_{\mathbf{x}} \max_{i \in \mathcal{X}} \ r(\mathbf{x} \mid \mathbf{P}_i, \mathbf{u}_i), \quad \text{where } r(\mathbf{x} \mid \mathbf{P}_i, \mathbf{u}_i) = \left\| \mathbf{u}_i - \frac{\mathbf{P}_i^{1:2}\tilde{\mathbf{x}}}{\mathbf{P}_i^3\tilde{\mathbf{x}}} \right\|_p, \quad (1)$$

$$\text{subject to} \quad \mathbf{P}_i^3\tilde{\mathbf{x}} > 0, \ \forall i \in \mathcal{X}$$

Here, $\mathbf{P}_i^{1:2}$ and $\mathbf{P}_i^3$ respectively denote the first-2 rows and 3rd row of $\mathbf{P}_i$, and $\tilde{\mathbf{x}}$ is $\mathbf{x}$ in homogeneous coordinates. The chirality constraints $\mathbf{P}_i^3\tilde{\mathbf{x}} > 0$ ensure that the estimated point lies in front of all the cameras. The reprojection error

$$r(\mathbf{x} \mid \mathbf{P}_i, \mathbf{u}_i) = \left\| \mathbf{u}_i - \frac{\mathbf{P}_i^{1:2}\tilde{\mathbf{x}}}{\mathbf{P}_i^3\tilde{\mathbf{x}}} \right\|_p = \frac{\left\| \left(\mathbf{u}_i\mathbf{P}_i^3 - \mathbf{P}_i^{1:2} \right)\tilde{\mathbf{x}} \right\|_p}{\mathbf{P}_i^3\tilde{\mathbf{x}}} \quad (2)$$

is basically the distance between the observed point $\mathbf{u}_i$ and the projection of $\mathbf{x}$ onto the i-th image plane. We have left p in the reprojection error undefined, since it remains a "design choice"; the optimisation problem above, therefore, is dubbed (ℓ_∞, ℓ_p) to reflect this choice. Typical values of p are 1, 2 and ∞.

The (ℓ_∞, ℓ_p) triangulation problem can be re-expressed as

$$\min_{\mathbf{x}} \quad \delta$$

$$\text{subject to} \quad \frac{\left\|\left(\mathbf{u}_i\mathbf{P}_i^3 - \mathbf{P}_i^{1:2}\right)\tilde{\mathbf{x}}\right\|_p}{\mathbf{P}_i^3\tilde{\mathbf{x}}} \leq \delta, \tag{3}$$

$$\mathbf{P}_i^3\tilde{\mathbf{x}} > 0, \quad \forall i \in \mathcal{X}$$

where the optimal δ^* is precisely the minimised maximum reprojection error.

2.1 Bisection and Generalised Fractional Programs

Although Eq. (2) is not convex, it is quasiconvex in the region $\mathbf{P}_i^3\tilde{\mathbf{x}} > 0$. Since the objective function in Eq. (1) is the maximum of a set of quasiconvex functions, it is also quasiconvex. This implies that for any *fixed* (and non-negative) δ, the following feasibility problem is convex for $p \geq 1$:

$$\text{does there exist} \quad \mathbf{x}$$

$$\text{such that} \quad \frac{\left\|\left(\mathbf{u}_i\mathbf{P}_i^3 - \mathbf{P}_i^{1:2}\right)\tilde{\mathbf{x}}\right\|_p}{\mathbf{P}_i^3\tilde{\mathbf{x}}} \leq \delta, \tag{4}$$

$$\mathbf{P}_i^3\tilde{\mathbf{x}} > 0, \quad \forall i \in \mathcal{X}$$

The method of bisection exploits the convexity of Eq. (4) to efficiently solve Eq. (3) [13]. Specifically, a binary search is conducted to find the minimum δ. In each step, the feasibility test Eq. (4) (that involves all of the data) is conducted.

Bisection can be viewed as an instance of generalised fractional programming (GFP) [2]. More efficient GFP algorithms exist, such as Dinkelbach's method [6] and Gugat's method [11]. However, all these methods share the similarity that a convex problem involving all of the data must be solved at each iteration. This represents a significant computational bottleneck for large problems [7].

2.2 Generalised Linear Programs

For $p \geq 1$, (ℓ_∞, ℓ_p) problem also belongs to a broader class of problems called generalised linear programs (GLP) [3]. Two useful properties of GLPs, stated in the context of (ℓ_∞, ℓ_p), are as follows.

Property 1 (Monotonicity). For any $\mathcal{C} \subseteq \mathcal{X}$,

$$\min_{\mathbf{x}} \max_{i \in \mathcal{C}} \ r(\mathbf{x} \mid \mathbf{P}_i, \mathbf{u}_i) \leq \min_{\mathbf{x}} \max_{i \in \mathcal{X}} \ r(\mathbf{x} \mid \mathbf{P}_i, \mathbf{u}_i) \tag{5}$$

given the appropriate chirality constraints on both sides.   □

Property 2 (Support set). Let $\mathbf{x}^*$ and δ^* respectively be the minimiser and minimised objective value of (ℓ_∞, ℓ_p). There exists a subset $\mathcal{B} \subseteq \mathcal{X}$ with $|\mathcal{B}| \leq 4$, such that for any $\mathcal{C}$ that satisfies $\mathcal{B} \subseteq \mathcal{C} \subseteq \mathcal{X}$, the following holds

$$\delta^* = \min_{\mathbf{x}} \max_{i \in \mathcal{B}} \ r(\mathbf{x} \mid \mathbf{P}_i, \mathbf{u}_i) = \min_{\mathbf{x}} \max_{i \in \mathcal{C}} \ r(\mathbf{x} \mid \mathbf{P}_i, \mathbf{u}_i) = \min_{\mathbf{x}} \max_{i \in \mathcal{X}} \ r(\mathbf{x} \mid \mathbf{P}_i, \mathbf{u}_i) \tag{6}$$

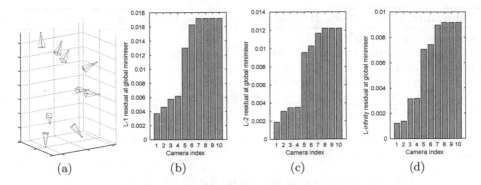

Fig. 1. Triangulating a point **x** observed in 10 views by solving (ℓ_∞, ℓ_p). The red '+' in panel (a) is the ℓ_∞ solution $\mathbf{x}^*$ for $p = 2$ in the reprojection error. The reprojection error of all the data at the solution $\mathbf{x}^*$ for respectively $p = 1$, 2 and ∞ are shown in panels (b), (c) and (d). Observe that there are at most 4 data with the same residual at $\mathbf{x}^*$; these data are the support set $\mathcal{B}$ of the respective problems. (Color figure online)

given the appropriate chirality contraints. In fact, the three problems in Eq. (6) have the same minimiser $\mathbf{x}^*$. Further,

$$r(\mathbf{x}^* \mid \mathbf{P}_i, \mathbf{u}_i) = \delta^* \quad \text{for any } i \in \mathcal{B} \tag{7}$$

The subset $\mathcal{B}$ is called the "support set" of the problem.   □

See [3,15,23] for details and proofs related to the above properties. Intuitively, Eq. (7) states that, at the solution of (ℓ_∞, ℓ_p), the minimised maximum error occurs at the support set $\mathcal{B}$; Fig. 1 illustrates. Further, Eq. (6) states that solving (ℓ_∞, ℓ_p) amounts to solving the same problem on $\mathcal{B}$. Many classical algorithms in computational geometry [4,16,20,22] exploit this property to solve GLPs.

One of the most basic algorithms is due to Sharir and Welzl [22]. Li [15] more recently applied this algorithm to triangulation. Algorithm 1 summarises the method, in the context of (ℓ_∞, ℓ_p).

The algorithm finds the support set $\mathcal{B}$ by iterating through the data, and checking for violations to the current $\mathcal{B}$ (Step 8). If a datum i violates the current $\mathcal{B}$, a new support set is calculated from the data indexed by $\mathcal{B}$ and i (Steps 3 and 10), which implicitly also obtains the current estimate $\mathbf{x}^*$ and objective value δ^* from problem based on data $\mathcal{B}$ (Steps 4 and 11). Support set updating is conducted using a *primitive solver* - we refer the reader to [22] for details. The algorithm terminates when none of the data violate the current $\mathcal{B}$, which implies that $\mathcal{B}$ is the support set of the whole input data.

Matoušek et al. [16] proved that the algorithm of Sharir and Welzl [22] has sub-exponential runtime. Observe that, unlike the GFP algorithms (Sect. 2.1), each update iteration of Algorithm 1 involves only small subset of the data ($\mathcal{B} \cup \{i\}$). However, as we will show in Sect. 4, the runtime of Algorithm 1 can still be significant on large scale 3D reconstruction datasets.

Algorithm 1. Subexponential-time algorithm [22] for solving (ℓ_∞, ℓ_p).

Require: Input data $\{\mathbf{P}_i, \mathbf{u}_i\}_{i=1}^N$.
 1: Randomly permute the order of $\{\mathbf{P}_i, \mathbf{u}_i\}_{i=1}^N$, and define $\mathcal{X} = \{1, \ldots, N\}$.
 2: $\mathcal{B} \leftarrow \{1, 2, 3, 4\}$.
 3: $\mathcal{B} \leftarrow$ Support set of data indexed by $\mathcal{B} \cup \{i\}$.
 4: $(\mathbf{x}^*, \delta^*) \leftarrow$ Solution of Eq. (3) on data indexed by $\mathcal{B}$.
 5: **while** true **do**
 6:     $v \leftarrow 0$.
 7:     **for** $i = 1, \ldots, N$ **do**
 8:         **if** $r(\mathbf{x}^* \mid \mathbf{P}_i, \mathbf{u}_i) > \delta^*$ **then**
 9:             $v \leftarrow v + 1$.
10:             $\mathcal{B} \leftarrow$ Support set of data indexed by $\mathcal{B} \cup \{i\}$.
11:             $(\mathbf{x}^*, \delta^*) \leftarrow$ Solution of Eq. (3) on data indexed by $\mathcal{B}$.
12:             Optionally rearrange $\mathcal{X}$ by moving i to the first position.
13:         **end if**
14:     **end for**
15:     **if** $v = 0$ **then**
16:         Break.
17:     **end if**
18: **end while**
19: **return** $\mathbf{x}^*$ and δ^*.

---

3   Meta-Algorithm for ℓ_∞ Triangulation

The primary source of inefficiency in Algorithm 1 is that multiple passes over all N of the data are usually required. This is due to the fact that the algorithm keeps track of only at most 4 of the data at once (i.e., $\mathcal{B}$). Thus, a large number of updates are required before convergence. Secondly, the choice of the datum i for updating $\mathcal{B}$ is determined randomly (via the random permutation of $\mathcal{X}$ during initialisation). To hasten convergence, the selection should be made more strategically. Figure 2(a) provides an analogy of each iteration of Algorithm 1.

We propose Algorithm 2 as a more efficient technique for ℓ_∞ triangulation [21]. First, Algorithm 2 is a meta-algorithm, since it requires an underlying (ℓ_∞, ℓ_p) solver to carry out the updates (see Steps 3 and 10). Any of the previous algorithms [2,5,7,9,13,14,19] can be embedded. Thus, although Algorithm 2 appears simple, a lot of complexity due to solving (ℓ_∞, ℓ_p) has been abstracted away (in this sense, Algorithm 1 is also a meta-algorithm, since there remains the primitive solver routine whose specification is independent of the main structure).

Conceptually, instead of explicitly finding the support set $\mathcal{B}$ directly, Algorithm 2 seeks a representative subset of the data $\mathcal{C}$, which is no smaller than 4. The algorithm incrementally expands $\mathcal{C}$, by choosing the data q that most violates the current solution $\mathbf{x}^*$. This helps to accelerate convergence, since the "radius" of $\mathcal{C}$ (i.e., the minimised maximum reprojection error of the data in $\mathcal{C}$) is expanded quickly; see Fig. 2(b). Contrast this to Algorithm 1, whose selection of the "pivot" datum i is achieved effectively by random selection.

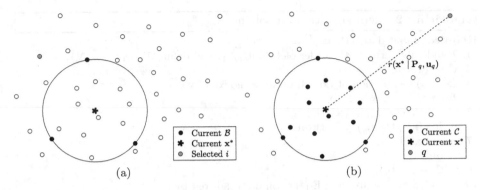

Fig. 2. Illustration of conceptual difference between Algorithms 1 and 2 on the minimum bounding circle problem, which is also a GLP and thus analogous to ℓ_∞ triangulation. (a) Algorithm 1 seeks the support set $\mathcal{B}$ of the data, and in each iteration, $\mathcal{B}$ is updated using the primitive solver [??] from the current $\mathcal{B}$ and a (randomly) selected datum i that violates $\mathcal{B}$. (b) Algorithm 2 seeks a representative subset $\mathcal{C}$ of the data, and in each iteration, $\mathcal{C}$ is updated using an (ℓ_∞, ℓ_p) solver [2,5,7,9,13,14,19] from the current $\mathcal{C}$ and the most violating datum q.

Algorithm 2 terminates when $\mathcal{C}$ equals $\mathcal{X}$, or when $\mathcal{C}$ contains the support set $\mathcal{B}$ of the whole input data. A formal statement is provided as follows.

Theorem 1. *Algorithm 2 finds* $\mathbf{x}^*$ *in at most N iterations.*

Proof. Given the current $\mathcal{C}$ with solution $\mathbf{x}^*$ and value δ^*, let q be obtained according to Step 5 in Algorithm 2.

- If $q \in \mathcal{C}$, then, by how $\mathbf{x}^*$ and δ^* were calculated in Step 10, the condition in Step 6 must be satisfied and $\mathbf{x}^*$ is the global minimiser.
- If $q \notin \mathcal{C}$ and the condition in Step 6 is satisfied, then Eq. (6) is implied and $\mathbf{x}^*$ is the global minimiser.
- If $q \notin \mathcal{C}$ and the condition in Step 6 is not satisfied, then Algorithm 2 will insert q into $\mathcal{C}$. There are at most N of such insertions (including the initial four insertions in the initialisation). In the worst case all of $\mathcal{X}$ will finally be inserted, and $\mathcal{C}$ converges to $\mathcal{X}$. $\qquad\square$

Clearly, for Algorithm 2 to be more efficient than executing an (ℓ_∞, ℓ_p) solver directly on the full input data, $\mathcal{C}$ must be expanded in a way that incorporates $\mathcal{B}$ into $\mathcal{C}$ quickly. This enables $\mathcal{C}$ and the number of iterations to be small, and, equally importantly, the embedded (ℓ_∞, ℓ_p) solver need only be invoked on small subsets $\mathcal{C}$ of $\mathcal{X}$. Section 4 demonstrates that, in practice, Algorithm 2 can in fact find the global minimiser much more efficiently than invoking an (ℓ_∞, ℓ_p) solver [2,5,7,9,13,14,19] in "batch mode" on the whole data.

3.1   Connections to Coreset Method

Algorithm 2 is in fact an instance of a coreset algorithm [1] (the representative subset $\mathcal{C}$ can be interpreted as a coreset of $\mathcal{X}$). Coreset methods are applied

---

Algorithm 2. Meta-algorithm for solving (ℓ_∞, ℓ_p).

Require: Input data $\{\mathbf{P}_i, \mathbf{u}_i\}_{i=1}^N$.
1: Randomly permute the order of $\{\mathbf{P}_i, \mathbf{u}_i\}_{i=1}^N$, and define $\mathcal{X} = \{1, \ldots, N\}$.
2: $\mathcal{C} \leftarrow \{1, 2, 3, 4\}$.
3: $(\mathbf{x}^*, \delta^*) \leftarrow$ Solution of Eq. (3) on data indexed by $\mathcal{C}$.
4: **for** $t = 1, \ldots, N - 4$ **do**
5:     $q \leftarrow \arg max_{i \in \mathcal{X}} r(\mathbf{x}^* \mid \mathbf{P}_i, \mathbf{u}_i)$.
6:     **if** $r(\mathbf{x}^* \mid \mathbf{P}_q, \mathbf{u}_q) \leq \delta^*$ **then**
7:         Exit for loop.
8:     **end if**
9:     $\mathcal{C} \leftarrow \mathcal{C} \cup \{q\}$.
10:     $(\mathbf{x}^*, \delta^*) \leftarrow$ Solution of Eq. (3) on data indexed by $\mathcal{C}$.
11: **end for**
12: **return** $\mathbf{x}^*$ and δ^*.

---

frequently in discrete geometry to obtain ϵ-approximation solutions for *extent* problems, such as minimum enclosing balls. However, the bounds arising from current theoretical results are usually too loose to be of practical use (e.g., to obtain a small approximation error ϵ, the maximum number of iterations predicted is often far larger than available data in real-life problems).

Nonetheless, Algorithm 2 remains a very efficient meta-algorithm for (ℓ_∞, ℓ_p), as we will demonstrate in the next section.

4   Experiments

We conducted experiments to investigate the performance of Algorithm 2 as a global minimiser to the ℓ_∞ triangulation problem. We used a standard machine with 3.2 GHz processor and 16 GB main memory.

4.1   Datasets and Initialisation

We tested on publicly available datasets for large scale 3D reconstruction, namely, Vercingetorix Statue, Stockholm City Hall, Arc of Triumph, Alcatraz, Örebro Castle [8,18], and Notre Dame [24]. The *a priori* estimated camera poses and intrinsics supplied with these datasets were used to derive camera matrices. For triangulation, the size of an instance is the number of observations of the target 3D point. To avoid excessive runtimes, we randomly sampled 10% of the scene points in each dataset - this reduces the number of problem instances, but not the size of each of the selected instances. Figure 3 illustrates the distribution of problem sizes in each of the datasets used.

As shown in the respective pseudo-codes, Algorithms 1 and 2 were initialised by randomly choosing four data to instantiate $\mathbf{x}^*$ by solving (ℓ_∞, ℓ_p).

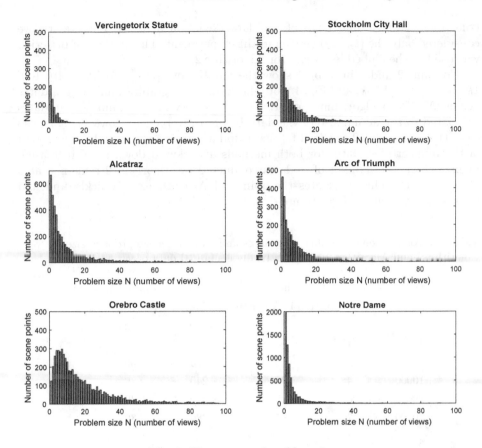

Fig. 3. Histograms of problem size.

4.2   Comparing Algorithm 1 and Algorithm 2

As mentioned above, Algorithms 1 and 2 can both be regarded as meta-algorithms, in that they require an embedded solver (the primitive solver [22] for Algorithms 1 and an (ℓ_∞, ℓ_p) solver for Algorithm 2), and the main structure of the algorithm is independent from the specification of the embedded solvers. Further, the solvers are only executed on a subset of the data in each update. In this section, we compare Algorithms 1 and 2 in terms of the number of updates (i.e., the number of calls to the embedded solver) required before convergence.

Synthetic Data. First, we generated synthetic data of varying sizes N (recall that the size of a triangulation problem is the number of views/measurements of the same 3D point). The camera centres were distributed uniformly around the 3D point, with the camera orientation pointing roughly towards the 3D point; Fig. 1 illustrates an instance of the synthetically generated data with $N = 10$. The observed point coordinates in each view were obtained by perturbing the true point with Normal noise of standard deviation 3 pixels. For brevity, we considered only the (ℓ_∞, ℓ_2) version in this experiment. The method of bisection was used as the embedded solver for Algorithm 2.

Columns 2 and 3 in Table 1 show the number of updates for $N = 10$, 100, $1k$, $10k$, $20k$ and $50k$, and Fig. 4 plots the results in a graph (number of updates versus N). Both algorithms performed only a very small number of updates compared to the actual size of the data before convergence. However, it is also clear that Algorithm 2 required far fewer updates than Algorithm 1. The growth of the number of updates for both methods also slowed down as N increased. In particular, the number of updates conducted by Algorithm 2 remained at 5 after $N = 1k$. This illustrates the ability of Algorithm 2 to quickly find and incorporate the support set $\mathcal{B}$ into $\mathcal{C}$.

Table 1. Comparison of number of updates and cumulative subproblem size between Algorithm 1 and Algorithm 2 on synthetically generated data of varying size N.

Data size N	Number of updates		Cumul. subproblem size	
	Algorithm 1	Algorithm 2	Algorithm 1	Algorithm 2
10	8	3	40	22
100	15	4	75	30
1000	35	5	175	39
10000	48	5	240	39
20000	62	5	310	39
50000	76	5	380	39

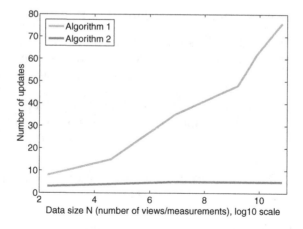

Fig. 4. Number of updates as a function of input size N (synthetic data) for Algorithm 1 and Algorithm 2.

A comparison of the number of updates will not be complete without comparing the total computational effort required. Instead of recording the actual runtime, however, which is dependent on the specific solver used and the maturity of its implementation, we compare the cumulative size of the subproblems that needed to be solved across the updates — intuitively, this is the amount of data that the algorithm accessed for triangulation. For Algorithm 1, since each update is invoked on the data of size 5 (i.e., $|\mathcal{B} \cup \{i\}|$), the cumulative subproblem size is simply 5 times the number of updates. For Algorithm 2, the subproblem size depends on the size of $\mathcal{C}$; the relationship between the cumulative subproblem size and the number of updates/iterations is

$$4 \text{ (initialisation)} + 5 \text{ (iter 1)} + 6 \text{ (iter 2)} + 7 \text{ (iter 3)} + \ldots \tag{8}$$

Columns 4 and 5 in Table 1 show the cumulative subproblem sizes for both algorithms. Clearly Algorithm 2 remains superior in this respect, thus we expect Algorithm 2 to be faster than Algorithm 1 in actual applications.

Real Data. The experiment above was repeated on real data; here, the three variants of (ℓ_∞, ℓ_p) with $p = 1$, 2 and ∞ were tested. The Dinkelbach's method was used as the embedded (ℓ_∞, ℓ_p) solver in Algorithm 2. Tables 2 and 3 show respectively the *total* number of updates and *total* cumulative subproblem size of both algorithms across all triangulation instances in each dataset. The same conclusion can be drawn, i.e., Algorithm 2 is far more efficient in terms of number of updates and total amount of measurements accessed.

Table 2. Number of updates.

Dataset	(ℓ_∞, ℓ_1)		(ℓ_∞, ℓ_2)		$(\ell_\infty, \ell_\infty)$	
	Algorithm 1	Algorithm 2	Algorithm 1	Algorithm 2	Algorithm 1	Algorithm 2
Vercingetorix	2883	1065	2788	1069	2914	1104
Stockholm	14766	5811	13624	5350	14956	5915
Arc of Triumph	20796	7430	19147	7069	20676	7524
Alcatraz	36943	11028	34881	12533	36094	11966
Örebro Castle	63613	22506	57806	20912	61978	22380
Notre Dame	62577	10743	44758	13451	66469	12358

Table 3. Cumulative subproblem size.

Dataset	(ℓ_∞, ℓ_1)		(ℓ_∞, ℓ_2)		$(\ell_\infty, \ell_\infty)$	
	Algorithm 1	Algorithm 2	Algorithm 1	Algorithm 2	Algorithm 1	Algorithm 2
Vercingetorix	11532	4958	11152	4974	11656	5213
Stockholm	59064	30327	54496	26875	59824	30913
Arc of Triumph	83184	38970	76588	36236	82704	39616
Alcatraz	147772	57902	139524	66303	144376	63980
Örebro Castle	254452	126963	231224	114711	247912	126260
Notre Dame	250308	49034	179032	64200	265876	58721

4.3   Relative Speed-Up of Meta-Algorithm over Batch Solution

We compared running Algorithm 2 with a specific solver as a sub-routine, and the direct execution of the same solver in "batch mode" on the whole data. Since the runtime of Algorithm 2 depends on the efficiency of embedded solver, the key performance indicator here is the *relative speed-up* achieved by the meta-algorithm over batch execution.

Choice of Solvers. In this experiment, we tested the three variants of (ℓ_∞, ℓ_p), specifically (ℓ_∞, ℓ_1), (ℓ_∞, ℓ_2) and $(\ell_\infty, \ell_\infty)$.

Table 4. Runtime comparison between Algorithm 2 and batch execution for (ℓ_∞, ℓ_1), (ℓ_∞, ℓ_2) and $(\ell_\infty, \ell_\infty)$. Points: number of scene points in the dataset. Views: total number views involved. For Algorithm 2, the number in parentheses indicates the percentage amongst all triangulation instances where Algorithm 2 was faster than batch execution.

				Total runtime (s)	
(ℓ_∞, ℓ_1)	Dataset	Points	Views	Batch	Algorithm 2
	Vercingetorix	594	68	1.61	1.58 (13%)
	Stockholm	2176	43	10.49	9.87 (19%)
	Arc of Triumph	2744	173	20.57	12.00 (49%)
	Alcatraz	4431	419	166.34	40.12 (73%)
	Örebro Castle	5943	761	1011.96	56.01 (94%)
	Notre Dame	7149	715	1535.57	116.22 (87%)
				Total runtime (s)	
(ℓ_∞, ℓ_2)	Dataset	Points	Views	Batch	Algorithm 2
	Vercingetorix	594	68	23.65	17.43 (26%)
	Stockholm	2176	43	109.51	74.02 (32%)
	Arc of Triumph	2744	173	204.40	89.91 (56%)
	Alcatraz	4431	419	452.73	239.55 (47%)
	Örebro Castle	5943	761	1440.97	351.25 (76%)
	Notre Dame	7149	715	2399.22	582.64 (76%)
				Total runtime (s)	
$(\ell_\infty, \ell_\infty)$	Dataset	Points	Views	Batch	Algorithm 2
	Vercingetorix	594	68	1.82	1.83 (−1%)
	Stockholm	2176	43	12.20	11.77 (4%)
	Arc of Triumph	2744	173	23.57	14.15 (40%)
	Alcatraz	4431	419	146.26	28.93 (80%)
	Örebro Castle	5943	761	887.27	52.20 (94%)
	Notre Dame	7149	715	865.35	31.72 (96%)

For (ℓ_∞, ℓ_1) and (ℓ_∞, ℓ_2), we chose Dinkelbach's method [6] (equivalent to [19]). Although the best performing technique in [2] was Gugat's algorithm [11], our experiments suggested that it did not outperform Dinkelbach's method on the triangulation problem — in any case, the performance metric of interest here is the relative speed-up of the meta-algorithm (Algorithm 2) over batch execution. If a faster solver was used, it would likely improve both Algorithm 2 and batch solution by the same factor. SeDuMi [25] was used to solve the convex subproblems (LP for (ℓ_∞, ℓ_1) and SOCP for (ℓ_∞, ℓ_2)) in Dinkelbach's method. For $(\ell_\infty, \ell_\infty)$, we used the state-of-the-art polyhedron collapse solver [7].

Results and Analysis. Table 4 shows the total runtime of Algorithm 2 and batch execution on the large scale 3D reconstruction datasets described in Sect. 4.1. Clearly on most of the datasets, embedding the (ℓ_∞, ℓ_p) solver into Algorithm 2 significantly cuts down the total runtime. On datasets where the computational gains were not evident, this was because the datasets were too

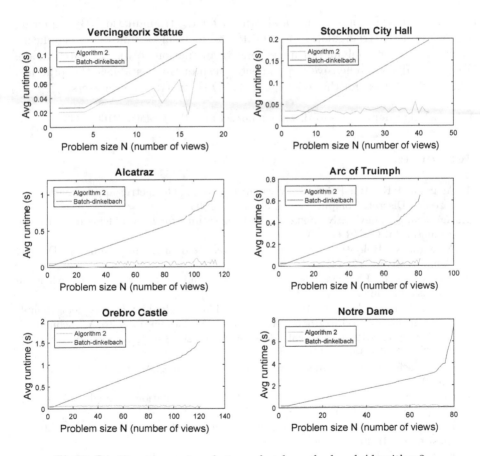

Fig. 5. Runtime comparison between batch method and Algorithm 2.

small (in terms of the number of triangulation instances and the size of the instances) for the benefit of Algorithm 2 to be exhibited.

For the case of (ℓ_∞, ℓ_2), the average runtimes of Algorithm 2 and batch solution as a function of problem size N (number of views/measurements) are plotted in Fig. 5. Observe that the runtime of batch increased linearly and then exponentially, whilst Algorithm 2 exhibited almost constant runtime — the latter observation is not surprising, since Algorithm 2 usually terminated at ≤ 10 iterations regardless of the problem size, as established in Sect. 4.2.

Of course, on all of the datasets, most of the triangulation instances are small, as summarised in the histograms in Fig. 3. However, in these datasets, there are sufficient numbers of moderate to large problem instances, such that the total runtime of Algorithm 2 is still much smaller than the total runtime of batch, as evidenced in Table 4.

5   Conclusions

In this paper, we propose a meta-algorithm for ℓ_∞ triangulation. By exploring the fact that ℓ_∞ triangulation is a GLP, the meta-algorithm offers significant acceleration over applying an underlying triangulation solvers in batch mode. We provided comprehensive experimental results that establish the practical value of the meta-algorithm on large scale 3D reconstruction datasets.

Acknowledgement. This work was supported by ARC Grant DP160103490.

References

1. Agarwal, P.K., Har-Peled, S., Varadarajan, K.R.: Geometric approximation via coresets. Discret. Comput. Geom. **52**, 1–30 (2005)
2. Agarwal, S., Snavely, N., Seitz, S.: Fast algorithms for l_∞ problems in multiview geometry. In: CVPR (2008)
3. Amenta, N.: Helly-type theorems and generalized linear programming. Discret. Comput. Geom. **12**, 241–261 (1994)
4. Clarkson, K.L.: Las Vegas algorithms for linear and integer programming when the dimension is small. J. ACM **42**, 488–499 (1995)
5. Dai, Z., Wu, Y., Zhang, F., Wang, H.: A novel fast method for L_∞ problems in multiview geometry. In: Fitzgibbon, A., Lazebnik, S., Perona, P., Sato, Y., Schmid, C. (eds.) ECCV 2012. LNCS, vol. 7576, pp. 116–129. Springer, Heidelberg (2012). doi:10.1007/978-3-642-33715-4_9
6. Dinkelbach, W.: On nonlinear fractional programming. Manag. Sci. **13**, 492–498 (1967)
7. Donné, S., Goossens, B., Philips, W.: Point triangulation through polyhedrom collapse using the l_∞ norm. In: ICCV (2015)
8. Enqvist, O., Olsson, C., Kahl, F.: Stable structure from motion using rotational consistency. Technical report (2010)
9. Eriksson, A., Isaksson, M.: Pseudoconvex proximal splitting for l_∞ problems in multiview geometry. In: CVPR (2014)

10. Furukawa, Y., Ponce, J.: Accurate, dense, and robust multi-view stereopsis. IEEE TPAMI **32**, 1362–1376 (2010)
11. Gugat, M.: A fast algorithm for a class of generalized fractional programs. Manag. Sci. **42**, 1493–1499 (1996)
12. Hartley, R.I., Schaffalitzky, F.: l_∞ minimization in geometric reconstruction problems. In: CVPR (2004)
13. Kahl, F.: Multiple view geometry and the l_∞ norm. In: ICCV (2005)
14. Ke, Q., Kanade, T.: Quasiconvex optimization for robust geometric reconstruction. In: ICCV (2005)
15. Li, H.: Efficient reduction of l_∞ geometry problems. In: CVPR (2009)
16. Matoušek, J., Sharir, M., Welzl, E.: A subexponential bound for linear programming. Algorithmica **16**, 498–516 (1996)
17. Mur-Artal, R., Tardós, J.D.: Probabilistic semi-dense mapping from highly accurate feature-based monocular SLAM. In: RSS (2015)
18. Olsson, C., Enqvist, O.: Stable structure from motion for unordered image collections. In: Heyden, A., Kahl, F. (eds.) SCIA 2011. LNCS, vol. 6688, pp. 524–535. Springer, Heidelberg (2011). doi:10.1007/978-3-642-21227-7_49
19. Olsson, C., Eriksson, A., Kahl, F.: Efficient optimization for l_∞ problems using pseudoconvexity. In: ICCV (2007)
20. Seidel, R.: Small-dimensional linear programming and convex hulls made easy. Discret. Comput. Geom. **6**, 423–434 (1991)
21. Seo, Y., Hartley, R.I.: A fast method to minimize l_∞ error norm for geometric vision problems. In: ICCV (2007)
22. Sharir, M., Welzl, E.: A combinatorial bound for linear programming and related problems. In: Finkel, A., Jantzen, M. (eds.) STACS 1992. LNCS, vol. 577, pp. 567–579. Springer, Heidelberg (1992). doi:10.1007/3-540-55210-3_213
23. Sim, K., Hartley, R.: Removing outliers using the l_∞ norm. In: CVPR (2006)
24. Snavely, N., Seitz, S.M., Szeliski, R.: Modeling the world from internet photo collections. IJCV **80**, 189–210 (2007)
25. Sturm, J.F.: Using SeDuMi 1.02, a Matlab toolbox for optimization over symmetric cones. Optim. Methods Softw. **11–12**, 625–653 (1999)

Synchronization Error Compensation of Multi-view RGB-D 3D Modeling System

Ju-Hwan Lee, Eung-Su Kim, and Soon-Yong Park[⊠]

School of Computer Science and Engineering,
Kyungpook National University, Daegu 41566, Korea
sasinhwan@nate.com, jsm80607@gmail.com, sypark@knu.ac.kr

Abstract. Finding 3D transformation relationship between multiple RGB-D cameras accurately is necessary to generate a complete 3D model from depth images. In this paper, we propose a convenient method to find the 3D transformation between multiple RGB-D cameras using centroids of an independently moving spherical object. In a multiple camera system, we should always consider the synchronization problem. Therefore, we first define how we managed to compensate for this problem. Next, we introduce a method to find the centroids of the moving sphere using the RANSAC algorithm and refine them based on the movement distance values. Finally, we calculate the 3D transformation relationship using the refined centroids and evaluate the robustness of the proposed calibration method by comparing 3D reconstruction results.

1 Introduction

3D scanning technology has been researched for a long time in the field of computer vision and graphics. Simultaneously obtaining object shape information and color information is essential for restoring the shape of the object accurately. Microsoft Kinect, which consists of an RGB camera and a depth camera, first introduced in 2010 and released a V2 version with increased performance in 2012. In general, these Kinect sensors are called 'RGB-D Cameras' and have been used in advanced research areas such as object restoring [1–3], 3600 full-body scanning [4,5]. Representing the coordinate system of each RGB-D camera into a common coordinate system is essential for restoring the entire shape of an object using depth and color information taken from 360° directions.

In our previous research paper, we introduced a method to conveniently find the calibration relationship between each RGB-D camera using the centroid of a moving sphere object [4]. However, this approach uses all the 3D points on the surface to determine the centroid of the sphere. Therefore, calculated centroid has an error because of the object is not an idle sphere and also the captured 3D data has noise. Furthermore, the accuracy of the centroid is affected by the synchronization errors when using data from multiple Kinects.

In this paper, we introduce a new method to accurately find the 3D coordinate of the sphere center avoiding the aforementioned problems and improved the calibration accuracy. First, we segment the captured 3D points of the sphere

© Springer International Publishing AG 2017
C.-S. Chen et al. (Eds.): ACCV 2016 Workshops, Part II, LNCS 10117, pp. 162–174, 2017.
DOI: 10.1007/978-3-319-54427-4_13

surface into five regions and, select four arbitrary 3D points from each region (except one area) to calculate the centroid using the RANSAC algorithm [6]. Then we compare distances between centroids between two consecutive frames to minimize the error. Finally, we use the centroid of the sphere to find the calibration relationship between each RGB-D camera.

2   Centroid Extraction of the Sphere Object

Figure 1 depicts the experimental setup that consists of four Kinect V2 sensors used in our proposed multi-view calibration method. The IR sensor of each Kinect device is denoted as $V_{i(i=1,2,3,4)}$ and data acquisition time between each sensor is synchronized. A yellow color spherical shaped object is moved in-between the four Kinects freely in an independent paths and N number of frames are captured simultaneously. Figure 2 shows an example set of captured 3D points of the sphere from the four sensors.

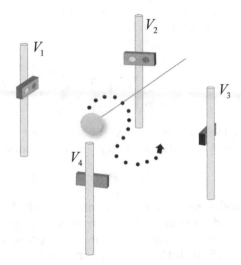

Fig. 1. Experiment setup used for data acquisition in the proposed multi-view calibration method. (Color figure online)

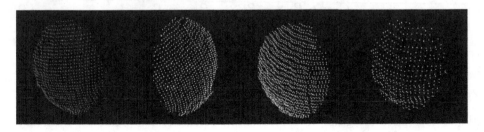

Fig. 2. A set of captured 3D points of the sphere from four Kinects.

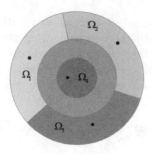

Fig. 3. The divided areas of the sphere surface to extract four arbitrary points. (Color figure online)

2.1   Extraction of Four Points on the Ball Surface

The first step of our proposed multi-view calibration method is determining the centroids of the spheres that are captured using the four Kinects as explained in Sect. 2. We can estimate the centroid of each sphere using four arbitrary 3D points on the surface according to Eq. 1.

$$\begin{bmatrix} x \\ y \\ z \end{bmatrix} = \frac{1}{2} \begin{bmatrix} x_2 - x_1 \; y_2 - y_1 \; z_2 - z_1 \\ x_3 - x_2 \; y_3 - y_2 \; z_3 - z_2 \\ x_4 - x_1 \; y_4 - y_1 \; z_4 - z_1 \end{bmatrix}^{-1} \begin{bmatrix} x_2^2 - x_1^2 + y_2^2 - y_1^2 + z_2^2 - z_1^2 \\ x_3^2 - x_2^2 + y_3^2 - y_2^2 + z_3^2 - z_1^2 \\ x_4^2 - x_1^2 + y_4^2 - y_1^2 + z_4^2 - z_1^2 \end{bmatrix} \quad (1)$$

Equation 2 defines the number of ways to select four arbitrary 3D points where Ω is the total number of 3D points on the sphere surface.

$$C = {}_{\Omega}P_4 \quad (2)$$

The general method requires a large number of calculations as it has to check all these cases, and reducing these unnecessary computations is required to reduce the computation time. Therefore, this paper proposes an area division method as shown in Fig. 3, such that an arbitrary point from each area (except blue area) are extracted. Using this approach, we can reduce the number of ways (c') to select the four points, which is calculated as in Eq. 3.

$$C' = {}_{\Omega_1}P_1 \times {}_{\Omega_2}P_1 \times {}_{\Omega_3}P_1 \times {}_{\Omega_4}P_1 \quad (3)$$

As the distance between the selected four points is far, not only the calculated centroid has a smaller error but also has a faster computational speed.

2.2   Finding Ball Centroid Using RANSAC Algorithm

3D data acquired using the Kinect IR sensors consist of noise, and it is necessary to remove these noise before estimating the centroid of the sphere. We use the well-known RANSAC algorithm for removing noise [6]. We can extract a random point from each area as shown in Fig. 3 and create a sphere using them. Then we can calculate distance of all 3D points of the ball surface from the centroid of sphere. If the distance from the centroid of the sphere to each 3D point of ball surface satisfies the real radius of ball and radius error, then it is considered as an inlier point, otherwise as an outlier point.

If the total number of inlier points is over a certain percentage, the calculated centroid can be considered as the true centroid of the sphere. Figure 4 shows the whole process. Then we can take the centroids of all captured frames of each Kinect as shown in Fig. 5.

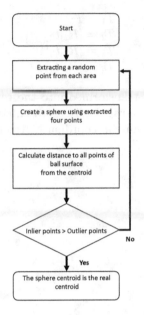

Fig. 4. Flowchart of the whole process.

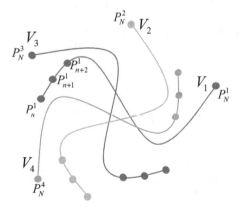

Fig. 5. Extracted centroids of the sphere trajectory of each Kinect.

3   View Calibration Using Center Points of the Sphere

Theoretically, 3D transformation relation T_i between two cameras, which consist of rotation matrix R_i and translation vector t_i, can be estimated using three 3D points [4]. In the proposed method, we estimate the 3D transformation T_i between each V_i using 3D coordinates of the spheres centroid P^i. Here we use N number of centroids $(P^i_{n(n=1,2,3,N)})$ that represent the trajectory of the sphere centroid to calculate the transformation. Considering V_1 as the reference, the 3D transformation matrix T_i from each V_i to V_1 can be represented as follows:

$$
\begin{aligned}
T_1 &= I \\
T_2 &= [R_2 | t_2] \\
T_3 &= [R_3 | t_3] \\
T_4 &= [R_4 | t_4]
\end{aligned}
\tag{4}
$$

Then each centroid P^i_n is transformed into the reference camera coordinate system using Eq. 5 and can be graphically represented as in Fig. 6, where all the points are converted into a common coordinate system.

$$
P'^i_n = T_i P^i_n
\tag{5}
$$

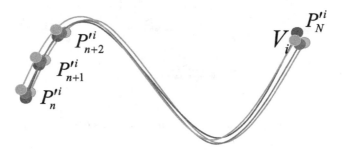

Fig. 6. Converted centroids of the sphere trajectories into a single coordinate system using transformation relationship. V_1 is selected as the reference.

4   Distance Control of the Centroids

We find the centroids of the sphere using 3D points on its surface that we acquired from each Kinect. Then we define the distance between centroids on two consecutive frames (d^i_n) in each V_i as mentioned in Eq. 6.

$$
d^i_n = ||P^i_n - P^i_{n+1}||
\tag{6}
$$

If the data acquisition time between each V_i is perfectly synchronized and acquired 3D point data of the sphere from each V_i does not contain any noise, the movement distance d_n^i from each V_i should be similar.

$$d_n^1 = d_n^2 = d_n^3 = d_n^4 \qquad (7)$$

In real world situations, these d_n^i values are not always identical (Fig. 7) due to synchronizing errors of Kinects, geometric deformities of the object (not being an ideal sphere), and the noise in captured 3D data. Considering it as a fact, we introduce a new idea to counterbalance this problem. In our proposed method, we first calculate movement distance d_n^i, between P_n^i and $P_{(n+1)}^i$ for each i and find the median distance l_n.

$$l_n = median(d_n^1, d_n^2, d_n^3, d_n^4) \qquad (8)$$

Then we correct the centroid coordinate of the $P_{(n+1)}^i$ using Eq. 9 where unit vector of movement direction $(\hat{t}_n^i)$ is calculated as in Eq. 10.

$$Q_{n+1}^i = P_n^i + l_n \hat{t}_n^i \qquad (9)$$

$$\hat{t}_n^i = P_{n+1}^i - P_n^i \qquad (10)$$

We follow this process for each $n_{(n=1,2,3,N)}$ to correct the centroid in all frames. Figure 8 shows the distance control method of the centroid, whereas the corrected centroids are shown in Fig. 9. The following pseudocode summarizes the whole distance control process.

Algorithm 1: Distance controlling process

---

1 **while** N **do**
2   **if** $n==1$ **then**
3          $Q_n^i = P_n^i$;
4   n=n+1;
5   **else**
6          $d_{n-1}^i = ||\mathrm{P}_n^i - Q_{n-1}^i||$;
7          $l_{n-1} = median(d_{n-1}^1, d_{n-1}^2, d_{n-1}^3, d_{n-1}^4)$;
8          $\hat{t}_n - 1^i = \mathrm{P}_n^i - Q_{n-1}^i$;
9          $Q_n^i = Q_{n-1}^i + l_{n-1}\hat{t}_n - 1^i$;
10   n = n+1;
11   **end**
12 **end**

Instead of using P_n^i, we use the corrected centroid Q_n^i to calculate the new 3D transformation matrix T_i' using Eq. 1. Once the new transformation relationship is estimated, we can convert 3D point data of any object into a common coordinate system. In our experiments, we convert the 3D point clouds into 3D meshes using the Poisson surface reconstruction algorithm [7].

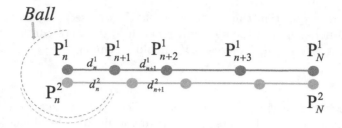

Fig. 7. The movement distance from P_n^i to $P_{(n+1)}^i$.

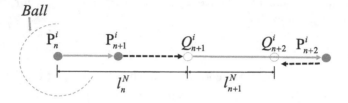

Fig. 8. The method of distance control for correcting the centroid.

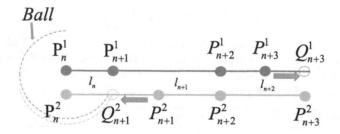

Fig. 9. Corrected centroids of the sphere using proposed distance control method.

5   Experiments and Implementations

Figure 10 represents the overall system configuration we used in our experiment setup. Here, we used four Kinect V2 sensors and mounted them on steel poles

Fig. 10. The test field and the ball for the test. (Color figure online)

inside a $3\,\mathrm{m} \times 3\,\mathrm{m}$ test area. Each Kinect sensor is connected to an individual computer, which runs ROS, and all four individual computers are connected to a separate common computer. A yellow color ball with a radius of $102.5\,\mathrm{mm}$ is used as the spherical object. We freely moved the ball in an independent path and captured its shape from each Kinect sensor in real-time.

5.1   Comparison of Centroid Extraction Methods

We extracted the centroids of the ball using captured 3D points on the ball surface from each V_i. Figure 11 shows the trajectory of the ball centroid P_n^i and the trajectory of the corrected ball centroid Q_n^i. Figure 12 shows the trajectories of $P_n^{'i}$ and $Q_n^{'i}$, which are transformed in to a common coordinate system considering $V1$ as the reference. Magnified view confirmed that the distribution of the centroids $(Q_n^{'i})$ is less in the method which uses the distance controlling technique.

Figure 13 shows an example of how the centroids $P_n^{'i}$ and $Q_n^{'i}$ are distributed in the 3D space. As a quantitative analysis, we calculated both the average distance and variance from the center point to each four centroids. The results are summarized in Tables 1 and 2 for three different experiment setups. Less variance values of the corrected centroids confirmed the accuracy of the proposed calibration process.

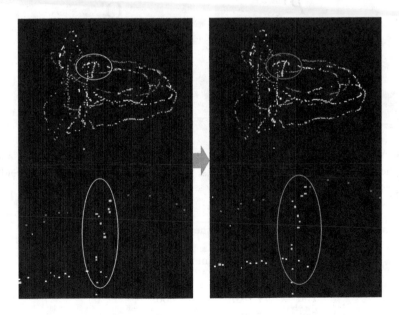

Fig. 11. The centroid trajectory and the centroid trajectory of moved the centroid.

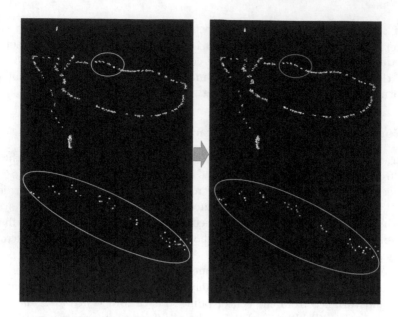

Fig. 12. Converted the centroid trajectory and converted the centroid trajectory of moved the centroid a single coordinate system.

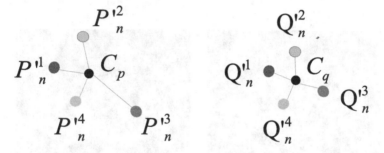

Fig. 13. An example showing how $P_n^{\prime i}$ and $Q_n^{\prime i}$ are distributed in 3D space. C_p and C_q denote the center points of two centroid sets, respectively. Converted $Q_n^{\prime i}$ centroids are distributed closer than $P_n^{\prime i}$ and consist of less variance.

Table 1. Average and variance of the distance from center to each $P_n^{\prime i}$

Data set	Average (mm)	Variance (mm)
1st	7.24	9.14
2nd	7.15	9.06
3rd	7.69	9.80

Table 2. Average and variance of the distance from center each Q'^i_n

Data set	Average (mm)	Variance (mm)
1st	6.39	6.51
2nd	6.64	7.74
3rd	6.23	7.79

5.2   Reconstruction Accuracy

We scanned a plaster model of Brutus to compare the accuracy of the two transformation relationships (T_i and T'_i). Figure 14 depicts the plaster model (left) and the 3D laser scanning point cloud result (right) that we used as the ground truth. Then, we reconstructed the object using the proposed multi Kinect system by positioning it in the middle of the test field. The distance between each Kinect and the object is about 1.8 m. We created two 3D point clouds using the two transformations and compared them with the ground truth (Fig. 15). Table 3 summarizes the comparison results and confirms the robustness of our proposed method. Figure 16 shows two different views of a mesh model created using our proposed method (using the Transformation matrix T'_i).

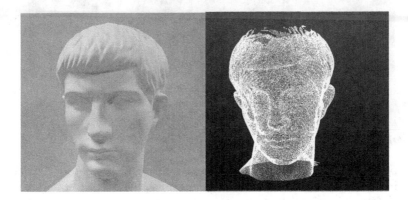

Fig. 14. The plaster model and the 3D laser scan point cloud that used as the ground truth.

Table 3. Mean distance and standard deviation (STD) of the distance from a plaster model to each reconstructed model

Data set	Average (mm)	Variance (mm)
1st	6.39	6.51
2nd	6.64	7.74
3rd	6.23	7.79

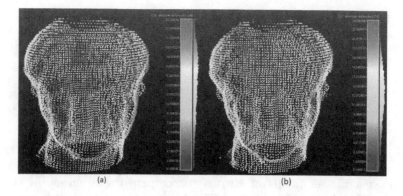

Fig. 15. Comparing point clouds with the ground truth. (a) point cloud created using the normal centroid trajectory. (b) point cloud created using the proposed method.

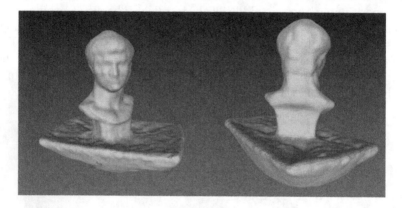

Fig. 16. Mesh model created using our proposed method.

6   Conclusion

This paper proposed a method to calibrate a multiple Kinect sensor setup conveniently using the RANSAC algorithm. We used a spherical object and extracted 3D points from its surface. To reduce the total computation time, we first divided the surface area of the sphere into five sections and randomly extracted four 3D points from four sections. We used these points to find centroids of the sphere in each frame and compared the distances to generate a refined trajectory of the sphere. Then, we used these refined points to calculate the 3D transformation relationship between the cameras. To compare the robustness of the proposed calibration method, we performed several 3D reconstruction tests. We created a ground truth point cloud using laser scanning technique and compared it with the point clouds we created using the calculated transformation relationship. Some other point cloud results along with mesh models are depicted in Figs. 17 and 18 respectively. As future work, we are planning to apply color values to these created mesh models using high tech DSLR cameras for more detailed representations.

Fig. 17. 3D point cloud of a moving human.

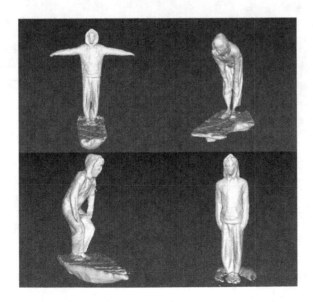

Fig. 18. 3D reconstruction results of a moving person.

Acknowledgement. This work was supported by the National Research Foundation of Korea funded by the Korean Government (2014R1A1A2059784).

References

1. Shum, H.P., Ho, E.S., Jiang, Y., Takagi, S.: Real-time posture reconstruction for Microsoft Kinect. IEEE Trans. Cybern. **43**, 1357–1369 (2013)
2. Nguyen, C.V., Izadi, S., Lovell, D.: Modeling kinect sensor noise for improved 3D reconstruction and tracking. In: 2012 Second International Conference on 3D Imaging, Modeling, Processing, Visualization and Transmission, pp. 524–530. IEEE (2012)
3. Lim, Y.W., Lee, H.Z., Yang, N.E., Park, R.H.: 3-D reconstruction using the kinect sensor and its application to a visualization system. In: 2012 IEEE International Conference on Systems, Man, and Cybernetics (SMC), pp. 3361–3366. IEEE (2012)
4. Park, S.Y., Choi, S.I.: Convenient view calibration of multiple RGB-D cameras using a spherical object. KIPS Trans. Softw. Data Eng. **3**, 309–314 (2014)
5. Tong, J., Zhou, J., Liu, L., Pan, Z., Yan, H.: Scanning 3D full human bodies using kinects. IEEE Trans. Vis. Comput. Graph. **18**, 643–650 (2012)
6. Fischler, M.A., Bolles, R.C.: Random sample consensus: a paradigm for model fitting with applications to image analysis and automated cartography. Commun. ACM **24**, 381–395 (1981)
7. Kazhdan, M., Bolitho, M., Hoppe, H.: Poisson surface reconstruction. In: Proceedings of the Fourth Eurographics Symposium on Geometry Processing, vol. 7 (2006)

Can Vehicle Become a New Pattern for Roadside Camera Calibration?

Yuan Zheng[1(✉)] and Wenyong Zhao[2]

[1] Harbin Institute of Technology Shenzhen Graduate School, Shenzhen, China
zhengyuan@hitsz.edu.cn
[2] Shenzhen Institute of Information Technology, Shenzhen, China

Abstract. Roadside camera calibration is essential to intelligent traffic surveillance and still an unsolved problem. The commonly used pattern-based calibration methods are suitable for the laboratory environment rather than real traffic environment, since the calibration patterns (e.g., checkerboards) generally do not exist in traffic scenarios. In view of this, we propose a new framework for roadside camera calibration where the vehicle moving on the roadway is first introduced as a calibration pattern. Considering that the vehicles are main monitoring targets and inevitably appear in traffic scenarios, the proposed calibration method has a wide use range and is not limited to the structure information of traffic scenarios. Inspired by the traditional pattern-based calibration methods that utilize the matching of 3D-2D point correspondences, we utilize the 3D-2D vehicle matching for camera calibration. The key insight is to convert the camera calibration problem into a vehicle matching problem. To improve the accuracy of calibration results, a new measure function is provided to evaluate the vehicle matching degree and a dynamic calibration method using multi-frame information is proposed to correct camera parameters. Experiments on real traffic images demonstrate the effectiveness and practicability of the proposed calibration framework.

1 Introduction

With the development of society and economy, the range of traffic monitoring is constantly expanding, and the number of the roadside cameras used for traffic monitoring is explosively growing. Accordingly, the traditional manual monitoring no longer is an optimal way and the intelligent traffic surveillance has already become an inevitable trend. By combining the video surveillance technology, the intelligent traffic surveillance exploits the computers to complete the various monitoring tasks in order to save the human resources. Camera calibration is to provide a mapping relationship between 3D scene and 2D image and thus is the foundation for accomplishing various computer vision tasks. Once a roadside camera is calibrated, it can be widely used for traffic scene measurement, object analysis and traffic scene reconstruction. The accurate calibration results would help improve the accuracy of traffic surveillance. Obviously, roadside camera calibration is essential to intelligent traffic surveillance.

© Springer International Publishing AG 2017
C.-S. Chen et al. (Eds.): ACCV 2016 Workshops, Part II, LNCS 10117, pp. 175–188, 2017.
DOI: 10.1007/978-3-319-54427-4_14

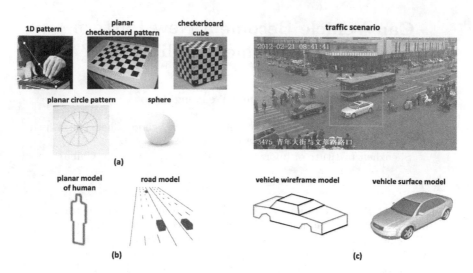

Fig. 1. Calibration patterns. (a) Traditional calibration patterns used for laboratory environment. (b) Some calibration patterns used in practical applications. (c) Our proposed calibration pattern (i.e., the vehicle on the road plane) used for traffic scenarios.

It is well known that camera calibration is a fundamental problem in computer vision and has a wide range of applications. Over the past decades, a large number of camera calibration methods emerged. Up to now, the pattern-based calibration methods are still commonly used that employ the information provided by a calibration pattern to perform calibration. In the existing methods, common calibration patterns contain 1D pattern, planar checkerboard pattern, checkerboard cube, planar circle pattern and sphere (see Fig. 1(a)). Unfortunately, these calibration patterns generally do not exist in traffic scenarios. Moreover, returning to traffic scenarios and placing these calibration patterns is unrealistic and infeasible, which would consume a large amount of human resources, time and financial inputs. Accordingly, the traditional pattern-based calibration methods are not suitable for traffic scenarios and they are more suitable for man-made laboratory environment.

In recent years, roadside camera calibration has become an active research area due to its wide application range. Up to now, the mainstream methods are the vanishing-point-based calibration methods. Vanishing point is the projection of a point at infinity and it is closely related to camera parameters. Evidently, the vanishing-point-based calibration methods strongly depend on the accuracy and reliability of vanishing point. For various traffic scenarios, the vanishing point generally can be obtained by static scene structure (e.g., lane markings, road boundaries and light pole on the roadside), walking humans and moving vehicles. According to the number of the used vanishing points, the existing methods can be categorized into three types: based on one vanishing point [1–3], based on two vanishing points [2,4–7] and based on three vanishing points [8]. More vanishing points are able to estimate more camera parameters. Consequently, three-vanishing-points-based method is supposed to be the best. However, this kind of methods requires that three mutually orthogonal vanishing points are available and this requirement is too strict for most traffic scenarios. The limitation of the vanishing-point-based methods resides in the strong dependence on the structure information of traffic scenarios. If the traffic scenario is capable of providing more and better vanishing points, a more accurate calibration result can be obtained, and vice versa. Obviously, this limitation severely affects the generality of the vanishing-point-based methods.

Our goal is to propose a generic roadside camera calibration method that can be suitable for most traffic scenarios and does not strongly depend on the structure information of traffic scenario. Fortunately, we observed a phenomenon that the vehicles definitely appear in traffic scenario since they are main monitoring targets. Accordingly, a natural idea is to make full use of the vehicle information to accomplish camera calibration. Based on this idea, we propose a novel framework for roadside camera calibration. Our main contributions are listed as follows: (1) considering that the roadside camera to be calibrated inevitably captures the images of the vehicles, we take the vehicle moving on the roadway as a calibration pattern and convert the camera calibration problem into a vehicle matching problem; (2) as for the evaluation of vehicle matching degree, we provide a new measure function that is more robust to clutter and occlusion by combining the local measure (i.e., image-gradient-based measure) with the global measure (i.e., image-silhouette-based measure); (3) we present a dynamic calibration process that exploits the vehicle information at different times (i.e., different image frames) to update and correct camera parameters.

2   Related Work

Up to now, the pattern-based camera calibration methods are still commonly used. This kind of methods exploits the information provided by a calibration pattern and thus requires that the geometric structure of the calibration pattern is known beforehand.

Checkerboard is the most common pattern for camera calibration. When putting the checkerboard pattern onto a cube, a 3D calibration pattern, namely checkerboard cube, is formed. Based on this checkerboard cube, the calibration methods [9,10] establish the equations of camera parameters through the relationship between the 3D points on the checkerboard cube and the corresponding 2D points on the image. Tsai's method [9] is the representative method that uses at least seven 3D-2D point correspondences to estimate camera parameters including the radial distortion coefficients. When putting the checkerboard pattern onto a plane, a 2D calibration pattern, namely planar checkerboard pattern, is formed. Similarly, the calibration methods based on planar checkerboard pattern [11–14] also use the relationship between the 3D points on the calibration pattern and the corresponding 2D points on the image to estimate camera parameters. The typical method is Zhang's method [12] that requires several images of planar checkerboard pattern from different viewpoints and then uses the point correspondences among these images for camera calibration.

Except for 3D calibration patterns and 2D calibration patterns, 1D calibration patterns has also attracted the researcher's attention. 1D calibration pattern consists of several collinear points and the distances between these points are known. Generally, the calibration methods based on 1D calibration pattern [15–17] require several images of 1D calibration pattern from different viewpoints or under its different poses, and then use the point correspondences and the distance information to establish the equations of camera parameters. In 2004, Zhang [15] first proposed 1D-pattern-based calibration method. The later works [16,17] have further improved this kind of calibration methods.

From the beginning of this century, the calibration methods based on circle and sphere have emerged, where the planar circle pattern and the sphere are viewed as a 2D calibration pattern and a 3D calibration pattern respectively. Unlike the calibration methods mentioned above, this kind of methods does not require known 3D-2D point correspondences. In the perspective geometry, the image of circular point is closely related to camera intrinsic parameters. The circle-based calibration methods [18–21] use the image of circular point obtained by the projection of planar circle pattern to estimate camera parameters. In the perspective geometry, the image of absolute conic is also closely related to camera intrinsic parameters. Similarly, the sphere-based calibration methods [22–25] use the image of absolute conic obtained by the projection of sphere for camera calibration.

Generally, these pattern-based calibration methods are only suitable for the laboratory environment, since the calibration patterns mentioned above can be manually placed in the laboratory. In practical applications such as indoor surveillance and outdoor surveillance, these calibration patterns do not exist in scenarios and thus the calibration methods using them are not suitable. In [26], the authors took the planar model of human as a calibration pattern, and then utilized the matching between the projection of the planar model and the corresponding region on the image to estimate camera parameters. In [27], the road model is viewed as a calibration pattern and Markov Chain Monte Carlo method is used to fit the road model and the corresponding image information. Once the geometry information of the road (e.g., road width and the length of lane markings) was obtained, the authors employed the vanishing point extracted from the road information to calibrate camera.

3   Camera Model

Camera model describes the mapping relationship between 3D space domain and 2D image domain, which can be expressed as

$$\lambda \begin{pmatrix} u \\ v \\ 1 \end{pmatrix} = K \begin{bmatrix} R\,t \end{bmatrix} \begin{pmatrix} X \\ Y \\ Z \\ 1 \end{pmatrix}, \tag{1}$$

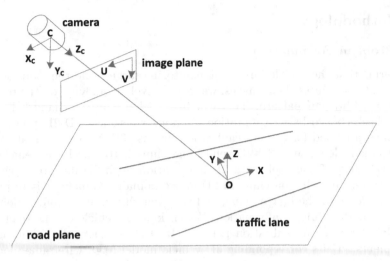

Fig. 2. Geometry relationship of roadside camera and roadway, where X-, Y- and Z-axes denote world coordinate system, X_C-, Y_C- and Z_C-axes denote camera coordinate system and U- and V-axes denote image coordinate system.

where (X, Y, Z) denotes a 3D point in world coordinate system, (u, v) denotes the corresponding 2D image point in image coordinate system and λ is a scaling factor. Figure 2 shows the geometric relationship between roadside camera and roadway.

In Eq. (1), K, R and t are camera intrinsic parameters matrix, rotation matrix and translation vector respectively, which can be expanded as

$$K = \begin{pmatrix} rf & s & u_0 \\ 0 & f & v_0 \\ 0 & 0 & 1 \end{pmatrix}, \tag{2}$$

$$R = \begin{pmatrix} 1 & 0 & 0 \\ 0 & \cos\varphi & -\sin\varphi \\ 0 & \sin\varphi & \cos\varphi \end{pmatrix} \cdot \begin{pmatrix} \cos\gamma & 0 & \sin\gamma \\ 0 & 1 & 0 \\ -\sin\gamma & 0 & \cos\gamma \end{pmatrix} \cdot \begin{pmatrix} \cos\theta & -\sin\theta & 0 \\ \sin\theta & \cos\theta & 0 \\ 0 & 0 & 1 \end{pmatrix}, \tag{3}$$

$$t = \begin{pmatrix} t_1 \\ t_2 \\ t_3 \end{pmatrix} = -R \begin{pmatrix} X_{cam} \\ Y_{cam} \\ h \end{pmatrix}, \tag{4}$$

where f is focal length, (u_0, v_0) is principal point, r is aspect ratio, s is skew factor, $(\varphi, \gamma, \theta)$ are three rotation angles, (X_{cam}, Y_{cam}) is camera's position on the ground and h is camera's height above the ground. Accordingly, camera calibration is to determine camera parameters including intrinsic parameters, rotation angles and spatial location.

4   Methodology

4.1   Problem Formulation

Considering that the vehicles are main monitoring targets and the roadside camera inevitably captures their images, we take the vehicle moving on the roadway as a new calibration pattern. This will bring what kind of change? Inspired by the checkerboard-based calibration methods where the 3D-2D point correspondences are used for camera calibration, we use 3D-2D vehicle matching to calibrate roadside camera. 3D-2D vehicle matching is to match the vehicle's 3D information (e.g., 3D model) with 2D information (e.g., 2D image). In general, it is converted into the matching in 2D image domain. Namely, it is to project the 3D model onto the image plane, and then match the model projection with the corresponding image information. Which kind of vehicle on the road plane would be chosen as a calibration pattern? It should satisfy the following two requirements: (1) its corresponding 3D vehicle model (e.g., wireframe model or surface model) is available that is used for 3D-2D vehicle matching; (2) its GPS information is available, namely its location information in 3D scene is known. With the development of modeling technology, 3D model of the vehicle can be easily obtained.

In the checkerboard-based calibration methods, the camera parameters correspond to a best matching of 3D-2D point correspondences. Similarly, when the vehicle is taken as a calibration pattern, the camera parameters correspond to a best vehicle matching. We provide a measure function, also called fitness function, to evaluate the degree of vehicle matching. The better the projection of vehicle model fits the corresponding image region, the greater the value of fitness function is. That is to say, the camera parameters correspond to the maximum value of fitness function. Accordingly, the roadside camera calibration problem is converted into an optimization problem of fitness function and the camera parameters can be estimated by maximizing the fitness function, which is expressed as

$$\hat{p} = \max_{p} E(p; I, M), \tag{5}$$

where p denotes camera parameters, I denotes vehicle image, M denotes vehicle model and $E(\cdot)$ denotes fitness function.

4.2   A New Measure for Vehicle Matching

The measure function (i.e., fitness function) is used to evaluate 3D-2D vehicle matching. The existing methods exploit the image information such as edge, silhouette, pixel intensity and pixel gradient. In order to improve the accuracy and robustness, we study the vehicle matching problem from both local and global perspectives, and exploit image gradient information as well as image silhouette information (see Fig. 3). Here we choose the vehicle wireframe model that is widely used due to its simplicity. Generally, the wireframe model is composed of several 3D line segments that describe the vehicle outline and some high contrast borders (see Fig. 1(c)).

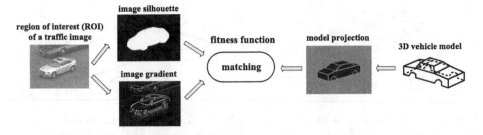

Fig. 3. Illustration of 3D-2D vehicle matching, where fitness function is used to evaluate vehicle matching degree that exploits both image gradient information and silhouette information.

From the local perspective of vehicle matching, a good matching means that the projected lines of the wireframe model coincide with the corresponding image edges. At this time, the image pixels lying on the projected lines have the maximum gradient value in the direction normal to the projected lines. Accordingly, we exploit the pixel gradient information around model projection. For every projected line, we introduce a rectangular neighborhood (see Fig. 4) and give a matching evaluation within it by

$$e\left(l_i\right) = \frac{\sum\limits_{s_j \in S_{rect}} w_{s_j} \cdot G_{\perp l_i}(s_j)}{L_i}, \tag{6}$$

where l_i denotes the i-th projected line, S_{rect} denotes the rectangular neighborhood of l_i, s_j denotes the j-th image pixel within S_{rect}, $G_{\perp l_i}(s_j)$ denotes the gradient magnitude of s_j in the normal direction of l_i, w_{s_j} denotes the weight of s_j and L_i denotes the length of l_i in pixels. Evidently, the image pixels that are closer to the projected line make greater contribution. Consequently, w_{s_j} obeys the following Gaussian distribution

$$w_{s_j} = \frac{1}{\omega\sqrt{2\pi}} exp\left(-\frac{d^2}{2\omega^2}\right), \tag{7}$$

where d is the distance from s_j to l_i in pixels and ω is the half width of S_{rect}.

We observe that there is the discrepancy between real vehicle and the wireframe model, since the streamlined design of real vehicle is to substitute the smooth surfaces for the wireframes. In view of this discrepancy, a natural idea is to group the model's wireframes (see Fig. 5) and treat them differently. Based on this idea, we utilize a weighting strategy for the fitting of the model's wireframes, namely

$$m(l_i) = \begin{cases} \alpha \cdot e\left(l_i\right) & l_i \text{ is primary wireframe} \\ (1-\alpha) \cdot e\left(l_i\right) & l_i \text{ is secondary wireframe} \end{cases}, \tag{8}$$

where α is a weight coefficient. Since the primary wireframes delineate the top, middle and bottom of the vehicle, they generally well fit the corresponding image edges. We set α as a constant that is close to 1, which means that the fitting

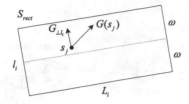

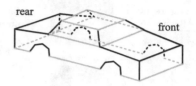

Fig. 4. Rectangular neighborhood S_{rect} for the projected line l_i.

Fig. 5. Grouping of the model's wireframes, where the primary wireframes are marked in red color and the rest is the secondary wireframes. (Color figure online)

of the primary wireframes will be emphasized. Taking all visible projected lines into account, we give a gradient-based measure of vehicle matching by

$$E_g = \frac{1}{n} \sum_{i=1}^{n} m(l_i), \tag{9}$$

where n is the number of the visible projected lines.

Except for local perspective, we study the vehicle matching problem from global perspective. Intuitively, if the projected contour of the wireframe model exactly covers the image silhouette of target vehicle, a good matching is obtained. Accordingly, we utilize the overlap rate to evaluate the similarity between the projected contour and image silhouette. The silhouette-based measure for vehicle matching is given by

$$E_s = \frac{A_{model} \cap A_{image}}{A_{model} \cup A_{image}}, \tag{10}$$

where A_{model} denotes the area of model projection and A_{image} denotes the area of image silhouette. It can be seen from Eq. (10) that the range of E_s is from 0 to 1.

In order to improve the accuracy and robustness, the proposed fitness function E is to combine the gradient-based measure E_g (i.e., local measure) and the silhouette-based measure E_s (i.e., global measure), which can be expressed as

$$E(E_g, E_s) = E_g \cdot E_s. \tag{11}$$

The advantage of incorporating global measure into local measure is to compensate for the weakness of local measure and improve the accuracy of vehicle matching.

Once the fitness function E is available, Eq. (5) can be rewritten as

$$\hat{p} = \max_{p} E(E_g, E_s). \tag{12}$$

It is well known that the selection of optimization algorithm is strongly dependent on the property of the objective function. Considering that the proposed fitness function is not smooth and the camera parameters to be optimized are seven, we adopt the evolutionary optimization algorithms (e.g., the estimation of distribution algorithms (EDAs)). Since the optimization result is strongly affected by the initial values of the parameters, we next discuss how to obtain the initial values of camera parameters.

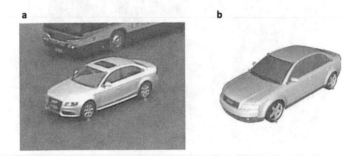

Fig. 6. Anchor points used for the calculation of initial values. (a) Anchor points on the image, marked in red color. (b) The corresponding points on the vehicle model, marked in blue color. (Color figure online)

In the model-based face recognition methods, anchor points are commonly used that provide several known 3D-2D point correspondences. Similarly, we utilize anchor points to calculate the initial values of camera parameters. First, we choose several points on the vehicle as anchor points, such as the window corners and the contact point between the wheel and the road plane, as shown in Fig. 6. Then, we compute the initial values through minimizing the distance between the anchor points and the corresponding projection points, namely

$$p_0 = \arg\min_p \sum_k \|q_k - M(Q_k, p)\|^2, \tag{13}$$

where q_k denotes the k-th anchor point, Q_k denotes the corresponding point on the vehicle model and $M(Q_k, p)$ denotes the projection point of Q_k under camera parameters p. For the nonlinear optimization problem in Eq. (13), the Levenberg-Marquardt algorithm is adopted.

4.3 Dynamic Calibration

The camera calibration process using a single image is described above. By exploiting the vehicle information at different times (i.e., different image frames), the multiple estimates of camera parameters can be obtained. Notice that these estimates are helpful for camera calibration and supposed to be used to correct camera parameters. Accordingly, a dynamic calibration process can be given by

$$\bar{p} = \frac{\sum_m w_m \cdot p^m}{\sum_m w_m}, \tag{14}$$

Fig. 7. Test video of real traffic scenario, where a white SUV is chosen as a calibration pattern.

where p^m is the m-th estimate of camera parameters and w_m is a weight assigned for p^m. The strength of the dynamic calibration is that it would help improve the accuracy and reliability of calibration results.

5   Experiments

To verify the performance of the proposed calibration framework, we conduct experiments on real traffic images. We choose a representative traffic video as test video, as shown in Fig. 7. It can be seen from this figure that there are no traditional calibration patterns in such scenario and the information of vanishing points obtained from such scenario is insufficient for camera calibration. Accordingly, the traditional pattern-based calibration methods and the vanishing-point-based calibration methods are not suitable. However, our calibration framework using the vehicle information is suitable for such scenario. In the test video, a white SUV is chosen as a calibration pattern (see Fig. 7). We develop a software platform using OpenCV library and OpenSceneGraph (OSG) library. By means of this platform, we can simulate roadside camera using the estimated calibration results and further obtain the ground truths of camera parameters.

Here we adopt the assumptions that the aspect ratio is one, the skew factor is zero and the principal point is the image's center. These assumptions are reasonable and are used in most roadside camera calibration methods [2]. Accordingly, the camera parameters to be estimated are reduced to focal length, three rotation angles and the spatial location on the ground, namely $p = (f, \varphi, \gamma, \theta, X_{cam}, Y_{cam}, h)$. As for the initial values of camera parameters, we choose at least seven anchor points to calculate it using Eq. (13). For more reliable calibration results, we exploit the multi-frame images of the vehicle (more

Fig. 8. Distance measurements. (a) Length measurement, where the lane marking to be measured is marked in red color. (b) Wheelbase measurement, where the vehicle's wheelbases to be measured are from five frame images. (Color figure online)

than 20 frames) to correct camera parameters using Eq. (14), where the weight w_m is set as 1. Table 1 gives the calibration results using our calibration framework. In this table, the first row lists the camera parameters to be estimated; the second row gives the initial values obtained from the 2329 frame of test video; the third row gives the estimates obtained from the multi-frame images; the fourth row gives the ground truths obtained via our software platform; the last row gives the relative error between the estimates and the ground truths. As can be seen from this table, the relative error is less than 8%. This fact reveals that the calibration results obtained by our calibration method are accurate and reliable.

Table 1. Calibration results using our calibration framework, where the focal length, rotation angle and spatial location are measured in pixel, degree and millimeter, respectively.

Parameters	f	φ	γ	θ	X_{cam}	Y_{cam}	h
Initial value	1215	102	−2	−72	29880	−8330	7230
Estimate	1344	101	0	−75	31053	−7973	6440
Ground truth	1400	100	0	−76	33000	−8200	6000
Relative error	4.0%	1.0%	-	1.3%	5.9%	2.8%	7.3%

Next, we evaluate the accuracy of calibration results quantitatively. To do this, we recover the length of lane marking from test image (see Fig. 8(a)) and the vehicle's wheelbase from five frame images (see Fig. 8(b)), and then compare the recovered values with real world measurements. Tables 2 and 3 give the recovered values of lane marking and wheelbase using our calibration results, respectively. As can be seen from these two tables, the relative errors between the recovered values and ground truths are all less than 10%. This fact also demonstrates the accuracy and reliability of our calibration results.

Table 2. Comparison of the recovered value and ground truth for length measurement in Fig. 8(a), where the length of lane marking is measured in millimeter.

Length	Recovered value	Ground truth	Relative error
l_1	4398	4500	2.3%
l_2	4627	4500	2.8%
l_3	4772	4500	6.0%
l_4	4741	4500	5.4%
l_5	4916	4500	9.2%

Table 3. Comparison of the recovered value and ground truth for wheelbase measurement in Fig. 8(b), where the vehicle's wheelbase is measured in millimeter.

Frame	Recovered value	Ground truth	Relative error
2329	3178	3000	5.9%
2332	3135	3000	4.5%
2335	3020	3000	0.7%
2338	3078	3000	2.6%
2341	3080	3000	2.7%

6   Discussion

Compared with traditional checkerboard-based calibration methods, the proposed calibration method based on vehicle information may be less accurate. Unfortunately, the checkerboard-based calibration methods are only suitable for the laboratory environment (i.e., man-made scenarios) rather than real traffic scenarios. Although the calibration results using our method are relatively coarser than the ones using the checkerboard-based methods, they are reliable and acceptable for some applications in traffic scenarios, such as distance measurements. The experimental results mentioned above also confirm this point.

7   Conclusion

This paper presents a novel calibration framework for roadside camera. Considering that the vehicles inevitably appear in traffic scenarios, we make full use of the vehicle information for roadside camera calibration. The key idea is to take the vehicle moving on the roadway as a calibration pattern and use 3D-2D vehicle matching to calibrate roadside camera. As for the vehicle matching, we provide a new measure function that combines the gradient-based local measure with the silhouette-based global measure. To further improve the accuracy of calibration results, we provide a dynamic calibration method to correct camera parameters that exploits the multi-frame information of the vehicle. Experiments on real traffic images confirm the effectiveness and practicability of the proposed calibration framework.

Acknowledgement. This study was supported by National Natural Science Foundation of China (No. 61502119) and China Postdoctoral Science Foundation (No. 2015M571414).

References

1. Song, K., Tai, J.: Dynamic calibration of pan-tilt-zoom cameras for traffic monitoring. IEEE Trans. Syst. Man Cybern. Part B: Cybern. **36**, 1091–1103 (2006)
2. Kanhere, N., Birchfield, S.: A taxonomy and analysis of camera calibration methods for traffic monitoring applications. IEEE Trans. Intell. Transp. Syst. **11**, 441–452 (2010)
3. Álvarez, S., Llorca, D., Sotelo, M.: Hierarchical camera auto-calibration for traffic surveillance systems. Expert Syst. Appl. **41**, 1532–1542 (2014)
4. Lv, F., Zhao, T., Nevatia, R.: Camera calibration from video of a walking human. IEEE Trans. Pattern Anal. Mach. Intell. **28**, 1513–1518 (2006)
5. Lee, S., Nevatia, R.: Robust camera calibration tool for video surveillance camera in urban environment. In: 2011 IEEE Computer Society Conference on Computer Vision and Pattern Recognition Workshops (CVPRW), pp. 62–67. IEEE (2011)
6. Hodlmoser, M., Micusik, B., Kampel, M.: Camera auto-calibration using pedestrians and zebra-crossings. In: 2011 IEEE International Conference on Computer Vision Workshops (ICCV Workshops), pp. 1697–1704. IEEE (2011)
7. Zheng, Y., Peng, S.: A practical roadside camera calibration method based on least squares optimization. IEEE Trans. Intell. Transp. Syst. **15**, 831–843 (2014)
8. Zhang, Z., Li, M., Huang, K., Tan, T.: Practical camera auto-calibration based on object appearance and motion for traffic scene visual surveillance. In: 2008 IEEE Conference on Computer Vision and Pattern Recognition, CVPR 2008, pp. 1–8. IEEE (2008)
9. Tsai, R.: A versatile camera calibration technique for high-accuracy 3D machine vision metrology using off-the-shelf TV cameras and lenses. IEEE J. Robot. Autom. **3**, 323–344 (1987)
10. Weng, J., Cohen, P., Herniou, M.: Camera calibration with distortion models and accuracy evaluation. IEEE Trans. Pattern Anal. Mach. Intell. **14**, 965–980 (1992)
11. Sturm, P.F., Maybank, S.J.: On plane-based camera calibration: a general algorithm, singularities, applications. In: IEEE Computer Society Conference on Computer Vision and Pattern Recognition, vol. 1. IEEE (1999)
12. Zhang, Z.: A flexible new technique for camera calibration. IEEE Trans. Pattern Anal. Mach. Intell. **22**, 1330–1334 (2000)
13. Ueshiba, T., Tomita, F.: Plane-based calibration algorithm for multi-camera systems via factorization of homography matrices. In: Proceedings of the Ninth IEEE International Conference on Computer Vision, pp. 966–973. IEEE (2003)
14. Penate-Sanchez, A., Andrade-Cetto, J., Moreno-Noguer, F.: Exhaustive linearization for robust camera pose and focal length estimation. IEEE Trans. Pattern Anal. Mach. Intell. **35**, 2387–2400 (2013)
15. Zhang, Z.: Camera calibration with one-dimensional objects. IEEE Trans. Pattern Anal. Mach. Intell. **26**, 892–899 (2004)
16. Wu, F., Hu, Z., Zhu, H.: Camera calibration with moving one-dimensional objects. Pattern Recogn. **38**, 755–765 (2005)
17. de França, J.A., Stemmer, M.R., de M França, M.B., Alves, E.G.: Revisiting Zhang's 1D calibration algorithm. Pattern Recogn. **43**, 1180–1187 (2010)

18. Chen, Q., Wu, H., Wada, T.: Camera calibration with two arbitrary coplanar circles. In: Pajdla, T., Matas, J. (eds.) ECCV 2004. LNCS, vol. 3023, pp. 521–532. Springer, Heidelberg (2004). doi:10.1007/978-3-540-24672-5_41

19. Abad, F., Camahort, E., Vivó, R.: Camera calibration using two concentric circles. In: Campilho, A., Kamel, M. (eds.) ICIAR 2004. LNCS, vol. 3211, pp. 688–696. Springer, Heidelberg (2004). doi:10.1007/978-3-540-30125-7_85

20. Colombo, C., Comanducci, D., Bimbo, A.: Camera calibration with two arbitrary coaxial circles. In: Leonardis, A., Bischof, H., Pinz, A. (eds.) ECCV 2006. LNCS, vol. 3951, pp. 265–276. Springer, Heidelberg (2006). doi:10.1007/11744023_21

21. Chen, X., Zhao, Y.: A linear approach for determining camera intrinsic parameters using tangent circles. Multimed. Tools Appl. **74**, 5709–5723 (2015)

22. Agrawal, M., Davis, L.S.: Camera calibration using spheres: a semi-definite programming approach. In: Proceedings of the Ninth IEEE International Conference on Computer Vision, pp. 782–789. IEEE (2003)

23. Zhang, H., Wong, K.Y., Zhang, G.: Camera calibration from images of spheres. IEEE Trans. Pattern Anal. Mach. Intell. **29**, 499–502 (2007)

24. Wong, K.Y., Zhang, G., Chen, Z.: A stratified approach for camera calibration using spheres. IEEE Trans. Image Process. **20**, 305–316 (2011)

25. Staranowicz, A.N., Brown, G.R., Morbidi, F., Mariottini, G.L.: Practical and accurate calibration of RGB-D cameras using spheres. Comput. Vis. Image Underst. **137**, 102–114 (2015)

26. Micusik, B., Pajdla, T.: Simultaneous surveillance camera calibration and foothead homology estimation from human detections. In: 2010 IEEE Conference on Computer Vision and Pattern Recognition (CVPR), pp. 1562–1569. IEEE (2010)

27. Dawson, D.N., Birchfield, S.T.: An energy minimization approach to automatic traffic camera calibration. IEEE Trans. Intell. Transp. Syst. **14**, 1095–1108 (2013)

4th ACCV Workshop on e-Heritage

Digital Longmen Project: A Free Walking VR System with Image-Based Restoration

Zeyu Wang[1]($\boxtimes$), Xiaohan Jin[1], Dian Shao[1],
Renju Li[1], Hongbin Zha[1], and Katsushi Ikeuchi[2]

[1] Key Laboratory of Machine Perception, Peking University, Beijing, China
1200012927@pku.edu.cn
[2] Microsoft Research Asia, Beijing, China

Abstract. Located in China's ancient capital Luoyang, Longmen Grottoes are one of the finest examples of Buddhist stone carving art. Nowadays, many caves do not have public access due to heritage preservation. In order to let people appreciate these relics, we setup a VR system with smartphones and helmets based on scanned models and textures. Motion capture system is also utilized to make the viewpoint not fixed so that users can walk freely as if in the cave. Moreover, since some sculptures have been heavily damaged, we propose a digital restoration framework to enhance exhibition contents. The framework includes general and detailed restoration from a single old image by shape from shading and landmark driven mesh deformation respectively. In practice, we develop this system for the representative Middle Binyang Cave, with interactions such as gesture recognition exploited to provide satisfactory user experience, which can ease the conflict between tourism and preservation.

1 Introduction

Cultural heritage represents the splendid civilization our ancestors created, including outstanding and irreproducible historical relics, architecture, sites, and intangible cultural heritage. As masterpieces of both nature and human beings, they are invested with high values from historical, cultural, and scientific aspects. China is a country with proud history of five thousand years, and has 50 world cultural and/or natural sites approved by UNESCO since the 40th World Heritage Committee session. The Forbidden City, Mogao Caves, Longmen Grottoes and many others in the list demonstrate people's creativity, wisdom, artistic accomplishments, as well as ideologies, aesthetic customs, and religious beliefs, thus are valuable treasures of all mankind. Take Longmen Grottoes as an example, almost 110,000 Buddha statues were built from the late 5th century with exquisite stone carving crafts. Among them Middle Binyang Cave is one of the most well engraved, decorated, and preserved constructions. As a royal project Emperor Xuanwu of the Northern Wei Dynasty initiated to bring honor to his parents, this cave was built by more than 800 thousand workers in 24 years. The cave has a theme of Trikalea Buddhas (Buddhas of the Past, Present, and Future), and is a typical representative of ancient Chinese style with spectacular statues and colorfully textured relief sculpture.

© Springer International Publishing AG 2017
C.-S. Chen et al. (Eds.): ACCV 2016 Workshops, Part II, LNCS 10117, pp. 191–206, 2017.
DOI: 10.1007/978-3-319-54427-4_15

However, Longmen Grottoes have experienced quite a few harsh times. The power of nature such as sunshine, damp, earthquakes, and weathering has made many parts deteriorated beyond recognition, not to mention unbridled sabotage, thefts, and smuggling from the beginning of the last century. Japanese scholar Tadashi Sekino wrote, "Most of the Buddha heads, as long as they can be dismantled, are cut off and sold to foreigners." Middle Binyang Cave alone lost four Bodhisattva heads and two precious reliefs named *Processions of Emperor and Empress in Worship*, which are respectively collected in Tokyo National Museum, Osaka City Museum of Fine Arts, Metropolitan Museum of Art, and Nelson-Atkins Museum of Art. Since Longmen Grottoes are famous for colossal stone sculptures, physical restoration requires strenuous labor and is even likely to damage remaining relics. Recently, preservation and restoration utilizing digital technology were conducted with the cooperation from both academia and industry. Many caves were digitalized in 3D mesh format with texture using Leica laser scanners. Once we have 3D models, automatic and manual methods can be developed to merge the head elsewhere with the remaining body. It has been proved effective in Guyang Cave. Compared with repair and assembly in the real world, there is no doubt that the operations on 3D models can avoid potential damage and save human labor.

Considering many relics in Longmen Grottoes have been already damaged more or less, the administrative committee suggests that the flow of tourists should be controlled to prevent heritage from deterioration caused by human activities. Some caves including Middle Binyang Cave are officially closed, so that textures can be better protected from direct sunlight, although meanwhile general visitors are not allowed to enter. There are several attempts on solving the conflict between tourism and preservation. In the past few years, people were able to take a virtual tour based on panoramas shot from fixed viewpoints instead of onsite visits, but they suffered from the sense of restriction. For Mogao Caves in Dunhuang, a large spherical theater was built in 2014 at an expense of nearly 45 million dollars, and it also cost a lot on filmmaking. To overcome these limitations, we propose a free walking VR system for heritage exhibition based on scanned models and textures. Inspiringly, cheap smartphone displays become popular and competent at providing a user-centric, active, immersive, interactive approach to restored heritage exhibition.

In order to provide more complete contents for VR exhibition, digital restoration is necessary for damaged relics, especially stolen Buddha heads. By reconstructing these missing parts, it becomes possible to digitally show the merged model together. In fact, one of the biggest challenges for restoration is lack of archaeological evidence, but luckily, we find some valuable old photos taken over 70 years ago. Images can be used as a powerful and convincing reference for restoration as they record the shape and unique characteristics of the original statues. We exploit shape from shading and landmark driven mesh deformation for image-based restoration. Compared to merging parts from different statues, we note the variance between statues of the same figure. The case shows that 3D reconstruction from a single image can be applied to heritage restoration.

Fig. 1. A glance at Middle Binyang Cave, Longmen Grottoes. Some heads remain lost.

In this paper, we make the following contributions:

1. Since Middle Binyang Cave does not have access to public tourists, we setup a free walking VR environment on popular smartphones. Realtime user localization is solved using both inertial sensors and motion capture systems.
2. For exhibition contents, we propose a restoration framework based on old images, including shape from shading and Buddha face reconstruction method by Poisson-based deformation driven by shape priors and facial features.
3. We release a Unity3D-powered app on Android with user interaction for restoration and enhancement, which is going to be promoted in Longmen (Fig. 1).

2   Related Work

The field of digital cultural heritage adopts many techniques in computer vision and graphics. In particular, 3D reconstruction has been playing an important role in digital museum, relic archiving, and other archaeological research. The EU funded 3D MURALE project established a multimedia database for displaying historical remains in Sagalassos, Turkey [1]. Current 3D scanning methods mainly include laser scanners, structured light scanners, and multi-ocular cameras. For large scenes such as the Bayon Temple, ballon scanners were proposed to overcome the inaccessibility from the top, and the 3D model of high quality was fused under the constraints of data distortion, image motion, and ballon motion [2]. Material, texture, and reflection model are also frequently discussed. For example, texture pattern in Kyushu ancient tombs was extracted and enhanced using color modeling, which helped prove the archaeological guesses [3].

Based on the scanned models, some research focused more on restoration algorithms. For instance, the Forma Urbis Romae project at Stanford University recovered the urban map of ancient Rome by matching and stitching 1,186 textured marble pieces with arbitrary shape [4]. Similarly, a team from Princeton University also digitalized and restored frescoes in the ruins of Akrotiri, Santorini, with a multi-way Iterative Closest Point (ICP) framework for organizing and stitching [5]. Curve-based interaction methods in a 3D pottery retrieval and classification system are useful for analysis and restoration as well [6]. Software tools for medieval manuscript recognition, sculpture shape analysis, and aging material modeling were developed by the graphics group at Yale University [7–9]. In this paper, we focus on tourist-oriented exhibition and Buddha head restoration.

VR/AR Exhibition. Virtual reality exhibition, including video mapping, telepresence, and Head Mounted Display (HMD), has brought tremendous changes in the way how people interact with the environment. As digital technology is entering the era of low cost and high efficiency, it becomes possible to adopt VR display in archaeology, art, and entertainment, such as movies, games, and theme parks [10]. Augmented reality systems have also been developed for various scenic areas technically considering geometric consistency and photometric consistency [11,12].

Face Reconstruction. 3D face reconstruction from a single 2D image is a difficult problem with many previous research work, such as shape from shading, shape from texture, shape from silhouettes, shape from focus and using shape priors, though the results are far from the goal of high quality and realistic reproduction. For face reconstruction, the main ideas include shape from shading [13,14] and 3D morphable models [15]. Shape from shading requires information about the reflectance properties and lighting. Using morphable models circumvents these requirements by representing input faces as combinations of hundreds of stored 3D models. But in Longmen dataset, the number of available 3D faces is very limited, so this algorithm does not apply. Our method performs Poisson-based mesh deformation following precise face landmark localization, which is novel and robust for our dataset, using only one 3D face template.

3   VR Environment Setup

Since Middle Binyang Cave is representative for stone carving art and does not have public access, we setup a virtual reality environment using scanned 3D models and textures. The mesh data is stored in 19 `.wrl` files, with texture data in 42 `.rgb` files for each connected region. Considering this VR system will ultimately be released on Samsung Galaxy Note 4 smartphone, the raw mesh data is downsampled using Geomagic Studio 11 to stay in accordance with limited computing and storage capabilities. The final size of mesh data imported into Unity3D is 31.6 MB, with 212,829 vertices and 409,917 facets. The texture mapping data is 250.6 MB and meets the workload requirements after testing.

3.1   Localization by IMU

Current integrated VR devices, such as Oculus Rift, usually include modules of optics, display, sensing, and computing, which is costly for common consumers.

Therefore, we take advantage of smartphones that everyone has, and 20-dollar Baofeng helmets offer necessary lenses. In the aforementioned 3D environment, the viewpoint is initialized in the center of the cave, and what we see in the first place is Sakyamuni. Generally speaking, we need to estimate the user's motion and pose in VR systems by various sensors, thus the new viewpoint can be updated and stay consistent with real life customs. The Inertial Measurement Unit (IMU), including accelerators and gyros, is crucial for pose estimation. Since the displacement is integrated twice from the acceleration, while the angle is integrated only once from the angular velocity, gyro data is utilized in our system to avoid considerable error accumulation.

As for user pose estimation, denote the raw, pitch, and yaw angles at time t_1 by $\alpha_r(t_1)$, $\alpha_p(t_1)$, and $\alpha_y(t_1)$ respectively. Then denote the raw, pitch, and yaw angular velocities at time t by $\omega_r(t)$, $\omega_p(t)$, and $\omega_y(t)$ respectively. Thus the new raw, pitch, and yaw angles at time t_2, i.e., $\alpha_r(t_2)$, $\alpha_p(t_2)$, $\alpha_y(t_2)$ can be calculated by:

$$\alpha_{r,p,y}(t_2) = \alpha_{r,p,y}(t_1) + \int_{t_1}^{t_2} \omega_{r,p,y}(t)\,dt \tag{1}$$

We implement these computation by C# script codes in Unity3D with the help of ALPSVR package, and the rotation of the viewpoint is measured by quaternions. As a form of extended complex numbers, quaternions make interpolation smoother, and avoid the problem of gimbal lock. Multiple rotations can be represented by quaternion multiplication. The viewpoint parameters in Unity3D will be updated after this computation. The Euler angles of roll α_r, pitch α_p, and yaw α_y can be converted into quaternions as follows:

$$q = \begin{bmatrix} q_0 \\ q_1 \\ q_2 \\ q_3 \end{bmatrix} = \begin{bmatrix} \cos\left(\frac{\alpha_y}{2}\right) \\ 0 \\ 0 \\ \sin\left(\frac{\alpha_y}{2}\right) \end{bmatrix} \begin{bmatrix} \cos\left(\frac{\alpha_p}{2}\right) \\ 0 \\ \sin\left(\frac{\alpha_p}{2}\right) \\ 0 \end{bmatrix} \begin{bmatrix} \cos\left(\frac{\alpha_r}{2}\right) \\ \sin\left(\frac{\alpha_r}{2}\right) \\ 0 \\ 0 \end{bmatrix} \tag{2}$$

3.2   Localization Design of Free Walking

IMU-based head tracking can achieve good results, because the new viewing orientation is consistently updated by angular velocity integral. However, there are drawbacks such as error accumulation and viewpoint immobilization. Tourists cannot appreciate the statues from different perspectives despite horizontal displacement. Therefore, we exploit TenYoun motion capture system with reflective markers on the helmet to realize virtual tours of high precision. This passive system consists of 12 near infrared cameras of high sensitivity capturing the reflective markers stuck on the moving object in real time. Motion capture of high precision is achieved by reconstructing trajectories of marker and object motion. In our implementation, four fixed markers (two on the X axis, one on the Y and Z axes) are stuck on the Baofeng helmet as a rigid body, thus 3D rotational and translational transformation matrices can be computed and applied to the coordinates of the viewpoint.

Triangulation in Multi-view Vision. The principle of motion capture system tracking marker balls is triangulation. Depth information is lost when capturing 2D images of a 3D scene, but 3D information can be reconstructed when multiple cameras capture images of the same scene. In fact, a point in 3D space is identical to the intersection of extension lines from optical centers to imaging points. In this multi-view vision system, denote the calibrated parameter matrices of 12 cameras by $M_1, M_2, \cdots, M_{12}$ and denote the imaging point coordinates of point (X_w, Y_w, Z_w) by $(u_1, v_1), (u_2, v_2), \cdots, (u_{12}, v_{12})$, then we have:

$$\begin{bmatrix} u_i \\ v_i \\ 1 \end{bmatrix} = M_i \begin{bmatrix} X_w \\ Y_w \\ Z_w \\ 1 \end{bmatrix}, \quad i = 1, 2, \cdots, 12 \tag{3}$$

In the motion capture system, reflective markers have strong responses in images captured by near infrared cameras of high sensitivity, so their pixel locations can be easily determined by intensity thresholding. Since the number and relative position of markers are fixed, multi-view matching can be achieved by enumeration. Theoretically, only two calibrated cameras are enough for 3D reconstruction, but more cameras eliminate occlusion and bring higher precision. Therefore, a realtime human motion capture system usually consists of 12 or 24 cameras.

3D Transformation Estimation. There are four reflective markers stuck on the Baofeng helmet for VR display, because the user's 3D transformation matrix can be estimated by the trajectories of these markers. Denote the current world coordinates of these four markers by (X_1, Y_1, Z_1), (X_2, Y_2, Z_2), (X_3, Y_3, Z_3), and (X_4, Y_4, Z_4). Denote the world coordinates in the next frame by (X'_1, Y'_1, Z'_1), (X'_2, Y'_2, Z'_2), (X'_3, Y'_3, Z'_3), and (X'_4, Y'_4, Z'_4). Then the 3D transformation between two frames can be represented by a 3×3 orthogonal rotation matrix R and a vector of 3D translation t:

$$\begin{bmatrix} X_j \\ Y_j \\ Z_j \\ 1 \end{bmatrix} = \begin{bmatrix} R & t \\ 0^T & 1 \end{bmatrix} \begin{bmatrix} X'_j \\ Y'_j \\ Z'_j \\ 1 \end{bmatrix}, \quad j = 1, 2, 3, 4 \tag{4}$$

Since the relative position of these four markers is fixed, and rigid transformation only has the degree of freedom of 6, two equations are theoretically sufficient for calculating R and t. Similarly, solving the over-determined equation set by least square method can improve the precision. Once we know the rigid transformation in real 3D space, we can apply the same transformation matrix to viewpoint in VR environment. Therefore, the user can appreciate the cave and experience consistent virtual tour as long as he or she is walking freely in the room equipped with our motion capture system.

Camera Calibration. From the above principle, we first needs to calibrate camera to get geometric parameters of the imaging model, which is generally the pinhole model to approximate the real nonlinear lens imaging model. We create four Descartes coordinate systems to describe the geometric relations in the imaging process: image pixel coordinate (u, v), image physical coordinate (x, y), camera coordinate (X_c, Y_c, Z_c), and world coordinate (X_w, Y_w, Z_w).

In the pinhole model, draw a line from an arbitrary point P in the 3D space to the camera's optical center, and their intersection p on the imaging plane is the imaging point. Denote the focal length by f. Let $\mathbf{R}_0$ be a 3×3 orthogonal rotation matrix, $\mathbf{t}_0$ be a vector of 3D translation. Denote the proportion of physical unit to column and row pixel unit by Δx and Δy respectively. If the image pixel coordinate of camera's optical center is (u_0, v_0), the image pixel coordinate of p, (u, v), can be calculated from the world coordinate of P, (X_w, Y_w, Z_w), as follows:

$$Z_c \begin{bmatrix} x \\ y \\ 1 \end{bmatrix} = \begin{bmatrix} \frac{1}{\Delta x} & 0 & u_0 \\ 0 & \frac{1}{\Delta y} & v_0 \\ 0 & 0 & 1 \end{bmatrix} \begin{bmatrix} f & 0 & 0 & 0 \\ 0 & f & 0 & 0 \\ 0 & 0 & 1 & 0 \end{bmatrix} \begin{bmatrix} \mathbf{R}_0 & \mathbf{t}_0 \\ \mathbf{0}^{\mathrm{T}} & 1 \end{bmatrix} \begin{bmatrix} X_w \\ Y_w \\ Z_w \\ 1 \end{bmatrix} \tag{5}$$

Put internal and external parameters together, and let $\mathbf{M}$ be an arbitrary 3×4 matrix. The objective of camera calibration is to determine this matrix, by capturing multiple images of unique calibration boards. It can be computed by least square method from the equation set of 3D-to-2D transformation. According to Eq. 5, $\mathbf{M}$ multiplied by a non-zero constant still represents the relation between (X_w, Y_w, Z_w) and (u, v), thus we might set an element to 1 as well. The degree of freedom is 11 in total, and we need at least 6 non-coplanar calibration points. All 12 calibrated cameras in the motion capture system should not be adjusted before running the VR display.

3.3   Display and Communications

Since the screen is very close to our eyes in VR display, there are convex lenses in the helmet between the screen and our eyes, but they also introduce image distortion, where the image magnification decreases when the distance to the optical axis increases. Barrel distortion correction based on Brown-Conrady model is usually used to counteract this effect [16].

$$(x_d, y_d) = (x_u, y_u) \left(1 + K_1 r^2 + K_2 r^4 + \cdots \right) \tag{6}$$

where (x_d, y_d) is the coordinate in distorted images, (x_u, y_u) is the coordinate in undistorted images, (x_c, y_c) is the coordinate of the distortion center, K_n is the nth parameter of radial distortion, and the distance from the pixel to the distortion center in undistorted images is $r = \sqrt{(x_u - x_c)^2 + (y_u - y_c)^2}$. The VR image flow delivered to the user from all viewpoints should be corrected against distortion in advance for people's perceptual comfort.

Fig. 2. The free walking VR environment using the motion capture system. The smartphone display is put in a helmet, on which four marker balls are stuck.

It remains an issue to transmit data from the motion capture system to the smartphone app in real time. There is a server for the motion capture system which returns current world coordinates of four marker balls. For communications between the server and the smartphone, they are connected to an identical WiFi under TCP/IP protocol, and then data can be transmitted in the form of Socket. We initialize marker positions in the VR environment and write a C# script listener to receive motion capture stream data from a certain IP address and port, thus estimated 3D transformation can be applied to the current viewpoint and new images can be rendered promptly. In our experiments, user's free walking is enabled in a 5 m × 5 m area, with an adequate frame rate of more than 30 fps for rendered display (Fig. 2).

4   Heritage Restoration

Currently, damaged Buddha models are available through 3D scanning, and old pictures of complete Buddha can be collected from photographers. In order to restore digital models for exhibition contents based on images, we first try an improved version of shape from shading, in which we extract reflectance map by minimizing the local variation of log-reflectance and minimizing the global entropy of log-reflectance. Then we introduced our method which is inspired by mesh deformation algorithms [17,18] in computer graphics field. We evaluate our method on existing statues. The proposed framework is shown in Fig. 3.

4.1   Shape, Illumination, and Reflectance from Shading

Shape from shading problem has been well surveyed since 1970s [19–21]. It deals with the recovery of shape from a gradual variation of shading in the image. The concept of "intrinsic image" comes up to complement the preprocessing of a natural image, and to decompose an image (I) into its shading (S) and reflectance (R) components. The algorithm of Shape, Illumination, and Reflectance from

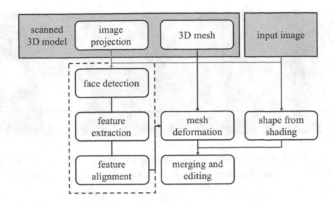

Fig. 3. The proposed restoration framework. Red arrows indicate the first method. Blue ones take two as input to the image processing module (dashed box), and then supervise mesh deformation. We can merge and edit these results in the end. (Color figure online)

Shading (SIRFS) [14], a generalization of an intrinsic image algorithm, infers the most-likely explanation of a single image, including the shape, surface normals, reflectance, shading, and illumination which produced that image. Shading (S) is explicitly parametrized as a function of shape (Z) and illumination (L).

$$I = R + S(Z, L) \tag{7}$$

The following priors are the core of SIRFS: reflectance images tend to be piecewise smooth and low-entropy, surfaces tend to be isotropic and bend infrequency, and illumination tends to be natural. So the goal is to

$$\min_{Z,L} g(I - S(Z, L)) + f(Z) + h(L) \tag{8}$$

The estimation of reflectance map is based on the assumption of smooth shading variation and low-entropy pattern influences. The smoothness level $g_s(R)$ of the input image patch becomes higher when the shape variation is small and that is when the pattern reflects shading information. Furthermore, as a soft constraint, the color distribution of a reflectance map is limited. We consider this factor in two parts, $g_e(R)$: parsimony of reflectance by minimizing the global entropy of log-reflectance; $g_a(R)$: color similarities and preferences.

$$g(R) = \lambda_s g_s(R) + \lambda_e g_e(R) + \lambda_a g_a(R) \tag{9}$$

The prior on shape consists of three parts. $f_k(Z)$: shapes tend to be smooth so that the variation of mean curvature tends to be small; $f_i(Z)$: shapes are equally likely to face in different directions; $f_c(Z)$: shapes tend to face outward at the occluding contour.

$$f(Z) = \lambda_k f_k(Z) + \lambda_i f_i(Z) + \lambda_c f_c(Z) \tag{10}$$

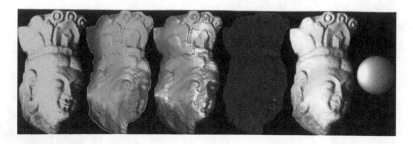

Fig. 4. The outputs of SIRFS method, including the shape (height map), orientation of normals, reflection map, shading image and orientation of illumination.

For illumination, we assume it is spherical-harmonic illumination, and train a multivariate Gaussian model in MIT intrinsic images dataset. $\boldsymbol{\mu}_L$ and $\boldsymbol{\Sigma}_L^{-1}$ are the parameters of the Gaussian we learned.

$$h(L) = \lambda_L (L - \boldsymbol{\mu}_L)^T \boldsymbol{\Sigma}_L^{-1} (L - \boldsymbol{\mu}_L) \tag{11}$$

Because the Gaussian parameters are trained on MIT intrinsic images but the texture of real-world Longmen statues are just fading and vague, so we have to lower the weights of reflectance $g(R)$. Given these priors and multi-scale optimization technique, this method produces reasonable reconstruction results on images of Longmen statues, but the resolution is quite low and we could only get a rough shape of the object (Fig. 4).

4.2 Landmark Driven Mesh Deformation

Although SIRFS needs only one image as input, it is restricted by many assumptions and sometimes produces unsatisfactory results. To improve this, it is highly desirable to warp a 3D template toward the true 3D shape of the target according to feature point correspondences, so that the subsequent deformation can have a better initialization. A generic face template mesh of Buddha's follower Ananda in Longmen Grottos is given. His statue has a moderate and standard appearance without much headwear or abnormal facial expression. The first step we take is face landmark localization in 2D space both on projected template image and input unknown image. We assume weak perspective camera projection and Lambertian reflection. Next, feature alignment by generalized procrustes analysis is intended to initialize the position of target feature point sets. After registration, we adopt discrete Poisson equation to manipulate the gradient fields and boundary conditions.

Face Landmark Localization. Proper 2D face alignment is vital in providing registration among images we input. Prior to our work, face feature extraction

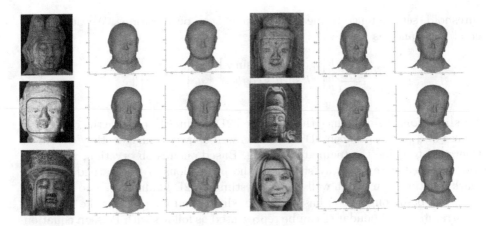

Fig. 5. Landmark driven face reconstruction results. By comparison, Poisson deformation (right columns) is better than Laplacian deformation (middle columns).

methods are mainly traditional methods such as Active Shape Model (ASM) [22] and Active Appearance Model (AAM) [23], their results are not very stable, so deformation driven by face landmarks is not reliable and there are few related work. But now this idea becomes practical with advance in feature extraction accuracy.

We employ a coarse-to-fine localization pipeline with deep convolutional networks cascade [24,25] to automatically fit q (= 83) facial landmarks onto each image. The first level networks predict the bounding boxes for the inner points and contour points separately. For the inner points, the second level predicted a initial estimation of the positions which are refined by the third level for each component. The fourth level is used to further improve the predictions of mouth and eyes by taking the rotated image patch as input. Two levels are used for contour points. The model is trained on Megvii Facial Landmark Database (MFLD). An example of landmark localization is given in the left columns in Fig. 5. Given an image $I(x, y)$, the landmark alignment returns a $2 \times q$ matrix W_i.

Feature Alignment. The initial template face is not nearly isometric to the individual face, e.g., the aspect ratio of the face may be different so that it will not fit closely to the images. Therefore, we take feature alignment as preprocessing step by Generalized Procrustes Analysis (GPA) [26]. It is a multivariate exploratory technique that involves transformations (e.g., translation, rotation, reflection, isotropic rescaling) of individual data matrices to provide optimal comparability. We already have a reference shape of point set as input. According to GPA algorithm, we superimpose another point set to current reference shape, then compute the mean shape of the current set of superimposed shapes. If the Procrustes distance between mean and reference shape is above a threshold, set reference to mean shape and keep on iterations. We define Procrustes distance as

$$P = \sum_{i=1}^{n} \left\| \begin{bmatrix} k\cos\theta & -k\sin\theta \\ k\sin\theta & k\cos\theta \end{bmatrix} \begin{bmatrix} x_i \\ y_i \end{bmatrix} - \begin{bmatrix} c_x \\ c_y \end{bmatrix} \right\|^2 \tag{12}$$

Mesh Deformation. Since the estimated 2D landmarks provide the correspondences of q points between 3D and 2D as well as across images, they should be leveraged to guide the template warping. Based on this observation, we aim to warp the template in a way such that the projections of the warped 3D landmark locations can match well with the estimated 2D landmarks. The technique we use is based on Poisson-based surface editing and adapted for the landmark constraints. The boundaries can be represented as follows with Poisson equation

$$\nabla^2 f = \nabla \cdot \mathbf{w}, f \mid_{\partial\Omega} = f^* \mid_{\partial\Omega} \tag{13}$$

where f is an unknown scalar function, $\mathbf{w}$ is a guidance vector field. f provides the desirable values on the boundary $\partial\Omega$. Specifically, in order to maintain the shape of the original template face while reducing the matching error from the 3D landmarks to the 2D landmarks, we minimize the scalar potential field ϕ

$$\int\int_{\Omega} \| \nabla\phi - \mathbf{w} \|^2 dA \tag{14}$$

$$Div(\nabla\phi) = Div\mathbf{w} \tag{15}$$

4.3   Fusion and Refinement

By SIRFS, we generate a coarse 3D reconstruction result of the shape estimated from an old facial image. On the same image, we utilize landmark driven mesh deformation to get a smoother and more accurate 3D face shape which shows high similarity with the original face in terms of the feature distribution of eyes, nose, mouth, and eyebrows. From these two methods, we obtain two depth maps both corresponding to the original image pixels. Thus, we do not need alignment in XY-plane. However, the depth scale estimated from these methods may be different. We manually align the two layers in Z-axis by the 3D editing tool ZBrush, so that the background and high-quality face shape can be fused together.

With the help of interactive software, we are also able to enhance the reconstruction based on the restoration result from our framework. Current 3D software has very strong functions, and some of them even encapsulate tools of crack detection, brush painting, and texture painting. Take the Bodhisattva on Maitreya's right side as example, once we reconstruct the background and the statue head, we can import, merge, and preprocess the models by enhancing the background texture using the carving brush, and reducing the noise using the smoothing brush. In addition, we can choose the proper point and stroke to

Fig. 6. Left: The old photo of a lost Buddha head. Middle: The refined model using Zbursh based on our automatic result. Right: Quantitative evaluation with another model of the same Buddha. Most parts are in the range of acceptable errors (green). (Color figure online)

restore the damaged parts, such as the headgear and hairstyle. Manual restoration is a work demanding aesthetic knowledge as well as patience, so our automatic reconstruction in the first place improves the efficiency to a great extent.

Except for qualitative evaluation, we conduct a quantitative one using a Buddha head with both scanned 3D model and pictures. Compared to the original model, most parts of our restored model are in the range of acceptable errors, which proves the effectiveness of this framework (Fig. 6).

5   Discussion

Our VR environment and restoration framework enable people to enjoy complete masterpieces of ancient Chinese Buddhist stone carving art. For user interaction, gesture recognition is designed using Hidden Markov Model (HMM). If tourists draw patterns such as circles and lines, then texture enhancement, relief restoration, and statue restoration can be triggered at an accuracy of 96.2%. Audio tour guide is also possible to be embedded into the smartphone app. Our accurate localization and natural interaction guarantee users to have a smooth and consistent experience. Therefore, very few cases of dizziness have been reported during our experiment. Most people would be very glad if such a system can be introduced to scenic spots and museums, because it allows the general public to appreciate original grottoes and relics conveniently.

The administrative committee of Longmen Grottoes has expressed their interests as well from the perspectives of archaeological research and tourism development. As the first "Internet+" scenic spot in China, Longmen Grottoes have WiFi and high speed 4G network access. Only needing some cheap helmets with simple optical components, a large number of tourists can use their own smartphones and download our app to enjoy impressive virtual tours (Fig. 7).

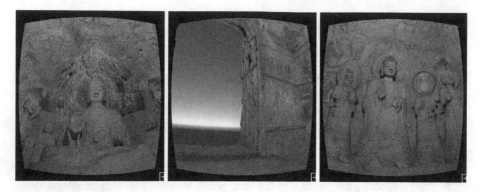

Fig. 7. Three views of rendered image after Barrel distortion correction for smartphone display (left channel). Left: The effect of texture enhancement. Middle: Restored relief mural from Internet images. Right: Restored head (circled) using our framework.

6   Conclusion

In this paper, we work on the Middle Binyang Cave, Longmen Grottoes, implement a VR system for heritage exhibition, and propose a feasible framework for digital restoration. Based on the digitized 3D model of Middle Binyang Cave, a virtual environment is set up, where user motion tracking is accomplished by the integration of IMU and the motion capture system, because this can solve the incapability of the viewpoint's translation and achieve a stable wandering experience. In order to reproduce the undamaged situation of Middle Binyang Cave, we firstly do a coarse restoration of the whole 3D structure and the background based on shape from shading, and then learn the facial features using 3D face reconstruction with shape priors. After this a new face model is produced through mesh deformation guided by feature points. The final step is to merge reconstruction results and add more details manually using current 3D editing software. Natural interactive methods such as gesture recognition are also designed using both static coordinates judgment and hidden Markov model to improve user experience. Our complete work on low-cost exhibition and efficient restoration contributes to the exploration of the digitalization of Longmen Grottoes, eases the conflict between tourism and preservation, and is hopeful to be applied to scenic spots and museums in the future.

There are some problems that need further study during related research. For example, we can investigate different statues of the same Buddha figure and analyze facial features and apparel characteristics to make our restoration framework more robust. Future work can also focus on considering how to apply the prototype system to Longmen Grottoes online and develop individualized augmented reality system for users.

References

1. Cosmas, J., Itegaki, T., Green, D., Grabczewski, E., Weimer, F., Vanrintel, D., Leberl, F., Grabner, M., Schindler, K., et al.: A novel multimedia system for archaeology. In: International Conference on Virtual Reality, Archeology, and Cultural Heritage, vol. 16 (2001)
2. Banno, A., Masuda, T., Oishi, T., Ikeuchi, K.: Flying laser range sensor for large-scale site-modeling and its applications in bayon digital archival project. Int. J. Comput. Vis. **78**, 207–222 (2008)
3. Ikeuchi, K., Miyazaki, D.: Digitally Archiving Cultural Objects. Springer Science & Business Media, Heidelberg (2008)
4. Koller, D., Levoy, M.: Computer-aided reconstruction and new matches in the forma urbis romae. Bullettino Della Commissione Archeologica Comunale di Roma **2** (2006)
5. Brown, B.J., Toler-Franklin, C., Nehab, D., Burns, M., Dobkin, D., Vlachopoulos, A., Doumas, C., Rusinkiewicz, S., Weyrich, T.: A system for high-volume acquisition and matching of fresco fragments: reassembling Theran wall paintings. ACM Trans. Graph. **27**, 84 (2008)
6. Koutsoudis, A., Pavlidis, G., Liami, V., Tsiafakis, D., Chamzas, C.: 3D pottery content-based retrieval based on pose normalisation and segmentation. J. Cult. Herit. **11**, 329–338 (2010)
7. Pintus, R., Yang, Y., Rushmeier, H.: Athena: automatic text height extraction for the analysis of text lines in old handwritten manuscripts. J. Comput. Cult. Herit. **8**, 1 (2015)
8. Kim, M.H., Rushmeier, H., Ffrench, J., Passeri, I., Tidmarsh, D.: Hyper 3D: 3D graphics software for examining cultural artifacts. J. Comput. Cult. Herit. **7**, 14 (2014)
9. Rushmeier, H.: Computer graphics techniques for capturing and rendering the appearance of aging materials. In: Martin, J.W., Ryntz, R.A., Chin, J., Dickie, R. (eds.) Service Life Prediction of Polymeric Materials, pp. 283–292. Springer, Heidelberg (2009)
10. Wojciechowski, R., Walczak, K., White, M., Cellary, W.: Building virtual and augmented reality museum exhibitions. In: International Conference on 3D Web Technology, pp. 135–144. ACM (2004)
11. Inaba, M., Banno, A., Oishi, T., Ikeuchi, K.: Achieving robust alignment for outdoor mixed reality using 3D range data. In: Symposium on Virtual Reality Software and Technology, pp. 61–68. ACM (2012)
12. Kakuta, T., Oishi, T., Ikeuchi, K.: Development and evaluation of asuka-kyo mr contents with fast shading and shadowing. In: International Conference on Virtual Systems and MultiMedia, pp. 254–260 (2008)
13. Horn, B.K., Brooks, M.J.: Shape from Shading. MIT Press, Cambridge (1989)
14. Barron, J.T., Malik, J.: Shape, illumination, and reflectance from shading. IEEE Trans. Pattern Anal. Mach. Intell. **37**, 1670–1687 (2015)
15. Blanz, V., Vetter, T.: A morphable model for the synthesis of 3D faces (1999)
16. Brown, D.C.: Decentering distortion of lenses. Photom. Eng. **32**, 444–462 (1966)
17. Lipman, Y., Sorkine, O., Cohen-Or, D., Levin, D., Rossi, C., Seidel, H.P.: Differential coordinates for interactive mesh editing. In: International Conference on Shape Modeling Applications, pp. 181–190. IEEE (2004)
18. Zhou, K., Yu, Y.: Mesh editing with Poisson-based gradient field manipulation. In: ACM Transactions on Graphics, pp. 641–648 (2004)

19. Horn, B.K.P.: Determining lightness from an image. Graph. Models Comput. Vis. Graph. Image Process. **3**, 277–299 (1974)
20. Ikeuchi, K., Horn, B.K.P.: Numerical shape from shading and occluding boundaries. Artif. Intell. **17**, 141–184 (1981)
21. Zhang, R., Tsai, P., Cryer, J.E., Shah, M.: Shape-from-shading: a survey. IEEE Trans. Pattern Anal. Mach. Intell. **21**, 690–706 (1999)
22. Cootes, T.F., Taylor, C.J., Cooper, D.H., Graham, J.: Active shape modelstheir training and application. Comput. Vis. Image Underst. **61**, 38–59 (1995)
23. Cootes, T.F., Edwards, G., Taylor, C.J.: Active appearance models. IEEE Trans. Pattern Anal. Mach. Intell. **23**, 681–685 (2001)
24. Zhou, E., Fan, H., Cao, Z., Jiang, Y., Yin, Q.: Extensive facial landmark localization with coarse-to-fine convolutional network cascade. In: IEEE International Conference on Computer Vision Workshops, pp. 386–391 (2013)
25. Huang, Z., Zhou, E., Cao, Z.: Coarse-to-fine face alignment with multi-scale local patch regression (2015). arXiv preprint arXiv:1511.04901
26. Gower, J.C.: Generalized procrustes analysis. Psychometrika **40**, 33–51 (1975)

Fast General Norm Approximation via Iteratively Reweighted Least Squares

Masaki Samejima$^{(\boxtimes)}$ and Yasuyuki Matsushita

Graduate School of Information Science and Technology,
Osaka University, Suita, Osaka, Japan
{samejima,yasumat}@ist.osaka-u.ac.jp

Abstract. This paper describes an efficient method for general norm approximation that appears frequently in various computer vision problems. Such a lot of problems are differently formulated, but frequently require to minimize the sum of weighted norms as the general norm approximation. Therefore we extend Iteratively Reweighted Least Squares (IRLS) that is originally for minimizing single norm. The proposed method accelerates solving the least-square problem in IRLS by warm start that finds the next solution by the previous solution over iterations. Through numerical tests and application to the computer vision problems, we demonstrate that the proposed method solves the general norm approximation efficiently with small errors.

1 Introduction

In various tasks in digitally archiving cultural heritages including 3D reconstruction [1,2], numerical optimization plays a central role. More specifically, we often optimize some ℓ_p-norm of the cost vector (equivalently, its p-th power ℓ_p^p) or the combination of different vector norms. For example, in compressive sensing, an unconstrained form of Lasso (least absolute shrinkage and selection operator) [3]

$$\min_x \|Ax - b\|_2^2 + \lambda \|x\|_1 \tag{1}$$

is used for reconstructing sparse signal x while ensuring data fitting. As another example, Tikhonov regularization (or ridge regression) [4],

$$\min_x \|Ax - b\|_2^2 + \lambda \|\Gamma x\|_2^2, \tag{2}$$

appears in image restoration, super-resolution, and image deblurring. These objective functions can be further augmented by additional ℓ_p-norm terms that represent further constraints for particular problems. For example, the elastic net [5] is defined as

$$\min_x \|Ax - b\|_2^2 + \lambda_1 \|x\|_1 + \lambda_2 \|x\|_2^2 \tag{3}$$

by regularizing the solution using both ℓ_1- and ℓ_2-norm terms. Some special cases are known to have closed-form solutions, *e.g.*, the minimizer of Tikhonov

© Springer International Publishing AG 2017
C.-S. Chen et al. (Eds.): ACCV 2016 Workshops, Part II, LNCS 10117, pp. 207–221, 2017.
DOI: 10.1007/978-3-319-54427-4_16

regularization (2) is given by $\hat{x} = (A^T A + \lambda \Gamma^T \Gamma)^{-1} A^T b$, or can be transformed into a simpler expression, $e.g.$, the elastic net problem (3) can be rewritten as an equivalent Lasso problem (1) with some augmentation [5].

Generally, these $\ell_p (p \geq 1)$ unconstrained minimization problems can be solved by any convex optimization methods[1]. To gain a greater computation performance, typically problem-specific structures are exploited to design a tailored solution method. For example, least-squares problems that consist of ℓ_2-norms can be solved analytically, or for a large-scale problem, conjugate gradient methods [6] are employed for faster computation. When the rank of the design matrix is small enough, randomized singular value decomposition (R-SVD) [7] may be employed for further acceleration. It is also understood that ℓ_1 minimization problems can be transformed into a linear programming problem [8], which can be efficiently solved by an interior point method [9]. On the other hand, it is still of broad interest to improve the performance of the general norm approximation problem, because in practical situations there is a strong need for testing with various formulations with different norms in designing computer vision applications. For example, one might initially formulate a regression problem with an ℓ_2-norm but later might add an ℓ_1 or ℓ_2 regularizer for stabilizing the solution.

This motivates us to develop a fast solver for the generalized norm approximation problem:

$$\min_{x} \sum_{k=1}^{K} \lambda_k \|A_k x - b_k\|_{p_k}^{p_k}, \qquad (4)$$

where $k = \{1, \cdots, K\}$ is the term index, $A_k \in \mathbb{R}^{m_k \times n}$ and $b_k \in \mathbb{R}^{m_k}$ the the design matrix and constant vector that define the k-th linear objective, and the overall objective function is defined as a linear combination of p_k-th power of ℓ_{p_k}-norm weighted by λ_k. Our method is built upon a simple yet powerful iteratively reweighted least-squares (IRLS) scheme. In the IRLS scheme, the problem can be reduced to iteratively solving a linear system that is derived as a normal equation of the sum of weighted squares of the terms.

In this paper, we present a fast method for deriving the approximate solution for this problem that outperforms the state-of-the-art solution methods such as [10,11]. Our method exploits the trait that the solution is gradually updated in an iterative manner in the IRLS scheme, and achieves acceleration by taking the previous estimate as an initial guess at each iteration for an LSQR solver [12]. The proposed method is faster and more stable than the previous state-of-the-art approaches as we are going to see in the experimental validation. In addition, the solution method for the general expression (4) has not been explicitly described in the literature that we are aware of, and we show in this paper that the general form can be solved in a unique manner regardless of the number of terms and diverse ℓ_p-norm objectives and benefit from the proposed method.

---

[1] While it may be still valid even when $p < 1$, the problem becomes non-convex when $p < 1$; thus they may be trapped by local minima.

2   Related Works

The early studies of IRLS can be found back in 1960's [13], developed for approximating a Chebyshev or ℓ_∞ norm. It has been later extended to approximate a general ℓ_p-norm term [14]. While the early studies focus on convex approximations with $p \geq 1$ mostly with a single term, later, the focus has been shifted to the case where $p < 1$ (non-convex cases). The original sparse recovery using the IRLS scheme has been known as FOCUSS [15] prior to that shift, and it has been known useful for robust estimation tasks. With the rise of compressive sensing and sparse recovery, norm approximation with $p < 1$ has been extensively studied. Chartrand and Yin [16] introduced a regularizer ϵ, for augmenting the IRLS weight, that varies from large to small over iterations so that it effectively smooths out the objective function and as a result avoids local minima. Daubechies et al. [10] proposed an alternative method for updating the weight over iterations and showed the convergence property in sparse recovery. Candès et al. [17] introduced an iteratively reweighted ℓ_1 minimization method, which repeatedly solves ℓ_1 minimization problem, for further enhancing sparsity. Wipf and Nagarajan [18] provided an extensive analysis on ℓ_2 and ℓ_1 reweighting schemes [16,17] and made distinction between separable (i.e, the weighting of a particular coefficient depends only that in previous iteration) and non-separable iterative reweighting schemes. The development of an effective numerical algorithm for sparse recovery is still an active research topic, and there is a broad interest in the area.

With these theoretical and algorithmic development, the IRLS scheme has expanded its application domain. It has been used for various signal processing applications, such as FIR filter design [19], image deblurring with ℓ_p-norm ($p = 0.8$) minimization [20,21], denoising based on TV regularization [22], and super-resolution [23]. The IRLS scheme has also been used for minimizing nuclear norm [24] and structured sparse recovery [25]. These new applications widen the use of IRLS scheme for even more diverse applications. This paper is motivated by the background that accelerating the general sum of ℓ_p^p terms is urgent because of its increasing need in various computer vision applications.

Because the problem of (4) is unconstrained and convex when $p_k \geq 1$, it can be also solved via a family of efficient quasi-Newton methods, such as limited-memory Broyden-Fletcher-Goldfarb-Shanno (L-BFGS) method, although such general convex optimizers are typically not optimal in terms of their performance.

3   Fast General Norm Approximation

This section describes the proposed method for general norm approximation using IRLS. In Sect. 3.1, we begin with briefly reviewing the IRLS and show how the generalized form of Eq. (4) can be solved via IRLS. We then describe an acceleration method using LSQR in Sect. 3.2.

3.1   IRLS for Norm Approximation Problems

Since 1960's, it has been understood that norm approximation problems can be solved via IRLS [13,14]. With the IRLS framework, norm approximation problems can be casted to iteratively solving weighted least-squares problem. Let us take an example of minimizing the p-th power of ℓ_p-norm of a real-valued vector $(Ax - b)$:

$$\min_x f(x), \quad f(x) = \|Ax - b\|_p^p = (Ax - b)^T W^T W (Ax - b). \tag{5}$$

The above can be expressed by a weighted squares of the vector as:

$$\min_x f(x), \quad f(x) = \|W(Ax - b)\|_2^2, \tag{6}$$

with a proper diagonal weight matrix W, whose elements are all non-negative. where W is a diagonal matrix of weights to be determined in IRLS. The problem of (6) is a quadratic programming; therefore the minimizer x^* is attained when

$$\frac{\partial}{\partial x} \|W(Ax - b)\|_2^2 = 2p \left(A^T W^T W A x - A^T W^T W b \right) = 0. \tag{7}$$

Therefore, the approximate solution x^* becomes

$$x^* = (A^T W^T W A)^{-1} A^T W^T W b. \tag{8}$$

In the IRLS scheme, the weight matrix W is iteratively refined for a more focal estimate. Let w_i and e_i denote the i-th diagonal element of W and the i-th element of the residual vector $e = Ax - b$, respectively. Since

$$\|Ax - b\|_p^p = \sum_i |e_i|^{p-2} |e_i|^2 = \sum_i w_i^2 |e_i|^2, \tag{9}$$

at each iteration, the weight matrix element w_i is updated by $w_i \leftarrow |e_i|^{(p/2-1)}$ if $e_i \neq 0$ or $w_i \leftarrow 1/\varepsilon$ otherwise where ε is a sufficiently small positive value. Typically, the weight matrix W is initialized as an identity matrix.

IRLS for General Norm Approximation. We have seen how ℓ_p-norm minimization for a single term is achieved by IRLS. We now describe its extension to the multiple terms for applying IRLS to the general norm approximation problem (4). For a general norm approximation problem

$$\min_x f(x), \quad f(x) = \sum_{k=1}^K \lambda_k \|A_k x - b_k\|_{p_k}^{p_k}, \tag{10}$$

there are K terms that are defined by ℓ_{p_k}-norm, where p_k may be different across the terms, each weighted by λ_k. With K weight matrices W_k, it can be approximated by

$$f(x) = \sum_{k=1}^K \lambda_k (A_k x - b_k)^T W_k^T W_k (A_k x - b_k), \tag{11}$$

in a similar manner to the single term case. From the normal equation of above, the approximate solution can be determined by differentiating $f(x)$ w.r.t. x as

$$\frac{\partial f(x)}{\partial x} = \sum_{k=1}^{K} 2p_k \lambda_k \left(A_k^T W_k^T W_k A_k x - A_k^T W_k^T W_k b_k \right) = 0. \tag{12}$$

Therefore, the minimizer x^* can be obtained by

$$x^* = \left(\sum_{k=1}^{K} p_k \lambda_k A_k^T W_k^T W_k A_k \right)^{-1} \left(\sum_{k=1}^{K} p_k \lambda_k A_k^T W_k^T W_k b_k \right). \tag{13}$$

The pseudo-code of IRLS for general norm approximation is summarized in Algorithm 1.

Procedure 1. IRLS for general norm approximation

Input: $A_k \in R^{m_k \times n}$ and $b_k \in R^{m_k}$
Output: Solution x
// Initialize the weight matrix
$W_1, W_2, \cdots, W_K \leftarrow I$
// Concatenate matrices and vectors; $A \in \mathbb{R}^{M \times n}, b \in \mathbb{R}^M, W \in \text{diag}\left(\mathbb{R}_+^M\right), M = \sum_k m_k$
$A \leftarrow [A_1^T, A_2^T, \cdots, A_K^T]^T$ // Matrix of vertically stacked $A_1, \cdots, A_K$
$b \leftarrow [b_1^T, b_2^T, \cdots, b_K^T]^T$ // Vector of vertically stacked $b_1, \cdots, b_K$
while x is not converged **do**
  $W \leftarrow \text{diag}(\sqrt{\lambda_1 p_1} W_1, \sqrt{\lambda_2 p_2} W_2, \cdots, \sqrt{\lambda_K p_K} W_K)$
  // Solve the weighted least-squares problem: $WAx = Wb$
  $x \leftarrow \text{LeastSquares}(WA, Wb)$
  $e \leftarrow Ax - b$
  // Update weight
  for all k, n **do**
    if $e_k(n) \neq 0$ **then**
      $W_k(n) = 1/|e_k(n)|^{(1-p_k/2)}$
    else
      $W_k(n) = 1/\epsilon$ // ϵ is a sufficiently small positive value.
    end if
  end for
end while

3.2   Acceleration of IRLS

The major bottleneck of the IRLS algorithm is its need for solving the weighted least-squares problem over iterations until convergence. In particular, when the size of matrix A_k is large, the computation cost significantly increases if an analytic solution method like Eq. (13) is used. To accelerate the whole algorithm, it is needed to efficiently solve the weighted least-squares problem. We exploit

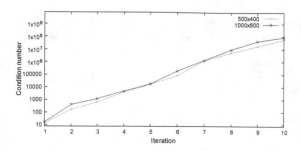

Fig. 1. Growth of the condition number of A in each iteration. The plots show the average condition numbers over ten trials for each setting.

the fact that the solution x and weight matrix W_k are gradually updated over iterations of IRLS and use the previous estimate of x for efficiently updating the solution x using LSQR.

Previously, conjugate gradient methods have been applied to the least-squares problem in an iterative framework [25, 26] in a similar spirit. They are effective when the design matrix A in $\|Ax - b\|_2^2$ is well-conditioned. Convergence of the conjugate gradient method is analyzed by the relative error [27]:

$$\frac{\left\|x - x^{(n)}\right\|_A}{\left\|x - x^{(0)}\right\|_A} \leq 2 \left(\frac{\sqrt{\kappa} - 1}{\sqrt{\kappa} + 1}\right)^n, \tag{14}$$

where $\|\cdot\|_A$ indicates A-norm, κ is matrix A's condition number calculated by $\kappa = \sigma_{\max}(A)/\sigma_{\min}(A)$, $\sigma_{\max}(A)$ and $\sigma_{\min}(A)$ are the maximum and the minimum singular values of A, respectively. When the relative error in the left side of Eq. (14) is large, convergence from $x^{(0)}$ to $x^{(n)}$ through n iterations takes much time. The upper bound of the relative error can be calculated with κ, and a greater κ makes the convergence slower. Unfortunately, for the weighted least-squares problem $WAx = Wb$ in IRLS, the condition number of matrix WA naturally increases as the iteration proceeds [26]. To depict this issue, we show a preliminary experiment of running IRLS for solving $\min_x \|Ax - b\|_2^2$ ten times with randomly generated matrices $A \in \mathbb{R}^{500 \times 400}$ with vector $b \in \mathbb{R}^{500}$ and also a larger scale setting, $A \in \mathbb{R}^{1000 \times 800}$ with vector $b \in \mathbb{R}^{1000}$. Figure 1 shows the variation of the average condition number over iterations plotted in a log scale. As seen in the figure, the condition number grows exponentially over iterations, which makes conjugate gradient methods slower and less stable in later iterations.

To overcome this issue, previous approaches use *preconditioning* to yield the equivalent least-squares problem with the small condition number of A by multiplying a matrix P called *preconditioner* [25, 26, 28]. However, the preconditioner P is problem-dependent [25]; therefore, it is not straightforward to incorporate the preconditioned conjugate gradient method into the generalized norm approximation problem.

To avoid these problems, we use the LSQR method [12], which is a stable iterative method for ill-conditioned least-squares problems [29,30]. In LSQR, the least-square problem is reduced to another least-square problem including a bidiagonal matrix, and the reduced problem is solved by QR factorization. To accelerate the LSQR computation in IRLS framework, we use a "warm start" strategy by taking the previous estimate of the solution as the initial guess for the next iteration. The pseudo-code of LSQR to find a solution $x^{(i+1)}$ at the $(i+1)$-th iteration from $x^{(i)}$ from the previous iteration in our context is shown in Algorithm 2.

Procedure 2. LSQR

Input: $A \in R^{M \times n}$, $x^{(i)} \in R^n$, and $b \in R^M$
Output: Solution $x^{(i+1)} \in R^n$
  // Initialization
  $\beta \leftarrow \|b - Ax^{(i)}\|_2$, $u \leftarrow (b - Ax^{(i)})/\beta$
  $\alpha \leftarrow \|A^T u\|_2$, $v \leftarrow A^T u/\alpha$
  $\bar{\rho} \leftarrow \beta$, $\bar{\phi} \leftarrow \alpha$
  while x is not converged **do**
    // Bidiagonalization
    $\beta \leftarrow \|Av - \alpha u\|_2$, $u \leftarrow (Av - \alpha u)/\beta$
    $\alpha \leftarrow \|A^T u - \beta v\|_2$, $v \leftarrow (A^T u - \beta v)/\alpha$
    // Construct and apply next orthogonal transformation
    $c \leftarrow \bar{\rho}/\sqrt{\bar{\rho}^2 + \beta^2}$, $s \leftarrow \beta/\rho$
    $\theta \leftarrow s\alpha$, $\phi \leftarrow c\bar{\phi}$
    $\bar{\rho} \leftarrow -c\alpha$, $\bar{\phi} \leftarrow s\bar{\phi}$
    // Update the solution and weight
    $x \leftarrow x + w\phi/\rho$
    $w \leftarrow v - w\theta/\rho$
  end while
  $x^{(i+1)} \leftarrow x$

---

4 Performance Evaluation

To evaluate the computational efficiency of the proposed method, we test the algorithm using the following weighted norms minimization problems:

$$(\text{Problem 1}) \qquad \min_x \|A_1 x - b_1\|_1, \qquad\qquad (15)$$

$$(\text{Problem 2}) \qquad \min_x \|A_2 x - b_2\|_2^2 + \|A_3 x - b_3\|_1, \qquad (16)$$

where $A_1 \in \mathbb{R}^{500 \times 400}, b_1 \in \mathbb{R}^{500}$ and $A_2, A_3 \in \mathbb{R}^{1000 \times 800}, b_2, b_3 \in \mathbb{R}^{1000}$. Matrices A_k and solution x are randomly generated, and vector b_k is computed by $b_k \leftarrow A_k x$. To add outliers to the systems, the signs of 10% of elements in b_k are flipped.

We implement the following methods to compare the computational times on a PC equipped with Intel Core i7 930 @2.8GHz and 24 GB Memory.

– L-BFGS
– IRLS by QR decomposition with column pivoting (IRLS-QR)
– IRLS by Jacobi-preconditioned Conjugate Gradient method (IRLS-CG)
– IRLS by Jacobi-preconditioned Conjugate Gradient method with warm start (IRLS-CG-WS)
– IRLS by LSQR (IRLS-LSQR)
– IRLS by LSQR with warm start (IRLS-LSQR-WS) (proposed method)

We use ALGLIB [31] for L-BFGS and Eigen [32] matrix library for matrix operations. Because L-BFGS does not converge well on Problem 1 and Problem 2 that include non-smooth ℓ_1-norms, we approximate $\|A_1 x - b_1\|_1$ to $\sqrt{(A_1 x - b_1)^2 + \gamma}$ with a sufficiently small positive value $\gamma(= 10e^{-8})$ when applying L-BFGS.

Table 1 summarizes the average computation times and residuals of ten trials, and Figs. 2 and 3 show computation times and residuals over iterations when solving Problems 1 and 2, respectively. As shown in Table 1, IRLS-based methods are faster and more accurate than L-BFGS, even though we use a relaxed tolerance for L-BFGS for faster convergence. Although IRLS-QR solves the smaller problem fast with the smallest residual, the computation time rapidly grows as the size of the matrix becomes larger.

The effectiveness of the warm start strategy is clearly seen in IRLS by Conjugate Gradient and LSQR, showing about 20 times faster convergence. As shown in Fig. 2, while computation times for methods without warm start drastically increase as the iteration proceeds, they are significantly reduced with warm start. It is also shown in Fig. 3 that the warm start strategy is effective in reducing the residual by providing a guide to solve least-squares with high condition numbers at later iterations.

Table 1. Computation times and residuals of each method applied to problems 1 and 2.

	Problem 1		Problem 2	
	Computation time (sec.)	Residual	Computation time (sec.)	Residual
L-BFGS	710*	741	> 1.0e^5	–
IRLS-QR	5.6	642	137[†]	2456[†]
IRLS-CG	43	763	990[†]	2468[†]
IRLS-CG-WS	7.0	664	30	2468
IRLS-LSQR	54	721	634[†]	2620[†]
IRLS-LSQR-WS (ours)	2.6	682	14	2475

*Due to the enormous computational time of L-BFGS, we set the tolerance greater $(1.0e^{-6})$ than usual $(1.0e^{-8})$.
[†]Because these methods did not converge, the computational times and residuals at 100-th iteration are shown.

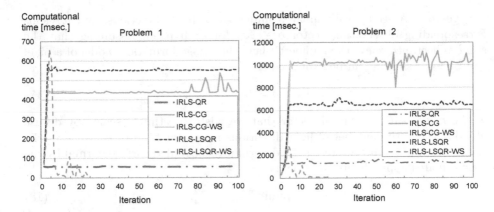

Fig. 2. Variation of computation times over iterations. Our method benefits from the warm start strategy and the computation time quickly drops at the early stage of iterations.

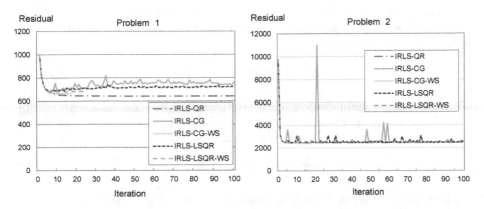

Fig. 3. Variation of residuals over iterations.

5   Applications

The expression of the general norm approximation (4) offers flexibility of treating diverse objective functions in a unified manner, and they can generally benefit from the proposed efficient computation method. As example use cases, we show two applications in this section: Photometric stereo in Sect. 5.1 and surface reconstruction from normals in Sect. 5.2.

5.1   Photometric Stereo

Photometric stereo is a method for estimating surface normal of a surface from its appearance variations under different lightings. Let us assume that we have an $f \times 3$ light direction matrix L that contains f distinct light directions as row vectors. A scene is illuminated under these light directions, and it yields corresponding observation vector $o \in \mathbb{R}_+^f$ for each pixel. Assuming that the scene reflectance obeys the Lambert's law, the image formation model of a pixel can be expressed as can be expressed as

$$o = Ln, \tag{17}$$

where n is a surface normal vector that we wish to estimate scaled by diffuse albedo. When the number of light directions is greater than three ($f > 3$), the Lambertian photometric stereo [33] determines the scaled surface normal by the least-squares approximate solution as

$$n^* = \underset{n}{\mathrm{argmin}} \, \|Ln - o\|_2^2. \tag{18}$$

The solution method for the above problem is rather straightforward and does not even require our method, while it can still be represented as a special case of (4).

In reality, scene reflectances may contain specularity that can be regarded as unmodelled outliers. It motivated the previous work [34, 35] to use an ℓ_1-norm minimization as robust estimation as

$$n^* = \underset{n}{\mathrm{argmin}} \, \|Ln - o\|_1. \tag{19}$$

It corresponds to minimization of the residual $e = o - Ln$, as above can be rewritten as the problem $\min_n \|e\|_1$ s.t. $e = o - Ln$. This hard constraint can be relaxed as a regularizer as depicted in [34, 35] as

$$n^* = \underset{n}{\mathrm{argmin}} \, \lambda_1 \|o - Ln - e\|_2^2 + \|e\|_1 \tag{20}$$

for allowing a certain magnitude of Gaussian error in the constraint.

In a special case, when we have an external method for determining surface normal, e.g., surface normal obtained from a measured depth by structured light, and can use it as prior for stabilizing the solution, the photometric stereo problem can be formulated as

$$n^* = \underset{n}{\mathrm{argmin}} \, \|Ln - o\|_1 + \lambda_2 \|n - n_g\|_2^2, \tag{21}$$

in which n_g is surface normal that are obtained by an external method.

These four different problem settings (18)–(21) can all be represented by general norm approximation (4), and our method is applicable to any of the settings. To demonstrate this, we render 40 images of *Bunny* and *Caesar* scenes under different light directions and solve these four problems. We set $\lambda_1, \lambda_2 = 10e^6$. n_g in (21) is generated by adding 1% of Gaussian noise to every normal elements in the ground truth normal vector. Figure 4 shows estimated surface normal and error maps of the proposed IRLS-LSQR-WS method. Putting aside that there are variations in errors because of different formulations, it shows that our method is applicable to diverse formulations.

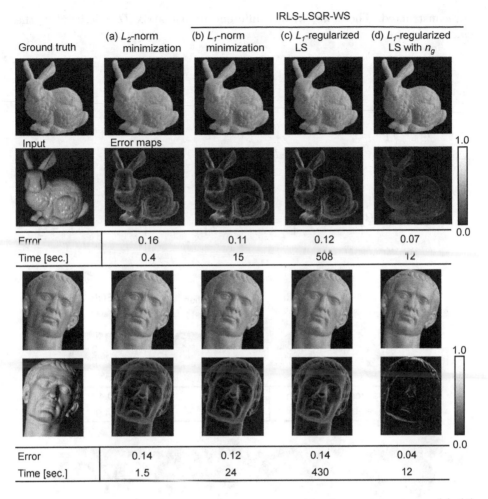

| | IRLS-LSQR-WS | | | |
	(a) L_2-norm minimization	(b) L_1-norm minimization	(c) L_1-regularized LS	(d) L_1-regularized LS with n_g
Error	0.16	0.11	0.12	0.07
Time [sec.]	0.4	15	508	12
Error	0.14	0.12	0.14	0.04
Time [sec.]	1.5	24	430	12

Fig. 4. Surface normal estimates and error maps with photometric stereo. (a)–(d) correspond to the settings (18)–(21). "Error" indicates the angular error in degrees, and "Time" corresponds to the computation time.

5.2 Surface Reconstruction from Normals

Once the surface normal is obtained by photometric stereo or shape from shading, one may like to reconstruct a surface from the normal, e.g., by [36]. From surface normal $n = (n_x, n_y, n_z)^T$ defined in the image coordinates (u, v), the gradients g_u and g_v are computed as

$$g_x(u, v) = \frac{n_x(u, v)}{n_z(u, v)}, \quad g_y(u, v) = \frac{n_y(u, v)}{n_z(u, v)}. \tag{22}$$

Let $G_u, G_v \in \mathbb{R}^{m \times n}$ denote matrices of g_u and g_v, respectively. These gradients corresponds to the 1-st order differentiation of the surface $Z \in \mathbb{R}^{m \times n}$ to be reconstructed. Therefore, with a differentiation matrix $D \in \mathbb{R}^{2mn \times mn}$, the relationship can be written as

$$Dz \approx g, \tag{23}$$

where

$$D = \frac{1}{h} \begin{bmatrix} D_x \\ D_y \end{bmatrix}, \quad z = \text{vec}(Z), \quad g = \begin{bmatrix} \text{vec}(G_u) \\ \text{vec}(G_v) \end{bmatrix},$$

$$D_x = \begin{bmatrix} B & & & \\ & B & & \\ & & \ddots & \\ & & & B \end{bmatrix}, \quad D_y = \begin{bmatrix} -I & I & & & \\ & -I & I & & \\ & & \ddots & \ddots & \\ & & & -I & I \\ & & & I & -I \end{bmatrix}, \quad B = \begin{bmatrix} -1 & 1 & & & \\ & -1 & 1 & & \\ & & \ddots & \ddots & \\ & & & -1 & 1 \\ & & & 1 & -1 \end{bmatrix},$$

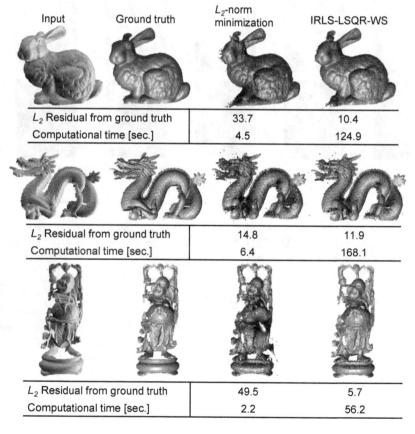

Fig. 5. Surface reconstruction from normal maps by the proposed method (IRLS-LSQR-WS)

with the identity matrix I and pixel size h. Similar to [37], we consider that there are outliers in normals that are 10 times larger than true normals and 10% of pixels are corrupted by them. In this setting, we consider the ℓ_1-residual minimization to reconstruct the surface in the presence of outliers as

$$\min_z \|Dz - g\|_1. \tag{24}$$

This problem, again, is a special case of the general norm approximation problem (4), and our method can be applied to derive the solution z. We use three different scenes, *Bunny*, *Dragon*, and *Happy Buddha* as target scenes for testing this scenario. For comparison, we show the result of surface reconstruction by ℓ_2-norm minimization as reference.

Figure 5 shows the reconstructed surfaces, ℓ_2-residual from the ground truth, and computation times. The accuracy indicates the strength of ℓ_1-residual minimization, but more importantly, our method is capable of handling any of these formulations because of the generalized form of norm minimization (4).

6   Conclusions

We presented a fast general norm approximation that can be applicable to diverse problem settings in computer vision. The proposed method (IRLS-LSQR-WS) is assessed in comparison to other state-of-the-art techniques and shows the favorable computation efficiency and accuracy at a time. In addition to the numerical tests, we show application scenarios by taking photometric stereo and surface reconstruction as examples to illustrate the usefulness of the general norm approximation.

During the experiments, we found that the proposed method is advantageous over IRLS-CG-WS in terms of stability, *i.e.*, IRLS-CG-WS occasionally becomes unstable when the condition number of the problem grows rapidly, while the proposed method does not suffer from this issue. On the other hand, IRLS-CG-WS tends to converge slightly faster when the design matrix A is sparse compared to the proposed method. We are interested in studying this trade-off by characterizing the problem by looking into the design matrix structure. In addition, further acceleration by preconditioning for LSQR [38] is another interesting venue to investigate.

Acknowledgement. This work was partly supported by JSPS KAKENHI Grant Numbers JP16H01732 and JP26540085.

References

1. Lu, M., Zheng, B., Takamatsu, J., Nishino, K., Ikeuchi, K.: In: 3D Shape Restoration via Matrix Recovery. Springer, Heidelberg (2011)
2. Futragoon, N., Kitamoto, A., Andaroodi, E., Matini, M.R., Ono, K.: In: 3D Reconstruction of a Collapsed Historical Site from Sparse Set of Photographs and Photogrammetric Map. Springer, Heidelberg (2011)

3. Tibshirani, R.: Regression shrinkage and selection via the lasso. J. R. Stat. Soc. B **58**, 267–288 (1996)
4. Tikhonov, A.N., Arsenin, V.Y.: Solution of Ill-posed Problems. Winston & Sons, Washington (1977). ISBN 0-470-99124-0
5. Zou, H., Hastie, T.: Regularization and variable selection via the elastic net. J. R. Stat. Soc. B **67**, 301–320 (2005)
6. Hestenes, M., Stiefel, E.: Methods of conjugate gradients for solving linear systems. J. Res. Natl. Bur. Stand. **49**, 409–436 (1952)
7. Halko, N., Martinsson, P.G., Tropp, J.A.: Finding structure with randomness: probabilistic algorithms for constructing approximate matrix decompositions. SIAM Rev. **53**, 217–288 (2011)
8. Gentle, J.E.: In: Matrix Algebra: Theory, Computations, and Applications in Statistics. Springer, New York (2007). ISBN 978-0-387-70872-0
9. Karmarkar, N.: A new polynomial-time algorithm for linear programming. Combinatorica **4**, 373–395 (1984)
10. Daubechies, I., DeVore, R., Fornasier, M., Gunturk, S.: Iteratively re-weighted least squares minimization: proof of faster than linear rate for sparse recovery. In: 42nd Annual Conference on Information Sciences and Systems, pp. 26–29 (2008)
11. Aftab, K., Hartley, R.: Convergence of iteratively re-weighted least squares to robust m-estimators. In: 2015 IEEE Winter Conference on Applications of Computer Vision, pp. 480–487 (2015)
12. Paige, C.C., Saunders, M.A.: LSQR: an algorithm for sparse linear equations and sparse least squares. ACM Trans. Math. Softw. **8**, 43–71 (1982)
13. Lawson, C.L.: Contributions to the theory of linear least maximum approximations. Ph.D. thesis, UCLA (1961)
14. Rice, J.R., Usow, K.H.: The lawson algorithm and extensions. Math. Comput. **22**, 118–127 (1968)
15. Gorodnitsky, I.F., Rao, B.D.: Sparse signal reconstruction from limited data using focuss: a re-weighted minimum norm algorithm. IEEE Trans. Signal Process. **45**, 600–616 (1997)
16. Chartrand, R., Yin, W.: Iteratively reweighted algorithms for compressive sensing. In: Proceedings of IEEE International Conference on Acoustics, Speech, and Signal Processing (ICASSP), pp. 3869–3872 (2008)
17. Candès, E.J., Wakin, M.B., Boyd, S.: Enhancing sparsity by reweighted ℓ_1 minimization. J. Fourier Anal. Appl. **14**, 877–905 (2008)
18. Wipf, D.P., Nagarajan, S.: Iterative reweighted ℓ_1 and ℓ_2 methods for finding sparse solutions. J. Sel. Top. Signal Process **4**(2), 317–329 (2010)
19. Burrus, C.S., Barreto, J., Selesnick, I.W.: Iterative reweighted least-squares design of fir filters. IEEE Trans. Signal Process. **42**, 2926–2936 (1994)
20. Levin, A., Fergus, R., Durand, F., Freeman, W.: Image and depth from a conventional camera with a coded aperture. ACM Trans. Graph. **26**, 70 (2007). Proceedings of SIGGRAPH
21. Joshi, N., Zitnick, L., Szeliski, R., Kriegman, D.: Image deblurring and denoising using color priors. In: Proceedings of IEEE Conference on Computer Vision and Pattern Recognition (CVPR) (2009)
22. Wohlberg, B., Rodríguez, P.: An iteratively reweighted norm algorithm for minimization of total variation functionals. IEEE Signal Process. Lett. **14**, 948–951 (2007)
23. Liu, C., Sun, D.: On Bayesian adaptive video super resolution. IEEE Trans. Pattern Anal. Mach. Intell. **36**, 346–360 (2014)

24. Mohan, K., Fazel, M.: Iterative reweighted algorithms for matrix rank minimization. J. Mach. Learn. Represent. **13**, 3441–3473 (2012)
25. Chen, C., Huang, J., He, L., Li, H.: Preconditioning for accelerated iteratively reweighted least squares in structured sparsity reconstruction. In: Proceedings of IEEE Conference on Computer Vision and Pattern Recognition (CVPR), pp. 2713–2720 (2014)
26. Fornasier, M., Peter, S., Rauhut, H., Worm, S.: Conjugate gradient acceleration of iteratively re-weighted least squares methods. Comput. Optim. Appl. **65**, 205–259 (2016)
27. Shewchuk, J.R.: An introduction to the conjugate gradient method without the agonizing pain. Technical report, Pittsburgh, PA, USA (1994)
28. Howell, G.W., Baboulin, M.: LU preconditioning for overdetermined sparse least squares problems. In: Wyrzykowski, R., Deelman, E., Dongarra, J., Karczewski, K., Kitowski, J., Wiatr, K. (eds.) PPAM 2015. LNCS, vol. 9573, pp. 128–137. Springer, Heidelberg (2016). doi:10.1007/978-3-319-32149-3_13
29. Benbow, S.J.: Solving generalized least-squares problems with LSQR. SIAM J. Matrix Anal. Appl. **21**, 166–177 (1999)
30. Nolet, G.: Solving Large Linearized Tomographic Problems: Seismic Tomography, Theory and Practice, pp. 227–247. Chapmanand Hall, London (1993)
31. Bochkanov, S., Bystritsky, V.: ALGLIB. http://www.alglib.net/
32. Guennebaud, G., Jacob, B., et al.: Eigen v3
33. Woodham, R.J.: Photometric method for determining surface orientation from multiple images. Opt. Eng. **19**, 191139–191139 (1980)
34. Wu, L., Ganesh, A., Shi, B., Matsushita, Y., Wang, Y., Ma, Y.: Robust photometric stereo via low-rank matrix completion and recovery. In: Kimmel, R., Klette, R., Sugimoto, A. (eds.) ACCV 2010. LNCS, vol. 6494, pp. 703–717. Springer, Heidelberg (2011). doi:10.1007/978-3-642-19318-7_55
35. Ikehata, S., Wipf, D., Matsushita, Y., Aizawa, K.: Robust photometric stereo via low-rank matrix completion and recovery. In: Proceedings of IEEE Conference on Computer Vision and Pattern Recognition (CVPR) (2012)
36. Harker, M., O'leary, P.: Regularized reconstruction of a surface from its measured gradient field. J. Math. Imaging Vis. **51**, 46–70 (2015)
37. Reddy, D., Agrawal, A.K., Chellappa, R.: Enforcing integrability by error correction using l1-minimization. In: Proceedings of IEEE Conference on Computer Vision and Pattern Recognition (CVPR), pp. 2350–2357 (2009)
38. Avron, H., Maymounkov, P., Toledo, S.: Blendenpik: supercharging lapack's least-squares solver. SIAM J. Sci. Comput. **32**, 1217–1236 (2010)

Radiometry Propagation to Large 3D Point Clouds from Sparsely Sampled Ground Truth

Thomas Höll$^{(\boxtimes)}$ and Axel Pinz

Institute of Electrical Measurement and Measurement Signal Processing,
Graz University of Technology, Graz, Austria
thomas.hoell@tugraz.at

Abstract. Good radiometry of a 3D reconstruction is essential for digital conservation and versatile visualization of cultural heritage artifacts and sites. For large sites, "true" radiometry for the complete 3D point cloud is very expensive to obtain. We present a method that is capable to reconstruct the radiometric surface properties of an entire scene despite the fact that we only have access to the "true" radiometry of a small part of it. This is done in a two stage process: First, we transfer the radiometry to spatially corresponding parts of the scene, and second, we propagate these values to the entire scene using affinity information. We apply our method to 3D point clouds and 2D images, and show excellent quantitative and visually pleasing qualitative results. This approach can be of high value in many applications where users want to improve phototextured models towards high-quality yet affordable radiometry.

1 Introduction

This paper presents a novel method to significantly improve the radiometric quality of image-based 3D models. Typically, such 3D models are generated from a vast amount of individual images, taken by either specific equipment, e.g. an unmanned aerial vehicle, UAV, equipped with a consumer camera [1], or from unordered photo collections taken by many different people, using many different cameras, under different seasonal, diurnal, illumination and weather conditions (e.g. [2]). In the course of the 3D reconstruction process, each individual 3D point is visible in many images and finally is assigned a colour value that is calculated from the various colours extracted from these images (e.g. average, median, etc. [3,4]). This processing results in a phototextured 3D point-cloud or triangulated mesh that can also be visualized under various illumination conditions, but will always exhibit the particular phototexture, as captured by the camera(s), under the particular illumination conditions at the time of capture. However, true radiometric surface properties would be highly desired for digital models to provide a conservation of the "true" geometry *and radiometry* at the time of recording, i.e. e-heritage, and to provide flexible realistic visualization of coloured 3D models under arbitrary illumination conditions. Applications range from architecture (indoor and outdoor), to large-scale 3D city models, forensic crime scene reconstruction, and cultural heritage.

© Springer International Publishing AG 2017
C.-S. Chen et al. (Eds.): ACCV 2016 Workshops, Part II, LNCS 10117, pp. 222–235, 2017.
DOI: 10.1007/978-3-319-54427-4_17

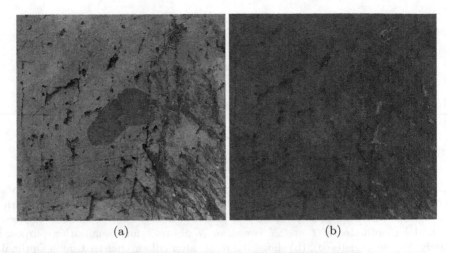

(a)   (b)

Fig. 1. 3D point clouds in an archaeological application. (a) shows part of a large rock panel that has been reconstructed at low spatial resolution using a UAV, plus a smaller area that has been scanned at high spatial resolution and true radiometry. We first transfer the true radiometry to the spatially registered subregion of the low resolution point cloud, and then propagate this radiometry to the remainder of the reconstruction, as shown in (b).

Imagine a large, phototextured 3D reconstruction *plus* sparsely sampled high quality radiometry data. Figure 1 illustrates the idea for an archaeological use case, where Fig. 1a shows part of a large rock panel that has been reconstructed at low spatial resolution using a UAV, plus a smaller area that has been scanned at high spatial resolution and true radiometry. In this paper, we present a method to (a) transfer the true radiometry to the spatially registered subregion of the low resolution point cloud, and (b) propagate this radiometry to the remainder of the reconstruction, as shown in Fig. 1b.

Figure 2 outlines our proposed method in more detail. First, we assume to have a possibly large-scale 3D reconstruction at a certain spatial resolution. Regarding its surface properties we assume that the images were taken under diffuse illumination conditions resulting in a reconstruction coloured with the photo texture of the images, but without direct, cast shadows. Second, we assume that we access to an accurate radiometric reconstruction of parts of the same scene, possibly captured at a different, more detailed spatial scale. Here, our concept builds on the idea, that several *radiometrically relevant* parts of the scene have been scanned - it would be unrealistic to try to solve the problem for surfaces exhibiting completely different radiometric properties than the samples taken. We map these accurate radiometric values to the entire 3D scene in a two-stage process: First, we transfer the "true" radiometric values to the photo texture of the spatially corresponding part of the scene, and second, we propagate these values to the entire scene using affinity information.

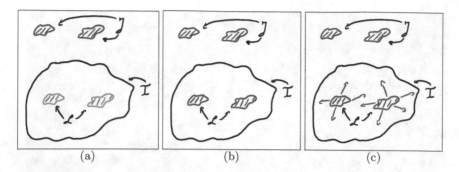

Fig. 2. Description of our method. (a) sketches the initial setting. We have two types of point clouds: $\mathcal{I}$, which is coloured with photo texture and $\mathcal{J}$, for which radiometric surface properties are available. The set of points in $\mathcal{L}$ is the set of matches between the spatial coordinates of $\mathcal{I}$ and $\mathcal{J}$ (we draw $\mathcal{J}$ above $\mathcal{I}$ for visualization purpose, actually they are registered). (b) shows the result after colour transfer trough Optimal Transport as described in Sect. 2. We transfer the colour distribution from point cloud $\mathcal{J}$ to the points in the set $\mathcal{L} \subset \mathcal{I}$. (c) sketches the colour propagation approach in which we propagate the known radiometric values to the areas where only photo texture is available. We describe this process in Sect. 3. (Color figure online)

In our experimental validation, we show that this method generates visually pleasing qualitative results, which, in quantitative evaluation, are very similar to ground truth measurements.

2   Colour Transfer Through Optimal Transport

Colour transfer was originally used to transfer the colour characteristic of one image to another one. Seminal work about colour transfer between two images was published by Reinhard *et al.* [5], in which the authors use a colour space that minimizes the correlation between colour channels [6] and simple statistical analysis to transfer the colour characteristic between two images. Pitié *et al.* [7,8] transform one N-dimensional colour distribution into another N-dimensional distribution by iterativly calculating their marginal distribution with subsequent histogram specification [9]. This provides more flexibility than [5] because one is not restricted to a certain distribution. In [10], Rabin *et al.* established a link between this prior work on distribution transfer and the *Wasserstein metric*.

This section explains how we transfer colour values between spatially overlapping areas were both is available: the "true" radiometric surface properties and the photo texture. This process must be able to handle inaccurate spatial registration as well as potentially different spatial scales of the two 3D point clouds. Therefore, we prefer not to depend on the spatial correspondence but instead transfer the global colour characteristic between the patch with "true" radiometry and the corresponding patch of the photo-textured data.

We define the set $\mathcal{I}$ as point cloud where only the photo texture is known, the set $\mathcal{J}$ as the point cloud where we have access to the radiometric surface properties, and the set $\mathcal{L} = \texttt{match}(\mathcal{I}, \mathcal{J})$ as the set of matches between the spatial coordinates of point cloud $\mathcal{I}$ and $\mathcal{J}$. This means that the set $\mathcal{L}$ contains the points of $\mathcal{I}$ ($\mathcal{L} \subset \mathcal{I}$) for which we know the radiometric surface properties but because of different scales and inaccurate registration, we have no point-wise match between $\mathcal{J}$ and $\mathcal{L}$. Next, we denote $\mathbf{r}_j = [r_r, r_g, r_b]$ as the "true" radiometry of a point $j \in \mathcal{J}$ expressed in RGB-colour space and $\mathbf{p}_i = [p_r, p_g, p_b]$ as the corresponding RGB-vector of the photo texture of a point $i \in \mathcal{I}$. We represent the two matrices $\mathbf{R} = \{\mathbf{r}_j\}_{j \in \mathcal{J}}$ and $\mathbf{P} = \{\mathbf{p}_i\}_{i \in \mathcal{L} \subset \mathcal{I}}$ as two distributions in three dimensional colour space and want to transform the distribution $\{\mathbf{p}_i\}$ such that it is as similar as possible to the distribution $\{\mathbf{r}_j\}$.

We utilize the concept of *Optimal Transport* (OT) [11] to achieve this desired colour transformation. To simplify the explanation of this concept let us assume that the number of points in $\{\mathbf{r}_j\}_{j \in \mathcal{J}}$ and $\{\mathbf{p}_i\}_{i \in \mathcal{L} \subset \mathcal{I}}$ is equal, where we denote L as the number of points. In OT one minimize the cost of transporting one distribution onto another distribution using an assignment σ:

$$\min_{\sigma \in \Sigma_L} \sum_{i \in \mathcal{L}} d_{Lab} \left(\mathbf{p}_i, \mathbf{r}_{\sigma(i)} \right)^2, \tag{1}$$

where Σ_L is the set of all permutations of L elements and d_{Lab} is the distance between two colour values in the *Lab*-colour space. The solution of this optimization problem is an assignment which map the photo texture $\mathbf{t}_i$ of point i to a radiometric surface colour $\mathbf{r}_{\sigma(i)}$, expressed as

$$\mathbf{p}_i \mapsto \mathbf{r}_{\sigma(i)} = \hat{\mathbf{r}}_i := [\hat{r}_r, \hat{r}_g, \hat{r}_b]. \tag{2}$$

We denote $\hat{\mathbf{r}}_i$ as the result of this mapping. One could use the flow obtained by calculating the *Earth Movers's Distance* [12] to find such a mapping but it suffers from the high ($O(n^3 \log n)$) computational complexity [13]. We therefore chose to compute the mapping using the *Sliced Wasserstein Distance* [10] approximation in which the *Wasserstein metric* is approximated by a set of random 1D projections of the distributions and calculating 1D assignments.

Rabin *et al.* [10] define the sliced Wasserstein distance as follows

$$\widetilde{W}(\mathbf{P}, \mathbf{R}) = \int_{\Theta} \underbrace{\min_{\sigma_\Theta} \sum_i \left(\langle \mathbf{p}_i - \mathbf{r}_{\sigma_\Theta(i)}, \Theta \rangle \right)^2}_{\mathcal{E}(\mathbf{P})} d\Theta, \tag{3}$$

in which Θ is a 1D line onto which $\mathbf{p}_i$ and $\mathbf{r}_{\sigma_\Theta(i)}$ are projected. In Eq. 1 we calculate the distance between colours in the *Lab*-colour space. For consistency, for the remainder of this section $\{\mathbf{p}_i\}$ and $\{\mathbf{r}_i\}$ are converted into this colour space. Our goal is to transform $\mathbf{P}$, making it as similar as possible to $\mathbf{R}$, to minimize $\widetilde{W}(\mathbf{P}, \mathbf{R})$, which approaches a minimum, if, for every Θ, $\mathcal{E}(\mathbf{P})$ is minimized. Because of the 1D projection in $\mathcal{E}(\mathbf{P})$ one can minimize this term w.r.t. $\mathbf{P}$ by calculating a *Histogram Specification* [9] of the projected points

$$\{r_{\Theta, i}\} = \texttt{hist\_spec}(\text{proj}_\Theta \mathbf{P}, \text{proj}_\Theta \mathbf{R}), \tag{4}$$

where the operator proj_Θ projects points onto the 1D line Θ, and hist_spec($\{x_i\}, \{y_i\}$) performs histogram specification such that $\{y_i\}$ matches $\{x_i\}$ as closely as possible (which also holds for different numbers of points in $\{x_i\}$ and $\{y_i\}$). Minimizing 3 with respect to $\mathbf{P}$

$$\mathbf{P}^\star = \{\hat{\mathbf{r}}_i\} = \underset{\mathbf{P}}{\operatorname{argmin}} \widetilde{W}(\mathbf{P}, \mathbf{R}), \tag{5}$$

can be done using *Stochastic Gradient Descent* [14] where we randomly generate the 1D line Θ. The update rule for $\mathbf{P}$ reads

$$\mathbf{P}^{t+1} = \mathbf{P}^t + \gamma \nabla_{\mathbf{P}} \mathcal{E}(\mathbf{P}), \tag{6}$$

where γ is the step size and $\nabla_{\mathbf{P}}\mathcal{E}(\mathbf{P})$ is the gradient of $\mathcal{E}(\mathbf{P})$. Computing this gradient

$$\nabla_{\mathbf{p}_i}\mathcal{E}(\mathbf{P}) = 2\Theta \left(\Theta^T \mathbf{p}_i - \underbrace{\Theta^T \mathbf{r}_{\sigma_\theta(i)}}_{r_{\Theta,i}} \right), \tag{7}$$

involves solving the histogram specification Eq. 4 and can be done very efficiently. The stochastic gradient descent algorithm terminates after a predefined, fixed number of iterations, or as soon as no further changes occur.

3   Radiometry Propagation

Why do we need to treat radiometry propagation different from colour transfer discussed in Sect. 2? Because we cannot rely on patches where we have both, "true" radiometry $\mathbf{r}$ and phototexture $\mathbf{p}$, but wish to propagate true radiometry into the unknown. In this case, we may observe similar phototexture, where it will make sense to propagate according radiometric values, but we also may observe significantly different phototexture, which has not been covered by the "true radiometric" scans at all. These values should not be included into the colour distributions used in Sect. 2. We therefore require a different method to solve this case and explain our solution, which is based on label propagation, below.

The result of colour transfer described in the previous section is a set $\{\hat{\mathbf{r}}_i\}_{i \in \mathcal{L}}$ of radiometric values expressed as RGB colour vectors. This set of vectors is associated with a subset of colour values of the initial photo textured point cloud $\mathcal{I}$. The goal is to propagate the available radiometric values to those areas $\{\mathbf{p}_i\}_{i \in \mathcal{I} \setminus \mathcal{L}}$ in the point cloud where we have no information about the radiometry. We use ideas from the field of semi-supervised learning, namely *Label Propagation* (LP) by Zhu and Ghahramani [15], to propagate the radiometric values to the entire scene. For this sake we first construct a matrix $\hat{\mathbf{R}}_l = \{\hat{\mathbf{r}}_i\}_{i \in \mathcal{L}}$ containing the known (labeled) radiometric values and second, a matrix $\hat{\mathbf{R}}_u = \{\mathbf{p}_i\}_{i \in \mathcal{I} \setminus \mathcal{L}}$ containing the photo texture (unlabeled). Next, we concatenate both matrices to obtain $\hat{\mathbf{R}} = (\hat{\mathbf{R}}_l, \hat{\mathbf{R}}_u)$. In the same way we order the colour values in the point cloud $\mathcal{I}$ according to the criterion, whether we can associate a radiometric colour value or not.

To set up our LP method we construct an affinity matrix $\mathbf{W}$ consisting of Gaussian kernels

$$w_{ij} = \exp\left(-\frac{d_{Lab}(\mathbf{p}_i, \mathbf{p}_j)^2}{2\sigma_{Lab}^2}\right),\tag{8}$$

in which i and j are points from the point cloud $\mathcal{I}$, d_{Lab} is the distance in the Lab-colour space, and σ_{Lab} is the width of the Gaussian kernel. To illustrate the concept behind $\mathbf{W}$, it can be interpreted as a graph whose nodes are the points in $\mathcal{I}$, and the edge weights are proportional to the colour similarity between points in $\mathcal{I}$. Because of the high storage requirements and computational burden of such an affinity matrix we limit the computation of the affinity to the k nearest neighbours in spatial and radiometric domain, by considering only a k-neighborhood for every point i

$$j \in \mathsf{N}_{k,Lab}(\mathbf{p}_i) \cup \mathsf{N}_{k,spatial}(\mathbf{p}_i),\tag{9}$$

where $\mathsf{N}_{k,Lab}$ denotes the set of radiometric neighbours in the Lab-colour space, $\mathsf{N}_{k,spatial}$ the set of spatial neighbours, and k is the number of neighbours to consider. The diagonal degree matrix $\mathbf{D}$ is constructed according to

$$d_{ii} = \sum_{j \in I} w_{ij}.\tag{10}$$

Using $\mathbf{W}$ and $\mathbf{D}$ we define the probabilistic transition matrix $\mathbf{T}$ as

$$\mathbf{T} = \mathbf{D}^{-1}\mathbf{W}.\tag{11}$$

The element t_{ij} of $\mathbf{T}$ corresponds to the probability of jumping from node j to node i. According to Zhu and Ghahramani [15] we can propagate our radiometric values (labels) through the graph by the following algorithm:

- Initialize: $\hat{\mathbf{R}}^t = (\hat{\mathbf{R}}_l^0, \hat{\mathbf{R}}_u)$ with $\hat{\mathbf{R}}_l^0 = \hat{\mathbf{R}}_l$.
- Iterate:
  1. Propagate: $\hat{\mathbf{R}}^{t+1} = \mathbf{T}\hat{\mathbf{R}}^t$.
  2. Persist in known radiometry: $\hat{\mathbf{R}}_l^{t+1} = \hat{\mathbf{R}}_l^0$.

Step 1 propagates the radiometric values to their neighbourhood, and step 2 ensures that the initial "true" radiometry does not fade out. This algorithm terminates either after a predefined, fixed number of iterations, or as soon as no further changes occur (i.e. $\hat{\mathbf{R}}_u^{t+1} = \hat{\mathbf{R}}_u^t$). Finally, the new, propagated colour values in $\hat{\mathbf{R}}_u^{t+1}$ are stored in the corresponding points in $\mathcal{I}$.

4   Experiments

Our main application in this paper deals with the radiometrically correct reconstruction of potentially large 3D rock-art sites. Section 4.1 provides an example that illustrates the difficulties of 3D data collection for a large rock-panel that contains many individual petroglyphs. On the one hand, the petroglyphs themselves need to be scanned at high spatial resolution and "true" radiometry, on the other hand, the whole rock panel is available only as a lower resolution point cloud with phototexture (see Figs. 3 and 4). In addition, we present further quantitative validation for a more colourful 3D point cloud in Sect. 4.2 and for radiometry propagation in 2D images in Sect. 4.3.

4.1 Radiometric Correction of a Large-Scale Cultural Heritage Reconstruction

For this experiment we consider a real world archaeological use case[1]. Recent advance in the field of autonomous image capture for photogrammetry, e.g. Mostegel *et al.* [1], allows the reconstruction of large-scale scenes with minimal effort. However, the colour values of such a reconstruction are still calculated based on the captured images and hence, no actual radiometric surface colour is available. On the other hand, for some parts of the scene one could obtain accurate radiometric values, using a special purpose scanner, e.g. Höll and Pinz [16]. Our approach can combine reconstructions within this setting. We consider this example a "large-scale cultural heritage reconstruction", because the rock panel itself has a diameter of approx. 19m and is reconstructed at the cm-scale leading to a 3D point cloud of 21 million points, but the individual petroglyphs (i.e. the rock-art itself) is reconstructed at much finer spatial resolution of 0.1 mm, leading to huge 3D point clouds of 100 million points for rather small portions of the rock-panel. Figure 3 illustrates this situation, and Fig. 4 provides an example of a single petroglyph reconstructed at highest spatial detail and "true" radiometry.

We have two types of reconstructions at hand: first, a large-scale reconstruction (`aerial`) based on images obtained by a micro-aerial vehicle and second, a set of radiometrically corrected small-scale, but high detail, reconstructions (A, B, C, D) (see Fig. 3). Given these reconstructions we perform first the colour transfer as explained in Sect. 2 and second, the radiometry propagation described in Sect. 3.

Table 1. Parameters for the experiments of Sect. 4.1.

σ_{Lab}	1.0
Neighborhood size of $N_{k,Lab}$	15
Neighborhood size of $N_{k,spatial}$	15
Number of iterations	2500

---

[1] We thank the authors of [1] for providing their data, and use our own high-resolution, "true" radiometry data [16].

Fig. 3. Images of the used dataset for the radiometric correction of 3D point clouds. (a) shows the large-scale photo textured `aerial` reconstruction. (b) shows the small-scale, high resolution, reconstructions with "true" radiometry (top row: A, B, C, bottom row D). (c) visualizes in red the registration of A, B, C and D w.r.t. `aerial`. The number of 3D points for the individual reconstructions is as follows: `aerial`: 21 million, A: 63 million, B: 23 million, C: 52 million, and D: 273 million points.

Fig. 4. 3D reconstruction of an individual petroglyph showing a hunter with bow. The point cloud consists of 21.7 million points and the petroglyph covers a tiny area of approx. $10\,cm^2$ on the rock.

Table 2. Leave-one-out evaluation of radiometry propagation on the 3D point cloud.

Omitted data:	A	B	C	D
d_{Lab}:	5.95	3.58	5.96	5.70

Figure 5 shows the initial point cloud `aerial` before (a) and after (c) radiometric correction (see parameter settings for this experiment in table 1). In addition to this qualitative, visually pleasing result, we provide a quantitative leave-one-out evaluation shown in table 2: We correct `aerial` by using just three out of four "true" radiometry reconstructions and use the remaining one to validate the performance. As performance measure we use the mean Euclidean distance in *Lab*-colour space.

(a)   (b)   (c)

Fig. 5. This figure shows the progress of the radiometric value propagation explained in Sect. 3. (a) shows the initial 3D point cloud. (b) shows an intermediate result (after 700 iterations) and (c) the final result at iteration 2500.

4.2   Radiometric Correction of a Colourful Painting

One might argue that the point clouds processed in Sect. 4.1 are not very colourful. Therefore, this section provides an example for the 3D point cloud of a colourful painting. Figure 6 shows our experimental setting. We have a photo-textured 3D point cloud of the whole painting, three point clouds representing subregions of the painting with radiometrically accurate colour values (see Fig. 6a for the phototextured pointcloud with the three subregions superimposed), and a complete, radiometrically accurate reconstruction of the painting that serves as ground truth to validate our approach (see Fig. 6c).

(a)   (b)   (c)

Fig. 6. Radiometric correction of a colourful painting. (a) shows the initial point cloud with three regions of "true" radiometry superimposed. (b) shows the final result after radiometry propagation and (c) the ground truth. The mean Euclidean distance in *Lab*-colour space between the final result and the ground truth is $d_{Lab} = 6.92$. (Color figure online)

As described in Sect. 2 we first transfer the radiometric values onto the photo-textured point cloud. Next, the radiometric values are propagated across the photo-textured reconstruction (as explained in Sect. 3). Our result is shown in Fig. 6b. Parameter settings for this experiments are as follows: $N_{k,Lab} = 15$, $N_{k,spatial} = 15$, and 2500 iterations. In comparison with the ground truth point cloud, we obtain a mean Euclidean distance in Lab-colour space of $d_{Lab} = 6.92$.

This experiment clearly demonstrates the benefits of our method. If small portions of all radiometrically meaningful colours of a scene are scanned, these "true" colours can successfully be propagated to a potentially much larger photo-textured 3D point cloud.

4.3 Quantitative Validation of Radiometry Propagation in the Image Domain

In this section, we provide further quantitative results for our radiometry propagation approach (Sect. 3) on the image dataset provided by Gehler $et~al.$ [17] and reprocessed by Lynch $et~al.$ [18]. This dataset consists of 482 3.2 MPixel images of outdoor and indoor scenes including a colour reference target. Based on this reference target, we apply colour correction [19] and define the corrected images as the set of images with "true" radiometry. For each of these images, we randomly sample a fraction of its pixel colours, except pixels within the colour reference target, and use them as the initial, "true" radiometric values (i.e., known labels, see Sect. 3). In Table 4, we show three different types of quantitative performance indicators for several fractions of sampled pixel values: first, $Structural~Similarity~SSIM$ [20] which is tailored to match the characteristics of the human visual system; second, $PSNR$ in dB; and third, Euclidean distance in Lab-colour space d_{Lab}. Pixels inside the colour reference target are omitted from this validation. Figure 7 shows radiometry propagation results for a sample image, and the parameters used in these experiments are provided in Table 3.

Table 3. Parameter settings for the random sampling experiment in Sect. 4.3.

σ_{Lab}	1.0
Neighborhood size of $N_{k,Lab}$	25
Neighborhood size of $N_{k,spatial}$	4
Number of iterations	120

We are fully aware that random sampling of individual points of the "true" radiometric values is not a realistic setting. However, we conducted this experiment to obtain statistically meaningful quantitative results on a large number of images, varying the sample size between 1% and 30% of the pixels. To conclude our experiments, we provide two examples for the more realistic scenario, where a supervisor decides which parts of an image need to be sampled to reconstruct the remaining part via radiometry propagation. We show experiments on two images of the dataset [17], IMG_0284 and IMG_0881, where we manually selected

(a) Original image   (b) "True" radiometry

(c) Result for a sample size of 30%.   (d) Result for a sample size of 5%.

Fig. 7. Radiometry propagation example on an image from [17]. For visualization purpose, the original (much darker) image (a) has been *Gamma* corrected. (b) shows the corresponding image with "true" radiometry, from which we randomly sample pixel values. Quantitative results for (c) are $PSNR = 35.45dB$, $SSIM = 0.92$, and $d_{Lab} = 3.14$. For (d), $PSNR = 30.89dB$, $SSIM = 0.852$, and $d_{Lab} = 5.56$. We omit the pixel values of the colour reference target for both, sampling and validation. (Color figure online)

Table 4. Results of radiometry propagation on the dataset provided by [17]. Sample size is given in % of pixels randomly sampled from the "true" radiometry image. Mean and standard deviation are calculated for the complete dataset of 482 images.

Sample size	$PSNR$		$SSIM$		d_{Lab}	
	Mean	Std	Mean	Std	Mean	Std
30%	35.05	3.55	0.92	0.05	3.41	1.41
20%	34.10	3.40	0.90	0.06	3.93	1.59
10%	32.42	3.27	0.88	0.07	4.80	1.84
5%	30.16	3.14	0.84	0.08	5.91	2.08
1%	15.68	3.62	0.51	0.12	20.44	6.58

two small areas of radiometrically "true" values and propagated them across the entire image. Figure 8 shows the selected images, sampled areas and the result of this user guided approach. Table 5 provides the validation results in terms of $PSNR$, $SSIM$, and d_{Lab} and the parameter settings used for radiometry propagation.

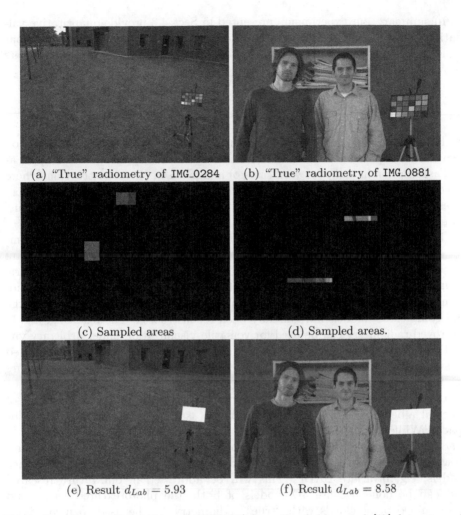

(a) "True" radiometry of IMG_0284   (b) "True" radiometry of IMG_0881

(c) Sampled areas   (d) Sampled areas.

(e) Result $d_{Lab} = 5.93$ (f) Result $d_{Lab} = 8.58$

Fig. 8. Two images (IMG_0284 and IMG_0881) from the Gehler *et al.* [17] dataset used for radiometry propagation. (a) and (b) show the "true" radiometry from which we sample values according to the areas shown in (c) and (d). (e) and (f) show the result of radiometry propagation and the distance in *Lab*-colour space (further quantitative measures can be found in Table 5). As expected, regions with colours that are not covered by the manually selected areas are not well reconstructed, *e.g.* the white T-shirt or the magazines on the book shelf. (Color figure online)

Table 5. Quantitative results of user guided radiometry propagation.

Image name	$PSNR$	$SSIM$	d_{Lab}	$N_{k,Lab}$	$N_{k,spatial}$	Number of iterations
IMG_0284	26.88	0.81	5.93	125	4	1000
IMG_0881	27.14	0.77	8.58	125	4	2500

In summary, our experiments presented in Sect. 4.3 provide quantitative validation of our radiometry propagation approach (Sect. 3). Our results show that "true" radiometry can be successfully propagated even if the sample size covers only a few percent of the data.

5   Conclusions

We have presented a method to estimate the radiometry of a large scene based on a sparsely sampled set of its "true" radiometry. This is a quite realistic application assumption in many relevant cases of image-based 3D reconstruction, where the complete sampling of the radiometry of a scene is prohibitive, for instance in cases of very large scenes, or when the available images are captured by various, potentially uncalibrated devices.

There are many benefits of the proposed method. In the case that the number of radiometrically relevant parts of a scene is managable, and when they are accessible for high quality scanning, the required sparsely sampled radiometric ground truth can be easily collected. Furthermore, this can be done on demand, even *after* a photocollection has been captured and when the potentially large and complex 3D models have already been reconstructed. The resulting, highly improved radiometry can be used for versatile visualization under arbitrary virtual illumination conditions, leading to highly realistic images, because unwanted bias regarding scene geometry and appearance of objects due to phototexture has been eliminated by our approach.

Our current algorithm propagates radiometric surface properties in terms of spectral albedo of individual surface points (as measured, e.g., in RGB color space). While this approach is fully valid only for Lambertian surfaces, it has, in practice, shown at least visually very plausible results for many natural scenes. However, the same concept will be extendable towards more complex reflectance models, e.g. bivariate BRDF. This will be addressed in our future research and can be done because 3D models of both, the phototextured scene, and the scanned point clouds with "true" radiometry can be used to incorporate information about surface normals, scanner illumination and viewing directions.

Acknowledgements. The research leading to these results was partly funded by the EC FP7 project 3D-PITOTI (ICT-2011-600545). The colourful painting in Sect. 4.2 is a 3D scan of a reproduction of a painting by August Macke. We thank ArcTron 3D GmbH (http://www.arctron.de) for providing us the data. We thank the Institute for Computer Graphics and Vision (ICG, TU Graz) for providing us the large-scale 3D reconstruction for our experiments in Sect. 4.1. We also thank MiBACT-SBA Lombardia and the Parco Archeologico Comunale di Seradina-Bedolina for permission to scan at Seradina I rock 12C. We appreciate the permission to use an academic license of the SURE software package [3] for dense 3D reconstruction.

References

1. Mostegel, C., Rumpler, M., Fraundorfer, F., Bischof, H.: UAV-based autonomous image acquisition with multi-view stereo quality assurance by confidence prediction. In: Proceedings of CVPR Workshop (2016)
2. Agarwal, S., Furukawa, Y., Snavely, N., Simon, I., Curless, B., Seitz, S., Szeliski, R.: Building rome in a day. Commun. ACM **54**, 105–111 (2011)
3. Rothermel, M., Wenzel, K., Fritsch, D., Haala, N.: SURE: photogrammetric surface reconstruction from imagery. In: Proceedings of LC3D Workshop (2012)
4. Furukawa, Y., Ponce, J.: Accurate, dense, and robust multiview stereopsis. IEEE Trans. Pattern Anal. Mach. Intell. **32**, 1362–1376 (2010)
5. Reinhard, E., Adhikhmin, M., Gooch, B., Shirley, P.: Color transfer between images. IEEE Comput. Graph. Appl. **21**, 34–41 (2001)
6. Ruderman, D.L., Cronin, T.W., Chiao, C.C.: Statistics of cone responses to natural images: implications for visual coding. J. Opt. Soc. Am. A **15**, 2036–2045 (1998)
7. Pitié, F., Kokaram, A.C., Dahyot, R.: N-dimensional probablility density function transfer and its application to colour transfer. In: Proceedings of ICCV (2005)
8. Pitié, F., Kokaram, A.C., Dahyot, R.: Automated colour grading using colour distribution transfer. Comput. Vis. Image Underst. **107**, 123–137 (2007)
9. Gonzalez, R.C., Woods, R.E.: Digital Image Processing, 3rd edn. Prentice-Hall Inc., Upper Saddle River (2006)
10. Rabin, J., Peyré, G., Delon, J., Bernot, M.: Wasserstein barycenter and its application to texture mixing. In: Bruckstein, A.M., Haar Romeny, B.M., Bronstein, A.M., Bronstein, M.M. (eds.) SSVM 2011. LNCS, vol. 6667, pp. 435–446. Springer, Heidelberg (2012). doi:10.1007/978-3-642-24785-9_37
11. Villani, C.: Topics in Optimal Transportation. Graduate Studies in Mathematics. American Mathematical Society, cop., Providence (2003)
12. Rubner, Y., Tomasi, C., Guibas, L.J.: A metric for distributions with applications to image databases. In: Proceedings of ICCV (1998)
13. Shirdhonkar, S., Jacobs, D.W.: Approximate earth mover's distance in linear time. In: Proceedings of CVPR (2008)
14. Bottou, L.: Online algorithms and stochastic approximations. In: Saad, D. (ed.) Online Learning and Neural Networks. Cambridge University Press, Cambridge (1998). Revised, October 2012
15. Zhu, X., Ghahramani, Z.: Learning from labeled and unlabeled data with label propagation. Technical report (2002)
16. Höll, T., Pinz, A.: Cultural heritage acquisition: geometry-based radiometry in the wild. In: Proceedings of 3D Vision (3DV) (2015)
17. Gehler, P.V., Rother, C., Blake, A., Minka, T., Sharp, T.: Bayesian color constancy revisited. In: Proceedings of CVPR (2008)
18. Lynch, S.E., Drew, M.S., Finlayson, G.D.: Colour constancy from both sides of the shadow edge. In: Proceedings of ICCV Workshop (2013)
19. Wolf, S.: Color correction matrix for digital still and video imaging systems. Technical report TM-04-406, National Telecommunications and Information Administration, Washington D.C. (2003)
20. Wang, Z., Bovik, A.C., Sheikh, H.R., Simoncelli, E.P.: Image quality assessment: from error visibility to structural similarity. IEEE Trans. Image Process. **13**, 600–612 (2004)

A 3D Reconstruction Method with Color Reproduction from Multi-band and Multi-view Images

Shuya Ito[1], Koichi Ito[1(⊠)], Takafumi Aoki[1], and Masaru Tsuchida[2]

[1] Graduate School of Information Sciences,
Tohoku University, Sendai 980-8579, Japan
ito@aoki.ecei.tohoku.ac.jp
[2] NTT Communication Science Laboratories, Atsugi 243-0198, Japan

Abstract. The purpose of digital archiving is the accurate rendering of authenticated contents in terms of color and shape. This paper proposes a novel 3D reconstruction method with color reproduction from multi-band and multi-view images. The proposed method consists of 5 steps: (i) feature tracking of multi-band and multi-view images, (ii) sparse 3D reconstruction using Structure from Motion (SfM), (iii) mesh generation from the sparse 3D point clouds, (iv) dense 3D reconstruction and (v) color reproduction. We create multi-band and multi-view image datasets of Japanese dolls with Kimono for quantitative performance evaluation. Through a set of experiments using our datasets, we confirm that our proposed method exhibits efficient performance of reconstructing a high-quality 3D model from multi-band and multi-view images compared with conventional methods.

1 Introduction

Digital archiving provides digital preservation of cultural resources and art by measuring them precisely for cultural, educational and scientific purposes. The goal of digital archiving is the accurate rendering of authenticated contents in terms of color and shape.

Accurate color of a target object can be reproduced from multi-band images which are images captured by different bands of wavelength [1]. An RGB image does not represent true color information, since the value of RGB varies depending on the circumstance of image acquisition. On the other hand, multi-band images, which are obtained by separating the wavelength band of visible light into more than 4 bands, make it possible to measure true color information from each reflectance distribution.

Accurate shape of a target object can be measured by 3D measurement systems with laser scanning or structured light projection. These systems require special measurement equipments depending on target objects and the application of these systems is limited, since the measurement equipments are relatively large and heavy. Addressing the above problem, we can utilize Multi-View Stereo

© Springer International Publishing AG 2017
C.-S. Chen et al. (Eds.): ACCV 2016 Workshops, Part II, LNCS 10117, pp. 236–247, 2017.
DOI: 10.1007/978-3-319-54427-4_18

(MVS), which is a technique used to reconstruct the 3D shape of an object using a set of images taken from different viewpoints [2–4].

So far, there are some methods for color reproduction from multi-band images [1,5] and for 3D reconstruction from multi-view images [3,4,6–8]. There is no method combined with color reproduction and 3D reconstruction from multi-band and multi-view images, although multi-band and multi-view images can be acquired by a camera and illumination. This paper proposes a novel 3D reconstruction method with color reproduction from multi-band and multi-view images. The proposed method consists of 5 steps: (i) feature tracking of a set of multi-band and multi-view images, (ii) sparse 3D reconstruction using Structure from Motion (SfM) [2,9], (iii) mesh generation from the sparse 3D point clouds, (iv) dense 3D reconstruction and (v) color reproduction.

The key technique for the proposed method is to use Phase-Only Correlation (POC) [10,11] for feature point tracking in step (i) and dense correspondence matching in step (iv). POC-based image matching is generally based on gray-scale images. To estimate accurate translational displacement from multi-band images, we modify POC-based image matching inspired by the approach proposed by Tsuchida et al. [12].

We create multi-band and multi-view image datasets of Japanese dolls with Kimono. Through a set of experiments using our datasets, we confirm that our proposed method exhibits efficient performance of reconstructing a high-quality 3D model from multi-band and multi-view images compared with conventional methods.

2   POC-Based Image Matching

This section briefly describes fundamentals of POC used for feature tracking and dense correspondence matching in the proposed method. We determine to use POC for MVS according to the report by Sakai et al. [13], where the use of POC made it possible to reconstruct more accurate 3D point clouds from multi-view images than those with Normalized Cross-Correlation (NCC) which is employed in most MVS methods.

Let $f(n_1, n_2)$ and $g(n_1, n_2)$ be the 2D image signals with $N_1 \times N_2$ pixels, where $-M_1 \leq n_1 \leq M_1$, and $-M_2 \leq n_2 \leq M_2$, $N_1 = 2M_1 + 1$ and $N_2 = 2M_2 + 1$. The normalized cross-power spectrum $R(k_1, k_2)$ is defined as

$$R(k_1, k_2) = \frac{F(k_1, k_2)\overline{G(k_1, k_2)}}{\left| F(k_1, k_2)\overline{G(k_1, k_2)} \right|}, \tag{1}$$

where $F(k_1, k_2)$ and $G(k_1, k_2)$ are the 2D Discrete Fourier Transforms (DFTs) of $f(n_1, n_2)$ and $g(n_1, n_2)$, $\overline{G(k_1, k_2)}$ denotes the complex conjugate of $G(k_1, k_2)$, $-M_1 \leq k_1 \leq M_1$ and $-M_2 \leq k_2 \leq M_2$. The 2D POC function $r(n_1, n_2)$ is given by 2D Inverse DFT (IDFT) of $R(k_1, k_2)$:

$$r(n_1, n_2) = \frac{1}{N_1 N_2} \sum_{k_1 k_2} R(k_1, k_2) W_{N_1}^{-k_1 n_1} W_{N_2}^{-k_2 n_2}, \tag{2}$$

where $\sum_{k_1 k_2}$ is $\sum_{k_1 = -M_1}^{M_1} \sum_{k_2 = -M_2}^{M_2}$, $W_{N_1} = \exp(-j\frac{2\pi}{N_1})$ and $W_{N_2} = \exp(-j\frac{2\pi}{N_2})$.

When two images are similar, their POC function gives a distinct sharp peak. When two images are not similar, the peak drops significantly. The height of the peak gives a good similarity measure for image matching, and the location of the peak shows the translational displacement between the images. As proposed by Takita et al. [14], there are some techniques for improving the accuracy of POC-based image matching: (i) function fitting for high-accuracy estimation of peak position, (ii) windowing to reduce boundary effects, (iii) spectral weighting for reducing aliasing and noise effects and (iv) correspondence matching with coarse-to-fine strategy using image pyramids. In this paper, we also employ the above techniques.

In the case of multi-band images, the accuracy of image matching using POC can be improved by combining POC functions calculated in each wavelength band as mentioned by Tsuchida et al. [12]. Let $f_i(n_1, n_2)$ and $g_i(n_1, n_2)$ be 2D image signals taken by an arbitrary band of wavelength, where i indicates i-th band, $i = 1, 2, \cdots, N$ and N is the number of wavelength bands. Let $R_i(k_1, k_2)$ be the normalized cross-power spectrum calculated for each band. The average normalized cross-power spectrum $\hat{R}(k_1, k_2)$ is calculated as a weighted mean of a set of $R_i(k_1, k_2)$ as follows:

$$\hat{R}(k_1, k_2) = \frac{\sum_{i=1}^{N} W_i(k_1, k_2) R_i(k_1, k_2)}{\sum_{i=1}^{N} W_i(k_1, k_2)}, \tag{3}$$

where $W_i(k_1, k_2)$ indicates a weight function for each band. In this paper, we employ an energy function of image signals as the weight function $W_i(k_1, k_2)$, which is defined by

$$W_i(k_1, k_2) = \left| F(k_1, k_2) \overline{G(k_1, k_2)} \right|. \tag{4}$$

This function represents a ratio of energy of frequency components of each band and affects a band with large S/N. POC function $\hat{r}(n_1, n_2)$ for multi-band images is calculated by IDFT of $\hat{R}(k_1, k_2)$.

3   3D Reconstruction from Multi-view Images

This section describes the proposed 3D reconstruction method, which consists of 4 steps: (i) feature tracking of a set of multi-band and multi-view images, (ii) sparse 3D reconstruction using SfM, (iii) mesh generation from the sparse 3D point clouds, (iv) dense 3D reconstruction as shown in Fig. 1. We explain the detail of each step in the following, where we assume that $f_i^l(n_1, n_2)$ indicates an image of l-th viewpoint and i-th band ($i = 1, \cdots, N$ and $l = 1, \cdots, L$).

(i) Feature Point Tracking

In this step, we repeat feature point detection and tracking from the initial image $f_i^1(n_1, n_2)$ to the last image $f_i^L(n_1, n_2)$ and then obtain the correspondence

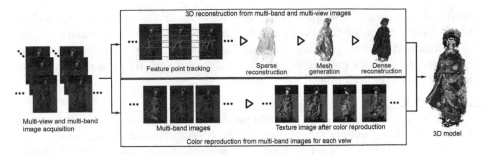

Fig. 1. An overview of the proposed method combined with 3D reconstruction and color reproduction. (Color figure online)

between each image, where this approach is inspired by Ishii et al. [15]. First, feature points are detected from image $f_s^l(n_1, n_2)$ with the center of wavelength bands, c, using the corner detection method proposed by Shi and Tomasi [16]. We introduce parameter D to control the density of extracted feature points, resulting in generating equivalent size of triangle meshes in step (iii). If feature points are extracted from the area within $\pm D$ pixels centered on a feature point, these points are removed. Next, we find the corresponding points in subsequent image $f_i^{l+1}(n_1, n_2)$ from the extracted feature points in current image $f_i^l(n_1, n_2)$ to track the feature points. We employ a correspondence matching method using POC for multi-band images as mentioned in Sect. 2 to find accurate corresponding points. If the corresponding point pair has a low correlation value of the POC function, we eliminate it as an outlier. Then, we extract feature points on $f_i^{l+1}(n_1, n_2)$ from the area without the feature point tracked from $f_i^l(n_1, n_2)$ for $f_i^{l+1}(n_1, n_2)$. By repeating the above processes until the last image $f_i^L(n_1, n_2)$, we can obtain a set of tracked feature points.

(ii) Sparse 3D Reconstruction

In this step, we estimate camera motion and reconstruct the sparse 3D point clouds of the object with SfM [2,9]. SfM repeats the linear solution and nonlinear optimization by sequentially adding images. The extrinsic camera parameters of l-th image $f_i^l(n_1, n_2)$ are estimated in the linear solution using the method proposed by Kneip et al. [17] from the geometric relation between the reconstructed 3D points and the coordinates of tracked feature points. Note that the extrinsic camera parameters of the first two images are estimated with the normalized five-point algorithm proposed by Nistér [18]. Note that the normalized five-point algorithm is used with RANdom SAmple Consensus (RANSAC) [19] to robustly estimate the parameters. The 3D points of tracked feature points are obtained using the estimated extrinsic parameters in the l-th image $f_i^l(n_1, n_2)$ according to triangulation. We refine the reconstructed 3D points and estimated camera parameters in nonlinear optimization by minimizing reprojection error using bundle adjustments [2,9]. The reprojection error is defined by the Euclidean distance $||\boldsymbol{m} - \boldsymbol{m}_{\mathrm{rep}}||^2$, where $\boldsymbol{m} = (u, v)$ is a feature point and $\boldsymbol{m}_{\mathrm{rep}}$ is a point obtaining by projecting a 3D point $\boldsymbol{M} = (X, Y, Z)$ onto the image using a projection matrix of a camera. We employ global and local bundle adjustments depending on the target range in this paper. Global bundle adjustment optimizes all the 3D points and camera parameters of all the images. Local bundle adjustment optimizes the camera parameters of the l-th image and 3D points observed in the position of the l-th image.

(iii) Mesh Generation

This step generates the 3D triangle meshes from the sparse 3D point clouds obtained in the previous step using the method proposed by Labatut et al. [20]. First, we generate a set of tetrahedrons from the sparse 3D point clouds using Delaunay triangulation, where each vertex corresponds to a 3D point. Next, we construct a graph whose vertices correspond to tetrahedrons, and edges correspond to triangular facets between adjacent tetrahedrons. Each tetrahedron is labeled as being inside or outside the scene by applying graph cuts to the constructed graph. Finally, the triangular meshes are then obtained by extracting the triangular facets between adjacent tetrahedrons having different labels.

(iv) Dense 3D Reconstruction

This step reconstructs accurate and dense 3D point clouds for each mesh using POC-based image matching with optimal viewpoint selection. We select an image located in front of the mesh as a reference camera position, C_{ref}, from a set of images having no occlusion. In this paper, we employ the method proposed by Havel and Herout [21] to detect occlusions. We select an image that has as wide a baseline as possible within the range of accurate stereo matching as the first target camera position, C_{tar}, where we employ the maximum convergence angle θ_{th}. We empirically determine $\theta_{\text{th}} = 15°$ in this paper. We also select an image located on the symmetrical position of C_{tar} as the second target camera position, C'_{tar}. We set two stereo camera pairs such as C_{ref}–C_{tar} and C_{ref}–C'_{tar} and carry out stereo matching with multi-band POC-based correspondence matching for the target triangular mesh. It may not obtain accurate correspondence between stereo images that have large perspective deformation using image matching-based approaches. Therefore, we employ the efficient approach of POC-based correspondence matching for small meshes as proposed by Ishii et al. [22]. This approach made it possible to achieve accurate stereo matching of POC by correcting the local deformation between images using projective transformation. The initial 3D points with constant density are placed on the mesh, and then the reference points for correspondence matching are obtained by projecting their points onto the reference frame of C_{ref} to obtain 3D points with constant density independent of the distance and relative angle between the mesh and C_{ref}. We obtain 3D points from corresponding point pairs between C_{tar} and C'_{tar} using stereo triangulation. If a corresponding point pair has a low correlation value of POC function and a large reprojection error to C_{ref}, such a point is eliminated as an outlier. We conduct stereo matching and 3D reconstruction again for such a reference point by changing the target camera position. We select the frame that has a maximum convergence angle from C'_{tar} or C_{tar} as the new C_{tar} or C'_{tar} from C_{match}, which has not yet been selected. We can reconstruct a 3D point from the optimal view for each reference point by the above iterative approach.

Finally, we obtain accurate 3D point clouds of the whole region by carrying out this procedure on all the meshes.

4   Color Reproduction from Multi-band Images

This section describes the color reproduction method from multi-band images for each viewpoint inspired by the procedure proposed by Tsuchida et al. [23]. Let $E(\lambda)$, $f(\lambda)$, N and $S_i(\lambda)$ be the illumination spectrum, spectral reflection, the number of bands and spectral sensitivity of i-th band, respectively, where λ indicates a wavelength. The signal value of i-th band can be represented by

$$g_i = \int S_i(\lambda)E(\lambda)f(\lambda)d\lambda, \tag{5}$$

where $i = 1, \cdots, N$. Note that the pixel coordinate is omitted in the above equation so as to discuss only one point on the image. Assuming $S_i(\lambda)E(\lambda) = h_i(\lambda)$, Eq. (5) can be written by

$$g_i = \int h_i(\lambda)f(\lambda)d\lambda. \tag{6}$$

Discretizing $h_i(\lambda)$ and $f(\lambda)$ with the fine interval of wavelength, Eq. (6) can be represented by

$$g = Hf, \tag{7}$$

where H is the $N \times P$ matrix whose element is $[H_{ip}] = h_i(\lambda_p)$, $f = (f(\lambda_1), \cdots, f(\lambda_p), \cdots, f(\lambda_P))$ and $g = (g_1, \cdots, g_N)^t$. P is the number of sampling points, e.g., $P = 401$ when sampling the range of wavelength 380 nm–780 nm by 1 nm. The spectral reflectance $\hat{f}$ is estimated using the Wiener estimation method [24] as follows:

$$\hat{f} = Mg, \tag{8}$$

where M is the Wiener estimation matrix obtained from H, which is defined by

$$M = RH^t \left(HRH^t\right)^{-1}. \tag{9}$$

R is *a priori* knowledge about the spectral reflectance of the object, which can be modeled by a first-order Markov process covariance matrix as

$$R = \begin{bmatrix} 1 & \rho & \rho^2 & \cdots & \rho^{L-1} \\ \rho & 1 & \rho & \cdots & \rho^{L-2} \\ \rho^2 & \rho & 1 & \cdots & \vdots \\ \vdots & \vdots & \vdots & \ddots & \vdots \\ \rho^{L-1} & \rho^{L-2} & \cdots & \cdots & 1 \end{bmatrix}, \tag{10}$$

where ρ represents a correlation value between adjacent elements. In this paper, we employ $\rho = 0.999$ from our experiments. Using the estimated spectral reflectance $\hat{f}$, the spectral power distribution of illumination for observation, and tone curves and chromaticity values of primary colors of display monitor, we can calculate the texture image after color reproduction.

5   Experiments and Discussion

This section describes performance evaluation of the proposed method. For this purpose, we make multi-band and multi-view image datasets of Japanese dolls with Kimono. We develop a system for acquiring multi-band and multi-view images as shown in Fig. 2. This system consists of a camera (Point Gray, Flea 3: FL3-U3-13E4C-C for RGB images and FL3-U3-13Y3M-C for multi-band images) with 1,280 × 1,024 pixels and a lens (μ-TRON, FV1022). The use of this system makes it possible to achieve automated image acquisition, since the rotating table, illumination and camera can be controlled by a PC. We use two dolls as a target object in the experiments as shown in Fig. 2. The target object is located on an automatic rotating table. The image is captured every 1° rotation with changing the wavelength of illumination. In this paper, we capture 8-band images whose center of wavelength is 450, 490, 530, 570, 610, 650, 690 and 730 nm. Figure 3 shows an example of multi-band images used in the experiment. To compare performance of the proposed method with the conventional method, RGB-color images are captured with white light. A 3D mesh model of the target object is measured with the laser scanner (KONICA MINOLTA VIVID) shown in Fig. 2 to quantitatively evaluate performance.

The intrinsic parameters of both camera are estimated by camera calibration using a planar checkerboard in advance, where we use Camera Calibration Toolbox for Matlab[1]. In the case of multi-band images, we take into account the chromatic aberration by changing parameters of lens distortion for each band.

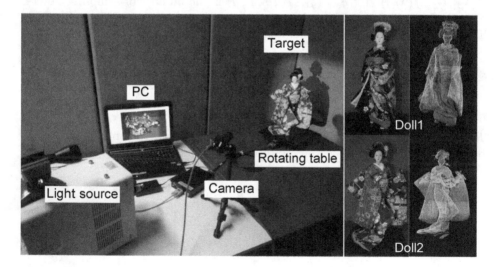

Fig. 2. The image acquisition system, target objects and their 3D mesh models. (Color figure online)

---

[1] http://www.vision.caltech.edu/bouguetj/calib_doc/.

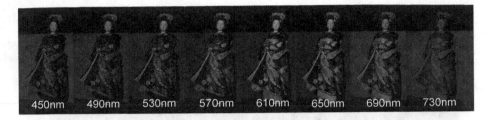

450nm   490nm   530nm   570nm   610nm   650nm   690nm   730nm

Fig. 3. Example of multi-band images of Doll 1.

We employ the conventional MVS method of VisualSFM [25] to estimate the camera positions and PMVS [7] to reconstruct 3D point clouds. To demonstrate effectiveness of multi-band images, we employ the proposed method with RGB-colored multi-view images.

We introduce the evaluation criteria $D_{accuracy}$ and $D_{complete}$, which are proposed by Jensen et al. [26]. $D_{accuracy}$ means a degree of correctness of the position of reconstructed points and is calculated as the distance from reconstructed points to the ground truth. $D_{complete}$ means a degree of coverage of reconstructed points and is calculated as the distance from the ground truth to reconstructed points. Note that $D_{accuracy}$ and $D_{complete}$ are calculated with points that have an error of less than 5 mm in the experiments.

Figure 4 shows reconstructed 3D point clouds, $D_{accuracy}$ and $D_{complete}$, while Table 1 shows Root Mean Square (RMS) of $D_{accuracy}$ and $D_{complete}$ and the number of 3D points for each method. As for the conventional method, VisualSFM+PMVS, the accuracy and completeness are worst and the number of reconstructed 3D points is less than others, although VisualSFM+PMVS is one of the accurate MVS methods. Comparing RGB-colored images and multi-band images, the accuracy and the completeness of the proposed method with multi-band images are better than that with RGB-colored images. The number of reconstructed 3D points for multi-band images is less than that for RGB-colored images. These results are depending on the shape of triangle meshes. In the case of RGB-colored images, the surface of the mesh model generated from sparse 3D point clouds is irregular. Therefore, the number of reference points for dense reconstruction is illegally increased. This results in producing a lot of 3D points with an error. On the other hand, in the case of multi-band images, the surface of the mesh model generated from sparse 3D point clouds is smoothed. The reference points are placed along the shape of the target object, resulting in producing accurate 3D point clouds. This result indicates that POC for multi-band images is effective for feature tracking and dense correspondence matching in the proposed method. In addition, the use of multi-band images makes it possible to reconstruct 3D shape even for black-colored regions of Kimono and hair, where other methods cannot reconstruct.

Figure 5 shows an RGB image and a texture image after color reproduction from multi-band images using the proposed method. Compared with both images, vivid color reproduction is achieved from multi-band images. The color information of the right image in Fig. 5 is the same as the actual looking by eye.

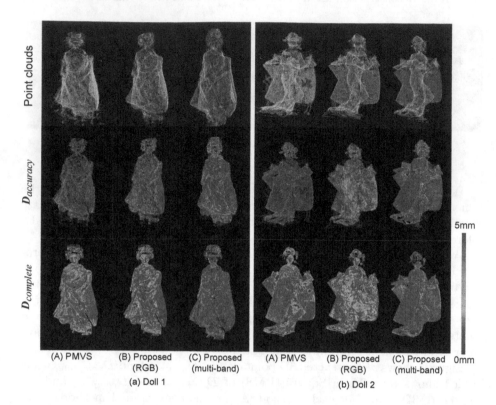

Fig. 4. Reconstructed 3D point clouds, $D_{accuracy}$ and $D_{complete}$ for (A) Visu-alSFM+MVS, (B) the proposed method with RGB-colored images and (C) the proposed method with multi-band images: (a) Doll 1 and (b) Doll 2. (Color figure online)

Table 1. RMS of $D_{accuracy}$ [mm], RMS of $D_{complete}$ [mm] and the number of 3D points for (A) VisualSFM+PMVS, (B) the proposed method with RGB-colored images and (C) the proposed method with multi-band images.

		(A) PMVS	(B) Proposed (RGB)	(C) Proposed (multi)
Doll 1	$D_{accuracy}$	1.58	1.53	1.30
	$D_{complete}$	1.61	1.22	1.17
	# of 3D points	78,731	236,769	214,119
Doll 2	$D_{accuracy}$	1.37	1.74	1.32
	$D_{complete}$	1.90	1.34	1.15
	# of 3D points	109,909	308,441	232,417

Figure 6 shows 3D point clouds with texture after color reproduction reconstructed by the proposed method. Almost the shape can be reconstructed and the color texture is clearly reconstructed. The region around the face is sparse compared with other regions. Corresponding points cannot be obtained from

Fig. 5. Example of an RGB image (left) and a texture image after color reproduction (right). (Color figure online)

Fig. 6. Reconstructed 3D point clouds with texture after color reproduction of Doll 1. (Color figure online)

such face region, since the texture of doll's face is constant and white color. On the other hand, the accurate mesh model can be reconstructed from the 3D point clouds so as to fill holes around the face region, since the accuracy of the reconstructed 3D points is high.

As observed above, the proposed method with multi-band and multi-view images exhibits efficient performance of creating 3D models for digital archiving.

6 Conclusion

This paper proposed the 3D reconstruction method with color reproduction from multi-band and multi-view images for digital archiving. The proposed method reconstructs accurate 3D shape from multi-band and multi-view images as well as accurate color information from multi-band images. Through a set of experiments, we demonstrated that the proposed method exhibits efficient performance of reconstructing high-quality 3D model from multi-band and multi-view images compared with conventional methods.

The target objects used in the experiments are hard to reconstruct their shape using MVS, since the shape of Japanese doll with Kimono is very complex. The dataset created in this paper is one of challenging datasets for MVS and is the first dataset of multi-band and multi-view images. The dataset will be available in our web page for performance evaluation of MVS algorithms.

References

1. Hardeberg, J.: Acquisition and Reproduction of Color Images: Colorimetric and Multispectral Approaches. Universal Publishers, Boca Raton (2001)
2. Szeliski, R.: Computer Vision: Algorithms and Applications. Springer-Verlag New York Inc., New York (2010)
3. Seitz, S.M., Curless, B., Diebel, J., Scharstein, D., Szeliski, R.: A comparison and evaluation of multi-views stereo reconstruction algorithms. In: Proceedings of the International Conference on Computer Vision and Pattern Recognition, pp. 519–528 (2006)
4. Strecha, C., von Hansen, W., Gool, L.V., Fua, P., Thoennessen, U.: On benchmarking camera calibration and multi-view stereo for high resolution imagery. In: Proceedings of the International Conference on Computer Vision and Pattern Recognition, pp. 1–8 (2008)
5. Vrhel, M., Trussell, H.: Color correction using principal components. Color Res. Appl. 17, 328–338 (1992)
6. Furukawa, Y., Curless, B., Seitz, S.M., Szeliski, R.: Towards internet-scale multiview stereo. In: Proceedings of the International Conference on Computer Vision and Pattern Recognition, pp. 1434–1441 (2010)
7. Furukawa, Y., Ponce, J.: Accurate, dense, and robust multiview stereopsis. IEEE Trans. Pattern Anal. Mach. Intell. 32, 1362–1376 (2010)
8. Wu, C., Liu, Y., Dai, Q., Wilburn, B.: Fusing multiview and photometric stereo for 3D reconstruction under uncalibrated illumination. IEEE Trans. Vis. Comput. Graph. 17, 1082–1095 (2011)
9. Hartley, R., Zisserman, A.: Multiple View Geometry. Cambridge University Press, Cambridge (2004)
10. Kuglin, C.D., Hines, D.C.: The phase correlation image alignment method. In: Proceedings of the International Conference on Cybernetics and Society, pp. 163–165 (1975)
11. Takita, K., Aoki, T., Sasaki, Y., Higuchi, T., Kobayashi, K.: High-accuracy subpixel image registration based on phase-only correlation. IEICE Trans. Fundam. **E86-A**, 1925–1934 (2003)

12. Tsuchida, M., Sakai, S., Ito, K., Kashino, K., Yamato, J., Aoki, T.: Efficient POC-based correspondence detection method for multi-channel images. In: Proceedings of Color and Imaging Conference, pp. 113–118 (2014)
13. Sakai, S., Ito, K., Aoki, T., Masuda, T., Unten, H.: An efficient image matching method for multi-view stereo. In: Lee, K.M., Matsushita, Y., Rehg, J.M., Hu, Z. (eds.) ACCV 2012. LNCS, vol. 7727, pp. 283–296. Springer, Heidelberg (2013). doi:10.1007/978-3-642-37447-0_22
14. Takita, K., Muquit, M.A., Aoki, T., Higuchi, T.: A sub-pixel correspondence search for computer vision applications. IEICE Trans. Fundam. **E87-A**, 1913–1923 (2004)
15. Ishii, J., Sakai, S., Ito, K., Aoki, T., Yanagi, T., Ando, T.: 3D reconstruction of urban environments using in-vehicle fisheye camera. In: Proceedings of the IEEE International Conference on Image Processing, pp. 2145–2148 (2013)
16. Shi, J., Tomasi, C.: Good features to track. In: Proceedings of the International Conference on Computer Vision and Pattern Recognition, pp. 593–600 (1994)
17. Kneip, L., Scaramuzza, D., Siegwart, R.: A novel parametrization of the perspective-three-point problem for a direct computation of absolute camera position and orientation. In: Proceedings of the International Conference on Computer Vision and Pattern Recognition, pp. 2969–2976. IEEE (2011)
18. Nistér, D.: An efficient solution to the five-point relative pose problem. IEEE Trans. Pattern Anal. Mach. Intell. **26**, 756–770 (2004)
19. Fischler, M.A., Bolles, R.C.: Random sample consensus: a paradigm for model fitting with applications to image analysis and automated cartography. Commun. ACM **24**, 381–395 (1981)
20. Labatut, P., Pons, J.P., Keriven, R.: Robust and efficient surface reconstruction from range data. Comput. Graph. Forum **28**, 2275–2290 (2009)
21. Havel, J., Herout, A.: Yet faster ray-triangle intersection (using SSE4). Trans. Vis. Comput. Graph. **16**, 434–438 (2010)
22. Ishii, J., Sakai, S., Ito, K., Aoki, T.: Wide-baseline stereo matching using ASIFT and POC. In: Proceedings of the IEEE International Conference on Image Processing, pp. 2977–2980 (2012)
23. Tsuchida, M., Sakai, S., Miura, M., Ito, K., Kawanishi, K., Kashino, K., Yamato, J., Aoki, T.: Stereo one-shot six-band camera system for accurate color reproduction. J. Electron. Imaging **23**, 033025-1–033025-12 (2013)
24. Pratt, W.K., Mancill, C.E.: Spectral estimation techniques for the spectral calibration of a color image scanner. Appl. Opt. **15**, 73–75 (1976)
25. Wu, C.: VisualSFM: A Visual Structure from Motion System. http://homes.cs.washington.edu/~ccwu/vsfm/
26. Jensen, R., Dahl, A., Vogiatzis, G., Tola, E., Anæs, H.: Large scale multi-view stereopsis evaluation. In: Proceedings of the IEEE Conference on Computer Vision and Pattern Recognition, pp. 406–413 (2014)

Multi-view Lip-Reading Challenges

Out of Time: Automated Lip Sync in the Wild

Joon Son Chung$^{(\boxtimes)}$ and Andrew Zisserman

Visual Geometry Group, Department of Engineering Science,
University of Oxford, Oxford, UK
`joon@robots.ox.ac.uk`

Abstract. The goal of this work is to determine the audio-video synchronisation between mouth motion and speech in a video.

We propose a two-stream ConvNet architecture that enables the mapping between the sound and the mouth images to be trained end-to-end from unlabelled data. The trained network is used to determine the *lip-sync error* in a video.

We apply the network to two further tasks: active speaker detection and lip reading. On both tasks we set a new state-of-the-art on standard benchmark datasets.

1 Introduction

Audio to video synchronisation (or lack of it) is a problem in TV broadcasting for the producer and the viewer. In television, a lip-sync error of up to several hundred milliseconds is not uncommon. The video usually lags the audio if the cause of the error is in the transmission. These errors are often noticeable – the threshold for detectability by an average viewer is around -125 ms (the audio lags the video) to $+45$ ms (the audio leads the video) [1].

In film production, audio to video synchronisation is a routine task, as the audio and the video are typically recorded using different equipment. Consequently, many solutions have been developed in this industry, the clapperboard being the most traditional one. Modern solutions use timecodes or sometimes time warping between the audio from the camera's built-in microphone and the external microphone, but it is not common to use the visual content as a guide to alignment.

Our objective in this work is to develop a *language independent* and *speaker independent* solution to the lip-sync problem, using only the video and the audio streams that are available to the TV viewer. The key contributions are the ConvNet architecture, and the data processing pipeline that enables the mapping between the sound and the mouth shapes to be learnt discriminatively from TV broadcast, without labelled data. To our knowledge, we are the first to end-to-end train a working AV synchronisation system.

This solution is of relevance to a number of different applications. We demonstrate that the method can be applied to three different tasks: (i) determining

© Springer International Publishing AG 2017
C.-S. Chen et al. (Eds.): ACCV 2016 Workshops, Part II, LNCS 10117, pp. 251–263, 2017.
DOI: 10.1007/978-3-319-54427-4_19

the *lip-sync error* in videos; (ii) detecting the speaker in a scene with multiple faces; and (iii) lip reading. The experimental performance on all of these tasks is extremely strong. In speaker detection and lip reading, our results exceed the state-of-the-art on public datasets, Columbia [4] and OuluVS2 [2].

1.1 Related Works

There is a large body of work on the audio to video synchronisation problem. The majority of these are based on methods that are not available to the television receiver (*e.g.* embedding timestamps in the transport stream); instead we focus on computer vision methods that only rely on the audio-visual data.

A number of papers have used *phoneme* recognition as a proxy task for solving the lip-sync problem. In Lewis [15], linear prediction is used to provide phoneme recognition from audio, and the recognised phonemes are associated with mouth positions to provide lip-sync video. Morishima *et al.* [19] classifies the face parameters into *visemes*, and uses the *viseme* to *phoneme* mapping to obtain the synchronisation. Although [13,18] do not explicitly classify the sounds into phonemes, their approaches are similar to those above in that they develop models by having the speaker record a set of vowels. Both [13,18] correlate face parameters such as jaw position to the FFT of the sound signal. Zoric and Pandzic [29] have used neural networks to tackle the problem. A multi-layer feedforward neural network is trained to predict the *viseme* from MFCC input vectors. A parametric face model is used for the visual processing. We do not make an intermediate classification of sounds and mouth shapes into vowels or *phonemes*.

More recent papers have attempted to find correspondence between speech and visual data without such labels. A number of approaches are based on canonical correlation analysis (CCA) [3,22] or co-inertia analysis (CoIA) [20] of audio and visual features (*e.g.* geometric parameters or 2D DCT features). The most related work to ours is that of Marcharet *et al.* [17] that uses a Deep Neural Network (DNN)-based classifier to determine the time offset based also on pre-defined visual features (speech class likelihoods, bottleneck features, etc.), whereas we learn the visual features directly.

Of relevance to the architectures developed in this paper are Siamese networks [6], in which similarity metrics are learnt for face classification without explicit class labels. [23,27] are also relevant in that they simultaneously train multi-stream networks in which the inputs are of different domains.

2   Representations and Architecture

This section describes the representations and network architectures for both the audio and the video inputs. The network ingests 0.2-s clips of each data type. In the dataset (Sect. 3), no explicit annotation (*e.g.* phonemes labels, or the precise time offset) is given for the audio-video data, however we make the assumption that in television broadcasts, the audio and the video are *usually* synced.

The network consists of two asymmetric streams for audio and video, each of which is described below.

2.1 Audio Stream

The input audio data is MFCC values. This is a representation of the short-term power spectrum of a sound on a non-linear mel scale of frequency. 13 mel frequency bands are used at each time step. The features are computed at a sampling rate of 100 Hz, giving 20 time steps for a 0.2-s input signal.

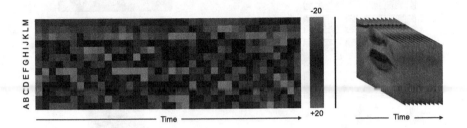

Fig. 1. Input representations. Left: temporal representations as heatmaps for audio. The 13 rows (A to M) in the audio image encode each of the 13 MFCC features representing powers at different frequency bins. Right: Grayscale images of the mouth area.

Representation. The audio is encoded as a heatmap image representing MFCC values for each time step and each mel frequency band (see Fig. 1). The top and bottom three rows of the image are reflected to reduce boundary effects. Previous work [9] has also attempted to train image-style ConvNet for similar inputs.

Architecture. We use a convolutional neural network inspired by those designed for image recognition. Our layer architecture (Fig. 2) is based on VGG-M [5], but with modified filter sizes to ingest the inputs of unusual dimensions. VGG-M takes a square image of size 224 × 224 pixels, whereas our input size is 20 pixels (the number of time steps) in the time-direction, and only 13 pixels in the other direction (so the input image is 13 × 20 pixels).

2.2 Visual Stream

Representation. The input format to the visual network is a sequence of mouth regions as grayscale images, as shown in Fig. 1. The input dimensions are $111 \times 111 \times 5$ ($W \times H \times T$) for 5 frames, which corresponds to 0.2-s at the 25 Hz frame rate.

Architecture. We base our architecture on that of [7], which is designed for the task of visual speech recognition. In particular, the architecture is based on the Early Fusion model, which is compact and fast to train. The *conv1* filter has been modified to ingest the 5-channel input.

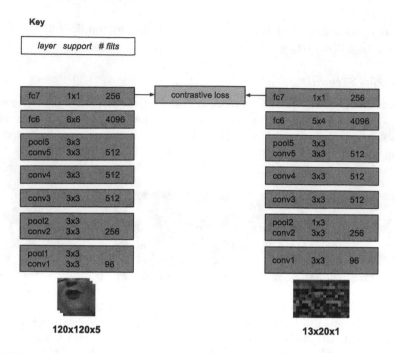

Fig. 2. Two-stream ConvNet architecture. Both streams are trained simultaneously.

2.3   Loss Function

The training objective is that the output of the audio and the video networks are similar for *genuine* pairs, and different for *false* pairs. Specifically, the Euclidean distance between the network outputs is minimised or maximised. We propose to use the contrastive loss (Eq. 1), originally proposed for training Siamese networks [6]. v and a are fc_7 vectors for the video and the audio streams, respectively. $y \in [0, 1]$ is the binary similarity metric between the audio and the video inputs.

$$E = \frac{1}{2N} \sum_{n=1}^{N} (y)\, d^2 + (1 - y) \max\left(margin - d, 0\right)^2 \tag{1}$$

$$d = ||v_n - a_n||_2 \tag{2}$$

An alternative to this would be to approach the problem as one of classification (on-sync/off-sync, *or* into different offset bins using synthetic data), however we were unable to achieve convergence using this method.

2.4   Training

The training procedure is an adaptation of the usual procedure for a single-stream ConvNet [14,24] and inspired by [6,23]. However our network is different in that it consists of non-identical streams, two independent sets of parameters and inputs from two different domains. The network weights are learnt using stochastic gradient descent with momentum. The parameters for both streams of the network are learnt simultaneously.

Data Augmentation. Applying data augmentation often improves validation performance and reduces overfitting in ConvNet image classification tasks [14]. For the audio, the volume is randomly altered in the range of $\pm 10\%$. We do not make changes to the audio playback speed, as this could affect the important timing information. For *false* examples only, we take random crops in time. For the video, we apply the standard augmentation methods used on the ImageNet classification task by [14,24] (*e.g.* random cropping, flipping, colour shift). A single transformation is applied to all video frames in a single clip.

Details. Our implementation is based on the MATLAB toolbox MatConvNet [26] and trained on a NVIDIA Titan X GPU with 12 GB memory. The network is trained with batch normalisation [10]. A learning rate of 10^{-2} to 10^{-4} is used, which is slower than that typically used for training a ConvNet with batch normalisation. The training was stopped after 20 epochs, or when the validation error did not improve for 3 epochs, whichever is sooner.

3   Dataset

In this section, we describe the pipeline for automatically generating a large-scale audio-visual dataset for training the lip synchronisation system. Using the methods described, we collect several hundred hours of speech from BBC videos, covering hundreds of speakers. We start from BBC News programs recorded between 2013 and 2016 (Fig. 3), given that a large number of different people appear in the news, in contrast to dramas with a fixed cast. The training, validation and test sets are divided in time, and the dates of videos corresponding to each set are shown in Table 1.

Fig. 3. Still images of BBC News videos.

The processing pipeline is summarised in Fig. 4. The visual part of the pipeline is based on the methods used by Chung and Zisserman [7], and we

Table 1. Dataset statistics: recording dates, and number of genuine (positive) and false lip-sync audio-video training samples, number of hours of facetrack.

Set	Dates	# pairs	# hours
Train	01/07/2013–31/08/2015	3,707K	606
Val	01/09/2015–31/12/2015	316K	42
Test	01/01/2016–31/05/2016	350K	47

give a brief sketch of the method here. First, shot boundaries are determined by comparing color histograms across consecutive frames [16]. The HOG-based face detection method of [12] is then performed on every frame, and the face detections are grouped across frames using a KLT tracker [25]. We discard any clips in which more than one face appears in the video, as the speaker is not known in this scenario.

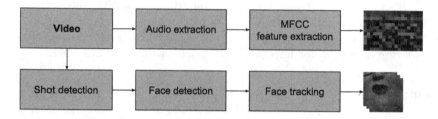

Fig. 4. Pipeline to generate the audio-visual dataset.

The audio part of the pipeline is straightforward. The Mel-frequency cepstral coefficient (MFCC) [8] features are used to describe the audio, which are commonly used in speech recognition systems. No other pre-processing is performed on the audio.

3.1 Compiling the Training Data

Genuine audio-video pairs are generated by taking a 5-frame video clip and the corresponding audio clip. Only the audio is randomly shifted by up to 2 s in order to generate synthetic *false* audio-video pairs. This is illustrated in Fig. 5. We take the audio from the same clip, so that the network learns to recognise the alignment, rather than the speaker.

Refining the Training Data. The training data generated using the proposed method is noisy in that it contains videos in which the voice and the mouth shapes do not correlate (*e.g.* dubbed videos) or are off-sync.

A network is initially trained on this noisy data, and the trained network is used to discard the false positives in the training set by rejecting positive pairs with distance over a threshold. A network is then re-trained on this new data.

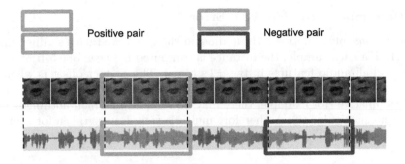

Fig. 5. The process of obtaining *genuine* and *false* audio-video pairs.

Discussion. The method does not require annotation of the training data, unlike some previous works that are based on phoneme recognition. We train on audio-video pairs, and the advantage of this approach is that the amount of available data is virtually infinite, and the cost of obtaining it is minimal (almost any video of speech downloaded from the Internet can be used for training). The key assumption is that the majority of the videos that we download are approximately synced, although some videos may have lip-sync errors. ConvNet loss functions and training are generally tolerant to the data being somewhat noisy.

4   Experiments

In this section we use the trained network to determine the lip-sync error in videos. The 256-dimensional fc_7 vectors for each stream are used as features representing the audio and the video. To obtain a (dis)similarity metric between the signals, the Euclidean distance of the features is taken. This is the same distance function that is used at training time. The histogram (Fig. 6) shows the distribution of the metric.

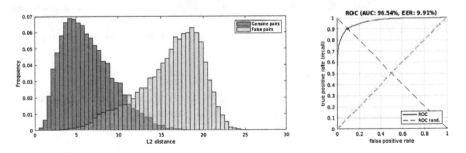

Fig. 6. The distribution of Euclidean distances for *genuine* and *false* audio-video pairs, using a single 0.2-s sample. Note that this is on the noisy validation data that may include clips of non-speakers or dubbed videos.

4.1 Determining the Lip-Sync Error

To find the time offset between the audio and the video, we take a sliding-window approach. For each sample, the distance is computed between one 5-frame video feature and all audio features in the ± 1 s range. The correct offset is when this distance is at a minimum. However as Table 2 suggests, not all samples in a clip are discriminative (for example, there may be samples in which nothing is being said at that particular time), therefore multiple samples are taken for each clip, and then averaged. Typical response plots are shown in Fig. 8.

Evaluation. The precise time offset between the audio and the video is not known. Therefore, the evaluation is done manually, where the synchronisation is considered successful if the lip-sync error is not detectable to a human. We take a random sample of several hundred clips from the part of the dataset that has been reserved for testing, as described in Sect. 3. The success rates are reported in Table 2.

Table 2. Accuracy to within human-detectable range.

Method	Accuracy
Single sample (0.2 s)	81%
Averaged over a clip	>99%

Experiments were also performed on a sample of Korean and Japanese videos (Fig. 7), to show that our method works across different languages. Qualitative results are extremely good, and will be available from our research page.

Fig. 7. Images of Korean and Japanese videos that were used for testing.

Performance. The data preparation pipeline and the network runs significantly faster than real-time on a mid-range laptop (Apple MacBook Pro with NVIDIA GeForce GT 750M graphics), with the exception of the face detection step (external application), which runs at around ×0.3 real-time.

4.2 Application: Active Speaker Detection

The problems of AV synchronisation and active speaker detection are closely related in that the correspondence between the video and the accompanying audio must be established. Therefore, the synchronisation method can be extended to determine the speaker in a scene where multiple faces are present. We define the confidence score of a time offset (synchronisation error) as the difference between the minimum and the median of the Euclidean distances (*e.g.* this value is around 6 to 7 for both plots in Fig. 8). In a multi-subject scene, the speaker's face is naturally the one with the highest correspondence between the audio and the video. A non-speaker should have a correlation close to zero and therefore also a very low score.

Unlike the uni-modal methods for active speaker detection that rely on the lip motion only, our method also can detect cases where the person is speaking, but is uncorrelated to the audio (*e.g.* in dubbed videos).

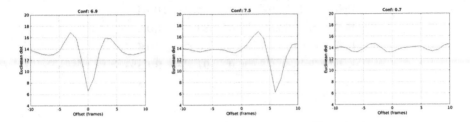

Fig. 8. Mean distance between the audio and the video features for different offset values, averaged over a clip. The actual offset lies at the trough. The three example clips shown here are for different scenarios. Left: synchronised AV data; Middle: the audio leads the video; Right: the audio and the video are uncorrelated.

Evaluation. We test our method using the dataset (Fig. 9) and the evaluation protocol of Chakravarty and Tuytelaars [4]. The objective is to determine who the speaker is in a multi-subject scene.

Fig. 9. Still images from the Columbia dataset [4].

The dataset contains 6 speakers, of which 5 (Bell, Bollinger, Lieberman, Long, Sick) are used for testing. A score threshold is set using the annotations on the remaining speaker (Abbas), at the point where the ROC curve intersects the diagonal (the equal error rate).

We report the F_1-scores in Table 3. The scores for each test sample are averaged over a 10-frame or 100-frame window. The performance is almost perfect for the 100-frame window. The disadvantage of increasing the size of the averaging window is that the method cannot detect examples in which the person speaks for a very short period; though that is not a problem in this case.

Table 3. F_1-scores on the Columbia speaker detection dataset. The results of [4] have been digitised from Fig. 3b of their paper, and are accurate to around ±0.5%.

Method	[4]		Ours	
Window	10	100	10	100
Bell	82.9%	90.3%	93.7%	**100%**
Bollinger	65.8%	69.0%	83.4%	**100%**
Lieberman	73.6%	82.4%	86.8%	**100%**
Long	86.9%	96.0%	97.7%	**99.8%**
Sick	81.8%	89.3%	86.1%	**99.8%**

4.3   Application: Lip Reading

Training a deep network for any task requires large quantities of data, but for problems such as lip reading, large-scale annotated data can be prohibitively expensive to collect. However, unlabelled spoken videos are copious and easy to obtain.

A useful by-product of the synchronisation network is that it enables very strong mouth descriptors to be learnt without any labelled data. We use this result to set the new state-of-the-art on the OuluVS2 [2] dataset. This consists of 52 subjects uttering the same 10 phrases (*e.g.* 'thank you', 'hello', etc.) or 10 predetermined digit sequences. It is assessed on a speaker-independent experiment, where 12 specified subjects are reserved for testing. Only the video stream is used for training and testing, *i.e.* this is a 'lip reading' experiment rather than one of audio-visual speech recognition.

Experimental Setup. A simple uni-directional LSTM classifier with one layer and 250 hidden units is used for this experiment. The setup is shown in Fig. 10. The LSTM network ingests the visual features (*fc7* activations from the ConvNet) of the 5-frame sliding window, moving 1-frame at a time, and returns the classification result at the end of the sequence.

Training Details. Our implementation of the recurrent network is based on the Caffe [11] toolbox. The network is trained with stocastic gradient descent, with a learning rate of 10^{-3}. The gradients are back-propagated for the full length of the clip. Softmax log loss is used, which is typical for a n-way classification problem. Here $n = 10$ for the 10 phrases or digit sequences. The loss is computed only at the final timestep.

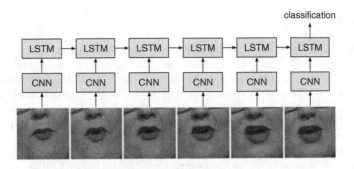

Fig. 10. Network configuration for the lip reading experiment. ConvNet weights are not updated at the time of LSTM training.

Table 4. Test set classification accuracy on OuluVS2, frontal view.

Method	Short phrases	Fixed digits
Zhou et al. [28]	73.5%	-
Chung and Zisserman [7]	93.2%	-
VGG-M + LSTM	31.9%	25.4%
SyncNet + LSTM	**94.1%**	**92.8%**

Evaluation. We compare our results to the previous state-of-the-art on this dataset; and also the same LSTM setup, but instead with a VGG-M [5] convolutional network pre-trained on ImageNet [21]. We report the results in Table 4. In particular, it is notable that our result beats that of [7], which is obtained using a network that has been pre-trained on a very large *labelled* dataset.

5   Conclusion

We have demonstrated that a two-stream ConvNet can be trained to synchronise audio to mouth motion, from natural videos of speech that are easy to obtain. A useful application of this method is in media players, where the lip-sync error can be corrected on a local machine at run-time. Furthermore, the approach can be extended to any problem where it is useful to learn a similarity metric between correlated data in different domains.

We have also shown that the trained network works effectively for the tasks of speaker detection in video, and lip reading.

Acknowledgements. We are very grateful to Andrew Senior for suggesting this problem; to Rob Cooper and Matt Haynes at BBC Research for help in obtaining the lip synchronisation dataset; and to Punarjay Chakravarty and Tinne Tuytelaars for supplying the Columbia dataset. Funding for this research is provided by the EPSRC Programme Grant Seebibyte EP/M013774/1.

References

1. Bt.1359: Relative timing of sound and vision for broadcasting. ITU (1998)
2. Anina, I., Zhou, Z., Zhao, G., Pietikäinen, M.: Ouluvs2: a multi-view audiovisual database for non-rigid mouth motion analysis. In: 11th IEEE International Conference and Workshops on Automatic Face and Gesture Recognition (FG), vol. 1, pp. 1–5. IEEE (2015)
3. Bredin, H., Chollet, G.: Audiovisual speech synchrony measure: application to biometrics. EURASIP J. Appl. Signal Process. **2007**(1), 179 (2007)
4. Chakravarty, P., Tuytelaars, T.: Cross-modal supervision for learning active speaker detection in video. arXiv preprint arXiv:1603.08907 (2016)
5. Chatfield, K., Simonyan, K., Vedaldi, A., Zisserman, A.: Return of the devil in the details: delving deep into convolutional nets. In: Proceedings of BMVC (2014)
6. Chopra, S., Hadsell, R., LeCun, Y.: Learning a similarity metric discriminatively, with application to face verification. In: Proceedings of the CVPR, vol. 1, pp. 539–546. IEEE (2005)
7. Chung, J.S., Zisserman, A.: Lip reading in the wild. In: Proceedings of ACCV (2016)
8. Davis, S.B., Mermelstein, P.: Comparison of parametric representations for monosyllabic word recognition in continuously spoken sentences. IEEE Trans. Acoust. Speech Signal Process. **28**(4), 357–366 (1980)
9. Geras, K.J., Mohamed, A.R., Caruana, R., Urban, G., Wang, S., Aslan, O., Philipose, M., Richardson, M., Sutton, C.: Compressing LSTMS into CNNS. arXiv preprint arXiv:1511.06433 (2015)
10. Ioffe, S., Szegedy, C.: Batch normalization: accelerating deep network training by reducing internal covariate shift. arXiv preprint arXiv:1502.03167 (2015)
11. Jia, Y.: Caffe: an open source convolutional architecture for fast feature embedding (2013). http://caffe.berkeleyvision.org/
12. King, D.E.: Dlib-ml: a machine learning toolkit. J. Mach. Learn. Res. **10**, 1755–1758 (2009)
13. Koster, B.E., Rodman, R.D., Bitzer, D.: Automated lip-sync: direct translation of speech-sound to mouth-shape. In: 1994 Conference Record of the Twenty-Eighth Asilomar Conference on Signals, Systems and Computers, vol. 1, pp. 583–586. IEEE (1994)
14. Krizhevsky, A., Sutskever, I., Hinton, G.E.: ImageNet classification with deep convolutional neural networks. In: NIPS, pp. 1106–1114 (2012)
15. Lewis, J.: Automated lip-sync: background and techniques. J. Vis. Comput. Anim. **2**(4), 118–122 (1991)
16. Lienhart, R.: Reliable transition detection in videos: a survey and practitioner's guide. Int. J. Image Graph. **1**, 469–486 (2001)
17. Marcheret, E., Potamianos, G., Vopicka, J., Goel, V.: Detecting audio-visual synchrony using deep neural networks. In: Sixteenth Annual Conference of the International Speech Communication Association (2015)
18. McAllister, D.F., Rodman, R.D., Bitzer, D.L., Freeman, A.S.: Lip synchronization of speech. In: Audio-Visual Speech Processing: Computational & Cognitive Science Approaches (1997)
19. Morishima, S., Ogata, S., Murai, K., Nakamura, S.: Audio-visual speech translation with automatic lip syncronization and face tracking based on 3-D head model. In: IEEE International Conference on Acoustics, Speech, and Signal Processing (ICASSP), vol. 2, p. II-2117. IEEE (2002)

20. Rúa, E.A., Bredin, H., Mateo, C.G., Chollet, G., Jiménez, D.G.: Audio-visual speech asynchrony detection using co-inertia analysis and coupled hidden markov models. Pattern Anal. Appl. **12**(3), 271–284 (2009)
21. Russakovsky, O., Deng, J., Su, H., Krause, J., Satheesh, S., Ma, S., Huang, S., Karpathy, A., Khosla, A., Bernstein, M., Berg, A., Li, F.: ImageNet large scale visual recognition challenge. IJCV **115**, 211–252 (2015)
22. Sargin, M.E., Yemez, Y., Erzin, E., Tekalp, A.M.: Audiovisual synchronization and fusion using canonical correlation analysis. IEEE Trans. Multimed. **9**(7), 1396–1403 (2007)
23. Simonyan, K., Zisserman, A.: Two-stream convolutional networks for action recognition in videos. In: NIPS (2014)
24. Simonyan, K., Zisserman, A.: Very deep convolutional networks for large-scale image recognition. In: International Conference on Learning Representations (2015)
25. Lucas, B.D., Kanade, T.: An iterative image registration technique with an application to stereo vision, Vancouver, BC, Canada (1981)
26. Vedaldi, A., Lenc, K.: Matconvnet - convolutional neural networks for MATLAB. CoRR abs/1412.4564 (2014)
27. Zhong, Y., Arandjelović, R., Zisserman, A.: Faces in places: compound query retrieval. In: British Machine Vision Conference (2016)
28. Zhou, Z., Hong, X., Zhao, G., Pietikäinen, M.: A compact representation of visual speech data using latent variables. IEEE Trans. Pattern Anal. Mach. Intell. **36**(1), 1 (2014)
29. Zoric, G., Pandzic, I.S.: A real-time lip sync system using a genetic algorithm for automatic neural network configuration. In: 2005 IEEE International Conference on Multimedia and Expo, pp. 1366–1369. IEEE (2005)

Visual Speech Recognition Using PCA Networks and LSTMs in a Tandem GMM-HMM System

Marina Zimmermann[1(✉)], Mostafa Mehdipour Ghazi[2],
Hazım Kemal Ekenel[3], and Jean-Philippe Thiran[1]

[1] Signal Processing Laboratory (LTS5),
Ecole Polytechnique Fédérale de Lausanne (EPFL), Lausanne, Switzerland
{marina.zimmermann,jean-philippe.thiran}@epfl.ch
[2] Faculty of Engineering and Natural Sciences, Sabanci University, Istanbul, Turkey
mehdipour@sabanciuniv.edu
[3] Department of Computer Engineering,
Istanbul Technical University (ITU), Istanbul, Turkey
ekenel@itu.edu.tr

Abstract. Automatic visual speech recognition is an interesting problem in pattern recognition especially when audio data is noisy or not readily available. It is also a very challenging task mainly because of the lower amount of information in the visual articulations compared to the audible utterance. In this work, principle component analysis is applied to the image patches — extracted from the video data — to learn the weights of a two-stage convolutional network. Block histograms are then extracted as the unsupervised learning features. These features are employed to learn a recurrent neural network with a set of long short-term memory cells to obtain spatiotemporal features. Finally, the obtained features are used in a tandem GMM-HMM system for speech recognition. Our results show that the proposed method has outperformed the baseline techniques applied to the OuluVS2 audiovisual database for phrase recognition with the frontal view cross-validation and testing sentence correctness reaching 79% and 73%, respectively, as compared to the baseline of 74% on cross-validation.

1 Introduction

Visual speech recognition has seen increasing attention in recent decades. The research interest in this topic arises from several factors; first, visual speech recognition can be used in automatic audio-visual speech recognition together with the audio data [23]. It has been shown that highly noisy audio data can thus be supported and higher recognition rates can be achieved in such scenarios [23]. In these cases, the supporting visual information helps similarly to its contribution in human-human interaction, where a listener's concentration on the lip movement increases in noisy environments. Secondly, visual speech recognition (VSR) is an interesting topic with varied applications. To name just a few, cybersecurity (pronounced passwords) [12], sign language (accompanying

© Springer International Publishing AG 2017
C.-S. Chen et al. (Eds.): ACCV 2016 Workshops, Part II, LNCS 10117, pp. 264–276, 2017.
DOI: 10.1007/978-3-319-54427-4_20

mouthings) [24], speech production [2] and in general human machine interfaces have an interest in VSR.

While audio-based speech recognition has improved significantly over the past decades and is nowadays applicable in many real-life scenarios, visual speech recognition still mostly focuses on speech produced in controlled lab conditions. However, there is a lot of interest to address, for example, the problem of head pose, which is a large hindrance in the application to real-world scenarios. Various works have already addressed this problem by taking different view angles into account [8,18]. To this end, several databases have been recorded simultaneously with cameras at different angles [11,17]. The recently published OuluVS2 database [1] aims at being a comparative dataset to allow a comprehensive comparison of approaches on multi-view data where cameras at five different angles record a subject simultaneously.

Efforts to bring visual speech recognition up to date with novel techniques used in both audio speech recognition and computer vision direct researchers to utilize deep learning techniques. Deep neural networks (DNNs) are widely employed in audio-based automatic speech recognition resulting in the current baseline accuracies [9]. DNNs have also become the standard techniques in computer vision to set baselines in recognition or analysis tasks [6,7]. However, one big problem in applying these networks to visual speech data is the fact that visual speech databases are not comparable to audio databases in terms of their sizes and number of speakers, meaning that insufficient amounts of training data are available. This is an important drawback since having a large amount of data is a necessity for training deep learning frameworks for complex acoustic models and complete recognition chains used for continuous speech. Although a few larger audio-visual databases such as TCD-TIMIT and OuluVS2 [1,11] have been published recently, the problem still remains highly challenging.

In this paper, we propose a visual speech recognition approach based on a two-stage PCA-based convolutional network [6] followed by a layer of long short-term memories (LSTMs) to extract a set of unsupervised spatiotemporal visual features. These features are then used in a tandem GMM-HMM system for speech recognition. Our contribution is two fold, with a major focus on feature extraction. First, we use principal component analysis in a multi-stage convolutional network to extract the optimal unsupervised learning lip representations. Secondly, we apply recurrent neural networks (RNNs) with LSTM cells to lip representations to extract spatiotemporal features. This approach does not only find the time-series dependencies within the video frame sequences, but also decreases the lips feature set dimension for further processing with the GMM-HMM scheme. Using this system, we were able to improve the baseline cross-validation results for phrase recognition for this workshop from a frontal and 30° side view with a large margin of roughly 5%, reaching 79% of all sentences being recognized correctly for each of these views. Combining these two views leads to an even higher recognition rate of 83% of all sentences.

The rest of this paper is organized as follows. Section 2 briefly reviews the related work and state-of-the-art approaches. Section 3 explains the details of the proposed method for visual speech recognition based on a PCA network,

LSTMs, and the GMM-HMM system. Section 4 describes the utilized dataset, experiments, and obtained results. Finally, Sect. 5 concludes the paper with a summary and discussions.

2   Related Work

Visual speech recognition requires a series of steps to process the video and extract relevant features. First of all, a region of interest (ROI) around the mouth, which contains the largest amount of information about the utterance, has to be extracted [23]. This can be done by hand or with the help of a face tracker. The latter is more common nowadays even though manual corrections are still sometimes applied. The ROI is later used to extract the features.

In general three types of features are used: texture-based features, shape-based features, or a combination of both [5,23]. Texture-based features exploit the pixel values in a ROI — usually closely around the mouth or including the jaws [23]. Typically, this is done by applying a transformation such as the discrete cosine transform (DCT) and/or a dimensionality reduction technique such as the linear discriminant analysis (LDA) to the ROI, possibly in combination with a principle component analysis (PCA) or a maximum-likelihood linear transform (MLLT) [23]. A common feature post-processing technique involves a chain of LDAs and MLLTs on concatenated frames, the so-called HiLDA [22].

Shape-based features, on the other hand, try to extract information about the shape of the mouth. This can be done for example with the help of snakes, taking into account the outer contours of the mouth, or by computing the geometrical distances between certain points of interest around the mouth [23]. In recent works these feature points are generally extracted with the help of a face or mouth tracker. Some researchers also directly use these points or shapes and extract information by applying a PCA to them. This technique is, for example, the case for the use of active appearance models (AAMs) [3,5].

The next step in the recognition system is the classification of the utterance, traditionally performed through a system composed of Hidden Markov Models (HMMs) with Gaussian Mixture Models (GMMs). The GMMs model the acoustics, i.e. the phonemes, or visemes in visual speech, while the states of an HMM model the time evolution within a phoneme and the overall evolution within and between words [23].

Even though there are still many studies working on texture-based features such as DCT, DCT-HiLDA, or scattering [19] and, similarly, many researchers still work with GMM-HMM recognition systems, recently, more focus is being put on deep learning techniques. These networks are widely spread both in audio speech recognition and visual recognition tasks to extract features, construct acoustic models, or replace the complete recognition chain.

In recent literature, deep network based approaches have consistently shown superior performances over traditional methods. Deep Boltzmann machines have been used as stacked autoencoders for feature extraction [20] or post-processing of local binary patterns from three orthogonal planes (LBP-TOP) [25]. These features are then classified using Support Vector Machines (SVMs) [20], where all utterance lengths have to be normalized, or using a tandem system [13], where the features are passed into a GMM-HMM recognizer [15,21,25]. Similarly, feature extraction has been performed by convolutional neural networks (CNNs) [16,21] and deep belief networks (DBNs) [15]. The outputs of these networks can be used as an acoustic model in the so-called hybrid approach, where the posterior probability outputs are passed directly to the HMM [4]. Finally, the recognition system itself can be replaced by DNNs, either in the form of bilinear [19] or recurrent neural networks [26]. In the former case, DNNs are used to classify texture-based features while in the latter case the whole processing chain is replaced by a LSTM network.

Comparing these approaches to our proposed method, one speciality of the PCA network is the effectiveness with which it extracts information from the given frames without needing any prior knowledge — it is unsupervised. The two-stage projection onto the leading principle components allows to capture the main variations within the image patches, while also extracting higher level features through the concatenation of these in two stages. The subsequent binarization and extraction of histograms leads to an indexing and pooling of these, a non-linear step. The following LSTM network then reduces the feature size dramatically by taking into account the temporal information between different frames and the classes assigned to these. As a result, the posterior probabilities serve as good spatiotemporal descriptors and can be utilized in a tandem system as features for a GMM-HMM recognizer.

3   The Proposed Method

In this section, novel feature extraction methods are explored for visual speech recognition. More specifically, a two-stage PCA-based convolutional network [6] followed by a layer of LSTMs [14] extracts features from the cropped mouth images. The obtained spatiotemporal features are then processed in a tandem system with a GMM-HMM basis for speech recognition.

Feature extraction is performed in a sequential fashion as shown in Fig. 1. First, a two-stage PCA network is applied to each video frame (see Fig. 2). The first layer network weights are learned by applying PCA to concatenated square patches — which are extracted from the mouth video frames and then vectorized.

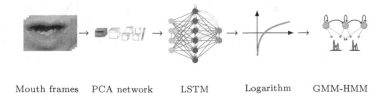

Mouth frames   PCA network   LSTM   Logarithm   GMM-HMM

Fig. 1. The proposed method for visual speech recognition from the mouth video frames.

We use eight principal components as the networks' first layer filter bank and convolve these with the input images. In a cascaded scheme, a similar procedure is applied to the filtered patches to obtain the second layer filter bank. After convolution, the output maps are binarized with a Heaviside step function and every eight binary images are stacked together to compose an 8-bit image — similar to the first layer outputs. Finally, block histograms — with 256 bins — are extracted from the obtained maps and concatenated, resulting in a long feature vector for each video frame. In this work, we extract 16 block histograms which result in feature vectors of length $16 \times 256 \times 8 = 32{,}768$.

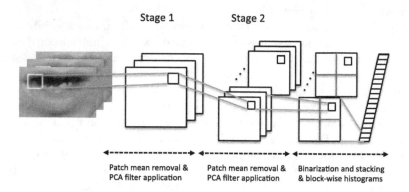

Fig. 2. The PCA network used in the first stage of the proposed method.

Secondly, an LSTM network is connected to the outputs of the PCA network to extract more abstract representations while taking the time-series dependencies between the video frames into account. This type of RNN is composed of memory cells to store the past values or ignore the dependencies when needed. Therefore, each cell has an input, an output, and a forget gate that can be activated at different levels. This architecture results in three cases: accepting the new input value, forgetting the existing value, or outputting a value at the given level [10]. Since we label each video frame in the phrase recognition subset based on the audio phonemes, there are 28 output nodes in our LSTM network.

Last but not least, the posterior probabilities received from the LSTMs are passed as spatiotemporal features concatenated with their delta and acceleration components into a GMM-HMM based speech recognition system, the so-called tandem approach. This system is implemented using the Hidden Markov Model Toolkit (HTK) [27]. However, since the outputs of the LSTM network show small variations, we first take the logarithm of these features to make them more discriminative. Our tandem system contains GMMs with 15 Gaussian mixtures per observation and 4 states per word.

4 Performance Analysis

In this section, we review details of the utilized dataset, evaluation metrics, and the conducted experiments. We present the validation and test results and discuss them in detail.

4.1 The Dataset

We use the phrase recognition subset of the OuluVS2 database [1] in our experiments. This dataset contains video clips of 52 subjects from five different views: frontal and four side views at 30°, 45°, 60°, and 90° (the profile). During each recording session, the subjects were asked to utter 10 daily short English phrases shown on a computer monitor. Each phrase was repeated three times resulting in 30 video recordings (utterances) per subject per view. The recording was performed in an ordinary office environment with varying lighting conditions and background noises producing a more real-world audio-visual dataset. Each of these videos was recorded with a resolution of 1920 × 1080, at 30 fps, and with an audio bit rate of 128 kbps. The challenge organizers also provided aligned and cropped mouth videos along with the original videos and fixed the training and test subsets: 40 out of 52 subjects are assigned for training and the rest are used for testing.

4.2 Experimental Results

In our experiments, we use the provided cropped mouth videos by first extracting and converting all video frames to grayscale images of size 60 × 90 pixels. PCA is applied to all image patches of size 7 × 7 pixels to learn eight filter banks in a two-stage cascaded PCA network. We add a max-pooling layer to the output of this network to obtain a more abstract representation before histogram pooling. Finally, 16 block histograms are extracted and concatenated to obtain a 32,768-dimensional feature vector for each frame.

For spatiotemporal recognition using the LSTM network, we need to obtain frame-based labels using phoneme level transcription. For this purpose, the audio data is first aligned to the sentence transcriptions using a standard GMM-HMM system with MFCCs trained on the training subset. These transcriptions are then used as labels for the obtained feature set from the PCA network. We train a one-layer LSTM network with a Sigmoid activation function in the gates and cells. The learning rate, weight decay penalty, and momentum value are set to 0.5, 0.001, and 0.8, respectively. Moreover, we use a random batch size and train the network until 10,000 iterations.

Three metrics are used to present the results: the accuracy and correctness at the word level, and the percentage of correct sentences. The word accuracy and correctness are defined as follows

$$\text{Accuracy} = \frac{H - I}{N} \cdot 100\% \tag{1}$$

$$\text{Correctness} = \frac{H}{N} \cdot 100\% \tag{2}$$

where H, I, and N are the number of correct words, number of erroneous words (insertion error), and the total number of words, respectively. The number of correct words is equal to the number of all words minus the total number of ignored words (deletion error) and the number of wrongly recognized words (substitution error), i.e. $H = N - D - S$.

To adjust our system parameters, we use a leave-one-out cross-validation scheme on the given training set. Later on, we apply the system in a leave-one out cross-validation scheme on the whole data, similar to the baseline[1], and, trained on the training set, to the test set for the final recognition at the word or phrase levels. Figures 3 and 4 show our cross-validation and test results on the OuluVS2 dataset for phrase recognition.

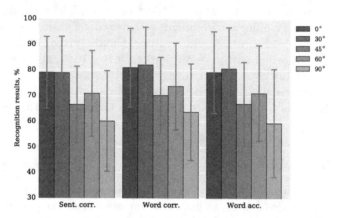

Fig. 3. Mean phrase recognition results on the multi-view dataset of OuluVS2 using our proposed method with the cross-validation technique and the standard deviation across subjects.

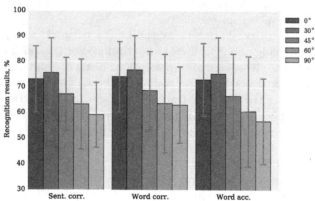

Fig. 4. Mean phrase recognition results on the multi-view dataset of OuluVS2 using our proposed method on the given test set and the standard deviation across subjects.

---

[1] The baseline results can be found at http://ouluvs2.cse.oulu.fi/preliminary.html.

Table 1. Phrase recognition results in % on the multi-view dataset of OuluVS2 using our proposed method on the given test set per speaker and the corresponding means and standard deviations across speakers (with SC = Sentence correctness, WC = Word correctness and WA = Word accuracy).

Spkr.	0°			30°			45°			60°			90°		
	SC	WC	WA	SC	WC	WA	SC	WC	WA	SC	WC	WA	SC	WC	WA
6	73.3	73.3	73.3	70.0	70.0	73.3	53.3	60.0	56.0	40.0	40.0	37.3	50.0	57.3	52.0
8	56.7	54.7	54.7	66.7	66.7	62.7	66.7	74.7	73.3	66.7	66.7	66.7	63.3	66.7	60.0
9	43.3	45.3	41.3	53.3	53.3	56.0	56.7	54.7	50.7	43.3	45.3	40.0	36.7	36.0	24.0
15	73.3	77.3	76.0	63.3	63.3	58.7	53.3	54.7	53.3	40.0	40.0	34.7	40.0	41.3	30.7
26	76.7	74.7	74.7	90.0	90.0	89.3	63.3	62.7	58.7	70.0	65.3	65.3	80.0	85.3	82.7
30	73.3	80.0	78.7	76.7	76.7	77.3	90.0	92.0	92.0	73.3	82.7	80.0	73.3	84.0	74.7
34	96.7	97.3	97.3	86.7	86.7	90.7	80.0	85.3	85.3	80.0	80.0	78.7	63.3	61.3	56.0
43	73.3	78.7	76.0	83.3	83.3	81.3	63.3	62.7	56.0	56.7	50.7	41.3	70.0	70.7	68.0
44	80.0	81.3	81.3	86.7	86.7	88.0	80.0	78.7	78.7	93.3	97.3	97.3	50.0	52.0	48.0
49	86.7	88.0	86.7	80.0	80.0	80.0	93.3	93.3	93.3	83.3	86.7	86.7	63.3	69.3	68.0
51	66.7	60.0	60.0	53.3	53.3	50.7	50.0	42.7	41.3	43.3	41.3	33.3	53.3	56.0	48.0
52	76.7	78.7	76.0	96.7	96.7	94.7	56.7	62.7	60.0	70.0	68.0	65.3	66.7	77.3	69.3
Mean	73.1	74.1	73.0	75.6	76.8	75.2	67.2	68.7	66.6	63.3	63.7	60.6	59.2	63.1	56.8
SD	12.9	13.8	14.2	13.6	13.4	14.3	14.3	15.3	16.6	17.6	19.2	21.5	12.7	14.9	16.7

Single-View Experiments. As the obtained results show, on average roughly 81% of the words are correctly recognized during the cross-validation approach for the frontal view. In addition, we have achieved a word recognition correctness of around 74% on the test set. Also, we can see that 79% of sentences are correctly recognized during cross-validation while the performance on the test set is 73%. The small differences between the word correctness and accuracies indicate that there are only few insertion errors. Comparing our obtained results with the baseline cross-validation results on the same dataset reveals that we have improved the performance with a large margin of roughly 5%.

The average phrase recognition results for the 30° view show similar improvements over the baseline provided. Almost 79% of all sentences are classified correctly for the cross-validation data — approximately 3% more than the baseline — and around 76% on the test data. Similarly, for the test set 77% of all words are correct and the accuracy reaches 75%, while on the cross-validation these values reach 82% and 81%. The other views do not show improvements over the baseline.

Looking into the standard deviation indicated in the figures or the individual test results in Table 1, we can see, however, that there is a large margin between the performance of the best speaker and the worst. This hints at a common problem in visual speech recognition where the variability between speakers is very large.

The frame recognition accuracy is shown in Table 2. The per frame results on a phoneme and a viseme basis for the training and test sets are displayed here. The observations are two-fold: First, it can be seen that on a frame level the differences between the different view angles does not seem very big, however, the combination of successive frames proves more successful for the frontal views as described above. Secondly, the phoneme and viseme-based classification show a big difference between the 28 phoneme classes and 12 viseme classes (defined according to [11]) due to the similarity between various phonemes represented only by the shape of the lips.

Table 2. Frame recognition accuracy results in % on the multi-view dataset of OuluVS2 using the LSTM output of our proposed on the given train and test sets across all speakers for phonemes and visemes (visemes defined according to [11]).

		0°	30°	45°	60°	90°
Train	Phoneme	19.5	20.1	18.0	17.7	15.7
	Viseme	34.4	32.9	33.8	33.7	30.8
Test	Phoneme	17.2	17.7	17.1	17.2	16.1
	Viseme	30.8	30.4	32.0	31.4	29.6

Multiple-View Experiments. In order to fully benefit from the multi-view recordings, further analyses are performed on combinations of different views. To this end the feature vectors obtained from the LSTM are concatenated and then processed similarly to the single views with their delta and acceleration components in a tandem GMM-HMM system. These multi-view experiments show interesting results (see Figs. 5 and 6). Various combinations of the frontal view with each of the four side views were tested, as well as the ensemble of all views together.

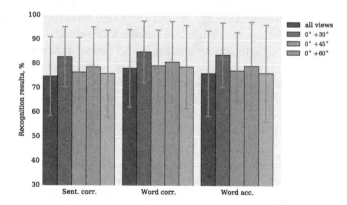

Fig. 5. Mean phrase recognition results on the combination of different views of the multi-view dataset of OuluVS2 using our proposed method with the cross-validation technique and the standard deviation across subjects.

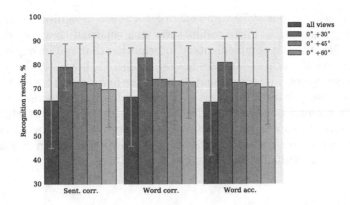

Fig. 6. Mean phrase recognition results on the combination of different views of the multi-view dataset of OuluVS2 using our proposed method on the given test set and the standard deviation across subjects.

Table 3. Phrase recognition results in % on the combination of different views of the multi-view dataset of OuluVS2 using our proposed method on the given test set per speaker and the corresponding means and standard deviations across speakers (with SC = Sentence correctness, WC = Word correctness and WA = Word accuracy).

Spkr.	All views			$0° + 30°$			$0° + 45°$			$0° + 60°$			$0° + 90°$		
	SC	WC	WA	SC	WC	WA	SC	WC	WA	SC	WC	WA	SC	WC	WA
6	30.0	29.3	25.3	76.7	80.0	77.3	63.3	68.0	64.0	40.0	44.0	41.3	50.0	53.3	52.0
8	56.7	61.3	58.7	70.0	74.7	70.7	60.0	64.0	61.3	53.3	57.3	57.3	50.0	54.7	54.7
9	36.7	44.0	38.7	70.0	76.0	73.3	43.3	44.0	41.3	60.0	64.0	62.7	43.3	54.7	45.3
15	70.0	65.3	65.3	70.0	76.0	73.3	66.7	66.7	65.3	70.0	68.0	68.0	70.0	69.3	66.7
26	60.0	56.0	50.7	80.0	84.0	80.0	86.7	88.0	88.0	66.7	62.7	61.3	66.7	66.7	64.0
30	76.7	82.7	81.3	86.7	90.7	90.7	83.3	88.0	88.0	90.0	93.3	93.3	83.3	90.7	88.0
34	96.7	98.7	98.7	86.7	90.7	90.7	96.7	98.7	98.7	96.7	97.3	97.3	90.0	90.7	89.3
43	76.7	80.0	77.3	76.7	81.3	78.7	86.7	89.3	89.3	73.3	80.0	77.3	86.7	85.3	85.3
44	70.0	72.0	72.0	83.3	88.0	88.0	76.7	81.3	78.7	100.0	100.0	100.0	70.0	70.7	70.7
49	80.0	82.7	82.7	86.7	92.0	89.3	86.7	88.0	86.7	90.0	93.3	92.0	83.3	88.0	86.7
51	40.0	37.3	33.3	63.3	61.3	58.7	46.7	33.3	32.0	40.0	36.0	30.7	56.7	56.0	54.7
52	86.7	88.0	86.7	100.0	100.0	100.0	76.7	77.3	76.0	86.7	81.3	81.3	86.7	92.0	88.0
Mean	65.0	66.4	64.2	79.2	82.9	80.9	72.8	73.9	72.4	72.2	73.1	71.9	69.7	72.7	70.4
SD	19.9	20.6	22.2	9.7	9.8	10.8	16.2	18.8	19.5	20.1	20.3	21.4	15.8	15.2	15.8

The multiple-view results show very good improvements especially for the combination of the frontal and the 30°-side view. On the cross-validation set a sentence accuracy of nearly 83% is achieved, while word correctness and word accuracy are around 85% and 84% respectively. Thus this amounts to improvements of around 3–10% over the separate results for these views. Similar improvements can be seen on the test set, where for the same combination the recognition of sentences is at 79% and 83% of words are recognised correctly. The word accuracy lies at 81%. These results show that especially between the frontal and the 30°-view there is complementary information that can be exploited. The improvements for the other views are not as significant, however, there could be further improvements. The concatenation of all the feature vectors from all views shows a particularly bad result. This is probably due to the increase in dimensionality, which could be aided by prior dimensionality reduction techniques.

Furthermore, again a large variability in the performance between the different speakers can be observed from the standard deviation shown in Figs. 5 and 6 as well as the individual speaker results in Table 3.

5   Conclusion

In this paper, we have proposed a visual speech recognition system that utilizes a two-stage cascaded PCA network to extract unsupervised learning based lip representations together with a layer of LSTM networks to obtain a set of spatiotemporal visual features. These features have later been used in a tandem GMM-HMM system for speech recognition. As the results indicate, the proposed method has outperformed the baseline technique with a large margin. They also show interesting results in a multiple-view recognition scenario, indicating the complementary information contained in the different views.

In this study only a limited dataset with a small vocabulary has been explored to point out the benefits of using PCA networks in combination with LSTMs. Future works should thus extend this approach to other available datasets such as TCD-TIMIT [11] that allow phoneme classification and provide a larger vocabulary. In addition, the influence of the different views and their complementary nature within the framework of these spatiotemporal features could be explored in a more detailed multiple-view visual speech recognition study.

Acknowledgement. This work was supported by TUBITAK project number 113E067 and by a Marie Curie FP7 Integration Grant within the 7th EU Framework Programme.

References

1. Anina, I., Zhou, Z., Zhao, G., Pietikainen, M.: OuluVS2: A multi-view audiovisual database for non-rigid mouth motion analysis. In: 2015 11th IEEE International Conference and Workshops on Automatic Face and Gesture Recognition (FG) (2015)
2. Badin, P., Bailly, G., Revéret, L., Baciu, M., Segebarth, C., Savariaux, C.: Three-dimensional linear articulatory modeling of tongue lips and face, based on MRI and video images. J. Phonet. **30**(3), 533–553 (2002)
3. Biswas, A., Sahu, P., Chandra, M.: Multiple camera in car audio–visual speech recognition using phonetic and visemic information. Comput. Electr. Eng. **47**, 35–50 (2015)

4. Bourlard, H.A., Morgan, N.: Connectionist Speech Recognition. Springer Nature, Berlin (1994)
5. Bowden, R., Cox, S., Harvey, R., Lan, Y., Ong, E.J., Owen, G., Theobald, B.J.: Recent developments in automated lip-reading. In: Zamboni, R., Kajzar, F., Szep, A.A., Burgess, D., Owen, G. (eds.) Optics and photonics for counterterrorism crime fighting and defence IX and optical materials and biomaterials in security and defence systems technology X. In: SPIE-The International Society of Optics and Photonics (2013)
6. Chan, T.H., Jia, K., Gao, S., Lu, J., Zeng, Z., Ma, Y.: PCANet: A simple deep learning baseline for image classification? IEEE Trans. Image Process. **24**(12), 5017–5032 (2015)
7. Donahue, J., Hendricks, L.A., Guadarrama, S., Rohrbach, M., Venugopalan, S., Darrell, T., Saenko, K.: Long-term recurrent convolutional networks for visual recognition and description. In: 2015 IEEE Conference on Computer Vision and Pattern Recognition (CVPR) (2015)
8. Estellers, V., Thiran, J.P.: Multi-pose lipreading and audio-visual speech recognition. EURASIP J. Adv. Slg. Process. **2012**(1), 51 (2012)
9. Graves, A., Jaitly, N.: Towards end-to-end speech recognition with recurrent neural networks. In: Jebara, T., Xing, E.P. (eds.) Proceedings of 31st International Conference on Machine Learning (ICML-2014), JMLR Workshop and Conference Proceedings, pp. 1764–1772 (2014)
10. Graves, A., Mohamed, A., Hinton, G.: Speech recognition with deep recurrent neural networks. In: 2013 IEEE International Conference on Acoustics Speech and Signal Processing (2013)
11. Harte, N., Gillen, E.: TCD-TIMIT: An audio-visual corpus of continuous speech. IEEE Trans. Multimedia **17**(5), 603–615 (2015)
12. Hassanat, A.: Visual passwords using automatic lip reading. Int. J. Sci.: Basic Appl. Res. (IJSBAR) **13**(1) (2014)
13. Hermansky, H., Ellis, D., Sharma, S.: Tandem connectionist feature extraction for conventional HMM systems. In: Proceedings of 2000 IEEE International Conference on Acoustics Speech, and Signal Processing (2000)
14. Hochreiter, S., Schmidhuber, J.: Long short-term memory. Neural Comput. **9**(8), 1735–1780 (1997)
15. Huang, J., Kingsbury, B.: Audio-visual deep learning for noise robust speech recognition. In: 2013 IEEE International Conference on Acoustics Speech and Signal Processing (2013)
16. Koller, O., Ney, H., Bowden, R.: Deep learning of mouth shapes for sign language. In: 2015 IEEE International Conference on Computer Vision Workshop (ICCVW) (2015)
17. Lee, B., Hasegawa-Johnson, M., Goudeseune, C., Kamdar, S., Borys, S., Liu, M., Huang, T.: AVICAR: Audio-visual speech corpus in a car environment. In: 8th International Conference on Spoken Language Processing (2004)
18. Lucey, P., Potamianos, G., Sridharan, S.: An extended pose-invariant lipreading system. In: Proceedings of AVSP 2007: International Conference on Auditory-Visual Speech Processing. International Speech Communication Association (2007)
19. Mroueh, Y., Marcheret, E., Goel, V.: Deep multimodal learning for audio-visual speech recognition. In: 2015 IEEE International Conference on Acoustics Speech and Signal Processing (ICASSP) (2015)
20. Ngiam, J., Khosla, A., Kim, M., Nam, J., Lee, H., Ng, A.Y.: Multimodal deep learning. In: Proceedings of 28th International Conference on Machine Learning (ICML), pp. 689–696 (2011)

21. Noda, K., Yamaguchi, Y., Nakadai, K., Okuno, H.G., Ogata, T.: Audio-visual speech recognition using deep learning. Appl. Intell. **42**(4), 722–737 (2014)
22. Potamianos, G., Neti, C., Gravier, G., Garg, A., Senior, A.: Recent advances in the automatic recognition of audiovisual speech. Proc. IEEE **91**(9), 1306–1326 (2003)
23. Potamianos, G., Neti, C., Luettin, J., Matthews, I.: Audio-visual automatic speech recognition: An overview. In: Bailly, G., Vatikiotis-Bateson, E., Perrier, P. (eds.) Issues in Visual and Audio-Visual Speech Processing, pp. 1–30. MIT Press, Cambridge (2004). Chap. 10
24. Schmidt, C., Koller, O.: Using viseme recognition to improve a sign language translation system. In: International Workshop on Spoken Language Translation, Heidelberg, Germany, pp. 197–203 (2013)
25. Sui, C., Bennamoun, M., Togneri, R.: Listening with your eyes: Towards a practical visual speech recognition system using deep Boltzmann machines. In: 2015 IEEE International Conference on Computer Vision (ICCV) (2015)
26. Wand, M., Koutnik, J., Schmidhuber, J.: Lipreading with long short-term memory. In: 2016 IEEE International Conference on Acoustics Speech and Signal Processing (ICASSP) (2016)
27. Young, S., Evermann, G., Hain, T., Kershaw, D., Moore, G., Odell, J., Ollason, D., Povey, D., Valtchev, V., Woodland, P.: The HTK Book. Technical report (2002)

Concatenated Frame Image Based CNN
for Visual Speech Recognition

Takeshi Saitoh[1(✉)], Ziheng Zhou[2], Guoying Zhao[2], and Matti Pietikäinen[2]

[1] Kyushu Institute of Technology, Iizuka, Japan
saitoh@ces.kyutech.ac.jp
[2] University of Oulu, Oulu, Finland
{ziheng.zhou,guoying.zhao,mkp}@ee.oulu.fi

Abstract. This paper proposed a novel sequence image representation method called concatenated frame image (CFI), two types of data augmentation methods for CFI, and a framework of CFI-based convolutional neural network (CNN) for visual speech recognition (VSR) task. CFI is a simple, however, it contains spatial-temporal information of a whole image sequence. The proposed method was evaluated with a public database OuluVS2. This is a multi-view audio-visual dataset recorded from 52 subjects. The speaker independent recognition tasks were carried out with various experimental conditions. As the result, the proposed method obtained high recognition accuracy.

1 Introduction

In the field of visual speech recognition (VSR), one of the most important problems is the extraction of visual features. All the existing approaches can be roughly grouped into four categories: (1) image-based, (2) motion-based, (3) geometric-feature-based, and (4) model-based [1,2]. For the image-based approaches, a gray-scale image is either used directly or after some image transformation, such as PCA and DCT, as a feature vector [3,4]. Typical motion-based methods are based on optical flow [5]. The geometric-feature-based approaches measure certain geometric features of the mouth such as the width, height, area, and aspect ratio. The model-based approaches are based on the active appearance models that jointly characterize the shapes and textures of talking mouths [6–8] and model parameters are used as visual features.

Recently, deep learning techniques have been successfully applied to learn features from audio-visual data for the tasks of VSR and audiovisual speech recognition (AVSR). Ngiam et al. [9] proposed to build a multimodal deep autoencoder consisting of stacks of the Restricted Boltzmann Machines (RBMs) for learning modality-specific information. In their work, two public datasets, AVLetters and CUAVE were used for the supervised classification. Hu et al. [10] proposed a Recurrent Temporal Multimodal Restricted Boltzmann Machines to model audio-visual sequences in an unsupervised fashion. The joint representations across the generated features of two modalities were learned using multimodal RBMs. Two public datasets, namely, the AVLetters and AVLetters2, as well as

© Springer International Publishing AG 2017
C.-S. Chen et al. (Eds.): ACCV 2016 Workshops, Part II, LNCS 10117, pp. 277–289, 2017.
DOI: 10.1007/978-3-319-54427-4_21

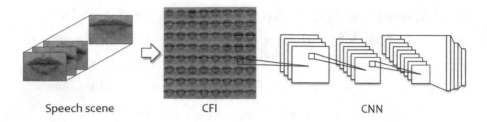

Speech scene   CFI   CNN

Fig. 1. Overview of proposed framework.

their collected dataset were used for the evaluation. Noda et al. [11] proposed a
visual feature extraction method for VSR utilizing a Convolutional Neural Network (CNN). Hidden Markov Models (HMM) with Gaussian mixtures were used
for a task of recognizing isolated word. The method was evaluated on an audio-
visual speech dataset comprising 300 Japanese words uttered by six speakers.
Amer et al. [12] proposed a hybrid model comprising of temporal generative and
discriminative models for classifying sequential data from multiple heterogeneous
modalities. Their method was evaluated on three datasets (AVEC, AVLetters,
and CUAVE). Takashima et al. [13] proposed a multi-modal feature extraction
method using a Convolutive Bottleneck Network (CBN), and applied to audio-
visual data. Extracted bottleneck audio and visual features were used as the
features input to the audio or visual HMMs and the recognition results then
integrated. Their method did not use the output labels of CBN and was evaluated on a work recognition task. They used 216 words as the test data and 2,620
words as the training data. Most of the above mentioned methods targeted the
problem of AVSR. Only [11] tackled the problem of VSR. Note that in their
method, CNNs were used for visual extraction features and the classification
was conducted by HMMs.

In this paper, we propose a novel image sequence representation, called the
concatenated frame image (CFI) and the data augmentation method for CFI. As
shown in Fig. 1, VSR is tackled by the CFI-based CNNs. We evaluate our approach on the newly collected public audiovisual database, OuluVS2 [14] and the
results show that our approach performs well in a speaker independent setting.

The rest of this paper is organized as follows: in Sect. 2 the proposed con-
catenated frame image is described. Section 3 provides details of the constructed
CNN. In Sect. 4, the OuluVS2 database and experimental results are described.
This paper concludes in Sect. 5.

2   Concatenated Frame Image

Let I_f denotes the f-th frame of a video sequence that records a certain utterance
and I'_f a resized image of I_f. Let the sequence length be F and the image sizes
of I_f and I'_f be $W \times H$ [pixels] and $W' \times H'$ [pixels], respectively. The proposed
concatenated frame image is an image that is formed by concatenating $\{I'_f\}$

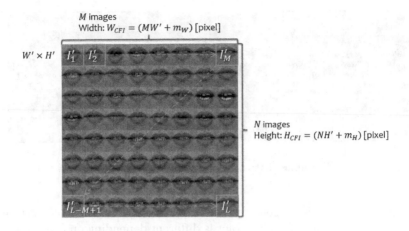

Fig. 2. Overview of base CFI.

following a specific rule that sample images with uniform intervals from $\{I'_f\}$ and re-organize them into M rows and N columns:

$$\mathrm{CFI} = \begin{pmatrix} I'_1 & \cdots & I'_M \\ \vdots & \ddots & \vdots \\ I'_{L-M+1} & \cdots & I'_L \end{pmatrix}, \tag{1}$$

where $L = M \times N$ is the number of sub-images in CFI. Here, F may be different across different videos. Each CFI is an image with a size of $W_{CFI} \times H_{CFI} = (MW' + m_W) \times (NH' + m_H)$ pixels, where m_W and m_H are the number of pixels surround sub-images. Figure 2 shows an overview of the construction of a CFI. In this CFI, $W' = 32$, $H' = 32$, $M = 8$, $N = 8$, $m_W = 0$, and $m_H = 0$. The left-top I'_f is the first frame image, and the right-bottom I'_f is the last frame image.

Figure 3 shows three samples of CFI. These CFIs are generated by the same speech scene. Five parameters of Fig. 3 are the same: $W = 228$ pixel, $H = 150$ pixel, $F = 146$, $W_{CFI} = 256$ pixel, and $H_{CFI} = 256$ pixel. However, the values of each L are different. L of the left-side CFI, middle-side CFI, and right-side CFI of Fig. 3 are 49, 64, and 81, respectively. The larger L indicates that CFI has the high time resolution.

2.1 Data Augmentation

Data augmentation (DA) techniques are effective for reducing overfitting on training datasets and therefore, improving generalization of the trained neural networks [15]. Typical DA methods for CNNs include applying some translation, rotation, mirror reverse, and color change to images. Here we propose two DA strategies for CFI to the tackle the problem of VSR:

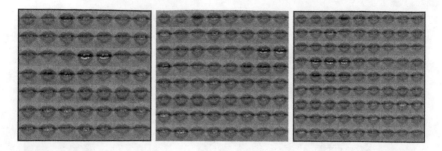

Fig. 3. CFI samples (left: $L = 49$, middle: $L = 64$, right: $L = 81$).

The first strategy is to generate CFIs by applying Gamma correction for brightness changes. Since the skin color is different depending on the gender and the race, Gamma correction is effective to this problem. Given an input image I, its pixel intensities are first scaled from the range $[0, 255]$ to $[0, 1]$. We then obtain a gamma corrected image I_{gamma} through:

$$I_{gamma}(r, c) = I(r, c)^{1/\gamma}, \tag{2}$$

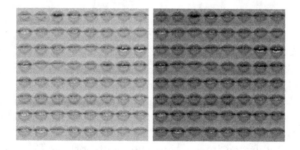

(a) Gamma correction (left: $\gamma = 0.6$, right: $\gamma = 1.4$)

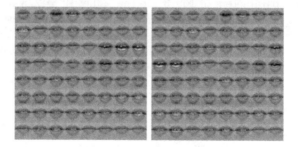

(b) temporal shift (left: $\alpha = -2$, right: $\alpha = 2$)

Fig. 4. Data augmentation samples.

where (r, c) denotes the pixel value and γ is the gamma value. Finally, I_{gamma} is scaled back to the range $[0, 255]$. Here, $\gamma < 1$ shifts an input image towards the darker end of the spectrum while $\gamma > 1$ makes the image appear lighter.

The second strategy is applied to the temporal domain. Utterance speed is different depending on individuals, and utterance time is different depending on the utterance content. Then, temporal shift is applied. The standard CFI as previously defined is built by L frames sampled from a video sequence of F frame length. For DA, $(L-\alpha)$ frames are sampled and temporal shifted CFI is generated with $(L - \alpha)$ frames. This means that the sampling interval $(= F/(L - \alpha))$ of the temporal shifted CFI is shorter than the sampling interval $(= F/L)$ of the standard CFI. To obtain L images for making the CFI, the first or last frame is replicated α times on the left-top or right-bottom of the CFI.

Figure 4 demonstrates the two DA strategies. Two samples of Fig. 4(a) are CFIs with the Gamma correction applied. Two of samples of Fig. 4(b) are CFIs which the temporal shift applied.

3   Convolutional Neural Network

Recent developments in deep learning technologies have greatly advanced the performance of the state of the art of many visual recognition tasks. In particular, convolutional neural networks (CNNs) have been established as a powerful class of models for image recognition tasks [15]. CNNs consist of alternating convolutional layers and pooling layers. Convolution layers take inner product of the linear filter and the underlying receptive field followed by a nonlinear activation function, such as rectifier, sigmoid, tanh, at every local portion of the input.

There have been a number of pre-trained CNN models available. In this research, we build our CNNs for VSR based on three well-known models: Network In Network (NIN) [16], AlexNet [15], and GoogLeNet [17].

NIN is proposed by Lin et al. [16]. NIN consists of mlpconv layers which use multilayer perceptrons to convolve the input and a global average pooling layer as a replacement for the fully connected layers in conventional CNN. NIN used in this research consists of four mlpconv layers, and the mlpconv layers are followed by a spatial max pooling layer which down-samples the input image by a factor of three. To reduce overfitting in the fully connected layers, regularization method called dropout is applied on the outputs of the last mlpconv layers.

AlexNet is proposed by Krizhevsky et al. [15]. This model consists of five convolutional layers, some of which are followed by max pooling layers, and three fully connected layers.

GoogLeNet is proposed by Szegedy et al. [17]. This model is based on using a sparsely connected architecture in order to avoid computational bottlenecks and improve computational efficiency over the entire network as they go deeper and wider. The sparsely connected architecture is called inception modules that construct a sparser representation of the convolution networks by clustering the neurons with the highest correlation and uses an extra 1×1 convolutional layer as dimensionality reduction. This model though much deeper than AlextNet.

4   Experiments and Results

4.1   Dataset

The OuluVS2 database[1] [14] is one of the largest dataset for VSR. It is a multi-view audio-visual dataset for non-rigid mouth motion analysis. The dataset contains video recording from 52 subjects (39 males and 13 females) speaking three types of utterances: continuous digit sequences, short phrases and TIMIT sentences. The lists of the utterance content of first two types are shown in Table 1. In phase 1, a subject was asked to utter continuously ten fixed digit sequences. Each sequence consisted of ten randomly generated digits and was repeated three times during recording. In phase 2, the subject was asked to speak ten daily-use short English phrases. The same set of phrases was used in the OuluVS dataset [18]. Every phrase was uttered three times.

Table 1. Lists of utterance content.

Phase 1: digit sequences		Phase 2: phrases	
p1-01	1735162667	p2-01	Excuse me
p1-02	4029185904	p2-02	Good bye
p1-03	1907880328	p2-03	Hello
p1-04	4912118551	p2-04	How are you
p1-05	8635402112	p2-05	Nice to meet you
p1-06	2390016764	p2-06	See you
p1-07	5271613670	p2-07	I am sorry
p1-08	9744435587	p2-08	Thank you
p1-09	6385398565	p2-09	Have a good time
p1-10	7324019950	p2-10	You are welcome

In OuluVS2, each utterance were filmed with six cameras placed around a subject. The six cameras included five GoPro Hero3 Black Edition cameras and a PuxeLink PL-B774U camera. The former five cameras are called HD cameras. The image size of these cameras is 1920×1080 pixels, and its frame rate is 30fps. On the other hand, the image size recorded by the latter camera is 640×480 pixels, and its frame rate is 100fps. As regards the five HD cameras, HD1, HD2, HD3, HD4, and HD5 are located in the following positions: $0°$ (frontal view), $30°$, $45°$, $60°$, and $90°$ (profile view) to the subject's right hand side. The recording was made in an ordinary office environment with three extra lights placed behind the camera to illuminate the subject's face.

The original images of OuluVS2 are covered subject's whole face, however, OuluVS2 provides the ROI image around the talking mouth. Figure 5 shows the examples of the provided ROI images by extracting from five HD cameras at each

---

[1] http://ouluvs2.cse.oulu.fi/.

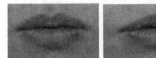

Fig. 5. ROI images (HD1, HD2, HD3, HD4, and HD5).

Table 2. Statistics of OuluVS2 (ROI).

		Digit sequences			Phrase sequences		
		min	max	ave	min	max	ave
HD1	Width [pixel]	162	294	208	158	262	201
	Height [pixel]	64	218	125	64	196	116
HD2	Width [pixel]	138	240	183	130	214	175
	Height [pixel]	64	190	122	64	186	114
HD3	Width [pixel]	130	230	180	124	210	171
	Height [pixel]	64	198	123	64	184	114
HD4	Width [pixel]	100	204	146	94	184	137
	Height [pixel]	64	186	118	64	168	109
HD5	Width [pixel]	64	164	97	64	136	89
	Height [pixel]	86	190	133	78	188	126
Frame number		83	297	161	8	20	36

frame. The ROI sizes of each scene are different. The statistics of OuluVS2 are summarized in Table 2. The ROI of near frontal view is horizontally long shape, and the ROI of near profile view is vertically long shape. As seen in Table 2, the number of frame of phrase sequences is smaller than digit sequences.

4.2 Experimental Settings

We tested our system in a speaker-independent VSR setting. We used 12 subjects (s06, s08, s09, s15, s26, s30, s34, s43, s44, s49, s51, and s52, 10 males and 2 females) for testing, and remaining 40 subjects for training. Note that digit strings and phrases were recognized as a whole instead of being modelled by visemes.

Moreover, we generated eight types of CFIs for DA by choosing four γ values (0.6, 0.8, 1.2, and 1.4) with $\alpha = 0$ as well as four α values (-2, -1, 1, and 2) with $\gamma = 1.0$. As a result, the test data included 360 ($= 10$ phrases $\times$ 12 subjects $\times 3$ samples) CFIs. The training data contained 1,200 ($= 10$ phrases $\times 40$ subjects $\times 3$ samples) CFIs without DA and the number increased to 10,800 ($= 10$ phrases $\times$ 40 subjects $\times$ 3 samples $\times$ 9 types) CFIs after DA. To generate CFIs, some parameters are required. In the experiment, we chose three sets of parameters as shown in Table 3.

Table 3. Three conditions and parameters for generating CFIs.

$M \times N$	L [frame]	W_{CFI} [pixel]	H_{CFI} [pixel]	W' [pixel]	H' [pixel]	m_W [pixel]	m_H [pixel]
7×7	49	256	256	36	36	4	4
8×8	64	256	256	32	32	0	0
9×9	81	256	256	28	28	4	4

As for the CNN model, three well-known models: NIN [16], AlexNet [15], and GoogLeNet [17], were used. We used Chainer [2], a flexible framework of deep learning for creating CNN model and training. We trained all models using stochastic gradient descent with 0.9 momentum, mini-batches of size 32, and the learning rate is initialized at 0.01. The softmax with the cross-entropy loss was used as a classifier. In the experiment, we used a personal computer with an Intel Core i7-3770 processor (3.4 GHz), 16 GB RAM, and a single NVIDIA GeForce GTX970 graphic processing unit with 6 GB on-board graphics memory.

4.3   Single-View Lip-Reading

In our experiments, training and test data were recorded by the same video camera. Since the OuluVS2 database has five camera views (HD1, HD2, HD3, HD4, and HD5), we carried out the recognition experiment for each of the five views.

Experimental results for recognizing digit strings are shown in Table 4. Table 5 shows the experimental results for recognizing short phrases. In either table, the upper half of the table shows the recognition results without DA, and the lower half the recognition results with DA. It is clear that our system achieved higher recognition accuracy when using DA than when not using DA. Regarding the three pre-trained models, there was no large difference among them in terms of system performance. Considering the parameter L, it can be seen that high recognition accuracy was obtained at $L = 49$ when recognizing digit strings, and at $L = 64$ when recognizing phrases.

Next, we selected the parameter and model settings with the highest recognition rates (those marked as bold in Tables 4 and 5) and have a closer look at the system performance for each digit string and short phrase. Tables 6 and 7 show the recognition results. Regarding viewing angle, frontal or near frontal viewing angle obtained high recognition accuracy. When recognizing the digit strings, the lowest recognition accuracy was 80.6%. It was 58.3% when recognizing short phrases. It shows that the system tends to recognize longer video sequences of a talking mouth better.

At last, we discuss the recognition result for each test subject. Figure 6 shows the average recognition results at each viewing angle. The blue and red bars are

---

[2] http://chainer.org/.

Table 4. Recognition results by various conditions (digit).

	CNN model (L)	HD1	HD2	HD3	HD4	HD5	ave.
Without DA	NIN (49)	45.8	10.1	13.5	44.6	73.5	37.5
	NIN (64)	52.1	11.3	13.8	63.2	57.9	39.6
	NIN (81)	61.7	9.7	13.5	56.1	70.1	42.2
	AlexNet (49)	12.1	9.6	45.1	70.0	59.3	39.2
	AlexNet (64)	10.7	10.6	10.4	9.6	44.9	17.2
	AlexNet (81)	9.9	9.2	11.5	42.9	33.8	21.4
	GoogLeNet (49)	11.5	9.4	10.3	47.2	52.8	26.3
	GoogLeNet (64)	9.9	11.9	9.7	8.5	14.2	10.8
	GoogLeNet (81)	12.2	8.8	8.9	10.3	28.3	13.7
With DA	NIN (49)	85.6	74.4	73.1	89.7	85.0	81.6
	NIN (64)	87.2	85.6	85.6	85.6	86.4	86.1
	NIN (81)	80.3	70.0	78.1	88.1	81.1	79.5
	AlexNet (49)	89.2	82.2	**90.8**	**91.7**	**87.5**	88.3
	AlexNet (64)	10.3	90.6	**90.8**	85.3	11.9	57.8
	AlexNet (81)	85.0	68.9	9.4	89.2	80.6	66.6
	GoogLeNet (49)	**89.4**	**92.5**	86.7	87.5	86.4	**88.5**
	GoogLeNet (64)	**89.4**	91.7	83.1	89.4	84.7	87.7
	GoogLeNet (81)	86.9	90.6	85.3	85.3	85.6	86.7

Table 5. Recognition results by various conditions (phrase).

	CNN model (L)	HD1	HD2	HD3	HD4	HD5	ave.
Without DA	NIN (49)	19.3	25.3	41.9	63.8	68.6	43.8
	NIN (64)	11.9	18.8	21.4	69.7	64.4	37.3
	NIN (81)	34.0	11.0	46.7	43.8	65.6	40.2
	AlexNet (49)	36.5	8.9	9.7	61.4	36.8	30.7
	AlexNet (64)	8.9	25.1	10.0	36.5	72.4	30.6
	AlexNet (81)	9.3	11.3	27.5	63.1	69.3	36.1
	GoogLeNet (49)	68.5	12.8	10.4	61.8	62.9	43.3
	GoogLeNet (64)	63.5	24.9	49.4	59.0	58.5	51.1
	GoogLeNet (81)	66.3	17.4	25.7	58.8	57.8	45.2
With DA	NIN (49)	77.5	77.5	78.9	71.1	74.7	75.9
	NIN (64)	81.1	79.7	**82.5**	81.9	74.7	80.0
	NIN (81)	75.3	81.7	77.8	77.8	73.3	77.2
	AlexNet (49)	82.8	75.6	80.6	80.8	79.2	79.8
	AlexNet (64)	81.7	**82.5**	81.9	**83.3**	75.3	80.9
	AlexNet (81)	73.6	74.4	75.8	76.9	75.3	75.2
	GoogLeNet (49)	83.6	81.7	81.9	78.3	76.7	80.4
	GoogLeNet (64)	**85.6**	79.7	80.8	**83.3**	**80.3**	**81.9**
	GoogLeNet (81)	83.1	76.7	78.6	79.7	78.6	79.3

Table 6. Recognition results in detail (digit).

Phrase	HD1	HD2	HD3	HD4	HD5
p1-01	86.1	88.9	86.1	83.3	80.6
p1-02	97.2	94.4	91.7	91.7	80.6
p1-03	91.7	97.2	94.4	91.7	91.7
p1-04	83.3	83.3	86.1	91.7	86.1
p1-05	100.0	94.4	97.2	100.0	91.7
p1-06	80.6	97.2	88.9	91.7	83.3
p1-07	80.6	94.4	94.4	86.1	91.7
p1-08	97.2	94.4	94.4	97.2	97.2
p1-09	91.7	86.1	86.1	91.7	88.9
p1-10	86.1	94.4	88.9	91.7	83.3
ave.	89.4	92.5	90.8	91.7	87.5

Table 7. Recognition results in detail (phrase).

Phrase	HD1	HD2	HD3	HD4	HD5
p2-01	88.9	91.7	94.4	94.4	80.6
p2-02	97.2	94.4	94.4	94.4	91.7
p2-03	80.6	66.7	72.2	69.4	69.4
p2-04	83.3	80.6	80.6	86.1	83.3
p2-05	100.0	94.4	91.7	91.7	88.9
p2-06	83.3	80.6	66.7	69.4	66.7
p2-07	94.4	88.9	100.0	91.7	88.9
p2-08	58.3	75.0	66.7	69.4	63.9
p2-09	91.7	86.1	94.4	97.2	91.7
p2-10	77.8	66.7	63.9	69.4	77.8
ave.	85.6	82.5	82.5	83.3	80.3

average recognition accuracies for the digit strings and short phrases, respectively. The recognition accuracies for speakers s06 and s51, especially from near profile viewing angle, are lower than the accuracies for other speakers. Speakers s30 and s44 obtained the highest recognition accuracy among all.

4.4   Comparison with Other Methods

The OuluVS2 database is a newly collected dataset, and there are only a few baseline recognition results provided in [14]. In their work, for feature extraction, 2D DCT features from each image were computed and PCA applied to reduce the feature dimension to 100. For recognition, a whole-word HMM was constructed

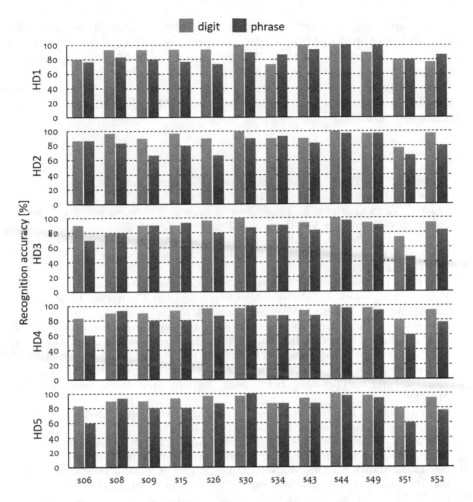

Fig. 6. Individual recognition results. (Color figure online)

for classification. In their experiment, leave-one-speaker-out cross validation was applied to the performance evaluation. The best recognition rates of 47% was obtained from HD4 camera image. This evaluation protocol is not the same as the protocol of this paper. It is clear that our method outperformed their baseline system by a large margin.

5   Conclusion

In this paper, we proposed a novel sequence image representation, namely, CFI, and a CFI-based CNN for VSR. The proposed system was evaluated on a public dataset, OuluVS2. In the experiments, the speaker independent setting was carried out with various parameter settings evaluated. In the current experiment, we did not compare the recognition accuracy with other state-of-the-art methods, such as 3D-ConvNet+LSTM [19] in the same protocol. In future, we plan to evaluate our result with other methods. We also consider to add further experiments with other datasets.

Acknowledgement. This work was supported by JSPS KAKENHI Grant Number 15K12601 and 16H03211.

References

1. Dupont, S., Luettin, J.: Audio-visual speech modeling for continuous speech recognition. IEEE Trans. Multimed. **2**, 141–151 (2000)
2. Zhou, Z., Zhao, G., Hong, X., Pietikainen, M.: A review of recent advances in visual speech decoding. Image Vis. Comput. **32**, 590–605 (2014)
3. Bregler, C., Konig, Y.: "Eigenlips" for robust speech recognition. In: IEEE International Conference on Acoustics, Speech, and Signal Processing (ICASSP 1994), pp. 669–672 (1994)
4. Lucey, P.J., Potamianos, G., Sridharan, S.: Patch-based analysis of visual speech from multiple views. In: Proceedings of International Conference on Auditory-Visual Speech Processing (AVSP 2008), pp. 69–73 (2008)
5. Shiraishi, J., Saitoh, T.: Optical flow based lip reading using non rectangular ROI and head motion reduction. In: 11th IEEE International Conference on Automatic Face and Gesture Recognition (FG2015) (2015)
6. Matthews, I., Cootes, T.F., Bangham, J.A., Cox, S., Harvey, R.: Extraction of visual features for lipreading. IEEE Trans. Pattern Anal. Mach. Intell. **24**, 198–213 (2002)
7. Shin, J., Lee, J., Kim, D.: Real-time lip reading system for isolated Korean word recognition. Pattern Recogn. **44**, 559–571 (2011)
8. Saitoh, T.: Efficient face model for lip reading. In: International Conference on Auditory-Visual Speech Processing (AVSP), pp. 227–232 (2013)
9. Ngiam, J., Khosla, A., Kim, M., Nam, J., Lee, H., Ng, A.Y.: Multimodal deep learning. In: 28th International Conference on Machine Learning, pp. 689–696 (2011)
10. Hu, D., Li, X., Lu, X.: Temporal multimodal learning in audiovisual speech recognition. In: IEEE Conference on Computer Vision and Pattern Recognition (CVPR), pp. 3574–3582 (2016)
11. Noda, K., Yamaguchi, Y., Nakadai, K., Okuno, H.G., Ogata, T.: Lipreading using convolutional neural network. In: INTERSPEECH, pp. 1149–1153 (2014)
12. Amer, M.R., Siddiquie, B., Khan, S., Divakaran, A., Sawhney, H.: Multimodal fusion using dynamic hybrid models. In: IEEE Winter Conference on Applications of Computer Vision (WACV), pp. 556–563 (2014)
13. Takashima, Y., Kakihara, Y., Aihara, R., Takiguchi, T., Araki, Y., Mitani, N., Omori, K., Nakazono, K.: Audio-visual speech recognition using convolutive bottleneck networks for a person with severe hearing loss. IPSJ Trans. Comput. Vis. Appl. **7**, 64–68 (2015)
14. Anina, I., Zhou, Z., Zhao, G., Pietikainen, M.: OuluVS2: a multi-view audiovisual database for non-rigid mouth motion analysis. In: IEEE International Conference on Automatic Face and Gesture Recognition (FG) (2015)
15. Krizhevsky, A., Sutskever, I., Hinton, G.E.: ImageNet classification with deep convolutional neural networks. In: Advances in Neural Information Processing Systems (2012)

16. Lin, M., Chen, Q., Yan, S.: Network in network. In: International Conference on Learning Representations (ICLR) (2014)
17. Szegedy, C., Liu, W., Jia, Y., Sermanet, P., Reed, S.: ImageNet classification with deep convolutional neural networks. In: IEEE Conference on Computer Vision and Pattern Recognition (CVPR) (2015)
18. Zhao, G., Barnard, M., Pietikainen, M.: Lipreading with local spatiotemporal descriptors. IEEE Trans. Multimed. **11**, 1254–1265 (2009)
19. Baccouche, M., Mamalet, F., Wolf, C., Garcia, C., Baskurt, A.: Sequential deep learning for human action recognition. In: International Workshop on Human Behavior Understanding (HBU 2011) (2011)

Multi-view Automatic Lip-Reading Using Neural Network

Daehyun Lee[1,2]([✉]), Jongmin Lee[1], and Kee-Eung Kim[1]

[1] School of Computing, KAIST, Daejeon, Korea
daehyun.lee@kaist.ac.kr
[2] Visual Display Division, Samsung Electronics, Suwon, Korea

Abstract. It is well known that automatic lip-reading (ALR), also known as visual speech recognition (VSR), enhances the performance of speech recognition in a noisy environment and also has applications itself. However, ALR is a challenging task due to various lip shapes and ambiguity of visemes (the basic unit of visual speech information). In this paper, we tackle ALR as a classification task using end-to-end neural network based on convolutional neural network and long short-term memory architecture. We conduct single, cross, and multi-view experiments in speaker independent setting with various network configuration to integrate the multi-view data. We achieve 77.9%, 83.8%, and 78.6% classification accuracies in average on single, cross, and multi-view respectively. This result is better than the best score (76%) of preliminary single-view results given by ACCV 2016 workshop on multi-view lip-reading/audio-visual challenges. It also shows that additional view information helps to improve the performance of ALR with neural network architecture.

1 Introduction

Human understands speech from not only acoustic information but also visual clues. An extreme case is shown in the McGurk effect [1], a visual information of /ga/ with an acoustic of /ba/ is perceived as /da/. Similar phenomenon also appears in machines using deep learning approach [2]. These observations present that visual information gives significant clues to the speech recognition in both human and machine.

Automatic lip-reading (ALR), also known as visual speech recognition (VSR) or speech reading, is understanding speech using only visual information (movements of the lips, face, tongue and so on) without acoustic information. In audio-visual speech recognition (AVSR), ALR has an important role in enhancing the performance of speech recognition through audio-visual data fusion. Furthermore, stand-alone ALR is a useful component of various applications such as a visual password, silent speech interface, and forensic video analysis.

ALR performance is measured by word or phrase classification error rate of utterance in the form of lip image sequence. ALR is a challenging task because of many visual factors such as various mouth shapes, changes of illumination, and head poses variation [3].

© Springer International Publishing AG 2017
C.-S. Chen et al. (Eds.): ACCV 2016 Workshops, Part II, LNCS 10117, pp. 290–302, 2017.
DOI: 10.1007/978-3-319-54427-4_22

There are several types of experimental setups in ALR task; speaker dependent (SD) or speaker independent (SI), single-view or multi-view. In SD setting, one speaker data are used for both training and evaluation. On the other hand, evaluation is performed to the unseen speaker(s) in SI setting. In the multi-view setting, we use the data recorded from the multiple cameras with various angles simultaneously [4] while the data in a single-view setting is recorded from one camera angle.

Classical approaches are based on visual feature extraction methods, such as principal component analysis, discrete wavelet transform, discrete cosine transform, active appearance model [5], local binary pattern [6], optical flow [7], Eigenlips [8], histograms of oriented gradients [9], internal motion histograms, motion boundary histograms, and their mixed models [8,10]. For multi-viewpoint lip-reading, [11] adopt a minimum cross-pose variance analysis technique.

Thanks to recent successful achievements in deep neural network approach in machine learning community, people started to apply neural network to ALR and AVSR. For example, feature extraction has been done by deep autoencoder [2] without massive hand-crafted engineering work. Recently, [12] show that with end-to-end neural network architecture, they can achieve a state-of-the-art performance in speaker dependent setting, compared to classical feature-based approaches.

In this paper, we tackle the ALR task using an end-to-end neural network approach without hand-crafted feature extraction in multi-viewpoint SI setting experiment. Our work motivated by recent works of video and image recognition researchers that use convolutional neural network (CNN) with long short-term memory (LSTM) for understanding video and image data [13–15].

2 Background

2.1 Problem Specification

We deal with three ALR tasks given by ACCV 2016 workshop on multi-view lip-reading/audio-visual challenges[1] (MLAC 2016) as follows:

- Single-view ALR: Train and test on data recorded from a single camera view.
- Cross-view ALR: Learn and transfer knowledge from a source view (e.g., the frontal view) to enhance learning for a target view (e.g., the profile view).
- Multiple-view ALR: Train and test on synchronized data recorded from multiple camera views.

2.2 Convolutional Neural Network

CNN is a biologically inspired variant of multi-layer perceptron containing small sub-regions of a visual field called receptive field [16]. Unlike fully connected layered network, CNN has sparse connectivity and shared weights for the purpose

---

[1] http://ouluvs2.cse.oulu.fi/ACCVW.html.

of increasing computational efficiency and global representation power. CNN is now the most popular and effective selection for learning visual features in computer vision and machine learning fields.

We obtain a feature map at layer h with input x pixel at coordinates (i, j) as the following equation:

$$h_{ij} = a((W * x)_{ij} + b), \tag{1}$$

where weight matrix W and bias vector b is the filter of this feature map, a is activation function for non-linearities.

2.3   Long Short-Term Memory

Recurrent neural network (RNN) is designed for processing sequential data by sharing weights across several time steps. Due to its vanishing gradient problem that appears to long-sequence training data, its variations including LSTM [17] become popularized in practical applications. LSTM consists of memory cells connected recurrently to each other, which is replacing hidden units of standard RNN. End-to-end learning architecture with LSTM is the typical model when dealing with a sequence dataset.

We update LSTM hidden state h_t at every timestep t as follows:

$$
\begin{aligned}
i_t &= \sigma(W_i x_t + U_i h_{t-1} + b_i) \\
f_t &= \sigma(W_f x_t + U_f h_{t-1} + b_f) \\
c_t &= i_t * tanh(W_c x_t + U_c h_{t-1} + b_c) + f_t * c_{t-1} \\
o_t &= \sigma(W_o x_t + U_o h_{t-1} + V_o c_t + b_o) \\
h_t &= o_t * tanh(c_t),
\end{aligned}
\tag{2}
$$

where x_t is an input at time t and W_i, W_f, W_c, W_o, U_i, U_f, U_c, U_o, V_o are weight matrices, b_i, b_f, b_c, b_o are bias vectors, subscripts represent input(i), forget(f), cell(c) and output(o) variables [18,19].

3   Dataset

3.1   OuluVS2 Database

OuluVS2 database, publicly provided by CMVS[2], contains video recordings from 52 speakers with five different camera views at the same time. It has three collections of data - ten continuous digit strings, ten daily-use short English phrases, and five randomly selected TIMIT sentences.

OuluVS2 provides region of interest (ROI) videos, which were preprocessed by segmenting individual utterances and cropping off ROIs, for digit strings and phrases collection. We use visual part of phrases collection, which are

---

[2] The Center of Machine Vision Research, Department of Computer Science and Engineering, University of Oulu, Finland.

Table 1. Data separation according to speakers. The test set is given by MLAC 2016, validation set is randomly chosen as the same size of test samples. (Total 7,800 samples with 52 speakers, ID 1 to 53 except 52.)

Data set	Speaker IDs	# samples
Training	$1, 2, 3, 10, 11, 12, 13, 18, 19, 20, 21, 22, 23, 24, 25,$ $27, 33, 35, 36, 37, 38, 39, 45, 46, 47, 48, 50, 53$	4,200
Validation	$4, 5, 7, 14, 16, 17, 28, 31, 32, 40, 41, 42$	1,800
Test	$6, 8, 9, 15, 26, 30, 34, 43, 44, 49, 51, 52$	1,800

same as the preliminary experiments conducted by MLAC 2016[3]. Every phrase was uttered three times in this collection, the total number of samples is $52(speakers) \cdot 5(views) \cdot 3(utterances) \cdot 10(phrases) = 7,800$, where maximum length of utterance is 36.

The preliminary experiments use three different methods including HiLDA [20], RAW + PLVM (raw pixel values classified by latent variable models [21]). All the experiments are single-view ALR tasks, and the best result is about 76% accuracy with 45° view data and RAW + PLVM.

OuluVS2 database is more suitable for SI experiment than SD setting because of relatively large number of speakers and a small number of the utterance of each speaker. It is considerable, though out of the scope of this paper, to evaluate with a semi-SD setting that trains with multiple speakers and test unseen utterance among them (Fig. 1).

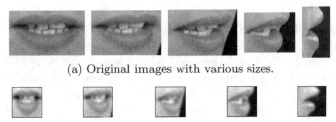

(a) Original images with various sizes.

(b) After pre-processing with fixed size and maximized contrast.

Fig. 1. Examples of original and pre-processed lip images for a frontal, 30°, 45°, 60°, profile view image from the left to the right.

3.2 Data Pre-processing

First of all, we extract image frames from each ROI video by FFmpeg[4] command line tool with option *qscale:v = 1*. After that, we resize all images of each view into the same size since the size of each image varies even in the same viewpoint,

---

[3] http://ouluvs2.cse.oulu.fi/preliminary.html.
[4] https://ffmpeg.org/about.html.

which enables us to conduct multi-viewpoint experiment easily. Square shape is a reasonable choice because profile view ROI image has longer height than width, unlike the others.

We have three options for pre-processing on the image itself; (1) Color: original RGB color image (2) Gray[†]: convert into grayscale with maximized contrast. (3) Color[†]: RGB color with maximized contrast.

We conduct experiments for selecting both size and image option, and the result is shown in Table 2. Experiments performed by 2D-CNN + LSTM architecture details in Sect. 4. As a result, we use 20 by 20 pixel color image with maximized the contrast, i.e. all pixel values in each channel are normalized as mapped into $[0, 1]$ interval.

Table 2. Comparison among various pre-processing options. All the results are obtained from validation set. Resizing into 20 by 20 pixels with contrast-maximized color image is the best option. All the later experiments are performed with this pre-processed data. [†]: Contrast is maximized.

Image size	Color	Gray[†]	Color[†]	Average
25×25	75.5%	77.1%	76.5%	76.4%
20×20	77.6%	78.6%	**80.8%**	**79.0%**
15×15	76.8%	74.9%	78.2%	76.6%
10×10	71.8%	75.9%	75.1%	74.3%
Average	75.4%	76.6%	**77.6%**	

4   Proposed Method

We combine visual model and temporal model based on CNN, LSTM for the basic neural network architecture in Fig. 2. We define visual models with various settings as 2D-CNN and 3D-CNN for automatic feature extraction. We also define temporal model as LSTM that learns the temporal features of the image sequence and classify the utterance into a phrase (Fig. 3).

Fig. 2. Overall architecture which returns probabilities of each class from an input of image sequence of a single utterance.

4.1   Visual Model

2D-CNN. Two convolutional layers with 16 to 256 filters in the shape of $(3, 3)$ used for feature extraction of lip region images. Each convolutional layer has a

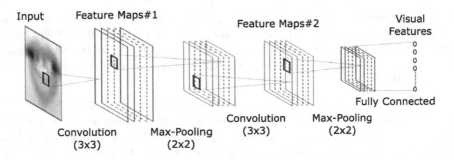

Fig. 3. 2D-CNN architecture

successive max-pooling layer for downscaling by the shape of $(2, 2)$. Last one fully connected layer with 8 to 64 dimension outputs used to the final output of image feature. We adopt dropout [22] technique for improving generalization power of the network. We use hyperopt [23] to find the optimal hyper-parameters (number of filters, output dimension of the fully connected layer) for our network.

3D-CNN. In the same way of Eq. (1), we generate the feature map from a 3D input where the x pixel at 3-dimensional coordinates (i, j, k) as following equation:

$$h_{ijk} = a((W * x)_{ijk} + b) \tag{3}$$

We construct 3D-CNN as almost same as 2D-CNN, the filter shapes are $(1, 3, 3)$ and $(2, 3, 3)$ for the first and the second convolutional layer respectively as shown in Fig. 4(a). The shape of pooling size is $(1, 2, 2)$ for every max-pooling layer, and the remaining parameters are same as 2D-CNN. Strictly speaking, the five view data is not a 3-dimensional data. However, we conduct an experiment with 3D-CNN to find out the possibilities of learning some features or not.

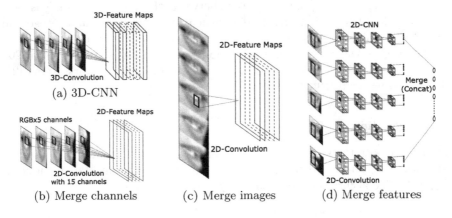

Fig. 4. Various visual model suggestions for the multi-view setting.

Merge Channels. We make an input image having 15 channels from the five view data with three (RGB) channels as shown in Fig. 4(b). The places in the same pixel position of each view data are the different locations in actual. Although it looks somewhat weird, we conduct this experiment as the same reason as 3D-CNN.

Merge Images: As shown in Fig. 4(c), we append five images from the different view at the same time into a single image as an input of the visual model. In this architecture, we expect to learn all the five view feature by 2D-CNN. While out of our experiment, a more elaborate configuration is that all five images avoid convolving each other along the edges.

Merge Features. The last variation is shown in Fig. 4(d). We merge the features that are generated by 2D-CNN from the images in the different view at the same time. We design this architecture to learn each view image separately and to combine them into a single feature for an input of the temporal model.

4.2 Temporal Model

We design the temporal model as a classifier that has inputs generated by the visual model and an output as probabilities of each class. It consists of two successive LSTM layer with 128 memory cells that lead to one FC layer having the same number of class, 10 output dimension. We select 128 by experiments using hyperopt which yield lowest validation loss between 8 to 256.

5 Experiments and Results

Basic protocol of experiments is SI setting, and the data is divided into the train, validation, and test sets as shown in Table 1. In order to improve generalization performance, we add dropout [22] layers between the layers. Learning is performed by Adam [24] optimizer with default learning rate 0.001, and categorical cross entropy objective loss function used. We evaluate the result on test data with the weights having minimum validation loss during the training until 200 epoch. All experiments are performed by deep learning framework Keras [25] with Theano backend [19] (Fig. 5).

5.1 Single-View ALR

Table 3 represents the single-view experiment result on test data. The baseline accuracies are given by MLAC 2016 as a preliminary experiments[5] in the form of a graph. The accuracies are approximated from the chart.

---

[5] http://ouluvs2.cse.oulu.fi/preliminary.html.

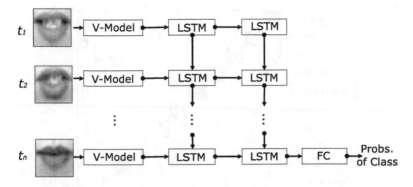

Fig. 5. Unfolded view of temporal model architecture, it returns probabilities of each class with inputs as $t_1, t_2, \ldots, t_n$ images in a single utterance. V-Model: Visual Model.

Table 3. Single-view word test accuracy results for each camera view. The best performance is obtained by profile view. [†]: Approximated accuracies from the graph image of preliminary single-view experiment given by MLAC 2016

| | Our Results (Accuracy of Test Data) | | | | | |
Training Data	(1)Frontal	(2)30°	(3)45°	(4)60°	(5)Profile	Average
(1)Frontal	81.1 %					
(2)30°		80 %				
(3)45°			76.9 %			77.9 %
(4)60°				69.2 %		
(5)Profile					82.2 %	

| | Single-View Baseline Results[†] (Accuracy of Test Data) | | | | | |
Method	(1)Frontal	(2)30°	(3)45°	(4)60°	(5)Profile	Average
DCT-PCA-HMM[†]	63%	62%	62%	63%	57%	61%
DCT-HiLDA-HMM[†]	74%	72%	73%	73%	68%	72%
RAW-PLVM[†]	73%	75%	76%	75%	70%	74%

The average accuracy is 5% higher than the average accuracy using RAW-PLVM method which is the best in the baseline. Moreover, all accuracies except 60° view are higher than the best score (76%) of the single-view baseline.

Our neural network produces the reasonable result of the prediction. The confusing phrases are similar to those of human. "Thank you" and "See you" pair is the most confusing as presented by confusion matrix in Fig. 6.

5.2 Cross-View ALR

As the first stage (cross-view), we conduct a cross-view experiment by training all the data in train set. In other words, we train frontal, 30°, 45°, 60°, profile view data all together, and test each view separately. In this approach, we get the average accuracy 82.6%, and all of the results are better than preliminary results.

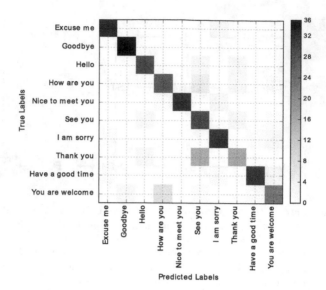

Fig. 6. An example of the confusion matrix. (Profile view result with test accuracy: 82.2%)

We infer from this result that neural networks learn and transfer knowledge from a view to the other view by simply learning with the data altogether without further work.

In the second stage (cross-view2), we perform ALR tasks with each view data using the neural network initialized by the weights from the first stage result, as similar as fine-tuning. We choose the weights for testing with min-validation-loss, note that there is no improvement with 30°, 45° view data. As a final result, we get 83.8% in average and 86.4% in the best (profile view), and this is outperformed the preliminary results. The confusion matrices of profile view show that the progress of increasing performance as shown in Fig. 7 (Table 4).

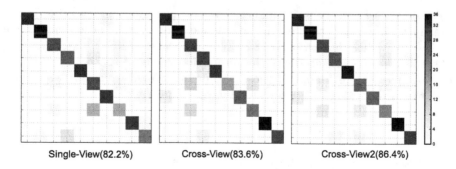

Single-View(82.2%)   Cross-View(83.6%)   Cross-View2(86.4%)

Fig. 7. The confusion matrices of test accuracies in profile view setting. It shows that a gradual improvement from single-view to cross-view. The x-axis is prediction classes; the y-axis is true classes with the same label in Fig. 6.

Table 4. Cross-view ALR test accuracy and validation-loss results for each camera view. CV (cross-view): train all the training data mixed, CV2 (cross-view2): train each view data individually after train all the training data mixed.

| | Training Data | Accuracy of Test Data | | | | | |
		(1)Frontal	(2)30°	(3)45°	(4)60°	(5)Profile	Average
CV	All	80.6 %	81.1 %	85 %	82.5 %	83.6 %	82.6 %
	All+(1)Frontal	82.8 %					
	All+(2)30°		81.1 %				
CV2	All+(3)45°			85 %			83.8 %
	All+(4)60°				83.6 %		
	All+(5)Profile					86.4 %	

| | Training Data | Loss of Validation Data | | | | | |
		(1)Frontal	(2)30°	(3)45°	(4)60°	(5)Profile	Average
CV	All	0.4372	0.4057	0.4109	0.3391	0.5606	0.4487
	All+(1)Frontal	0.3216					
	All+(2)30°		0.4957				
CV2	All+(3)45°			0.4109			0.424
	All+(4)60°				0.3371		
	All+(5)Profile					0.5546	

This result indicates that the test performance of target view has been improved by transferring knowledge of other source view data. For instance, in the profile view, the accuracy is 82.2% in single-view (train profile view data only) then 1.4% improved in cross-view (train five view data altogether). Finally, we get 86.4% accuracy in cross-view2 (initialize weights by the cross-view result and train profile view data once more) which is 4.2% higher than single-view.

5.3   Multiple-view ALR

We conduct four types of architecture explained in Sect. 4. In this experiments, we use hyper-parameters already tuned in the single-view experiment because we have not enough time to optimize each network one by one. Despite we adopt the hyper-parameters from single-view experiments, the multi-view result is better than the single-view result in average (Table 5).

Table 5. Multi-view results of prediction accuracy using various architectures.

Method	Test acc.
3D-CNN (MV-3D)	76.1%
Merge Channels (MV-MC)	78.9%
Merge Images (MV-MI)	80.0%
Merge Features (MV-MF)	79.4%

It infers that the more features (five times more than the single view) in a single frame help the network in the average performance.

On the other hand, the result is worse than cross-view in spite of using same data. The difference between them is the number of samples, the numbers of training sample are 4,200 (cross-view) vs. 840 (multi-view). We know that the number of samples impacts the neural network performance, more data is always better.

Table 6 shows the elapsed time in the average of each experiment. We use 1 GPU, and measure the time for 200 epoch. Cross-view experiment takes exactly 5 times more time than single-view because of the same architecture and the training data size. Note that the elapsed times in the multi-view are various in the range between 7 and 38 depending on the architecture. 3D-CNN takes 5.4 times more time than Merge Channels.

Table 6. The average computational times (minutes) of training data for 200 epochs. We use 1 GPU (Geforce GTX 1080) for each experiment.

Method	Time	# samples	Input dimension
Single-view	6	840	$20 \cdot 20 \cdot 3 = 1,200$
Cross-view	30	4,200	$20 \cdot 20 \cdot 3 = 1,200$
3D-CNN (MV-3D)	38	840	$5 \cdot 20 \cdot 20 \cdot 3 = 6,000$
Merge Channels (MV-MC)	7	840	$20 \cdot 20 \cdot 15 = 6,000$
Merge Images (MV-MI)	20	840	$100 \cdot 20 \cdot 3 = 6,000$
Merge Features (MV-MF)	23	840	$(20 \cdot 20 \cdot 3) \cdot 5 = 6,000$

All of the experiment results are summarized in Fig. 8.

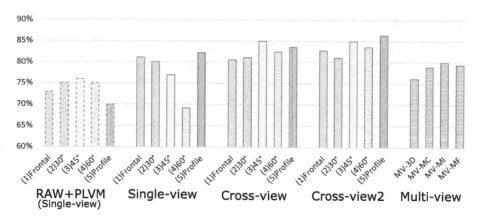

Fig. 8. The summary graph of the prediction accuracies. RAW + PLVM is the best result of the preliminary single-view experiment given by MLAC 2016. MV-3D: 3D-CNN, MV-MC: Merge Channel, MV-MI: Merge Images, MV-MF: Merge Features.

6   Conclusions

In this paper, we propose an end-to-end neural network architecture for speaker independent ALR task with multi-view data (OuluVS2 database). The evaluation is performed by experiments in single-view, cross-view, and multi-view setting. We report that the accuracies are 77.9%, 83.8%, and 78.6% in average; 82.2%, 86.4%, and 80.0% in the best respectively. All of the results are better than the best single-view result (76%) of the baseline. These results show that ALR performance is improved by multi-view data using neural network architectures.

We expect that the best score would come out in the multi-view result before experiments are performed, but the results are different as we reported. One of the reason is the lack of samples in the multi-view experiment in spite of more complex feature than a single view. Experiments with various data augmentation from multi-view data are future work.

OuluVS2 database provides the original high-resolution videos, experiment with this large size of the input video is another considerable future work.

Acknowledgement. This work was supported by Institute for Information and communications Technology Promotion (IITP) grant funded by the Korea government (MSIP) (No. B0101-16-0307, Basic Software Research in Human-level Lifelong Machine Learning (Machine Learning Center)).

References

1. McGurk, H., MacDonald, J.: Hearing lips and seeing voices. Nature **264**, 746–748 (1976)
2. Ngiam, J., Khosla, A., Kim, M., Nam, J., Lee, H., Ng, A.Y.: Multimodal deep learning. In: Proceedings of 28th International Conference on Machine Learning (ICML-2011), pp. 689–696 (2011)
3. Potamianos, G., Neti, C.: Audio-visual speech recognition in challenging environments. In: INTERSPEECH (2003)
4. Anina, I., Zhou, Z., Zhao, G., Pietikäinen, M.: Ouluvs2: a multi-view audiovisual database for non-rigid mouth motion analysis. In: 2015 11th IEEE International Conference and Workshops on Automatic Face and Gesture Recognition (FG), vol. 1, pp. 1–5. IEEE (2015)
5. Cootes, T.F., Edwards, G.J., Taylor, C.J., et al.: Active appearance models. IEEE Trans. Pattern Analysis Mach. Intell. **23**, 681–685 (2001)
6. Zhao, G., Barnard, M., Pietikainen, M.: Lipreading with local spatiotemporal descriptors. IEEE Trans. Multimedia **11**, 1254–1265 (2009)
7. Shaikh, A.A., Kumar, D.K., Yau, W.C., Azemin, M.C., Gubbi, J.: Lip reading using optical flow and support vector machines. In: 2010 3rd International Congress on Image and Signal Processing (CISP), vol. 1, pp. 327–330. IEEE (2010)
8. Bregler, C., Konig, Y.: Eigenlips for robust speech recognition. In: 1994 IEEE International Conference on Acoustics, Speech, and Signal Processing, ICASSP-1994, vol. 2, p. II-669. IEEE (1994)

9. Dalal, N., Triggs, B.: Histograms of oriented gradients for human detection. In: 2005 IEEE Computer Society Conference on Computer Vision and Pattern Recognition (CVPR 2005), vol. 1, pp. 886–893. IEEE (2005)
10. Rekik, A., Ben-Hamadou, A., Mahdi, W.: A new visual speech recognition approach for RGB-D cameras. In: Campilho, A., Kamel, M. (eds.) ICIAR 2014. LNCS, vol. 8815, pp. 21–28. Springer, Heidelberg (2014). doi:10.1007/978-3-319-11755-3_3
11. Pass, A., Zhang, J., Stewart, D.: An investigation into features for multi-view lipreading. In: 2010 IEEE International Conference on Image Processing, pp. 2417–2420. IEEE (2010)
12. Wand, M., Koutn, J., et al.: Lipreading with long short-term memory. In: 2016 IEEE International Conference on Acoustics, Speech and Signal Processing (ICASSP), pp. 6115–6119. IEEE (2016)
13. Venugopalan, S., Rohrbach, M., Donahue, J., Mooney, R., Darrell, T., Saenko, K.: Sequence to sequence-video to text. In: Proceedings of IEEE International Conference on Computer Vision, pp. 4534–4542 (2015)
14. Venugopalan, S., Xu, H., Donahue, J., Rohrbach, M., Mooney, R., Saenko, K.: Translating videos to natural language using deep recurrent neural networks (2014). arXiv preprint arXiv:1412.4729
15. Yao, L., Torabi, A., Cho, K., Ballas, N., Pal, C., Larochelle, H., Courville, A.: Describing videos by exploiting temporal structure. In: Proceedings of IEEE International Conference on Computer Vision, pp. 4507–4515 (2015)
16. Hubel, D.H., Wiesel, T.N.: Receptive fields and functional architecture of monkey striate cortex. J. Physiol. **195**, 215–243 (1968)
17. Hochreiter, S., Schmidhuber, J.: Long short-term memory. Neural Comput. **9**, 1735–1780 (1997)
18. Gers, F.A., Schmidhuber, J., Cummins, F.: Learning to forget: continual prediction with LSTM. Neural Comput. **12**, 2451–2471 (2000)
19. Bastien, F., Lamblin, P., Pascanu, R., Bergstra, J., Goodfellow, I., Bergeron, A., Warde-Farley, D., Bengio, Y.: Theano: new features and speed improvements (2012). arXiv preprint arXiv:1211.5590
20. Potamianos, G., Neti, C., Gravier, G., Garg, A., Senior, A.W.: Recent advances in the automatic recognition of audiovisual speech. Proc. IEEE **91**, 1306–1326 (2003)
21. Zhou, Z., Hong, X., Zhao, G., Pietikäinen, M.: A compact representation of visual speech data using latent variables. IEEE Trans. Pattern Anal. Mach. Intell. **36**, 1–1 (2014)
22. Srivastava, N., Hinton, G.E., Krizhevsky, A., Sutskever, I., Salakhutdinov, R.: Dropout: a simple way to prevent neural networks from overfitting. J. Mach. Learn. Res. **15**, 1929–1958 (2014)
23. Bergstra, J., Yamins, D., Cox, D.D.: Making a science of model search: hyperparameter optimization in hundreds of dimensions for vision architectures. ICML (1) **28**, 115–123 (2013)
24. Kingma, D., Ba, J.: Adam: a method for stochastic optimization (2014). arXiv preprint arXiv:1412.6980
25. Chollet, F.: Keras (2015). https://github.com/fchollet/keras

Lip Reading from Multi View Facial Images Using 3D-AAM

Takuya Watanabe[1(✉)], Kouichi Katsurada[2], and Yasushi Kanazawa[1]

[1] Department of Computer Science and Engineering,
Toyohashi University of Technology, Toyohashi, Japan
watanabe@img.cs.tut.ac.jp, kanazawa@cs.tut.ac.jp
[2] Department of Information Science, Tokyo University of Science,
Katsushika, Japan
katsurada@rs.tus.ac.jp

Abstract Lip reading is a technique to recognize the spoken words base on lip movement. In this process, it is important to detect the correct features of the facial images. However, detection is not easy in the real situations because the facial images may be taken from various angles. To cope with this problem, lip reading from multi view facial images has been conducted in several research institutes. In this paper, we propose a lip reading approach using the 3D Active Appearance Models (AAM) features and the Hidden Markov Model (HMM)-based recognition model. The AAM is a parametric model constructed from both shape and appearance parameters. The parameters are compressed into the combination parameters in the AAM, and are used in lip reading or some other facial image processing applications. The 3D-AAM extends the traditional 2D shape model to 3D shape model built from three different view angles (frontal, left, and right profile). It provides an effective algorithm to align the model with the RGB and the 3D range images obtained by the RGBD-camera. The benefit of using 3D-AAM in lip reading is that it enables to recognize the spoken words from any angle of the facial images. In the experiment, we compared the accuracy of lip reading using 3D-AAM with that of the traditional 2D-AAM on various angles of facial images. Based on the result, we confirmed that 3D-AAM is effective in cross view lip reading despite using only the frontal images in the HMM training phase.

1 Introduction

Lip reading is a technique to recognize spoken words from lip movement. This technique is generally used by a hearing-impaired person in communicating or in multimodal speech recognition when the audio information is corrupted by background noises. In general, in the multimodal speech recognition, lip reading obtains lip movement from the frontal facial images. Hence, the traditional lip reading approach [1–3] using frontal facial images is established as the standard method.

Although the traditional methods achieved high accuracy in lip reading from the frontal face, they did not work well for non-frontal facial images. To improve the accuracy for non-frontal facial images, some researchers have exploited lip reading

© Springer International Publishing AG 2017
C.-S. Chen et al. (Eds.): ACCV 2016 Workshops, Part II, LNCS 10117, pp. 303–316, 2017.
DOI: 10.1007/978-3-319-54427-4_23

from various angles of the facial images. Kumar et al. [4] evaluated the traditional frontal view approach on profile view the facial images. Lucey et al. [5] improved the accuracy in profile images using linear regression to find the transformation matrix from the synchronous frontal and profile images. Although these methods increase the accuracy in the profile view facial images, their accuracy significantly decreases when the views are not included in the training data. To enable lip reading from any view of facial images, Komai et al. [6] assumed that a three-dimensional head shape is on a sphere, and translated the non-frontal features obtained by the Active Appearance Models (AAM) to the frontal features using the coefficient of angles calculated in advance. This method achieved multi view recognition from −30° to 30°. However, the accuracy decreases when the angle is more than 30° because of self-occlusion. Lan et al. [7] obtained the optimal view for lip reading in the experiments and constructed the linear transformation model which converts the multi view features into the optimal view. Their results showed that feature transformation helps to increase the lip reading performance. However, this procedure may result in reduction of performance if the transformation matrix is not accurate enough. To avoid this problem, the use of a 3D model is suggested as one of the solutions.

In this paper, we propose a cross view lip reading approach using the angle-independent 3D-AAM [8]. The 3D-AAM extends the 2D shape model used in the traditional AAM to the 3D shape model. The models in the 3D-AAM are composed of an angle parameter and shape parameters. Since the shape parameters do not contain any angle information, they are not transformed when reading the lip movement. In our approach, we use the viseme-based Hidden Markov Model (HMM) to model the sequence of the lip movement. The other benefit of using 3D-AAM is that the HMM model can be constructed from any angle of the facial images because the 3D-AAM shape parameters do not contain any angle information. This means that our method can solve the problem caused by the angles of the images both in the HMM training phase and the recognition phase. In the experiment, we compare the performance of lip reading using traditional 2D-AAM [9] with the use of the 3D-AAM. Then, we use multi view images from 0°, 45°, and 90° as test images to evaluate the efficiency in the cross view lip reading.

In the next section, we describe the outline of the 3D-AAM construction and its fitting algorithm. Section 3 describes the methods to use HMM in lip reading. Section 4 details the experimental evaluation of the proposed approach. Finally, Sect. 5 summarizes and concludes this paper.

2   3D Active Appearance Models

The 3D-AAM is an extended model of the traditional 2D-AAM. The purpose of using 3D-AAM in lip reading is to extract the 3D facial feature from multi view images during lip reading. In this section, we describe how to construct the 3D-AAM and how to find the parameters when a facial image is given.

2.1   Construction of 3D-AAM

The 3D-AAM is constructed from the facial 3D shape model and the appearance model. In this section, we outline the formulation of these two models. The details of the construction process are given in literature [8].

The 3D Shape Model

The 3D shape parameter **s** is composed of n 3D feature points that are represented by (x_i, y_i, z_i) as shown in Eq. (1).

$$s = \begin{pmatrix} x_1 & x_2 & \ldots & x_n \\ y_1 & y_2 & \ldots & y_n \\ z_1 & z_2 & \ldots & z_n \end{pmatrix}. \tag{1}$$

The 3D shape must be modeled independently with the scale, rotation, and translation information; these information are removed before the model construction. Procrustes analysis [10] is conducted to normalize the 3D shapes and to calculate the mean 3D shape s_0. The principal component analysis (PCA) is applied to the normalized 3D shapes to find the principal components $\{s_1, s_2, \ldots s_m\}$. Consequently, a 3D shape is represented by the following equation:

$$s = s_0 + \sum_{i=1}^{m} p_i s_i, \tag{2}$$

where $p = \{p_i\}_{i=1}^{m}$ is a set of real number component scores (shape parameters) that control the variation of the shape. An example of a shape variation is shown in Fig. 1.

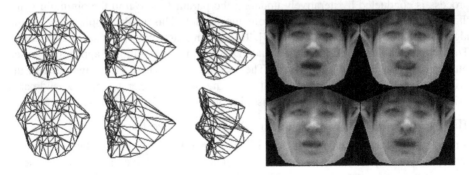

Fig. 1. Left: Examples of the 3D shapes with different shape parameters. The upper shows the mean shape and the below is created from a shape parameter. Right: Example of the appearance variation. Upper left is the mean appearance. Another three appearances are created from different parameters.

To obtain the facial shape from any angle, the translation parameter $t = (t_x, t_y, t_z)$ and head orientation parameter $\theta = (\alpha, \beta, \gamma)$ are included in the equation. The representation of a position is then

$$s' = R(\theta)s + t, \tag{3}$$

where R is a rotation matrix. A pose vector is defined as $q = (\theta, t)$.

The Appearance Model

The appearance model represents the color (or intensity) information of the target object. Given the reference shape, all training images are transformed into the shape-free images. The mean shape (also called the reference shape) is used to normalize the appearance of the different shapes of facial images. This normalization guarantees the dimension of texture parameters to be the same among images. Moreover, the difference of intensity among images because of lighting environment or camera setting is normalized before applying PCA. As a result, the appearance A is represented by the following equation:

$$A = A_0 + \sum_{i=1}^{n} \lambda_i A_i, \tag{4}$$

where A_0 is a mean appearance parameter, $\{A_i\}_{i=1}^{n}$ are the principal components, and $\lambda = \{\lambda_i\}_{i=1}^{n}$ is a set of appearance parameters. For example, the appearances calculated using different parameters are shown in Fig. 1.

2.2 3D-AAM Fitting Procedure

Fitting is a process to find the parameters when a target facial image is given. This process is conducted by iteratively updating the parameters to find the optimum value that can generate the target image from the model. This study employs the fitting algorithm proposed by Dopfer et al. [8]. Their fitting algorithm consists of two stages. The first stage, which is used in the usual 3D-AAM, is the traditional appearance based fitting to calculate the 2D constraint. The second stage, which is not included in the original 3D-AAM, is performed to align the model parameter to the 3D landmarks using the range image and the 2D constraint obtained in the first stage.

The First Stage: 3D-AAM Fitting with 2D Intensity Image Data

Since the target appearance in the first stage comprises intensity of image, we need to map the 3D shape obtained from the model to the 2D shape. For this purpose, a perspective camera model is used. The projection P is defined as follows:

$$P \begin{pmatrix} X \\ Y \\ Z \end{pmatrix} = \frac{f}{Z} \begin{pmatrix} X \\ Y \end{pmatrix} + \begin{pmatrix} o_x \\ o_y \end{pmatrix}, \tag{5}$$

where f is the focal length and (o_x, o_y) is the principal point location of projection image. The 3D shape is projected to corresponding 2D shape landmark by $\mathbf{P}$.

In the first stage, the fitting procedure is regarded as a problem of minimizing the norm between the appearance (facial image) calculated from the parameter and target facial image. The minimization function is written as

$$\min_{\lambda, p, q} \sum_{x \in s_0} \left(\mathbf{I}(\mathbf{W}(x; p, q)) - A_0(x) - \sum_{i=1}^{n} \lambda_i A_i(x) \right)^2, \tag{6}$$

where $\mathbf{x}$ belongs to the pixels that are not self-occluded. The function $\mathbf{W}(x; \mathbf{p}, \mathbf{q})$ is a warping that is used to map the pixels in the 3D model to their corresponding 2D image points. The notation $\mathbf{I}(\mathbf{x})$ is an intensity at a point in the image.

The parameters $\mathbf{p}, \mathbf{q}$ and λ are updated under the assumption of the first order approximation with respect to $\mathbf{p}$ and $\mathbf{q}$ in order to minimize the following cost function.

$$\sum_{x \in s_0} \left(\mathbf{I}(\mathbf{W}(x; p + \Delta p, q + \Delta q)) - A_0(x) - \sum_{i=1}^{n} (\lambda_i + \Delta \lambda_i) A_i(x) \right)^2. \tag{7}$$

The parameters $\Delta \mathbf{p}$, $\Delta \mathbf{q}$, and $\Delta \lambda_i$ are determined by least-square approach. The parameters $\Delta \mathbf{p}$ and $\Delta \mathbf{q}$ are used as temporal parameters in the next stage, while $\Delta \lambda_i$ is used to update λ_i as $\lambda_i = \lambda_i + \Delta \lambda_i$ in this stage. See literature [8] for more details about the calculation of $\Delta \mathbf{p}$, $\Delta \mathbf{q}$, and $\Delta \lambda_i$.

The Second Stage: 3D-AAM Fitting with Range Image Data

The alignment of the 3D-AAM with the range image is based on the ICP framework [8] for minimizing the distance between the model and the corresponding position of the target shape. The goal is to minimize the following function:

$$\min_{r} \left\{ \sum_{x \in s_0} u(x) f(x) \|\mathbf{W}_{3D}(x; r) - x'_{3D}\|_2^2 + \sum_{x \in s_0} g\left(x'_{2D}\right) \|\mathbf{W}(x; r) - x'_{2D}\|_2^2 \right\}, \tag{8}$$

where $r = (p, q)$. The function $\mathbf{W}_{3D}(x; r)$ is the 3D warp which maps a point of the shape model to the real 3D coordinates. The vector x'_{2D} is the result of the traditional 2D-based 3D-AAM fitting procedure described in the first stage. The vector x'_{3D} is the 3D position of x'_{2D}, which is obtained using the range image. The transformation $u(x)$ is a matrix to adapt the unit difference between the 3D and 2D data. $f(x)$ and $g(x'_{2D})$ are the weighting functions. $f(x)$ is a function whose value is between 0 and 1. The value of $f(x)$ becomes bigger if $\mathbf{W}_{3D}(x; r)$ is close to x'_{3D}.

Algorithm 1. Fitting a 3DAAM to range and intensity image

1: Initial estimate of model parameters, p_0 and λ_0, and pose parameter q_0
2: Set parameters $p \leftarrow p_0$, $\lambda \leftarrow \lambda_0$, and $q \leftarrow q_0$
3: Calculate gradient image ∇I of the input image I.
4: **while** not converged yet **do**
5:      Calculate the 3D-AAM on the intensity image to find Δp, Δq, and $\Delta \lambda$
6:      Set temporary parameter: $p' \leftarrow p + \Delta p$
7:      Set temporary parameter: $q' \leftarrow q + \Delta q$
8:      Update parameter: $\lambda' \leftarrow \Delta \lambda + \Delta \lambda$
9:      Find x'_{2D} using $W(x : p', q')$ from the temporary parameters.
10:     **while** not converged yet **do**
11:        Find x'_{3D} for each visible point on $W_{3D}(x; r)$
12:        Compute $u(x)$, $f(x)$, and $g(x)$.
13:        Compute Δp, and Δq according to Eq. (7).
14:        Update parameters: $p \leftarrow p + \Delta p$
15:        Update parameters: $q \leftarrow q + \Delta q$
16:     **end while**
17: **end while**

$g(x'_{2D})$ is also a function whose value is between 0 and 1. It has a bigger value if the point is close to an edge on the image. Since the landmarks are sometimes taken from the contour of the face, this function encourages the landmarks to be on the outline form of the face.

Two Stage Fitting Algorithm

The parameter $r = (p, q)$ is updated in the second stage for minimization (8). The obtained values p and q are returned to the first stage for the further elaboration of the parameters. Algorithm 1 shows the procedure of the two stage fitting. In this algorithm, the second stage (line 10 to 16) is included in the first stage (line 4 to 17). The iteration continues until the value of the minimization (6) is less than a certain threshold.

3   Lip Reading Using Hidden Markov Model

The Hidden Markov Model (HMM) is a statistical model of a sequence of data and is widely used in speech recognition. Lip reading is similar to speech recognition in terms of using the sequence data as features. Hence we use HMM in lip reading for handling the facial features extracted in the previous section.

In speech recognition, an HMM state usually represents a phoneme in the mono-phone model. However, in lip reading, some phonemes cannot be distinguished from one another because the shape of the mouth is sometimes the same among phonemes. Therefore, we use viseme instead of phoneme for the target of recognition. The relation between a phoneme and a viseme is shown in Table 1.

Table 1. The relation between a phoneme and a viseme

Phoneme	Viseme	Phoneme	Viseme	Phoneme	Viseme
/a/	‖a‖	/t/	‖t‖	/ky/	‖vfy‖
/a:/		/d/		/gy/	
/i/	‖i‖	/n/		/hy/	
/i:/		/ts/	‖s‖	/sh/	‖sh‖
/u/	‖u‖	/z/		/ch/	
/u:/		/s/		/dy/	‖ty‖
/e/	‖e‖	/w/	‖w‖	/ny/	
/e:/		/f/		/py/	‖py‖
/o/	‖o‖	/k/	‖vf‖	/my/	
/o:/		/g/		/by/	
/p/	‖p‖	/h/		/ry/	‖ry‖
/b/		/j/	‖y‖	/N/	‖N‖
/m/		/y/			
/r/	‖r‖	/q/	none		

For viseme recognition, left-to-right HMM with five states that contains input, output, and three emitting states are used. An HMM state is a mono-viseme model and is associated with the Gaussian Mixture Models. Figure 2 illustrates a sample of a five state HMM. A viseme is modeled by three states: the connection to the previous viseme, the center of the viseme, and the connection to the succeeding viseme. The feature vector includes the AAM parameters, their linear regression coefficients Δ, and ΔΔ. In the training, each HMM state is initialized using a flat start training. In the recognition, various insertion penalties are tested and the best parameter is selected in the evaluation.

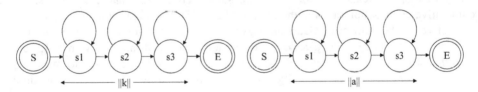

Fig. 2. The samples of left-right HMM with five states. Each state is a mono-viseme model and is associated with the Gaussian Mixture Models

4   Experiments and Results

4.1   Experimental Setup

The purpose of this experiment is to confirm that the effectiveness of facial feature given by the 3D-AAM in the cross view condition. We evaluate viseme recognition accuracy using the traditional 2D-AAM and 3D-AAM. The head angles used in the recognition are 0°, 45° and 90°.

We prepared 51 facial images of the frontal, left, and right views for AAM training, which are doubled using flip horizontal. All facial images are captured from a person using Microsoft Kinect V1. The image resolution is 640 × 480. The traditional 2D-AAM is built from 102 frontal images in which 73 landmarks (12 points from the eyebrow, 10 from the eye, 12 from the nose, 19 from the mouth, and 20 from the facial outline) are manually pointed for training the shape. The 3D-AAM is built from 306 images. A total of 91 landmarks are used to train the shape in 3D-AAM. A total of 18 points from the ear and the profile facial outline are newly added for the 3D-AAM construction. Figure 3 shows an example of the facial image and the landmarks used to construct 2D and 3D-AAMs.

Fig. 3. Examples of facial landmarks that are manually pointed. The 2DAAM is constructed from only the frontal landmarks. In the 3DAAM, the landmarks from three angle faces are integrated using the ICP framework and are used to construct the model.

For HMM training, eight sets of 258 utterances of Japanese words [11] are recorded from the frontal view. The words in the set are selected so that all Japanese tri-phone connections are included. The dimension of the feature vector is 90 which includes 12 shape parameters, 18 appearance parameters, their Δ, and their $\Delta\Delta$. In order for the dimension of the feature vector to be the same between 2D and 3D HMMs, the cumulative contribution ratios in the 2D-AAM and 3D-AAM are set at 95% and 80%, respectively. The number of Gaussian mixture component is set at 1, 2, 4, 8, or 16.

For evaluation, 216 utterances of ATR phoneme balanced Japanese words [12] are recorded from multiple views (0°, 45° and 90°). The lip reading performance is measured using the viseme recognition accuracy calculated by the following equation:

$$\text{acc}\% = \frac{H - I}{N} \times 100, \tag{10}$$

where N is the total number of viseme instances, H is the number of correctly recognized visemes, and I is number of insertion errors.

4.2   Experimental Results and Discussion

We first built two HMMs from the 2D-AAM features and 3D-AAM features that are obtained from only the frontal view images. Each HMM is evaluated with the testing utterances whose angles are 0°, 45° or 90°. The result is shown in Fig. 4. The horizontal and vertical axes are the variation of the Gaussian mixture components and the viseme recognition accuracy, respectively. The two numbers attached to the legend of each line represent the angles of training and testing data. The graph shows that the 3D-AAM achieves higher accuracy than the 2D-AAM at any parameters. For example, when the training and testing data are frontal images (0°) and the HMM mixture is 16, the accuracy of the 3D-AAM outperforms the 2D-AAM by 9%. Under the other conditions, the 3D-AAM is higher than the 2D-AAM by 1% to 15%. It is notable that the 3D-AAM can output the results when the test data is profile images (90°), while the 2D-AAM cannot output any results because the alignments failed for all data.

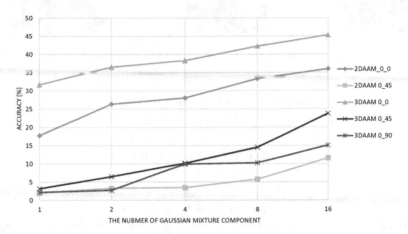

Fig. 4. The result of lip reading in different views where HMM is generated from only frontal image features.

To examine the reason why these results are obtained, we constructed the facial shapes from the AAM parameters. Figure 5 shows the aligned landmarks which are generated from the 2D-AAM and 3D-AAM parameters. It illustrates that the generated landmarks obtained from the 3D-AAM parameters are subjectively correct. On the other hand, the errors between the landmarks obtained from the 2D-AAM parameters and the real feature points are very large when the angles are different from 0°. Since the 2D-AAM pose parameter does not contain the information about the head rotation, it cannot output any correct position of the landmarks if there are some self-occlusions. In 3D-AAM, given that the pose parameter includes the 3D head rotation information as shown in Eq. (3), the generated landmarks obtained from the 3D-AAM parameters are more accurate than that from the 2D-AAM parameters.

For detailed analysis of the results, we investigated the accuracy of the recognition on each vowel as shown in Table 2. While the accuracy is high when the angle is 0° (same as the training data), the accuracy decreases when the angle is 45°, especially in the case of ||i|| and ||o||. This is because these vowels are mis-categorized into other vowels. Table 3 presents the confusion matrixes of viseme recognition. The rows and columns are the utteranced visemes and the recognized visemes, respectively. It shows

Fig. 5. Alignment results of each AAM. The top and bottom are the results using the traditional 2DAAM and 3DAAM, respectively.

Table 2. The result of vowel recognition.

[%]	$\|a\|$	$\|i\|$	$\|u\|$	$\|e\|$	$\|o\|$
0°	43.2	50.0	60.4	61.5	72.8
45°	17.0	0.8	35.2	43.8	16.9
90°	43.1	27.9	10.2	0	0

that $\|i\|$ and $\|o\|$ are mis-categorized into $\|e\|$ and $\|u\|$, respectively. This is caused by the similarity of the shapes of the lip when pronouncing $\|i\|$ and $\|e\|$ as shown in Fig. 6. The shape of visemes $\|o\|$ and $\|u\|$ are also similar. In addition, the alignment accuracy is lower due to self-occlusion when the angle is 45°. Therefore the average of viseme recognition accuracy becomes low.

When the angle is 90°, the average of the viseme recognition accuracy is lower than 45° in general because of self-occlusion. However, it is higher in $\|a\|$ and $\|i\|$ recognition because many visemes (including $\|a\|$ and $\|i\|$ themselves) are mis-categorized into $\|a\|$ and $\|i\|$ when the angle is 90°. Figure 7 shows the aligned landmarks which are generated from the 3D-AAM parameters. It indicates that the size of the open mouth of $\|e\|$ is placed between that of $\|a\|$ and $\|i\|$. Therefore, $\|e\|$ is mis-categorized into $\|a\|$ or $\|i\|$. $\|a\|$ and $\|i\|$ are not mis-categorized into $\|e\|$ because the number of occurrences of $\|a\|$ and $\|i\|$ are considerably more than $\|e\|$. Hence, $\|a\|$ and $\|i\|$ are trained more than $\|e\|$, which results in mis-categorizing only $\|e\|$ into $\|a\|$ or $\|i\|$. Table 3 also shows that the accuracies of $\|u\|$ and $\|o\|$ are low because the failed alignment of $\|u\|$ and $\|o\|$ around the mouth area as shown in Fig. 8. When pronouncing the vowels $\|u\|$ and $\|o\|$ in Japanese, the speaker has to pout the lip. Since there are not so many training data including pout shape and not enough feature points to express the shape, the 3D-AAM could not generate the correct 3D shapes. Figure 9 illustrates the facial image when pronouncing $\|u\|$ and its corresponding wireframe image generated from the 3D-AAM parameters. It shows that the shape around the mouth area could not be reconstructed correctly in detail because of the lack of feature points. The appropriate landmarks and the number of training data will have to be reconsidered for more accurate recognition.

Table 3. The confusion matrix of viseme recognition. (a), (b) and (c) are the results for angles 0°, 45° and 90°. The row and column represent the utteranced visemes and recognized visemes, respectively.

(a)

	a	i	u	e	b	vf	vfy	s	sh	t	ty	b	by	r	ry	w	y	N	Del
a	73	1	1	16	8	3	0	0	3	2	0	4	0	0	0	0	0	1	57
i	0	67	2	4	5	0	0	0	0	1	0	1	0	0	0	0	0	1	51
u	0	0	152	2	10	5	0	3	0	0	0	1	0	0	0	1	0	1	70
e	2	1	3	57	2	2	0	0	0	0	0	0	0	0	0	0	0	0	24
b	0	1	2	2	126	0	0	0	0	0	0	1	0	1	0	0	0	0	29
vf	2	2	4	4	3	20	0	5	1	2	0	2	0	0	0	3	0	0	78
vfy	0	0	2	0	3	2	0	2	0	0	0	0	0	0	0	0	0	0	24
s	0	3	0	1	2	6	0	8	1	3	0	1	0	0	0	0	0	0	32
sh	0	0	1	1	6	2	1	1	8	1	0	3	0	0	1	0	0	0	45
t	0	1	4	4	5	8	0	3	2	12	0	5	0	0	0	0	0	0	38
ty	0	0	0	0	1	0	0	0	0	1	0	0	0	0	0	0	0	0	7
b	0	0	1	0	4	1	0	0	1	2	0	18	0	0	2	0	1	0	45
by	0	0	0	0	4	0	1	0	0	1	0	4	0	1	0	0	0	1	15
r	0	0	2	3	3	2	0	3	1	3	0	1	0	0	0	3	0	0	40
ry	0	0	0	0	0	0	0	1	1	0	0	0	0	0	0	0	0	0	9
w	0	0	1	0	2	0	0	0	0	0	0	0	0	0	0	4	0	0	6
y	0	0	0	0	1	1	0	1	1	0	0	1	0	0	0	0	1	0	10
N	1	0	2	0	4	2	0	0	0	1	0	0	0	0	0	0	0	6	36
hs	0	1	4	1	8	4	0	1	0	6	0	6	0	0	0	1	0	1	

(b)

	a	i	u	e	b	vf	vfy	s	sh	t	ty	b	by	r	ry	w	y	N	Del
a	27	0	4	7	2	1	0	0	0	1	0	4	0	0	0	0	0	0	113
i	1	1	5	9	3	0	0	0	0	1	0	2	0	0	0	0	0	0	99
u	6	0	81	11	11	1	0	0	0	3	0	4	0	0	0	0	0	0	110
e	0	0	0	46	0	1	0	0	0	0	0	1	0	0	0	0	0	0	41
b	1	0	7	5	25	0	0	0	0	4	0	2	0	0	0	0	0	0	98
vf	0	0	0	4	3	4	0	0	0	2	0	0	0	0	0	0	0	0	98
vfy	0	0	1	1	0	0	0	0	0	1	0	1	0	0	0	0	0	0	23
s	0	0	0	2	1	0	0	0	0	0	0	0	0	0	0	0	0	0	52
sh	0	0	2	8	1	0	0	0	0	1	0	1	0	0	0	0	0	0	55
t	0	0	3	3	2	0	0	0	0	5	0	1	0	0	0	0	0	0	60
ty	0	0	0	0	0	0	0	0	0	0	0	0	0	0	0	0	0	0	8
b	1	0	0	2	2	0	0	0	0	0	0	7	0	0	0	0	0	0	57
by	0	0	0	2	0	1	0	0	0	0	0	0	0	0	0	0	0	0	21
r	1	0	3	2	3	0	0	0	0	1	0	1	0	0	0	0	0	0	47
ry	0	0	0	1	0	0	0	0	0	0	0	0	0	0	0	0	0	0	8
w	0	0	0	1	0	0	0	0	0	0	0	1	0	0	0	0	0	0	11
y	0	0	1	1	0	0	0	0	0	1	0	0	0	0	0	0	0	0	13
N	1	0	1	3	2	0	0	0	0	1	0	0	0	0	0	0	0	0	38
hs	0	0	1	7	1	1	0	0	0	2	0	2	0	0	0	0	0	0	

(c)

	a	i	u	e	b	vf	vfy	s	sh	t	ty	b	by	r	ry	w	y	N	Del
a	87	3	1	0	0	1	0	0	0	0	0	0	0	0	0	0	0	0	75
i	3	39	2	0	0	3	0	0	0	0	0	0	0	0	0	0	0	0	82
u	18	2	26	0	0	0	0	0	0	0	0	2	0	0	0	0	0	0	196
e	6	3	0	0	0	0	0	0	0	0	0	0	0	1	0	0	0	0	81
b	7	4	1	1	0	0	0	0	0	0	0	0	0	0	0	0	0	0	148
vf	3	1	0	0	0	2	0	0	0	0	0	0	0	0	0	0	0	0	115
vfy	0	0	0	0	0	0	0	0	0	0	0	0	0	0	0	0	0	0	32
s	0	3	0	0	0	1	0	0	0	0	0	1	0	0	0	0	0	0	53
sh	1	0	0	0	0	0	0	0	0	0	0	0	0	0	0	0	0	0	62
t	0	1	2	0	0	0	0	0	0	0	0	1	0	0	0	0	0	0	73
ty	0	0	0	0	0	0	0	0	0	0	0	0	0	0	0	0	0	0	10
b	0	0	1	0	0	0	0	0	0	0	0	3	0	0	0	0	0	0	66
by	0	0	0	0	0	0	0	0	0	0	0	0	0	0	0	0	0	0	28
r	0	0	0	0	0	0	0	0	0	0	0	0	0	0	0	0	0	0	59
ry	0	1	0	0	0	0	0	0	0	0	0	0	0	0	0	0	0	0	10
w	0	1	0	0	0	0	0	0	0	0	0	0	0	0	0	0	0	0	9
y	0	0	0	0	0	1	0	0	0	0	0	0	0	0	0	0	0	0	14
N	7	1	1	0	0	0	0	0	0	0	0	1	0	0	0	0	0	0	43
hs	15	3	1	0	0	0	0	0	0	0	0	0	0	0	0	0	0	0	

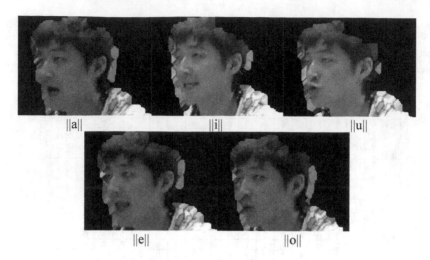

Fig. 6. Examples of the testing images (angle: 45°).

Fig. 7. Samples of profile images when ‖a‖, ‖e‖ and ‖i‖ are pronounced.

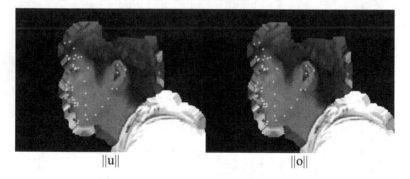

Fig. 8. Examples of the aligned landmarks in the testing image when ‖u‖ and ‖o‖ are pronounced.

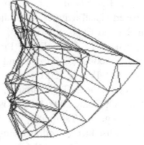

Fig. 9. Left: A profile image when ‖u‖ is pronounced. Right: A wireframe image of the aligned landmarks when ‖u‖ is pronounced.

5 Conclusion

In this study, we evaluate the efficiency of 3D-AAM in lip reading. We first built two HMMs from the 2D-AAM features and the 3D-AAM features that are obtained from only the frontal view images. Each HMM is evaluated with the testing utterances at 0°, 45° or 90° angles. The experimental result shows that the HMM constructed from the 3D-AAM can recognize the visemes even though the angle is 90°, while the 2D-AAM-based HMM cannot recognize any visemes because of alignment failure. However, the accuracy decreases when the angle is different from the training data. For example, when the angle is 90°, the recognition accuracy of vowel visemes ‖e‖, ‖u‖, and ‖o‖ are not good for practical use. This result is caused by the similarity of the shapes of the lip when pronouncing ‖a‖, ‖i‖ or ‖e‖ at 90°. Therefore, ‖e‖ is mis-categorized into ‖a‖ or ‖i‖. Another reason is that the shape around the mouth area when pronouncing ‖u‖ or ‖o‖ could not be correctly reconstructed in detail because of the lack of feature points.

Future work will focus in solving these problems. First, we will solve the problem of self-occlusion by assuming that the training face is a symmetry or using the robust error function in the fitting procedure [13]. In addition, we will redesign the location of landmarks around the mouth area for more accurate alignment of the lip movement.

Acknowledgement. This work has been supported by a Grant-in-Aid for Scientific Research (C) 16K00234, Scientific Research (B) 16H03211, and Scientific Research (C) 16K00251 by MEXT, Japan, and the Futaba Electronics Memorial Foundation.

References

1. Jang, K.S.: Lip contour extraction based on active shape model and snakes. J. Comput. Sci. **7** (10), 148–153 (2007)
2. Xiaopeng, H., Hongxun, Y., Yuqi, W. Rong, C.: A PCA based visual DCT feature extraction method for lip-reading. In: 2006 Proceedings of International Conference on Intelligent Information Hiding and Multimedia Signal Processing, IIH-MSP 2006, pp. 321–324 (2006)

3. Matthews, I., Potamianos, G., Neti, C., Luettin, J.: A comparison of model and transform-based visual features for audio-visual LVCSR. In: IEEE International Conference on Multimedia and Expo, ICME 2001, vol. 2, pp. 2–5 (2001)

4. Kumar, K., Chen, T., Stern, R.M.: Profile view lip reading. In: 2007 IEEE International Conference on Acoustics, Speech and Signal Processing - ICASSP 2007, Honolulu, HI, 2007, pp. IV-429–IV-432 (2007)

5. Lucey, P., Potamianos, G., Sridharan, S.: A unified approach to multi-pose audio-visual ASR. Interspeech Conf. **2007**(1–4), 809–812 (2007)

6. Komai, Y., Yang, N., Takiguchi, T., Ariki, Y.: Robust AAM-based audio-visual speech recognition against face direction changes. In: Proceedings of the 20th ACM International Conference on Multimedia – MM 2012, pp. 1161–1164 (2012)

7. Lan, Y., Theobald, B.J., Harvey, R.: View independent computer lip-reading. In: 2012 IEEE International Conference on Multimedia and Expo, Melbourne, VIC, 2012, pp. 432–437 (2012)

8. Dopfer, A., Wang, H., Wang, C.: 3D active appearance model alignment using intensity and range data. Robot. Auton. Syst. **62**(2), 168–176 (2014)

9. Cootes, T.F., Edwards, G.J., Taylor, C.J.: Active appearance models. IEEE Trans. Pattern Anal. Mach. Intell. **23**(6), 681–685 (2001)

10. Dryden, I.L., Mardia, K.V.: Statistical Shape Analysis. Wiley, New York (1998)

11. Matsuura, H., Nitta, T.: Speaker independent large vocabulary word recognition based on SMQ/HMM. IEICE Trans. **J76-D-2**(12), 2486–2494 (1993). (in Japanese)

12. Takeda, K., Sagisaka, Y., Katagiri, S., Kuwabara, H.: A Japanese speech database for various kinds of research purposes. Acoust. Sci. Technol. **44**(10), 747–754 (1998). (in Japanese)

13. Gross, R., Matthews, I., Baker, S.: Active appearance models with occlusion. Image Vis. Comput. **24**(6), 593–604 (2006)

Workshop on Facial Informatics (WFI)

Face Detection by Aggregating Visible Components

Jiali Duan[1(✉)], Shengcai Liao[2], Xiaoyuan Guo[3], and Stan Z. Li[2]

[1] School of Electronic, Electrical and Communication Engineering,
University of Chinese Academy of Sciences, Beijing, China
jli.duan@gmail.com
[2] Center for Biometrics and Security Research
and National Laboratory of Pattern Recognition,
Institute of Automation, Chinese Academy of Sciences, Beijing, China
{scliao,szli}@nlpr.ia.ac.cn
[3] School of Engineering Science, University of Chinese Academy of Sciences,
Beijing, China
xiaoyuanguo.ucas@gmail.com

Abstract. Pose variations and occlusions are two major challenges for unconstrained face detection. Many approaches have been proposed to handle pose variations and occlusions in face detection, however, few of them addresses the two challenges in a model explicitly and simultaneously. In this paper, we propose a novel face detection method called Aggregating Visible Components (AVC), which addresses pose variations and occlusions simultaneously in a single framework with low complexity. The main contributions of this paper are: (1) By aggregating visible components which have inherent advantages in occasions of occlusions, the proposed method achieves state-of-the-art performance using only hand-crafted feature; (2) Mapped from meanshape through component-invariant mapping, the proposed component detector is more robust to pose-variations (3) A local to global aggregation strategy that involves region competition helps alleviate false alarms while enhancing localization accuracy.

1 Introduction

Unconstrained face detection is challenging due to pose and illumination variations, occlusions, blur, etc. While illumination variations are handled relatively better due to many physical models, pose variations and occlusions are the most commonly encountered problems in practice[1]. Many approaches have been specifically proposed to solve pose variations [2–4] and occlusions [5–9], however, few of them addresses pose variations and occlusions in a model explicitly and simultaneously.

---

[1] Blur or low resolution is a challenging problem mainly in surveillance. Though many blur face images exist in current benchmark databases (e.g. FDDB [1]), they are intentionally made out of focus in background while the main focus is the center figures in news photography.

© Springer International Publishing AG 2017
C.-S. Chen et al. (Eds.): ACCV 2016 Workshops, Part II, LNCS 10117, pp. 319–333, 2017.
DOI: 10.1007/978-3-319-54427-4_24

Recently, a number of Convolutional Neutral Network (CNN) [10] based face detection methods [11–15] have been proposed due to the power of CNN in dealing with computer vision problems. However, CNN models generally deal with problems in face detection by learning from a large number of diverse training samples. Such data driven solutions may be good in dealing with various face variations, however, they usually result in very complex models that run slowly, which limits their application in practice, especially in embedding devices. On the other hand, Yang et al. [13] proposed a specific architecture called Faceness-Net, which considers facial component based scoring and their spatial configuration to explicitly deal with occluded face detection. This work inspires that explicit modeling of challenges in face detection is still required and more effective than pure data driven, though the fixed spatial configuration in Faceness-Net is still an issue, and the model is still expensive to apply.

Putting occlusions and large pose variations together, a common issue is that some facial components are invisible under either condition. This motivates us to only detect visible components that share some pose invariance property, and adaptively aggregate them together to form the whole face detection. Therefore, in this paper we propose a novel face detection method called Aggregating Visible Components (AVC), which addresses pose variations and occlusions simultaneously in a single framework.

Specifically, to handle pose variations, we define two pose-invariant (or pose-robust) components by considering half facial view, and a regression based local landmark alignment. Such a consistent component definition helps to reduce the model complexity. Accordingly, we train two component detectors, mirror them to detect the other half view, and introduce a local region competition strategy to alleviate false detections. To handle facial occlusions, we only detect visible facial components, and build a local to global aggregation strategy to detect the whole face adaptively. Experiments on the FDDB and AFW databases show that the proposed method is robust in handling pose variations and occlusions, achieving much better performance but lower model complexity compared to the corresponding holistic face detector.

The remaining parts of this paper are organized as follows. Section 2 gives a concise review of related works. Section 3 gives an overview of the proposed AVC detector. Section 4 introduces the pose-invariant component definition and the detector training. In Sect. 5, we present the local region competition strategy and the adaptive local to global aggregation strategy. Experimental results on AFW and FDDB are shown and discussed in Sect. 6 and we conclude the paper in Sect. 7.

2   Related Works

Given that the original Viola-Jones face detector [16] is limited to multi-view face detection, various cascade structures have been proposed to handle pose variations [2–4]. Today multi-view face detection by partitioning poses into discrete ranges and training independently is still a popular way to handle pose variations, for example, in recent works [12,17]. Zhu and Ramanan [18] proposed to jointly detect a face, estimate its pose, and localize face landmarks in the wild by a Deformable Parts-based Model (DPM), which was further improved in [19,20]. Ranjian et al. [21] proposed to combine deep pyramid features and DPM to handle faces with various sizes and poses in unconstrained settings. Chen et al. [22] proposed to combine the face detection and landmark estimation tasks in a joint cascade framework to refine face detection by precise landmark detections. Liao et al. [23] proposed to learn features in deep quadratic trees, where different views could be automatically partitioned. These methods are effective in dealing with pose variations, however, not occlusions simultaneously.

Face detection under occlusions is also an important issue but has received less attention compared to multi-view face detection, partly due to the difficulty of classifying arbitrary occlusions into predefined categories. Component-based face detector is a promising way in handling occlusions. For example, Chen et al. [8] proposed a modified Viola-Jones face detector, where the trained detector was divided into sub-classifiers related to several predefined local patches, and the outputs of sub-classifiers were re-weighted. Goldmann et al. [24] proposed to connect facial parts using topology graph. Recently, Yang et al. [13] proposed a specific architecture called Faceness-Net, which considers faceness scoring in generic object proposal windows based on facial component responses and their spatial configuration, so that face detection with occlusions can be explicitly handled. However, none of the above methods considered face detection with both occlusions and pose variations simultaneously in unconstrained scenarios.

Our work is also different from other part-based methods like [25–29] in that [25] describes an object by a non-rigid constellation of parts and jointly optimize parameters whereas we learn component detectors independently and apply an aggregation strategy to constitute a global representation. On the other hand, AVC define parts via component-invariant mapping, in contrast to [26] which defines parts by a search procedure while [27–29] deploy CNN structures.

Recently, the Convolutional Neutral Network (CNN) [10] based methods [11–15] have been proposed for face detection due to the power of CNN in dealing with computer vision problems. For example, Li et al. [11] proposed a cascade architecture based on CNN and the performance was improved by alternating between the detection net and calibration net. Most recently Zhang et al. [14] and Ranjan et al. [15] combined face detection with other vision tasks such as face alignment and involved multi-task loss into CNN cascade.

3   Overview of the Proposed Method

Figure 1 is an overview of the proposed AVC face detection method. It includes three main steps in the detection phase: visible component detection step, local region competition step, and the local to global aggregation step. AVC works by detecting only the visible components which would be later aggregated to represent the whole face. Two half-view facial component detectors are trained, and for this we introduce a pose-invariant component definition via a regression based local landmark alignment, which is crucial for training sample cropping and pose-invariant component detection. Then the two learned detectors are mirrored to detect the other half view of the facial components. Next, the detected visible facial components go through a local region competition module to alleviate false detections, and finally a local to global aggregation strategy is applied to detect the whole face adaptively.

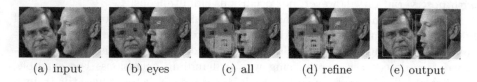

(a) input   (b) eyes   (c) all   (d) refine   (e) output

Fig. 1. The processing steps of the proposed AVC face detection method. (a) Input image. (b) Visible eye detection. (c) Detection of all visible components (Red: left eye; Blue: right eye; Green: left mouth; Pink: right mouth). (d) Refinement after local region competition. (e) Aggregated whole face detection. (Color figure online)

The intuition behind our component-based design is the fact that face images in real-world applications are often with large pose variations and occlusions. Consider for example, a face turning left over 60 degrees (see Fig. 2(a)), where the holistic face detector unavoidably includes unwanted backgrounds (see Fig. 2(b)).

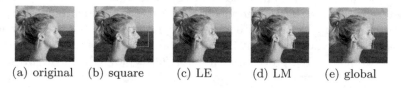

(a) original   (b) square   (c) LE   (d) LM   (e) global

Fig. 2. Illustration of holistic face detection and component-based face detection. (a) Input image. (b) Typical holistic face detection. (c) Left eye (LE) detection. (d) Left mouth (LM) detection. (e) Aggregating LE and LM to get a global detection.

However, a robust face detector should not only predict the number of faces but also give bounding boxes as tight as possible. The criteria on this performance was first introduced by FDDB [1], a face benchmark that employs both discrete metric and continuous metric for evaluation. While a typical face detector may fail to bound a profile face tightly and miss faces under occlusions, we discover however, that pose variations and occlusions can be jointly solved by locating and aggregating facial components. We trained two facial component detectors respectively for the detection of left eyebrow + left eye (denoted as LE Fig. 2(c)) and left nose + left mouth (denoted as LM Fig. 2(d)).

It's observed that although a face with large rotation towards left may lead to left eye invisible, we can still, under this circumstance, locate the right eye or mouth and nose etc. It also applies to occlusions where for example, the left half face is occluded by another person's shoulder, we can still locate the whole face by the detection of right eye. Furthermore, we only consider training two half facial view components, and mirror them to detect the other half view. This strategy not only reduces the training effort, but also enables us to deal with larger pose variations because for example, the left eye component appears to be invariant under 0–60° pose changes, and beyond this range the right eye or other component is usually detectable.

4 Pose-Invariant Component Detection

4.1 Pose-Invariant Component Mapping

As was indicated in AFLW [30], although there is largely an agreement on how to define anchor points and extents of rectangle for frontal faces, it's not so obvious for profile and semi-profile views, which makes it harder to get consistently annotated samples for training. Unlike the training input of a holistic face detector, facial part detector requires uniform eye patches and mouth patches as training set. This would not be made possible without pose-invariant component mapping.

Samples in AFLW consist of 21 landmarks. We first calculate the mean shape of the whole database with samples normalized and missing coordinates excluded. Region in the mean shape which we want to map i.e. left eyebrow and left eye for LE component is mapped directly to a new input sample by applying the transformation

$$a\bar{\mathbf{x}} + x_0 = \mathbf{x} \tag{1}$$

$$a\bar{\mathbf{y}} + y_0 = \mathbf{y} \tag{2}$$

Note that in (1) and (2) $\bar{\mathbf{x}}$ and $\bar{\mathbf{y}}$ are vectors representing x coordinates and y coordinates of mean shape while $\mathbf{x}$ and $\mathbf{y}$ representing those of a new sample. $\mathbf{E}$ is a nx1 vector with all elements being 1, x_0, y_0 are scalars that denote offsets and n is the number of landmarks used for regression. Closed form solution can be derived as the following

$$a = \frac{\bar{\mathbf{x}}^T \cdot \mathbf{x} + \bar{\mathbf{y}}^T \cdot \mathbf{y} - \frac{1}{n} \cdot (\bar{\mathbf{x}}^T \cdot \mathbf{E})(\mathbf{x}^T \cdot \mathbf{E}) - \frac{1}{n} \cdot (\bar{\mathbf{y}}^T \cdot \mathbf{E})(\mathbf{y}^T \cdot \mathbf{E})}{\bar{\mathbf{x}}^T \cdot \mathbf{x} + \bar{\mathbf{y}}^T \cdot \mathbf{y} - \frac{1}{n} \cdot (\bar{\mathbf{x}}^T \cdot \mathbf{E})^2 - \frac{1}{n} \cdot (\bar{\mathbf{y}}^T \cdot \mathbf{E})^2} \tag{3}$$

$$x_0 = \frac{1}{n} \cdot \mathbf{x}^T \cdot \mathbf{E} - a\frac{1}{n} \cdot \bar{\mathbf{x}}^T \cdot \mathbf{E} \tag{4}$$

$$y_0 = \frac{1}{n} \cdot \mathbf{y}^T \cdot \mathbf{E} - a\frac{1}{n} \cdot \bar{\mathbf{y}}^T \cdot \mathbf{E} \tag{5}$$

An intuitive visual interpretation is shown in Fig. 3. In Fig. 3(c), blue points are annotated landmarks while red points are mapped from meanshape. Positive samples extracted in this way retain excellent uniformity, which would be used for training LE and LM component detector. The pose-invariant component mapping method is also used for preparing negative samples for bootstrapping (see Fig. 4).

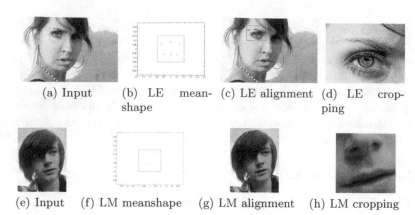

(a) Input   (b) LE mean-   (c) LE alignment   (d) LE crop-
             shape   ping

(e) Input   (f) LM meanshape   (g) LM alignment   (h) LM cropping

Fig. 3. Pose-invariant component mapping and cropping. (a) Input. (b) Meanshape of the LE component. (c) Regression based local landmark alignment of LE component. (d) Cropping of the LE component. (e) Input. (f) Meanshape of the LM component. (g) Regression based local landmark alignment of LM component. (d) Cropping of the LM component. (Color figure online)

Fig. 4. Positive and negative examples for components. The first and third rows show positive training samples of the LE and LM components respectively, while the second and forth rows show images for bootstrapping negative LE and LM samples respectively.

4.2 Why LE and LM?

In our paper, we trained two part-based detectors, namely LE (left eyebrow and left eye) and LM (left nose and left mouth) and Fig. 4 displays some positive and hard-negative training samples obtained using method of the last subsection. But why not eyes, noses or other patches? Our motivations are: (1) These patches are not defined arbitrarily or conceptually but based on the regression of local landmarks. As in Fig. 3, these landmarks are derived by LE/LM meanshape of AFLW to ensure that they retain invariance throughout the database

(2) Why 6 landmarks instead of 3 or 9? According to AFLW, a nose is defined by 3 landmarks, the width/height of these patches would then be too small for training and testing. While 9 landmarks would result with a facial area too broad thus vulnerable for occlusions.

4.3 Training Procedure

In this subsection, we give a brief introduction about the feature employed for facial representation as well as the work flow of the training algorithm.

Feature: We choose NPD [23] as our feature mainly for its two properties: illumination invariant and fast in speed because each computation involves only two pixels. For an image with size $p = w \times h$, the number of features computed is C_p^2 which can be computed beforehand, leading to superiority in speed for real world applications. With the scale-invariance property of NPD, the facial component detector is expected to be robust against illumination changes which is important in practice.

Training Framework: The Deep Quadratic Tree (DQT) [23] is used as weak classifier which learns two thresholds and is deeper compared to typical tree classifiers. Soft-Cascade [31] as well as hard-negative mining are applied for cascade training. While individual NPD [32] features may be "weak", the Gentle AdaBoost algorithm is utilized to learn a subset of NPD features organized in DQT for stronger discriminative ability.

5 Local to Global Aggregation

5.1 Symmetric Component Detection

Figure 5 shows some example outputs by LE and LM detector respectively. As can be seen, our component-based detector has the inherent advantages under occasions of occlusions (Fig. 5(a,h)) and pose-variations (Fig. 5(c,g)), where a holistic detector would normally fail. The detection of right eyebrow + right eye (RE) and right mouth + right nose (RM) can be achieved by deploying the detector of their left counterpart. Figure 6(a) to (d) illustrates how we locate RM and RE using the same detectors as LM and LE.

5.2 Local Region Competition

Adopting facial part detection also brings about many troublesome issues. If handled improperly, the performance will vary greatly. First, LE, LM, RE, RM detector for different facial parts will each produce a set of candidate positive windows with a set of confidence scores. But the goal for face detection is to locate faces each with a bounding box as tight as possible, so we need to merge these detections from different facial part detectors and remove duplicated windows. A common solution is Non-Maximum Suppression (NMS) [33] but issue

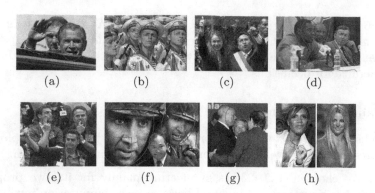

(a)   (b)   (c)   (d)

(e)   (f)   (g)   (h)

Fig. 5. Some example component detections by the proposed LE (upper row) and LM facial component detector.

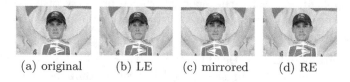

(a) original   (b) LE   (c) mirrored   (d) RE

Fig. 6. (a): Input image; (b): Left eye detection; (c): Left eye detection in mirrored image; (d): Right eye detection mapped back to the original image.

arises on how to do window merging with a good trade-off between high precision rate and high detection rate. Second, different benchmarks with different annotation styles could lead to biased evaluation. Noted in [20], this diversity becomes more prominent for profile faces. In this section, we address the above issues by exploiting the advantage of a competitive strategy.

Figure 1 illustrates the idea of the proposed local region competition. The core idea is to reject false alarms during merging (compete) while improving localization accuracy during aggregation (collaborate). In Algorithm 1 line 6 to line 11 first obtains candidate outputs of a specific facial region by LE, RE, LM, RM facial part detectors denoted as region_rects, see Fig. 1(c) which shows detection results of all components and Fig. 1(d) after competition as an example. In this example, left eye region may well contain the outputs of other facial part detectors such as RE (false alarms) other than LE and vice versa. It is through this competitive strategy that we ensure candidate windows of only one facial part detector are reserved for each region, rooting out the possibility of using false alarms for aggregation.

5.3   Aggregation Strategy

After deploying competitive strategy to exclude possible false positives, the task now is to ensure accurate localization of detection outputs. This is achieved by taking the full use of information from rects of different regions. We use rectangle as facial representation. Note that our proposed pipeline also applies to elliptical representation as the aforementioned workflow remains unchanged.

Algorithm 1. Detection framework with local region competition strategy

Require:

    Input and Model: LE and LM model; RGB or gray image I

    Options: eyeHeight, eyeWidth, mouthHeight, mouthWidth for scanning windows; overlap for IOU; minEyeSize; minMouthSize; numThreads for parallel computing

Ensure: outRect

  1: [LE,LM]=Scan(eyeModel,mouthModel,I,eyeHeight,eyeWidth,mouthHeight, mouthWidth,minEyeSize,minMouthSize,numThreads)

  2: Symmetrically detect RE and RM

  3: $LE \cup LM \cup RE \cup RM \subseteq R$

  4: predicate(i,j)=1 if IOU between R pair (i,j) > overlap

  5: [label,numCandi]=Partition(predicate)

  6: **for** i=1:numCandi **do**

  7:     Get region_rects{i} with rects labelled i

  8:     **for** r in region_rects{i} **do**

  9:         categorize r to LE, LM, RE, RM detector

10:     **end for**

11:     Reserve rects from the detector with the highest score

12:     Fitting rects to bounding boxes for the whole face

13: **end for**

14: predicate(i,j)=1 if IOU between rectangle pair (i,j) > overlap

15: [label,numCandi]=Partition(predicate)

16: **for** i=1:numCandi **do**

17:     Weight adjustment

18: **end for**

19: Elimination

20: **return** outRect

In Algorithm 1 line 12, winning rectangles from each region as illustrated in Fig. 5 are regressed directly to bounding boxes. Note that we only learn two sets of regression parameters (linear regression), because during inference the coordinates of RE/RM component are first mirrored, regressed and then mirrored back using the same parameters of their left counterparts. This is a local to global bottom up strategy because rects of different facial regions are mapped to global facial representations. In Algorithm 1 Line 15 to Line 18, these rects are then concatenated for partitioning using disjoint-set algorithm. Then the locations of partitioned rects are translated and adjusted by tuning their widths and heights according to their confidence scores (weights). Through this process, information of different regions are collaborated to get a more accurate localization of the whole face. Finally, NMS [33] is deployed to eliminate interior rects.

6 Experiments

6.1 Training Parameters:

Annotated Facial Landmarks in the Wild (AFLW) [1] is an unconstrained face benchmark that contains 25993 face annotations in 21997 real world images with large pose variations, occlusions, illumination changes as well as a diversity of ages, genders, and ethnicity. In total, we use 43994 images from AFLW together with its flipped counterpart as positive samples and 300000 background images for training. And an additional 12300 images of natural scenes are scraped from the Internet to mask face components for hard-negative mining. In training AVC, images of 15x20 pixels are assigned to LE component while images of 20x20 pixels are used for LM. Pose-invariant component mapping is deployed to crop positive training patches and prepare bootstrapping samples.

6.2   AFW Results:

Annotated Faces in the Wild (AFW) [18] contains 205 images collected from Flickr that contain images of cluttered scenes and different viewpoints.

To evaluate on AFW, we fit winning rects from local component detectors to rectangle representations of the whole face, which would be used for further aggregation. The fitting parameters are learned on AFLW using 10-cross validation and this also applies to the learning of elliptical fitting parameters for testing on FDDB.

We use the evaluation toolbox provided by [20]. The comparison of Precision-Recall curves generated by different methods is shown in Fig. 7(a). We compare AVC with both academic methods like DPM, HeadHunter, Structured Models and commercial systems like Face++ and Picasa. As can be seen from the figure, AVC outperforms DPM and is superior or equal to Face++ and Google Picasa. The precision of AVC is 98.68% with a recall of 97.13%, and the AP of AVC is 98.08%, which is comparable with the state-of-the-art methods. Example detection results are shown in the first row of Fig. 8, note that we output rectangle for evaluation on AFW.

6.3   FDDB Results:

Face Detection Data Set and Benchmark (FDDB) [1] contains 2845 images with 5171 faces, with a wide range of arbitrary poses, occlusions, illumination changes and resolutions. FDDB uses elliptical annotations and two types of evaluation metrics are applied. One is the discrete score metric which counts the number of detected faces versus the number of false alarms. A detected bounding box is considered true positive if it has an IoU of over 0.5 with ground truth. The other is the continuous score metric that measures the IoU ratio as the indicator for performance.

As FDDB uses ellipse for annotations, we fit the output rectangles to elliptical representations of the whole face. We use the evaluation code provided by Jain and Learned-Miller [1] and the results using discrete score metric are shown in Fig. 7. We compare our results with the latest published methods on FDDB including MTCNN, DP2MFD, Faceness-Net and Hyperface. Ours performs worse than MTCNN and DP2MFD which resort to powerful yet complex CNN features but is better than Faceness-Net, which is also component-based but with the help of CNN structure. AVC gets 84.4% detection rate at FP $= 100$,

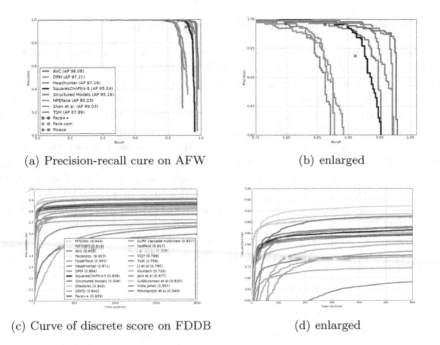

(a) Precision-recall cure on AFW   (b) enlarged

(c) Curve of discrete score on FDDB   (d) enlarged

Fig. 7. Experimental results on AFW and FDDB database. Best viewed in color.

Fig. 8. Qualitive results of AVC on AFW (first row using rectangle representations) and FDDB (second and third row using elliptical representations).

and a detection rate of 89.0% at FP = 300. Example detection results are shown in the second and third row of Fig. 8, where faces under poses changes and occlusions have been successfully located.

6.4 Does Component-Invariant Mapping Help?

We have tried two other methods when preparing facial-component patches for training component detectors. One is to define anchor points and extents of rectangle, the other is to project 3D landmarks back to 2D plane. However, unlike training holistic face detector that gets by with ordinary methods, the uniformity of component training-set under profile or semi-profile views deteriorates notably compared to those under frontal views. The resulting detectors that we have trained achieve at best 81% AP on FDDB. To the best of our knowledge, it remains a tricky issue on how to achieve consistency under profile views [30]. This motivates us to make new attempts and explore component-invariant mapping, whose performance is further boosted with the help of symmetric component detection because, when a face only exposes RE/RM component, LE/LM component detector would fail. Second, its likely that symmetric component detection presents a symmetric but unblocked or simpler view for detector. Third, symmetric detection obviates the need to train another two more detectors and regression parameters. Experiment shows that trained part-detectors using conventional cropped patches will decrease AP by about 8.2% on FDDB.

6.5 Model Complexity

As is shown in Table 1, different tree levels for training have been evaluated, leading to different training stages and number of weak classifiers. Training FAR indicates to what extent AVC has converged, but it can not reflect the performance of the model on test set. The complexity of the model is measured by aveEval, which means the average number of NPD features evaluated per detection window. The lower the value of aveEval, the faster the detector. For the sake of speed, this index is important for the choices of our component models.

Table 1. Comparison of model complexity between AVC and NPD

Model	Weaks	Features	aveEval
LE	200	6193	24.754
RE	200	6193	24.754
LM	300	3561	26.755
RM	300	3561	26.755
LE+RE+LM+RM	1000	19508	103.018
NPD	1226	46401	114.507

The aveEval in LE and LM are 24.754 and 26.755 respectively (See Table 1). So the total number of features per detection window that AVC has to evaluate is 103.018 with symmetric detection considered, which is faster than NPD holistic face detector implemented in [23] that has 46401 weak classifiers and an aveEval of 114.507. With regard to pose-variations and occlusions, AVC also outperforms

NPD detector by a notable margin on FDDB (See Fig. 7(c)). Another advantage of AVC is that storage memory required is low compared to CNN methods, which is crucial for real-world applications. The total model size of AVC is only 2.65 MB, smaller compared to NPD (6.31 MB) or a typical CNN model.

7   Conclusion

In this paper, we proposed a new method called AVC highlighting component-based face detection, which addresses pose variations and occlusions simultaneously in a single framework with low complexity. We show a consistent component definition which helps to achieve pose-invariant component detection. To handle facial occlusions, we only detect visible facial components, and build a local to global aggregation strategy to detect the whole face adaptively. Experiments on the FDDB and AFW databases show that the proposed method is robust in handling illuminations, occlusions and pose variations, achieving much better performance but lower model complexity compared to the corresponding holistic face detector. The proposed face detector is able to output local facial components as well as meanshape landmarks, which may be helpful in landmark detection initialization and pose estimation. We will leave it as future work for investigation.

Acknowledgement. This work was supported by the National Key Research and Development Plan (Grant No. 2016YFC0801002), the Chinese National Natural Science Foundation Projects #61672521, #61473291, #61572501, #61502491, #61572536, NVIDIA GPU donation program and AuthenMetric R&D Funds.

References

1. Jain, V., Learned-Miller, E.G.: FDDB: a benchmark for face detection in unconstrained settings. UMass Amherst Technical report (2010)
2. Wu, B., Ai, H., Huang, C., Lao, S.: Fast rotation invariant multi-view face detection based on real adaBoost. In: IEEE Conference on Automatic Face and Gesture Recognition (2004)
3. Li, S., Zhang, Z.: Floatboost learning and statistical face detection. IEEE Trans. Pattern Anal. Mach. Intell. **26**, 1112–1123 (2004)
4. Huang, C., Ai, H., Li, Y., Lao, S.: High-performance rotation invariant multiview face detection. IEEE Trans. Pattern Anal. Mach. Intell. **29**, 671–686 (2007)
5. Hotta, K.: A robust face detector under partial occlusion. In: International Conference on Image Processing (2004)
6. Lin, Y., Liu, T., Fuh, C.: Fast object detection with occlusions. In: Proceedings of the European Conference on Computer Vision, pp. 402–413 (2004)
7. Lin, Y., Liu, T.: Robust face detection with multi-class boosting (2005)
8. Chen, J., Shan, S., Yang, S., Chen, X., Gao, W.: Modification of the adaboost-based detector for partially occluded faces. In: 18th International Conference on Pattern Recognition (2006)

9. Goldmann, L., Monich, U., Sikora, T.: Components and their topology for robust face detection in the presence of partial occlusions. IEEE Trans. Inf. Forensics Secur. **2**, 559–569 (2007)
10. LeCun, Y., Bengio, Y.: Convolutional networks for images, speech, and time series. In: The Handbook of Brain Theory and Neural Networks, pp. 33–61 (1995)
11. Li, H., Lin, Z., Shen, X., Brandt, J., Hua, G.: A convolutional neural network cascade for face detection. In: Proceedings of the IEEE Conference on Computer Vision and Pattern Recognition, pp. 5325–5334 (2015)
12. Farfade, S.S., Saberian, M.J., Li, L.J.: Multi-view face detection using deep convolutional neural networks. In: Proceedings of the 5th ACM on International Conference on Multimedia Retrieval, pp. 643–650. ACM (2015)
13. Yang, S., Luo, P., Loy, C.C., Tang, X.: From facial parts responses to face detection: a deep learning approach. In: Proceedings of the IEEE International Conference on Computer Vision, pp. 3676–3684 (2015)
14. Zhang, K., Zhang, Z., Li, Z., Qiao, Y.: Joint face detection and alignment using multi-task cascaded convolutional networks. arXiv preprint arXiv:1604.02878 (2016)
15. Ranjan, R., Patel, V.M., Chellappa, R.: Hyperface: a deep multi-task learning framework for face detection, landmark localization, pose estimation, and gender recognition. arXiv preprint arXiv:1603.01249 (2016)
16. Viola, P., Jones, M.: Robust real-time object detection. Int. J. Comput. Vis. **4**, 34–47 (2001)
17. Yang, B., Yan, J., Lei, Z., Li, S.Z.: Aggregate channel features for multi-view face detection. In: 2014 IEEE International Joint Conference on Biometrics (IJCB), pp. 1–8. IEEE (2014)
18. Zhu, X., Ramanan, D.: Face detection, pose estimation, and landmark localization in the wild. In: 2012 IEEE Conference on Computer Vision and Pattern Recognition (CVPR), pp. 2879–2886. IEEE (2012)
19. Yan, J., Lei, Z., Wen, L., Li, S.: The fastest deformable part model for object detection. In: Proceedings of the IEEE Conference on Computer Vision and Pattern Recognition, pp. 2497–2504 (2014)
20. Mathias, M., Benenson, R., Pedersoli, M., Gool, L.: Face detection without bells and whistles. In: Fleet, D., Pajdla, T., Schiele, B., Tuytelaars, T. (eds.) ECCV 2014. LNCS, vol. 8692, pp. 720–735. Springer, Heidelberg (2014). doi:10.1007/978-3-319-10593-2_47
21. Ranjan, R., Patel, V.M., Chellappa, R.: A deep pyramid deformable part model for face detection. In: 2015 IEEE 7th International Conference on Biometrics Theory, Applications and Systems (BTAS), pp. 1–8. IEEE (2015)
22. Chen, D., Ren, S., Wei, Y., Cao, X., Sun, J.: Joint cascade face detection and alignment. In: Fleet, D., Pajdla, T., Schiele, B., Tuytelaars, T. (eds.) ECCV 2014. LNCS, vol. 8694, pp. 109–122. Springer, Heidelberg (2014). doi:10.1007/978-3-319-10599-4_8
23. Liao, S., Jain, A., Li, S.: A fast and accurate unconstrained face detector. IEEE Trans. Pattern Anal. Mach. Intell. **38**, 211–223 (2016)
24. Goldmann, L., Mönich, U.J., Sikora, T.: Components and their topology for robust face detection in the presence of partial occlusions. IEEE Trans. Inf. Forensics Secur. **2**, 559–569 (2007)
25. Azizpour, H., Laptev, I.: Object detection using strongly-supervised deformable part models. In: Fitzgibbon, A., Lazebnik, S., Perona, P., Sato, Y., Schmid, C. (eds.) ECCV 2012. LNCS, vol. 7572, pp. 836–849. Springer, Heidelberg (2012). doi:10.1007/978-3-642-33718-5_60

26. Bourdev, L., Maji, S., Brox, T., Malik, J.: Detecting people using mutually consistent poselet activations. In: Daniilidis, K., Maragos, P., Paragios, N. (eds.) ECCV 2010. LNCS, vol. 6316, pp. 168–181. Springer, Heidelberg (2010). doi:10.1007/978-3-642-15567-3_13

27. Zhang, N., Paluri, M., Ranzato, M.A.: Panda: Pose aligned networks for deep attribute modeling. In: Computer Vision and Pattern Recognition, pp. 1637–1644. IEEE, Springer, Berlin, Heidelberg (2014)

28. Zhang, N., Donahue, J., Girshick, R., Darrell, T.: Part-based R-CNNs for fine-grained category detection. In: Fleet, D., Pajdla, T., Schiele, B., Tuytelaars, T. (eds.) ECCV 2014. LNCS, vol. 8689, pp. 834–849. Springer, Heidelberg (2014). doi:10.1007/978-3-319-10590-1_54

29. Zhang, H., Xu, T., Elhoseiny, M., Huang, X., Zhang, S., Elgammal, A., Metaxas, D.: SPDA-CNN: Unifying semantic part detection and abstraction for fine-grained recognition. In: The IEEE Conference on Computer Vision and Pattern Recognition (CVPR) (2016)

30. Köstinger, M., Wohlhart, P., Roth, P.M., Bischof, H.: Annotated facial landmarks in the wild: a large-scale, real-world database for facial landmark localization. In: 2011 IEEE International Conference on Computer Vision Workshops (ICCV Workshops), pp. 2144–2151. IEEE (2011)

31. Bourdev, L., Brandt, J.: Robust object detection via soft cascade. In: IEEE Computer Society Conference on Computer Vision and Pattern Recognition, CVPR 2005, vol. 2, pp. 236–243. IEEE (2005)

32. Liao, S., Jain, A.K., Li, S.Z.: Unconstrained face detection. Technical report, MSU-CSE-12-15, Department of Computer Science, Michigan State University (2012)

33. Dalal, N., Triggs, B.: Histograms of oriented gradients for human detection. In: IEEE Computer Society Conference on Computer Vision and Pattern Recognition, CVPR 2005, vol. 1, pp. 886–893. IEEE (2005)

Deep Architectures for Face Attributes

Tobi Baumgartner$^{(\boxtimes)}$ and Jack Culpepper

Computer Vision and Machine Learning Group, Flickr, Yahoo, San Francisco, USA
tobi@yahoo-inc.com, jackcul@yahoo-inc.com

Abstract. We train a deep convolutional neural network to perform identity classification using a new dataset of public figures annotated with age, gender, ethnicity and emotion labels, and then fine-tune it for attribute classification. An optimal sharing pattern of computational resources within this network is determined by experiment, requiring only 1 G flops to produce all predictions. Rather than fine-tune by re-learning weights in one additional layer after the penultimate layer of the identity network, we try several different depths for each attribute. We find that prediction of age and emotion is improved by fine-tuning from earlier layers onward, presumably because deeper layers are progressively invariant to non-identity related changes in the input.

1 Introduction

We would like to efficiently compute a representation of faces that makes high level attributes, such as identity, age, gender, ethnicity, and emotion explicit. The deployed system must run within a computational budget, and we would like to maximize prediction accuracy. Like others, we believe that learning a deep representation that fuses multiple sources of label information is a promising avenue towards achieving this goal [13,20,21]. In this work, we approach the fusion idea stage-wise, and investigate the problem of how to properly fuse label information from tasks with conflicting invariances.

To illustrate the problem we address, consider the following straight-forward way of jointly solving identity, age, gender, ethnicity and emotion classification. First, train a deep network to solve a large-scale identity classification task. The representation learned by this network can then be adapted to another, related task through a process called fine-tuning [4,11,14,16]. To do so, extract features from the penultimate layer of the identity network, and use them to train classifiers for each attribute. This approach is still commonly used [10], but has a problem: identity is invariant to age and emotion. Thus, presumably the representation at the penultimate layer of the network trained to solve the identity task will reflect this and impair discrimination of ages and emotions. To address this, we varied the layer depth from which our feature was extracted and fine-tuned all layers of the network from that point forward, using cross-validation to pick the best depth. Our experiments show that fine-tuning from the penultimate layer leads to worse results than fine-tuning from an earlier layer onwards.

© Springer International Publishing AG 2017
C.-S. Chen et al. (Eds.): ACCV 2016 Workshops, Part II, LNCS 10117, pp. 334–344, 2017.
DOI: 10.1007/978-3-319-54427-4_25

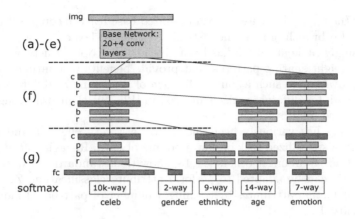

Fig. 1. Proposed network architecture with fine-tuned layers. The different face attribute sub task relearn different numbers of top layers. During the forward pass, i.e., in production, we can efficiently compute all outputs simultaneously (cf. Fig. 2 for full network and legend).

Our approach has several advantages over the more typical approach described above. A significant amount of the identity network is shared across attribute predictors, yet the invariance conflicts between tasks are resolved. Our approach is pragmatic when it comes to dealing with a high variance in the number of labels from one task to another – due to cross validation, tasks with more labels will benefit from them by sharing fewer weights with other tasks.

Figure 1 shows the architecture that we derive from experimentation. It shows the branching pattern that gives the best prediction accuracy for each task, and the order of depths gives a measure of the relationship of the task to identity classification. Admittedly, this relationship is slightly confounded by the variable number of training examples across tasks.

2   Related Work

Fusion of Multiple Information Sources. Liu *et al.* jointly fine-tune a location specialized network and a network for attribute detection and show that fusing information from attribute labels yields an improvement in the localization task [10]. Their attribute network (ANet) is pretrained using a massive identity database, and they use the representation from the penultimate layer to learn attribute predictors such as "is wearing sunglasses". If their identity database contained any individual pictured both with and without sunglasses, then their pretraining would have encouraged the penultimate layer to be invariant to this attribute. They did not utilize the technique we describe herein, which suggests that it is non-obvious and deserving of some attention.

Zhang *et al.* show that face attribute labels can improve landmark detection [20]. They simultaneously train with multiple attributes (*wearing glasses, smiling, gender, pose*) and landmark labels and introduce an early stopping criteria for each of the tasks. They use the same shared base feature computation for all tasks, but also branch only at the last fully connected layer.

Surprisingly little attention has been given to the topic of network architectures with weight sharing patterns that provide benefits to multiple tasks with conflicting invariances such as ours. The idea of branching at different layers of a deep network for different tasks is utilized in [12], but their branching pattern is not evaluated quantitatively.

Fine-tuning has variously been shown to be a very effective and powerful tool to learn a specialized task on low amounts of data. Razavian *et al.* [14] and Donahue *et al.* [4] amongst others [11,18] show that standard CNNs trained on ImageNet [15] can be adapted to perform attribute classification. Zhang *et al.* infer various human attributes by training a number of part-based models with pose-normalized CNNs [19].

Face Attribute Datasets. Research in *emotion recognition* is supported by 3 standard benchmarks. EmotiW [3] is a labeled set of movie clips, in which each clip portrays one of 7 emotions; hence, even though the dataset is fairly large with $400k$ images, it is not very diverse. In the MultiPie [1] dataset, a set of only 4 facial expression, as well as facial landmarks, are captured in very high detail with 15 cameras. The Toronto Face Database (TFD) by Susskind *et al.* [17] contains 7 emotions on a few thousand images.

The task of *age recognition* is addressed in the adience dataset by Levi *et al.* [9] and the cross-age celebrity dataset (CARC) by Chen *et al.* [2]. The former uses the yfcc dataset[1] and has $19k$ images. The latter implies celebrities' ages from photo timestamps and consists of $160k$ images.

To the best of our knowledge, there are no widely available diverse datasets for the task of *ethnicity recognition*.

The above datasets contain images from various ages/genders/ethnicities, but they all have limitations which make them unsuitable for our purposes: LFW and PubFig83 are too small; EmotiW, MultiPie, and Gallagher do not have enough diversity; EmotiW, TFD, and CARC are only available for academic use.

3 Model

In the following we describe our base network architecture for identity recognition, as displayed in Fig. 2, as well as the fine-tuned top layers, shown in Fig. 1.

3.1 Base Network

We first train a deep convolutional neural network (CNN) to solve the task of celebrity recognition, then fine-tune the upper layers of this network to solve the gender, ethnicity, age, and emotion tasks. The design for our CNN is inspired by the recent work on *residual networks* [6]. Like the networks in that paper, ours is built from a few simple functions: convolution, ReLU, pooling, and batch

---

[1] http://webscope.sandbox.yahoo.com/catalog.php?datatype=i\&did=67.

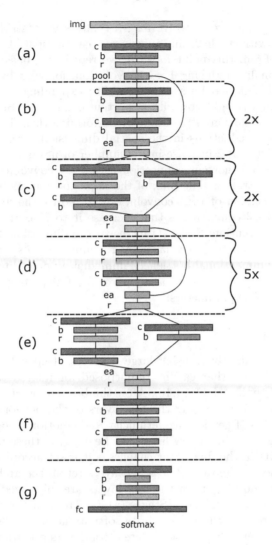

Fig. 2. The complete face-network. c = convolution, b = batchnorm, r = ReLU, ea = elementwise add, p = pool, fc = fully connected. Parts (b), (c) and (d) are repeated as indicated. Excluding the shortcut layers, there are 24 convolution layers.

normalization. We compose them together to build a complex, high capacity function with about 10M free parameters, using skip-connections to mitigate the degradation problem. We use the softmax function to convert the outputs of the topmost layer into a probability distribution and minimize the cross entropy between this and the label distribution. The total loss is the cross entropy averaged across all samples.

Figure 2 shows the structure of our base network. The input is a 224 × 224 pixel RGB image resulting from the output of our detection and alignment algorithms. In stage (a) we convolve 32 7×7 filters with stride 2 then apply 2×2 max pooling with stride 2; this quickly reduces the spatial resolution of the input to 56×56. From this point forward, we repeat the basic module shown in (b): a dimensionality reducing 1×1 convolution, followed by batch normalization and ReLU, followed by a dimensionality expanding 3×3 convolution, followed by element-wise addition with a shortcut connection. Stages (c) and (e) reduce the spatial resolution of the input by performing their 3×3 convolutions with stride 2. These reductions in the spatial dimensions are accompanied by a corresponding factor of 2 increase in the number of filters in the 3×3 convolutional layer, and the shortcut connection uses 1×1 convolutions to equalize dimensions, also at stride 2. The bulk of the flops and parameters are in stage (d), which is a succession of 1×1 convolutions to reduce the dimensionality to 128, followed by 3×3 convolutions that increase it to 256. Stage (d) operates at 14×14 spatial resolution. Stage (f) increases the dimensionality to 512, and stage (g) uses average pooling to collapse the remaining 7×7 spatial dimensions, and imposes a 320 dimensional bottleneck immediately preceding the final fully connected layer. In total, the network requires 0.9 Gflops to process an input image, and has 9M free parameters.

3.2   Fine Tuning

When trained discriminatively using large datasets, deep networks learn representations of the data that can be re-purposed for related tasks through a process called fine tuning [4,11,14,16]. After the basic network in Fig. 2 is fully trained, we fine-tune by replacing the top layers of the network, then use our datasets for the tasks of gender, age, ethnicity and emotion recognition to optimize only the weights in these new layers. To be clear, in these new layers, only the number of units in the final layer changes – it is set according to the number of target classes in the task. The other layers retain the architecture of the base network, but the weights in the new layers are initialized randomly and retrained. We depict this in Fig. 1. Stages (a)–(e) are summarized in the 'base network' box. We replace the $10k$-celebrity softmax layer ('fc'-box in Fig. 1(g)) with a task-specific softmax, and set different top-layers according to the depiction in Fig. 1. Once trained, the parameters of the base network are fixed and not altered during fine-tuning.

The task-specific architectures shown in Fig. 1 were determined experimentally, using a validation set. In Fig. 3 we show the results after fine-tuning for our tasks for a variety of branch depths. The cross-validation results give an ordered relationship between tasks, based on the branching depth: *emotion* branches earliest, then *age*, *ethnicity*, and, finally, *gender* performs best when we branch off the penultimate layer, as is typical.

Fine-tuning the network from earlier layers than the ones experimentally determined yields inferior results, likely due to overfitting caused by the limited number of labeled training samples for the attribute tasks. There is a trade-off between overfitting (backing off too far) and being too invariant to learn the task (not retraining enough layers of the network).

attribute layer	emotion	age	ethnicity	gender
conv17	0.44	0.22	0.60	0.98
conv19	**0.68**	0.28	0.68	0.94
conv21	0.30	0.24	0.50	0.98
conv22	0.34	**0.40**	0.55	0.95
conv-bn320	0.42	0.06	0.61	0.97
fc	0.44	0.24	**0.72**	**0.99**

Fig. 3. Grid of attribute fine-tuning experiments. Accuracy of face attribute tasks on validation set after convergence.

The depth of the best performing connectivity pattern has a noteworthy relationship to the invariances learned by each layer in the base network: due to the way it was trained, higher layers should be progressively more invariant to changes in the input that are unrelated to changes in the identity of the person pictured. That is, it makes perfect sense that the information required to discriminate emotional states should be unavailable at the top of the network – if it were, it would only degrade the capability of the network to identify a celebrity. Put another way, person identity is invariant to changes in emotion or age, so the features that are predictive of these concepts are found lower in the network, before the network has learned to distinguish identities. Gender and ethnicity are most closely related to the identity of a person, i.e. a person has a fixed ethnicity and an (almost) fixed gender, so it is natural that gender/ethnicity prediction relies on features that are more closely associated with identity.

Sharing the representation among related tasks significantly reduces the computational complexity. This allows us for fast evaluation of all tasks at test time in our production system, since the forward-pass through the base network only has to be computed once. Figure 4 summarizes the data and number of fine-tuned parameters for the specific tasks. In the following we describe the data collection, training process, and the evaluation results for each of the 5 tasks in more detail.

task	size	accuracy	n-way	#params
celebrity	2.5M	84%	$10k$	$10M$
emotion	280k	68%	7	$1,018,055$
age	290k	40%	14	$889,230$
ethnicity	197k	72%	9	642
gender	6M	99%	2	642

Fig. 4. This table summarizes the size of our newly collected datasets and the accuracy we achieve on this data. Each of the tasks has a different complexity, that can be described by the n-way classification. The trained layers for each task are illustrated in Fig. 1. The last column shows the number of free parameters that are learned during fine-tuning.

4   Experiments

In this section we show experimental results for our deep network. The applications and experiments in Sect. 4.1 uses the entire network and resulting feature vectors. In Sects. 4.4–4.7, we fine-tune the network for face attribute specific tasks. In Sect. 4.3 we propose an interpretation of our findings and relate them to the capacity of the deep net to learn levels of invariance.

4.1   Identifying Public Figures

In order to learn a mapping from pixels to a representation that is invariant to changes in the pixels that are unrelated to identity, we use a large dataset with many examples of many identities. As there was no suitable, freely available dataset, we collected our own proprietary dataset by analyzing the Yahoo image search engine query logs.

We started with a list of 40k public figures from wikipedia[2] and identified those for which we had at least 50 clicked images in the search logs (i.e. at least 50 positive training examples). For unbiased evaluation of our algorithm against the LFW benchmark, we also excluded identities that appear in LFW, resulting in about 10K identities and a total of 2.5M images.

We partitioned the dataset into training and validation, keeping 5 samples from each identity for validation. We initialize all weights in the network by drawing from a Gaussian distribution with mean 0 and variance 0.01, and optimize the cross entropy loss using stochastic gradient with momentum 0.9 and a batch size of 400. Initially, our learning rate is 0.1 and we reduce it by a factor of 4 every 10,000 minibatches. After training, our validation recognition accuracy on the 10K-way classification task is 84%.

4.2   Face Verification

In the LFW task, pairs of faces must be 'verified' as belonging either to the same person, or to two different people [7]. The dataset contains 1,680 people and 6,000 pairs of faces arranged in 10 splits with 300 'same' and 300 'not same' pairs each. We evaluated the 'unrestricted' protocol using the following leave one out evaluation paradigm. For each of the 6,000 pairs of faces, we run our detection and alignment, then extract the 320 dimensional features from the fully connected layer in Fig. 2(g). We calculate the cosine similarity between each pair of 320d vectors. We select the threshold that provides the best verification on 9 splits, then evaluate it on the remaining one. We reach an accuracy of 95.98% on LFW.

4.3   Attribute Tuning

After the base network is fully trained and reaches an accuracy of 84% on the 10k-way classification task, we fine-tune it starting at different layers for the

---

[2] http://wikipedia.org.

4 attribute tasks. For this, we initialize the same network with the pretrained weights and freeze the weights before the respective starting layer; during back-prop these weights will not be adjusted. We also replace the last fully connected layer with a slimmer fully connected layer that has the number of targets of the specific task. In Fig. 4 we show the number of labels for each of the attributes. All of the experiments have a 1-hot classification label and we summarize the accuracy on the test set in Fig. 3.

Gender and ethnicity prediction are closely related to identity prediction, and are well-suited to be learned by fine-tuning from the penultimate layer of an identity discriminating network. Age and emotion prediction benefit from branching earlier and retraining more layers. Furthermore, our results in Fig. 3 imply that age is more closely related to identity than emotion, which is likely explained by the prevalence in our identity database of photos from a certain time period in a public figure's life.

4.4   Predicting Gender

For *gender classification* we can bootstrap the aforementioned celebrity dataset; The gender of each celebrity can be extracted from the Yahoo Knowledge graph. In order to avoid overfitting, we separate the training and testing sets by celebrities, i.e., the same person will not appear in both the training and test sets. For gender prediction, we achieve accuracy of 99% on a dataset of 6M faces (note that the dataset is larger than for the 10k-way classification task as we do not need to perform the same filtering to have 50 training examples per celebrity).

4.5   Predicting Age

Similarly to the gender dataset, we bootstrap the data for *age recognition* from our celebrity dataset, again by using the Yahoo Knowledge graph, in this case to determine the birthdays of every celebrity. We then determine the capture date of every downloaded celebrity image from the exif-data, if available. Although this process is subject to some noise in the label collection, a similar approach was used in the Cross-Age Reference Coding benchmark [2].

Since not all the celebrity images have valid exif-dates, we end up with a dataset for age recognition of about 290k images. We split this data into 14 age bins, similar to the adience dataset [9]. With this, instead of perfectly predicting a persons age, the task now becomes to predict an age range. The adience dataset contains 19k images and Eidinger *et al.* [5] achieve an accuracy of 45.1%. In their experiment, they only consider 8 age-bins. The binning itself is the same as in our dataset, but they leave out intermediate bins, like 21–24. In our experiments we achieve and accuracy of 40%.

4.6   Predicting Emotion

For *emotion recognition*, we collect a new set of public images from Flickr[3] that have a high probability of containing emotions. We first filter the 15B images in

---

[3] http://flickr.com.

the Flickr corpus by the autotag "portrait". This tag has been computed using the face detector described in Sect. 3; which adds this tag if exactly 1 face is detected in an image and the size of the face is larger than 0.3 times the image height and width. We find all publicly available images that have a emotion-related user-generated tag or title associated with it. Our system distinguish between the following 7 emotions: happy, sad, angry, surprised, disgusted, scared and neutral. For each of these emotions, we compile a list of synonyms and facial expressions and arrive at a total set of $200 + 7$ terms for collecting user-labeled data. This emotion dataset consists of $280k$ images. Our best results yield a recognition accuracy of 68%. We also compare the performance of our network to the EmotiW [3] challenge, for which Kahou *et al.* [8] reported accuracy of 41.03% in the 2013 challenge on the task of emotion detection from images. With our method we achieve an accuracy of 30.23% on their data. As already mentioned in Sect. 2, the EmotiW data does not reflect a real-life task as presented in our scenario though.

4.7 Predicting Ethnicity

The task of *ethnicity detection* is not strictly well defined. There are different levels of granularity imaginable, e.g. Hispanic vs. Mexican, White vs. Central European. In this example, one term is based on region and the other on country borders. Ethnic groups identify by varying commonalities among their peers instead of just the color of their skin. These unifying characteristics are as diverse as: region (South Islanders), language (Gaels), nationality (Iraqis), ancestry (Afro-Brazilian) or religion (Sikhs)[4].

We adapt our celebrity dataset for this task by extracting celebrity ethnicity labels from `EthniCelebs`[5], which has a well curated set of ethnicity labels for over $15k$ people. We use this data to enhance our celebrity dataset with their family origins. The resulting labels are very specific, e.g., *austro-hungarian jewish* or *ulster-scots*. We next map these locations to 9 ethnicities in accordance with industry competitor Getty[6]. The resulting ethnicities we learn to distinguish between are: black, hispanic, east-asian, middle-eastern, pacific islander, south-asian, south-east-asian and white. Our dataset has a size of $197k$ images and we achieve an accuracy of 72%.

Obviously, the ethnicity of most people cannot be described in a single word. We have different backgrounds with parents from various places. To reflect this, we also trained a multi-target classifier for ethnicity detection; This means that the result of this model will be a 9-dimensional vector with probabilities that the tested person belongs to each of these ethnic groups. The evaluation of this task is done in terms of *true/false positive rate* (*tpr/fpr*). The *tpr* describes the percentage of actual ethnicity, that we correctly predict; a non-perfect score means that we miss to label a person with an ethnicity they identify with. More

---

[4] https://en.wikipedia.org/wiki/Ethnic_group.

[5] http://www.ethnicelebs.com.

[6] http://www.gettyimages.com/.

important for this sensitive task is the *fpr* though, since it measures the number of times we falsely assign an ethnicity to someone. People might take offense in errors of this kind. In our experiments, we chose an operating point with a *tpr* of 71.97% and *fpr* of only 1.03%; we err on the safe side, while still assigning some label to almost all test subjects: only 0.44% do not get a ethnicity prediction at all.

5 Discussion

The representations learned by deep networks trained in an identity discrimination task can be fine-tuned to perform well in the tasks of age, gender, ethnicity and emotion recognition. We expose limitations in the typical approach, and show benefits to careful architecture analysis via cross-validation. Our approach provides a straight-forward way to measure the level of similarity between two tasks, in terms of where the information required to solve them is represented inside the network. Capturing these ideas in a learning approach remains for future work.

Acknowledgments. We would like to thank Neil O'Hare for collaborating with us on the search engine query logs, and the entire Yahoo Vision and Machine Learning Team.

References

1. Baker, S.: Multi-PIE. In: Proceedings of the IEEE International Conference on Automatic Face and Gesture Recognition. IEEE Computer Society (2008)
2. Chen, B.-C., Chen, C.-S., Hsu, W.H.: Cross-age reference coding for age-invariant face recognition and retrieval. In: Fleet, D., Pajdla, T., Schiele, B., Tuytelaars, T. (eds.) ECCV 2014. LNCS, vol. 8694, pp. 768–783. Springer, Heidelberg (2014). doi:10.1007/978-3-319-10599-4_49
3. Dhall, A., Goecke, R., Joshi, J., Wagner, M., Gedeon, T.: Emotion recognition in the wild challenge. In: Proceedings of the 15th ACM on International Conference on Multimodal Interaction, ICMI (2013)
4. Donahue, J., Jia, Y., Vinyals, O., Hoffman, J., Zhang, N., Tzeng, E., Darrell, T.: DeCAF: a deep convolutional activation feature for generic visual recognition. CoRR, abs/1310.1531 (2013)
5. Eidinger, E., Enbar, R., Hassner, T.: Age and gender estimation of unfiltered faces. IEEE Trans. Inf. Forensics Secur. **9**(12), 2170–2179 (2014)
6. He, K., Zhang, X., Ren, S., Sun, J.: Deep residual learning for image recognition. In: CVPR (2016)
7. Huang, G.B., Ramesh, M., Berg, T., Learned-Miller, E.: Labeled faces in the wild: a database for studying face recognition in unconstrained environments. Technical report 07–49, University of Massachusetts, Amherst, October 2007
8. Kahou, S.E., Bouthillier, X., Lamblin, P., Gülçehre, Ç., Michalski, V., Konda, K.R., Jean, S., Froumenty, P., Dauphin, Y., Boulanger-Lewandowski, N., Ferrari, R.C., Mirza, M., Warde-Farley, D., Courville, A.C., Vincent, P., Memisevic, R., Pal, C.J., Bengio, Y.: EmoNets: multimodal deep learning approaches for emotion recognition in video. J. Multimodal User Interfaces, abs/1503.01800 (2015)

9. Levi, G., Hassner, T.: Age and gender classification using convolutional neural networks. In: CVPR workshops (2015)
10. Liu, Z., Luo, P., Wang, X., Tang, X.: Deep learning face attributes in the wild. In: ICCV (2015)
11. Oquab, M., Bottou, L., Laptev, I., Sivic, J.: Learning and transferring mid-level image representations using convolutional neural networks. In: CVPR (2014)
12. Pinto, L., Gandhi, D., Han, Y., Park, Y.-L., Gupta, A.: The curious robot: learning visual representations via physical interactions. In: ECCV (2016)
13. Ranjan, R., Patel, V.M., Chellappa, R.: Hyperface: a deep multi-task learning framework for face detection, landmark localization, pose estimation, and gender recognition. CoRR, abs/1603.01249 (2016)
14. Razavian, A.S., Azizpour, H., Sullivan, J., Carlsson, S.: CNN features off-the-shelf: an astounding baseline for recognition. In: CVPR workshops (2014)
15. Russakovsky, O., Deng, J., Su, H., Krause, J., Satheesh, S., Ma, S., Huang, Z., Karpathy, A., Khosla, A., Bernstein, M., Berg, A.C., Fei-Fei, L.: Imagenet large scale visual recognition challenge. Int. J. Comput. Vis. **115**(3), 211–252 (2015)
16. Sermanet, P., Kavukcuoglu, K., Chintala, S., LeCun, Y.: Pedestrian detection with unsupervised multi-stage feature learning. In: CVPR (2013)
17. Susskind, J.M., Anderson, A.K., Hinton, G.E.: The toronto face database. Department of Computer Science, University of Toronto, Toronto, ON, Canada. Technical report 3 (2010)
18. Zeiler, M.D., Fergus, R.: Visualizing and understanding convolutional networks. In: Fleet, D., Pajdla, T., Schiele, B., Tuytelaars, T. (eds.) ECCV 2014. LNCS, vol. 8689, pp. 818–833. Springer, Heidelberg (2014). doi:10.1007/978-3-319-10590-1_53
19. Zhang, N., Paluri, M., Ranzato, M., Darrell, T., Bourdev, L.D.: PANDA: pose aligned networks for deep attribute modeling. In: CVPR (2014)
20. Zhang, Z., Luo, P., Loy, C.C., Tang, X.: Facial landmark detection by deep multi-task learning. In: Fleet, D., Pajdla, T., Schiele, B., Tuytelaars, T. (eds.) ECCV 2014. LNCS, vol. 8694, pp. 94–108. Springer, Heidelberg (2014). doi:10.1007/978-3-319-10599-4_7
21. Zhang, Z., Luo, P., Loy, C.C., Tang, X.: Learning deep representation for face alignment with auxiliary attributes. IEEE Trans. Pattern Anal. Mach. Intell. **38**(5), 918–930 (2016)

Automatic Micro-expression Recognition from Long Video Using a Single Spotted Apex

Sze-Teng Liong[1]([✉]), John See[2], KokSheik Wong[1],
and Raphael Chung-Wei Phan[3]

[1] Faculty of Computer Science and Information Technology,
University of Malaya, Kuala Lumpur, Malaysia
`szeteng1206@hotmail.com, koksheik@um.edu.my`
[2] Faculty of Computing and Informatics,
Multimedia University, Cyberjaya, Malaysia
`johnsee@mmu.edu.my`
[3] Faculty of Engineering, Multimedia University, Cyberjaya, Malaysia
`raphael@mmu.edu.my`

Abstract. Recently, micro-expression recognition has seen an increase of interest from psychological and computer vision communities. As micro-expressions are generated involuntarily on a person's face, and are usually a manifestation of repressed feelings of the person. Most existing works pay attention to either the detection or spotting of micro-expression frames or the categorization of type of micro-expression present in a short video shot. In this paper, we introduced a novel automatic approach to micro-expression recognition from long video that combines both spotting and recognition mechanisms. To achieve this, the apex frame, which provides the instant when the highest intensity of facial movement occurs, is first spotted from the entire video sequence. An automatic eye masking technique is also presented to improve the robustness of apex frame spotting. With the single apex, we describe the spotted micro-expression instant using a state-of-the-art feature extractor before proceeding to classification. This is the first known work that recognizes micro-expressions from a long video sequence without the knowledge of onset and offset frames, which are typically used to determine a cropped sub-sequence containing the micro-expression. We evaluated the spotting and recognition tasks on four spontaneous micro-expression databases comprising only of raw long videos – CASME II-RAW, SMIC-E-HS, SMIC-E-VIS and SMIC-E-NIR. We obtained compelling results that show the effectiveness of the proposed approach, which outperform most methods that rely on human annotated sub-sequences.

1 Introduction

Micro-expression is a form of nonverbal communication that unconsciously reveals the true sentiments of a person. Micro-expressions are exhibited subtly and they typically occur very briefly, at a duration of about 1/5 to 1/25 of a

© Springer International Publishing AG 2017
C.-S. Chen et al. (Eds.): ACCV 2016 Workshops, Part II, LNCS 10117, pp. 345–360, 2017.
DOI: 10.1007/978-3-319-54427-4_26

second [1]. The intensity of the micro-expression is as small as twitching a tiny part of the facial muscles [2]. Thus, it is difficult to observe micro-expressions in real-time conversations due to the minuteness and the quickness of the motion. On the contrary, normal expressions, also known as macro-expressions, lasts between 3/4 of a second to 2 seconds, and can appear at multiple large areas of the facial regions [3], making them easier to be detected. Similar to macro-expressions, micro-expressions can be grouped into six basic expressions: happy, surprise, anger, sad, disgust and fear [4]. Various research groups, especially in the behavioral and computing fields, are interested to analyze micro-expressions mainly due to its usefulness in uncovering the emotional state of a person who is attempting to conceal it [5]. Hence, micro-expression recognition is useful in a wide range of applications, including clinical diagnosis, national security, and interrogation.

Micro-expression is a dynamic facial action which evolves in the following sequence of states: neutral-onset-apex-offset-neutral [6]. Starting from a neutral state, an onset frame indicates the beginning of a micro-expression where the facial muscles begin to undergo contraction, while offset frame is the end of the expression where the intensity of the muscles is reduced to zero. Apex frame is the instant when the micro-expression reaches its climax (the most intense movement). The apex is not necessarily located at the middle between the onset and offset frames, but it can be situated at any frame in the onset-offset range. In our work, we define the video sub-sequence that is composed only of frames from onset to offset as "*short video*". On the other hand, "*long video*" refers to the raw video sequence which may include the frames with micro-expressions and irrelevant motion that are present before the onset and after the offset. Figure 1 illustrates the short and long video sequences with onset-apex-offset frame annotations. Notice how a micro-expression sequence (frames 30–120) in a long video can be easily shrouded by frames outside the onset-offset range that contain eye blinks and head rotations (such as in frames 15 and 150). In current literature, most works categorized micro-expressions using the pre-cropped short videos. For these cases, the locations of the onset and offset frames are required. These annotations can be obtained from the ground-truth, which

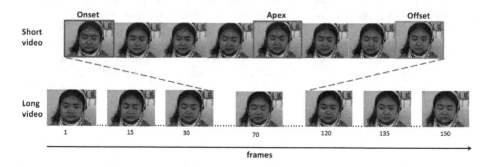

Fig. 1. An example of a long and short video with annotated ground-truth labels indicating the onset, apex and offset frames.

are manually marked and verified by psychologists or "coders". Nonetheless, precision of ground-truth labeling is highly dependent on the judgment of the psychologists, who decide on the onset and offset locations by frame-by-frame observation [7,8].

Apex frame contains vital information of a facial micro-expression as it is the best representative frame instant in the whole video sequence. A few recent works have been proposed for automatic apex frame spotting on the CASME II dataset, but all were tested on short videos with the luxury of onset and offset annotations. Yan et al. [9] demonstrated a pilot experiment by searching for the apex frame using two different feature extractors (i.e., Constraint Local Model (CLM) [10] and Local Binary Pattern (LBP) [11]) in 50 short video samples. The frame with the highest feature difference among the image frames is denoted as the apex. However, the CLM feature performed poorly as it was not able to annotate landmark points to a good degree of accuracy. Later on, Liong et al. [12] enhanced the work of [9] by employing Optical Strain (OS), a flow-based feature which was motivated by the work of Shreve et al. [3]. The authors also discovered that instead of considering the whole face for feature representation, the features from three regions-of-interest (i.e., "left eye + left eyebrow", "right eye + right eyebrow" and "mouth") provided more salient features with respect to micro-expressions. On a separate direction, several other works [3] attempted to detect a micro-expression sequence from a long video by detecting the frames that make up the sequence. "Peak frames" that show a hike in intensity indicate likely onset and offset frames. However, the success of these methods depends greatly on the choice of threshold parameters used to determine peak frames.

A recent work by Liong et al. [13] proved that the utilization of information from the apex frame alone is sufficient for micro-expression recognition. They validated their method on short videos from the CASME II and SMIC micro-expression databases. Features were obtained from the apex frame and the first frame of the short video (or onset frame) using the Bi-Weighted Oriented Optical Flow (Bi-WOOF) [13] feature extractor. However, the authors followed the assumption that the onset and offset frames are already annotated, and hence, constraints the finding of the apex to that particular range. This is unrealistic considering that long videos may contain many irrelevant motions that can be falsely spotted as micro-expressions.

As far as we are aware, there is only a single attempt [14] to work on long micro-expression videos in the literature, to realize a seamless automatic recognition system. They utilized two kinds of features, LBP and Histogram of Oriented Optical Flow (HOOF) [15], to characterize the frames in the sequence. A chi-square dissimilarity metric is used to compute the feature difference between each frame and a reference frame. Then, all spotted frames are thresholded and cross-checked against a pre-defined frame interval to determine the spotting accuracy. The spotting threshold was chosen (at true positive rate of 74.86%) to obtain the spotted micro-expression sequences which are fed to the recognition component. Although evaluation on the SMIC-E-VIS showed promising intent of such a scheme, the reliance on the annotated onset and offset frames, and use of a tunable threshold parameter warrants the need for manual intervention.

Evaluation of the micro-expression system on long videos is particularly challenging, primarily because of the presence of unwanted facial movements. These motions correspond to falsely detected micro-expressions, which may appear before the actual onset frame and after the offset frame. One common irrelevant facial movement that is unavoidable during the elicitation of micro-expression database is the eye blinking motion. Shreve et al. [3] suggested to remove the eye regions because eye blinking can adversely affect optical flow estimation, causing false detection of the micro- and macro-expressions. In their work, the boundaries of the eye regions were automatically marked using a land-mark annotator, unlike the work of [16] which was done manually.

In this paper, we present a novel approach that can automatically recognize the type of facial micro-expression given a long video without ground-truth annotations of the onset, apex and offset frames. A complete micro-expression system which combines both apex frame spotting and micro-expression recognition components that is capable of operating on long videos is introduced. In the apex frame spotting component, pre-processing is first performed to automatically mask the eye regions to prevent ambiguous eye behaviors (i.e., eye blinking). After which, optical strain magnitudes are computed and sum-aggregated for twelve facial blocks, and a max operator pinpoints the apex frame. In the recognition component, we employ the Bi-Weighted Oriented Optical Flow (Bi-WOOF) feature using only the spotted apex frame and a neutral reference frame (we take the first frame as the most neutral expression) to describe the video sequence. We validate the reliability of the system and the effectiveness of the proposed methods in four spontaneous micro-expression databases: CASME II-RAW, SMIC-E-HS, SMIC-E-VIS and SMIC-E-NIR. To the best of our knowledge, these are the only databases which contained long videos.

The structure of the paper is organized as follows: Sect. 2 explains the algorithms of the proposed micro-expression system in detail, Sect. 3 describes the databases, performance metrics and settings used in the experiments, Sect. 4 presents the results for both apex frame spotting and micro-expression recognition, with further analysis. Finally, Sect. 5 concludes the paper.

2   Proposed Approach

The micro-expression system proposed includes two main stages: apex frame spotting and micro-expression recognition. Firstly, the apex frame in a long video sequence is identified by applying Optical Strain feature extractor after eye masking and regions of interest selection techniques. The spotted apex frame is then fed into the micro-expression recognition stage, which is made up of a feature extractor and a classifier. The framework of the proposed algorithm is illustrated in Fig. 2, with detail of each stage elaborated as follows.

2.1   Apex Frame Spotting

In the apex frame spotting task, some of the frames in the long videos might contain irrelevant micro-expression movements, such as eye blinking action, which

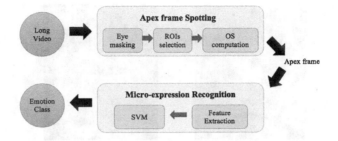

Fig. 2. Flow diagram of the proposed micro-expression system for long videos.

can possibly lead to erroneous apex frame spotting; and further to that, misclassification of micro-expressions. Therefore, we present a novel eye masking approach to address the eye blinking issue. Besides that, instead of using the whole face for feature extraction, we select a number of facial regions that contain meaningful micro-expression details, particularly at the eye and mouth regions [9,12]. The optical strain magnitudes are then computed for each region of interest (ROI); the frame with the highest sum of optical strain magnitudes (from any region) is chosen as the apex. We note that the eye masking and ROIs selection steps are fully automated and completely rely on the facial locations of the first frame of each video, that are marked by the landmark detector.

Eye Masking. Eye blinking is a natural motion of rapid opening and closing of the eyelids, and cannot be considered a micro-expression. Since the micro-expression databases are typically recorded at a high frame rate, the blinking action is clearly visible when displaying the video frame-by-frame; hence, it is significantly more intense compared to micro-expressions. Thus, it is a nagging issue that exists in some of the long video sequences. We overcome this issue by masking the left and right eye parts to reduce the false spotting of the apex frame. To ensure this is done automatically, the eye regions are removed based on the location of landmark points annotated by a robust landmark detector, Discriminative Response Map Fitting (DRMF) [17]. DRMF has shown to outperform state-of-the-art landmark detection methods [18,19], with lower computational time and real-time capabilities. The process of eliminating the eye regions is shown in Fig. 3. Landmark coordinates 37 to 42 indicate the boundary of the right eye region, while landmark coordinates 43 to 48 are the boundary points of the left eye region. To overcome potential inaccurate landmark annotation, a fifteen pixel margin is added to expand the eye boundaries.

ROI Selection. Subsequently, the choice of region to perform accurate spotting is crucial. In [12], features from two main facial regions that contribute important micro-expression information, i.e. "eye and eyebrow" and "mouth" regions, were considered rather than the whole face region. The cropping of these region-of-interests (ROIs) are done in a completely automatic way: (1) using the facial

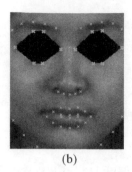

Fig. 3. The eye masking process: (a) There are six landmark coordinates which marked the boundaries of the left (landmark points 37, 38, 39, 40, 41 and 42) and right (landmark points 43, 44, 45, 46, 47 and 48) eye regions; (b) The eye regions are removed after adding some pixel margins.

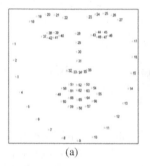
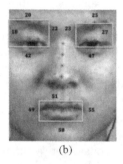
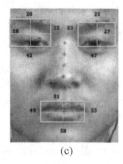

Fig. 4. Illustration of extraction of the three RoIs: (a) Sixty six landmark coordinates labeled by DRMF; (b) The four edges (i.e., top, bottom, left and right) are determined based on the landmark point locations; (c) Each ROI is partitioned into four blocks with the same size.

landmark points annotated earlier, the three ROIs are identified using rectangular bounding boxes determined based on the landmark locations; (2) the ROI bounding boxes are widened by a margin of ten pixels on all four edges to compensate for potentially imprecise landmark annotation; (3) each ROI is equally divided into four blocks to encode more local appearance features. Thus, there is a total of twelve facial region blocks in a frame. Figure 4 illustrates the steps involved in the ROIs selection.

Optical Strain Computation. Shreve et al. [3] employed optical strain magnitudes for macro- and micro-expression spotting. Inspired by this work, we adapted the idea to better characterize micro motions, by obtaining optical strain magnitudes based on a reference frame. Optical strain is the extension of optical flow, and is more effective than the latter in identifying the subtle deformable facial muscle within a time interval [20]. In this work, we use the

TV-L1 optical flow method [21], which is able to preserve flow discontinuities and arguably more robust compared to the classic Black and Anandan [22] optical flow method employed in [3].

Given a micro-expression video, $s_i = \{f_{i,j} | i = 1, \ldots, S; j = 1, \ldots, F_i\}$ of length F_i, we first compute the optical flow vector $\boldsymbol{p} = [p = \frac{u}{\Delta t}, q = \frac{v}{\Delta t}]$ between each frame in the video sequence (except the first frame) and the reference frame (first frame chosen as it contains the most neutral expression). Optical strain can be described by a two-dimensional displacement vector, $\boldsymbol{u} = [u, v]$ and its magnitude for each pixel can be calculated by taking the sum of squares of the normal and shear strain components, expressed as follows:

$$\begin{aligned} |\varepsilon| &= \sqrt{\varepsilon_{xx}{}^2 + \varepsilon_{yy}{}^2 + \varepsilon_{xy}{}^2 + \varepsilon_{yx}{}^2} \\ &= \sqrt{\frac{\partial u}{\partial x}^2 + \frac{\partial v}{\partial y}^2 + \frac{1}{2}(\frac{\partial u}{\partial x} + \frac{\partial u}{\partial x})^2} \end{aligned} \tag{1}$$

A more detailed discourse on optical strain can be found in [3,20].

As mentioned earlier, there are twelve facial blocks in each frame. The optical strain magnitudes are calculated for each of these regions after applying eye masking. The optical strain magnitudes in each block b are summed up and the frame with the highest block value (or sum of magnitudes) is designated as the spotted apex frame f^* for the sequence:

$$f^* = \arg\max_j \left\{ \sum_{j,b} |\varepsilon_{j,b}| \right\} \quad \text{for } j = [1, F_i - 1], b = [1, 12] \tag{2}$$

2.2   Micro-expression Recognition

To describe the features of the spotted apex frame, we employ the Bi-Weighted Oriented Optical Flow (Bi-WOOF) feature extractor in [13]. This is the only work that makes use of information from a single apex frame for representing a micro-expression video. For each video sequence, the orientation θ and magnitude ρ of the flow vector $\boldsymbol{p} = [p, q]$ computed from the spotted apex frame and a neutral reference frame (we choose the first frame) are calculated as follows:

$$\theta_{x,y} = tan^{-1} \frac{q_{x,y}}{p_{x,y}} \tag{3}$$

$$\rho_{x,y} = \sqrt{p_{x,y}{}^2 + q_{x,y}{}^2} \tag{4}$$

We utilize the first frame of each video instead of the onset frame used in [13] since we do not rely on any ground-truth annotations when processing long video sequences. Then, the orientation values are locally and globally weighted by the magnitude and optical strain values respectively to form the Bi-WOOF features. The flow magnitudes are used to weight the flow orientations at a finer bin level (local). The spatially-pooled optical strain magnitudes provide coarser (global) weighting to $N \times N$ equally-partitioned blocks in the frame.

3   Experiment

To assess the performance of the proposed approach to micro-expression spotting and recognition, experiments were performed on four publicly accessible spontaneous micro-expression datasets: CASME II-RAW [7], SMIC-E-HS [8], SMIC-E-VIS [8] and SMIC-E-NIR [8]. Note that all these datasets have been recorded under well-controlled laboratory conditions because an unconstrained environment poses a great challenge to the elicitation of emotions and further machine processing. For standardized experimentation, we first align the faces with DRMF [17], detect them using a standard face detector [23], and then resize the cropped faces to 170 × 140 pixels.

3.1   Datasets

CASME II-RAW. This database comprises of 246 micro-expressions from 26 subjects with a mean age of 22.03 years. The videos were collected using Point Grey GRAS-03K2C camera at a resolution of 640 × 480 pixels and a frame rate of 200 fps. The average frame length is 244 frames ($\sim$1.22s), with the longest being 1,024 frames ($\sim$5.12s) and the shortest being 51 frames ($\sim$0.26s). There are five main categories of micro-expressions, in the following distribution: 25 surprise videos, 27 repression videos, 32 happiness videos, 63 disgust videos and 99 other videos. The micro-expressions are elicited from the subjects by showing them some video clips and asking them to keep a poker face when watching the videos. The emotion types are labeled based on the action units decided by two coders, participants' self report and the content of the video shown. A reliability score of the action unit labeling is reported at 0.846. The ground-truths provided include the onset, apex, offset frame indices and the emotion class.

SMIC-E-HS. It consists of 157 micro-expression clips from 16 subjects (mean age of 28.1 years). The videos were recorded using PixeLINK PL-B774U camera with a temporal resolution of 100 fps and spatial resolution of 640 × 480 pixels. The average frame length is 590 frames ($\sim$5.9s); the longest is 1200 frames ($\sim$12s) while the shortest is 120 frames ($\sim$1.2s). There are three micro-expression classes: negative (66 videos), positive (51 videos) and surprise (40 videos). The micro-expression categories are determined by two coders based on the participants' self-report data. The ground-truths provided are onset, offset frame indices, and the emotion label. No apex frame indices are given.

SMIC-E-VIS. This dataset is made up of 71 micro-expression videos from 8 subjects. The videos were recorded using a standard visual camera with a frame rate of 25 fps at 640 × 480 pixels. The video clips have an average length of 150 frames ($\sim$ 6 s); the longest is 300 frames ($\sim$12s) and the shortest is 30 frames ($\sim$1.2s). It consists of three micro-expression classes: negative (24 videos), positive (28 videos) and surprise (19 videos). The procedure and the ground-truth labels provided are the same as that in SMIC-E-HS.

SMIC-E-NIR. It is a collection of 71 micro-expression video sequence obtained from 8 subjects. The videos were recorded with a near-infrared camera at a resolution of 640 × 480 pixels at $25fps$. The average frame length is 150 frames ($\sim$6 s); the longest is 300 frames ($\sim$12s) and the shortest is 30 frames ($\sim$1.2 s). The micro-expression videos are categorized into three classes: negative (23 videos), positive (28 videos) and surprise (20 videos). The process of the ground-truth acquisition is the same as that in SMIC-E-HS.

3.2 Performance Metrics

Spotting Task. The effectiveness of apex frame spotting can be determined using the Mean Absolute Error (MAE), which was also used in [9,12]. MAE indicates the average frame distance between the ground-truth and the spotted apex frame, and it can be computed by the following equation:

$$\text{MAE} = \frac{1}{N} \sum_{i=1}^{N} |e_i| \tag{5}$$

where N is the total number of video sequence in the database and e is the distance (in frames) between the ground-truth apex and the spotted apex. However, among the four databases used in the experiments, only CASME II-RAW provided ground-truth apex frame indices. Thus, we propose to evaluate the performance of apex frame spotting using another measurement, Apex Spotting Rate (ASR), which calculates the success rate in spotting apex frames within the onset and offset range given a long video. An apex frame is scored 1 if it is located between the onset and offset frames, and 0 otherwise:

$$\text{ASR} = \frac{1}{N} \sum_{i=1}^{N} \delta$$
$$\text{where } \delta = \begin{cases} 1, & \text{if } f^* \in (f_{i,onset}, f_{i,offset}) \\ 0, & \text{otherwise} \end{cases} \tag{6}$$

Recognition Task. The classifier adopted in all experiments reported in this paper is the Support Vector Machine (SVM) with linear kernel. To ensure subject independence in the classification process, a leave-one-subject-out (LOSO) cross validation is employed. In LOSO, for each k-fold (where k is the total number of the subjects), the video samples of one subject are held out as testing set, while the remaining video samples form the training set. Following the experiment settings in [13], the block size for the Bi-WOOF feature extractor is set to 8 × 8 for CASME II-RAW, and 5 × 5 for the SMIC databases. Since the video samples in different micro-expression classes are distributed unequally [24], we measure the recognition accuracy using the F1-score which conveys the balance by averaging the precision (exactness) and recall (completeness):

$$\text{F1-score} := 2\frac{\text{Precision} \times \text{Recall}}{\text{Precision} + \text{Recall}} \tag{7}$$

$$\text{Precision} := \frac{\text{TP}}{\text{TP} + \text{FP}} \tag{8}$$

$$\text{Recall} := \frac{\text{TP}}{\text{TP} + \text{FN}} \tag{9}$$

where TP, FN and FP are the true positive, false negative and false positive.

4   Results and Discussion

4.1   Results

Performance of Apex Frame Spotting. The MAE result for apex frame spotting task on the CASME II-RAW dataset shown in Table 1 compares the technique with and without applying eye masking on two types of feature extractors - LBP and OS. The LBP feature was utilized in [14] to spot the micro-expression frames while we propose the use of OS in this work. The lower the MAE (in frames), the closer the spotted apex frame is to the ground-truth apex frame, implying more accurate spotting. The spotting performance for OS outperforms the LBP. This result also emphasizes the importance of using eye masking (a more detailed look into the impact of this step can be found in Fig. 5). Eye masking improves the spotting accuracy with OS features by 36.38%.

On the other hand, Table 2 shows the apex spotting accuracy measured in terms of ASR. With eye masking, we observe important improvements of 20%, 41.68%, 20.02% and 31.58% on the CASME II-RAW, SMIC-E-HS, SMIC-E-VIS and SMIC-E-NIR databases respectively. From these results, we show that the elimination of eye regions (but not up to the extent of eyebrows) from consideration is able to increase the precision of searching for the apex frame. It is worth mentioning that the overall performance on the SMIC databases is still quite low (even with eye masking). We discuss this in detail in Sect. 4.2.

Performance of Micro-Expression Recognition. At this point of time, there are no papers in literature that reported the F1-score recognition performance on micro-expression long videos (i.e., CASME II-RAW, SMIC-E-HS, SMIC-E-VIS and SMIC-E-NIR). For reference purpose, we provide the state-of-the-art methods that worked on the short videos (i.e., CASME II, SMIC-HS, SMIC-VIS and SMIC-NIR), shown in Table 3. Whereas, the results for long videos are tabulated in Table 4. For the spotting task in Table 4, our methods

Table 1. Performance of apex frame spotting with and without eye masking on CASME II-RAW measured by MAE. Average frame number per video is 244.

Feature extractor	W/o eye mask	With eye masked	Improvement
LBP	51.86 frames	55.26 frames	−6.56%
OS	42.77 frames	**27.21 frames**	**36.38%**

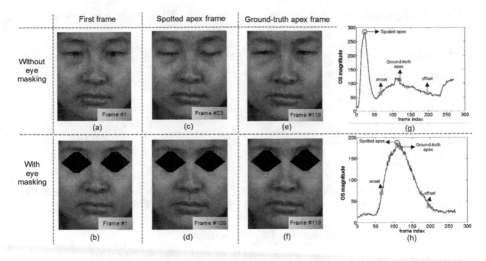

Fig. 5. Top row without eye masking, bottom row with eye masking: (a–b) The first frame in the video; (c–d) The spotted apex frame; (e–f) The ground-truth apex frame; (g–h) Plots of optical strain magnitudes across the video sequence. Relevant labeled frames are marked.

Table 2. Performance of apex frame spotting with and without eye masking on the CASME and SMIC databases measured by ASR.

Databases	W/o eye mask	With eye masked	Improvement
CASME II-RAW	0.6584	**0.8230**	**20.00%**
SMIC-E-HS	0.2229	**0.3822**	**41.68%**
SMIC-E-VIS	0.2253	**0.2817**	**20.02%**
SMIC-E-NIR	0.1831	**0.2676**	**31.58%**

Table 3. State-of-the-art recognition performance using F1-score in short videos databases.

#	Methods	CASME II	SMIC-HS	SMIC-VIS	SMIC-NIR
1	Baselines [7,8]	.39	.39	.39	.40
2	Le et al. [24]	.33	.47	-	-
3	Le et al. [25]	.51	-	-	-
4	Le et al. [26]	.51	.60	-	-
5	Wang et al. [27]	.40	.55	-	-
6	Liong et al. [28]	-	.45	-	-
7	Liong et al. [20]	.38	.54	-	-
8	Oh et al. [29]	.43	.35	-	-
9	Huang et al. [30]	.57	.58	-	-
10	Xu et al. [31]	.30	.54	**.60**	**.60**
11	Liong et al. [13]	**.61**	**.62**	.58	.58

Table 4. Comparison of recognition performance using F1-score in long videos databases.

#	Methods	CASME II-RAW	SMIC-E-HS	SMIC-E-VIS	SMIC-E-NIR
1	Spotting (random) + recognition	.36	.37	.33	.28
2	Spotting (w/o eye mask) + recognition	.46	.36	.44	.38
3	Spotting (with eye mask) + recognition	**.59**	**.47**	**.53**	**.43**

Table 5. Comparison of recognition accuracy (%) and methodology between the state-of-the-art method and the proposed method on the SMIC-E-VIS database.

Methods	Recognition (%)	Remarks
Li et al. [14]	56.67	Onset and offset frames used
		Only correctly spotted sequences used
Proposed method	**53.52**	No onset, offset, apex labels required
		All spotted apices are used

employed the OS feature. Method #1 randomly spots the apex frame in the video sequence; method #2 spots the apex frame without masking the eye regions; method #3 spots the apex frame after applying eye masking. We can see that our best approach (#3) generates the best performance in all the long video databases.

We also compared the proposed method (method #3, with eye masking) with the work of Li et al. [14], which is the only other work that implemented a micro-expression spotting and recognition system for long videos (Table 5). This could only be done using the Accuracy measure, and only on one database, SMIC-E-VIS. It is observed that our proposed method is comparable to that of Li et al. [14] but with several advantages. Our method does not rely on the ground-truth onset and offset labels, and has the computational benefit of only needing to find the apex frame for recognition purpose.

4.2 Discussion

Although we evaluated the proposed method (#3) in long videos, we obtained a superior result on CASME II-RAW in Table 4, compared to all other methods (#1 to #10) in Table 3 that were conducted on CASME II (short videos) despite not having onset, offset and apex frame labels. The performance of our proposed method on SMIC-E-HS, SMIC-E-VIS and SMIC-E-NIR databases are slightly mixed, compared to the results of SMIC-HS, SMIC-VIS and SMIC-NIR databases respectively, in Table 3. This may be due to the fact that most of

Table 6. Average number of frames in the short and long videos of the CASME II and three SMIC databases.

Databases	Short video	Long video	Frames with micro-expression
CASME II	67 frames	244 frames	~27%
SMIC-HS	33 frames	590 frames	~6%
SMIC-VIS	9 frames	150 frames	~6%
SMIC-NIR	9 frames	150 frames	~6%

the frames in these databases do not contain micro-expression-like motions. We show in Table 6 the average percentage of frames of the long videos that consists of micro-expressions. We note that only approximately 6% of the frames consists of micro-expressions in the three SMIC databases. In other words, 94% of the frames contain either neutral faces, macro-expressions or other forms of irrelevant motions such as head rotations and eyeball movements. In addition, SMIC has a much lower frame rate i.e. $100fps$, $25fps$ and $25fps$ in HS, VIS and NIR datasets respectively, compared to CASME II ($200fps$). This points to the possibility of macro-expression and irrelevant movements become more prominent while micro-expressions may occur only in a few frames. Hence, attempting to spot the apex frame in these circumstances is an arduous task. In future, better techniques can be designed to differentiate between these different states of emotions (macro and micro).

For method #1 in Table 4, the apex frame is spotted randomly, as a control method. As expected, the recognition performance is the poorest among all evaluated methods in all databases. This indicates the importance of obtaining the apex frame correctly. Both the spotting and recognition results (in Tables 2 and 4) prove that the eye masking technique enhances the micro-expression system by removing noises from eye blinking, resulting in more meaningful features. Figure 5 demonstrates the difference in the selection of spotted apex, with and without applying eye masking. Without the eye masking technique (the upper row in Fig. 5), the detected apex frame (frame 23) contains an eye closing motion, a falsely spotted micro-expression. This occurred because the facial movement is relatively more intense among all the frames in the video. On the contrary, the spotted apex frame with eye masking (frame 109) is much closer to the ground-truth apex frame (frame 119).

In Table 5, Li et al. [14] tested their full micro-expression system on the SMIC-E-VIS. Although their reported recognition performance is slightly better than that of our method, there are several glaring differences. Firstly, they utilized the ground-truth onset and offset frame labels to form a frame interval which was used to determine the spotted micro-expression sequence. Secondly, the incorrectly spotted micro-expression sequences are not considered for recognition. The authors pointed out that the reported performance was achieved using only the correctly spotted ME sequences (TPR = 74.86%) [14]. Hence, we believe that our proposed approach eliminates the need for human intervention;

Table 7. Confusion matrices for recognition task on the CASME-II-RAW database using our proposed method, measured by accuracy rate (%).

	Disgust	Happiness	Tense	Surprise	Repression
Disgust	**42.62**	8.20	44.26	1.64	3.28
Happiness	6.25	**56.25**	25.00	6.25	6.25
Others	23.47	9.18	**61.22**	1.02	5.10
Surprise	12.00	4.00	16.00	**68.00**	0
Repression	0	7.41	33.33	0	**59.26**

Table 8. Confusion matrices for recognition task on the SMIC-E-HS database using our proposed method, measured by accuracy rate (%).

	Negative	Positive	Surprise
Negative	**53.85**	32.31	13.85
Positive	39.22	**49.02**	11.76
Surprise	53.66	12.20	**34.15**

it does not make use of any hand-labeled ground-truth frames (i.e., onset, apex and offset). It also mimics a fully automatic and realistic system which considers the likelihood of a less-than-desirable spotted apex. For a closer inspection into how each class performed, we provide confusion matrices of the recognition task for CASME-II-RAW and SMIC-E-HS databases, tabulated in Tables 7 and 8.

5   Conclusion

A novel fully automatic micro-expression recognition system, which combines both apex frame spotting and micro-expression recognition, is proposed in this paper. We present the first known work that classifies the emotion class given a long micro-expression video without the knowledge of ground-truth onset and offset frames. In the spotting task, a major problem which exists in long videos is the presence of eye blinking motion, which can easily be mistaken as a micro-expression. We remove the eye regions by applying an automatic eye masking technique which depends entirely on detected landmark coordinates. Optical strain magnitudes of all frames are computed with the aid of eye masking and region selection in order to determine the apex frame. In the recognition task, we apply the recent Bi-WOOF feature extractor to capture discriminative features from each video, using only the spotted apex frame and a neutral reference frame (the first frame). Using a SVM classifier with linear kernel, we evaluated the proposed approaches on four spontaneous micro-expression databases that contain long videos. The proposed recognition approach outperforms all existing state-of-the-art methods on the CASME II database, achieving a promising F1-score of 59%.

References

1. Porter, S., ten Brinke, L.: Reading between the lies identifying concealed and falsified emotions in universal facial expressions. Psychol. Sci. **19**(5), 508–514 (2008)
2. Ekman, P., Friesen, W.V.: Nonverbal leakage and clues to deception. J. Interpers. Process. **32**, 88–106 (1969)
3. Shreve, M., Godavarthy, S., Manohar, V., Goldgof, D., Sarkar, S.: Towards macro- and micro-expression spotting in video using strain patterns. In: Applications of Computer Vision (WACV), pp. 1–6 (2009)
4. Ekman, P., Friesen, W.V.: Constants across cultures in the face and emotion. J. Personal. Soc. Psychol. **17**, 124 (1971)
5. Ekman, P.: Lie catching and microexpressions. The philosophy of deception, pp. 118–133 (2009)
6. Valstar, M.F., Pantic, M.: Combined support vector machines and hidden Markov models for modeling facial action temporal dynamics. Human Computer Interaction, pp. 118–127 (2007)
7. Yan, W.J., Wang, S.J., Zhao, G., Li, X., Liu, Y.J., Chen, Y.H., Fu, X.: CASME II: An improved spontaneous micro-expression database and the baseline evaluation. PLoS ONE **9**, e86041 (2014)
8. Li, X., Pfister, T., Huang, X., Zhao, G., Pietikainen, M.: A spontaneous micro-expression database: inducement, collection and baseline. In: Automatic Face and Gesture Recognition, pp. 1–6 (2013)
9. Yan, W.-J., Wang, S.-J., Chen, Y.-H., Zhao, G., Fu, X.: Quantifying micro-expressions with constraint local model and local binary pattern. In: Agapito, L., Bronstein, M.M., Rother, C. (eds.) ECCV 2014. LNCS, vol. 8925, pp. 296–305. Springer, Cham (2015). doi:10.1007/978-3-319-16178-5_20
10. Cristinacce, D., Cootes, T.: Automatic feature localisation with constrained local models. Pattern Recogn. **41**(10), 3054–3067 (2008)
11. Ojala, T., Pietikäinen, M., Harwood, D.: A comparative study of texture measures with classification based on featured distributions. Pattern Recogn. **29**(1), 51–59 (1996)
12. Liong, S.T., See, J., Wong, K., Le Ngo, A.C., Oh, Y.H., Phan, R.C.W.: Automatic apex frame spotting in micro-expression database. In: Asian Conference on Pattern Recognition (ACPR) (2015)
13. Liong, S.T., See, J., Wong, K., Phan, R.C.W.: Less is more: micro-expression recognition from video using apex frame. arXiv (2016)
14. Li, X., Hong, X., Moilanen, A., Huang, X., Pfister, T., Zhao, G., Pietikäinen, M.: Reading hidden emotions: spontaneous micro-expression spotting and recognition. arXiv preprint arXiv:1511.00423 (2015)
15. Dalal, N., Triggs, B., Schmid, C.: Human detection using oriented histograms of flow and appearance. In: European conference on computer vision, pp. 428–441 (2006)
16. Moilanen, A., Zhao, G., Pietikainen, M.: Spotting rapid facial movements from videos using appearance-based feature difference analysis. In: ICPR, pp. 1722–1727 (2014)
17. Asthana, A., Zafeiriou, S., Cheng, S., Pantic, M.: Robust discriminative response map fitting with constrained local models. In: Computer Vision and Pattern Recognition, pp. 3444–3451 (2013)
18. Saragih, J.M., Lucey, S., Cohn, J.F.: Deformable model fitting by regularized landmark mean-shift. Int. J. Comput. Vis. **91**(2), 200–215 (2011)

19. Zhu, X., Ramanan, D.: Face detection, pose estimation, and landmark localization in the wild. In: Computer Vision and Pattern Recognition, pp. 2879–2886 (2012)
20. Liong, S.-T., See, J., Phan, R.C.-W., Ngo, A.C., Oh, Y.-H., Wong, K.S.: Subtle expression recognition using optical strain weighted features. In: Jawahar, C.V., Shan, S. (eds.) ACCV 2014. LNCS, vol. 9009, pp. 644–657. Springer, Heidelberg (2015). doi:10.1007/978-3-319-16631-5_47
21. Pérez, J.S., Meinhardt-Llopis, E., Facciolo, G.: Tv-l1 optical flow estimation. Image Process. Line **3**, 137–150 (2013)
22. Black, M.J., Anandan, P.: The robust estimation of multiple motions: parametric and piecewise-smooth flow fields. Comput. Vis. Image Underst. **63**(1), 75–104 (1996)
23. Viola, P., Jones, M.: Rapid object detection using a boosted cascade of simple features. In: IEEE CVPR, vol. 1, pp. I–511 (2001)
24. Ngo, A.C., Phan, R.C.-W., See, J.: Spontaneous subtle expression recognition: imbalanced databases and solutions. In: Cremers, D., Reid, I., Saito, H., Yang, M.-H. (eds.) ACCV 2014. LNCS, vol. 9006, pp. 33–48. Springer, Heidelberg (2015). doi:10.1007/978-3-319-16817-3_3
25. Le Ngo, A.C., Liong, S.T., See, J., Phan, R.C.W.: Are subtle expressions too sparse to recognize? In: Digital Signal Processing (DSP), pp. 1246–1250 (2015)
26. Le Ngo, A.C., See, J., Phan, R.C.W.: Sparsity in dynamics of spontaneous subtle emotions: analysis and application. Trans. Affect. Comput. (2016)
27. Wang, Y., See, J., Phan, R.C.-W., Oh, Y.-H.: LBP with six intersection points: reducing redundant information in LBP-TOP for micro-expression recognition. In: Cremers, D., Reid, I., Saito, H., Yang, M.-H. (eds.) ACCV 2014. LNCS, vol. 9003, pp. 525–537. Springer, Cham (2015). doi:10.1007/978-3-319-16865-4_34
28. Liong, S.T., Phan, R.C.W., See, J., Oh, Y.H., Wong, K.: Optical strain based recognition of subtle emotions. In: International Symposium on Intelligent Signal Processing and Communication Systems, pp. 180–184 (2014)
29. Oh, Y.H., Le Ngo, A.C., See, J., Liong, S.T., Phan, R.C.W., Ling, H.C.: Monogenic Riesz wavelet representation for micro-expression recognition. In: Digital Signal Processing, pp. 1237–1241. IEEE (2015)
30. Huang, X., Wang, S.J., Zhao, G., Pietikainen, M.: Facial micro-expression recognition using spatiotemporal local binary pattern with integral projection. In: International Conference on Computer Vision Workshops, pp. 1–9 (2015)
31. Xu, F., Zhang, J., Wang, J.: Microexpression identification and categorization using a facial dynamics map. Trans. Affect. Comput. (2016)

Failure Detection for Facial Landmark Detectors

Andreas Steger and Radu Timofte[(⊠)]

Computer Vision Laboratory, D-ITET, ETH Zurich, Zürich, Switzerland
stegeran@ethz.ch, radu.timofte@vision.ee.ethz.ch

Abstract. Most face applications depend heavily on the accuracy of the face and facial landmarks detectors employed. Prediction of attributes such as gender, age, and identity usually completely fail when the faces are badly aligned due to inaccurate facial landmark detection. Despite the impressive recent advances in face and facial landmark detection, little study is on the recovery from and the detection of failures or inaccurate predictions. In this work we study two top recent facial landmark detectors and devise confidence models for their outputs. We validate our failure detection approaches on standard benchmarks (AFLW, HELEN) and correctly identify more than 40% of the failures in the outputs of the landmark detectors. Moreover, with our failure detection we can achieve a 12% error reduction on a gender estimation application at the cost of a small increase in computation.

1 Introduction

Face allows a non-invasive assessment of the human identity and personality. By looking to a face one can determine a large number of attributes such as identity of a person, biometrics (such as age, ethnicity, gender), facial expression, and various other facial features such as wearing lipstick, eyeglasses, piercings, etc. Therefore, in computer vision the automatic detection of faces from images and the estimation of such attributes are core tasks.

The past decades have shown tremendous advances in the field of object detection. More than a decade ago, Viola and Jones [1] were among the first to real-time accurately detect faces using cascaded detectors. Subsequent research led to large performance improvements for detection of a broad range of object classes [2,3], including the articulated and challenging classes such as pedestrian [4]. Nowadays, the solutions to vision tasks are expected to be both highly accurate and with low time complexity.

For face attribute prediction the common pipeline is to first detect the face in the image space, then to detect facial landmarks or align a face model, and finally model such landmarks to predict facial attributes. Each step is critical. Inaccurate face localization can propagate to erroneous detection of facial landmarks as the landmark detectors assume the presence of a full face in the input image region. Also, the accuracy of the facial landmarks strongly affects the performance of the attribute predictor as it relies on features extracted from (aligned and normalized) facial landmark regions.

© Springer International Publishing AG 2017
C.-S. Chen et al. (Eds.): ACCV 2016 Workshops, Part II, LNCS 10117, pp. 361–376, 2017.
DOI: 10.1007/978-3-319-54427-4_27

In this paper we study the detection of failures for facial landmark detectors and propose solutions at individual facial landmark level and at the whole face level. Detection of failures allows for recovery by using more robust but computationally demanding face detectors and/or alignment methods such that to reduce the errors of the subsequent processing steps (such as attribute predictors).

Our main contributions are:

1. We are the first to study failure detection in facial landmark detectors, to the best of our knowledge;
2. We propose failure detection methods for individual and groups of facial landmarks;
3. We achieve significant improvements on a gender prediction application with recovery from facial landmark failures at only small increase in computational time.

1.1   Related Work

Despite the dramatic propagation of errors in the forward face processing pipeline very few works focus on failure detection such that to recover and improve the performance at the end of the pipeline.

Most literature on failure detection concerns video and image sequence processing usually in the context of visual tracking where inaccurate localizations in each frame can lead to drifting and loose of track. One common model for increasing the stability and diminishing the drift or error accumulation is the forward-backward two-way checkup that effectively applies the tracking method twice in both directions from one frame to another to get the stable and reliable track [5,6]. On single image processing the failure detection approaches from visual tracking are generally not applicable as they employ temporal information.

Usually, in face detection and facial landmark literature the reduction of failures is the direct result of trading off the time complexity/running time of the methods. More complex models and intensive computations might allow for more robust performance. However, the costs for such reductions within the original models can be prohibitive for practical applications. For example, one might use a set of specialized detectors (or components) deployed together instead of a generic detector, and while the performance potentially can be improved, the time and memory complexities varies with the cardinality of the set [7–9].

The remainder of the paper is structured as follows. Section 2 briefly describes the datasets and facial landmark detectors from the experimental setup of our study. Section 3 studies the detection of failures for landmark detectors and introduces our methods. Section 4 validates our failure detection on gender estimation, a typical facial attribute prediction application. Section 5 concludes the paper.

2   Experimental Setup

Before proceeding with our study we first introduce the experimental setup, that is the two fast and effective recent facial landmark detectors we work with

(Uricar [9] and Kazemi [10]) and the two of the most used recent datasets of face images with annotated facial landmarks (AFLW [11] and HELEN [12]).

2.1 Facial Landmark Detectors

In order to validate our findings we use two fast and effective landmark detectors that are publicly available:

Uricar Detector. The real-time multi-view facial landmark detector learned by the Structured Output SVM of Uricar *et al.* [9] is the main facial landmark detector employed in our study. It uses the SO-SVM [13] to learn the parameters of a Deformable Part Model [2]. We directly use the authors' implementation "uricamic/clandmark" from github[1] and the joint learned models for the front view which provides us 21 landmarks as shown in Fig. 2. In order to have a comparable result to Kazemi and Sullivan [10] we also applied the dlib face detector. After using the same transformation method as shown in Fig. 3 we obtain the normalized landmarks as depicted in Fig. 2(b).

Kazemi Detector. To validate the generality of our study and derived failure detection methods we also use the highly efficient Ensemble of Regression Trees (ERT) algorithm of Kazemi and Sullivan [10]. ERT can run in a millisecond

a)   b)   c)   d)

Fig. 1. The images (a) and (c) show examples for the original annotations from AFLW [11] and HELEN [12]. Their normalized form which was used for all further computations is depicted in (b) and (d).

a)   b)   c)   d)

Fig. 2. Landmark detections on an AFLW [11] image. The original output from Uricar [9] and Kazemi Detector [10] are shown in (a) and (c). Their normalized form which was used for all further computations is depicted in (b) and (d).

---

[1] https://github.com/uricamic/clandmark.

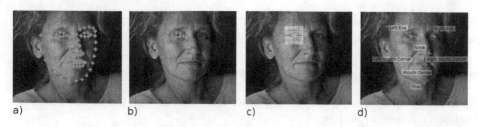

Fig. 3. Image (a) shows the landmarks predicted by Kazemi Detector [10], (b) shows an example for a landmark group, (c) shows the average of the group of landmarks from the left eye and the corresponding grid for features extraction, (d) shows the predicted landmarks marked with red + and the ground truth landmarks marked with green x and their names. (Color figure online)

for accurate face alignment results. It uses gradient boosting for learning an ensemble of regression trees [14]. As in [10] we use the HELEN dataset for our experiments and the authors' dlib implementation[2] that provides 68 landmarks. The detected landmarks are depicted in Fig. 3(a). We group them as shown for the left eye in (b) and then take the mean of each group as coordinate for that landmark (c). The picture in (d) shows the names of the used landmarks. We neglected the ears, because there are too few annotations (Sect. 2.2).

2.2   Datasets

We use two popular facial landmark datasets: AFLW [11] and HELEN [12].

AFLW [11]. It consists of 21k images with 24k annotated faces. It comes with a database that contains information about each face. The landmarks that we used from this database are: Left Eye Center, Right Eye Center, Nose Center, Mouth Left Corner, Mouth Right Corner, Mouth Center, Chin Center. From those we chose 5.5k faces that are aligned with pitch and yaw smaller than 15°. We used a test and validation set, each of 10%, while the remaining 80% was used for training. We did not use the ear landmarks, because about 40% of them are missing. The miss rate for the chosen landmarks is below 3%. Figure 1 shows an example for the original annotations of an AFLW image (a) and the annotations that we picked (b).

HELEN [12]. It consists of 2330 images which are splitted into a training set of 2000 images and a test set of 330 images. Moreover we split a validation set off the original training set with size of 10% of the total number of images. Each image is annotated with 194 points. From those points we extracted the same landmarks as from AFLW. That extraction was done by choosing a group of annotations and taking the mean position like introduced in Sect. 2.1. An example for that is shown in Fig. 1(c) and (d). This method did not work well for the chin and the mouth center because annotation points with the same index are skewed for faces that are not in an aligned frontal view.

---

[2] http://www.csc.kth.se/~vahidk/face_ert.html.

2.3   Train, Test and Validation Set

When reporting results on data that were already used for training the model or choosing parameters, then it would be overoptimistic. Therefore, we split both datasets into training, validation and test set. Their sizes are listed in Sect. 2.2. We trained the models for landmark confidence and gender prediction only on the training set. The model parameters were chosen in a 5 fold cross validation implemented in scikit-learn [15]. To choose the feature parameters, combination of features and combination of landmarks, we use the reported TrueCorrect95 on the validation set. The threshold for the predicted Confidence C is tuned on a disjunct set on the validation set. The final result in the application section is then reported on the test set.

2.4   Measures

Mean Absolute Error (MAE). To measure the (pixel) accuracy of n predicted landmark positions $\hat{Y}$ we calculate the standard mean absolute error (MAE) between the predictions and the known ground truth positions Y:

$$MAE = \frac{1}{n} \sum_{i=1}^{n} \sqrt{(\hat{Y}_{x,i} - Y_{x,i})^2 + (\hat{Y}_{y,i} - Y_{y,i})^2} \qquad (1)$$

The errors are generally small but there are some outliers and MAE is very sensitive with them. The following measure addresses this sensitivity.

Confidence (C). We need a transformation of the error in a way that small deviations of the predicted landmark from the ground truth have the highest dynamics. For this we use the Gaussian as depicted in Fig. 4, where $C \in (0, 1]$ and σ is the standard deviation:

$$C = e^{-\frac{(\hat{Y}-Y)^2}{2\sigma^2}} \qquad (2)$$

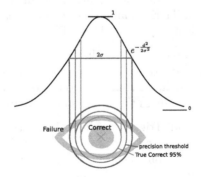

Fig. 4. Dependency of Confidence (C) from the distance of ground truth (green X). The standard deviation σ determines the width of the fitting function (blue), ground truth precision threshold (purple) and TrueCorrect95 (red). (Color figure online)

Threshold (T). Considering C measure for a fixed standard deviation σ we can threshold it to obtain the plots from our experiments. A zero error level, corresponding to perfectly aligned landmarks, get the highest confidence $C = 1$, while the failures get C values closer to 0. The ground truth precision threshold in Fig. 4 (purple) classifies the ground truth landmark confidence into correct and failure.

Coefficient of Determination (R^2). For training all models we used the Coefficient of determination also called R^2.

$$R^2(y, \hat{y}) = 1 - \frac{\sum_{i=1}^{n}(y_i - \hat{y}_i)^2}{\sum_{i=1}^{n}(y_i - \overline{y})^2} \tag{3}$$

TrueCorrect95. Rate of failed landmarks that are detected as a failure at an operating point where the rate of correct landmarks that are detected as correct is above 95%. It determines the threshold for the predicted confidence that is depicted in Fig. 4 (red). It classifies the detected landmarks into detected as correct and detected as failure. When reporting the TrueCorrect95 we split 20% off the set on which we report to tune the threshold for the predicted Confidence C.

Runtime. All runtimes were measured on an Intel Core i7-5500U @ 2.4 GHz processor when using 2 threads for calculation.

3 Failure Detection

3.1 Facial Features

Before extracting the features the face is rotated and cropped according to the predicted eyes annotations. We use a patch size of 128×128 pixels for the face plus a border of 48 pixels in each direction to be able to apply descriptors also near the edge. In order to avoid the trade off between descriptiveness and speed that CNN features have, we used handcrafted features instead. Therefore we used a combination of SIFT, HoG and LBP in order to get very descriptive features that can be efficiently calculated on CPUs.

PCA. For all descriptors listed below that have a feature vector larger than 1500, we apply a PCA that is learned from the training data to reduce the runtime and memory usage of the predictor training. To do so we use the scikit-learn [15] PCA implementation that reduces the dimensionality to 1500.

HoG. Navneet Dalal and Bill Triggs [16] showed that the combination of a grid of histogram of oriented gradients descriptors and an SVM can outperform other feature sets for human detection. Inspired by that we use the implementation of HoG form scikit-image to apply the same principle to a patch around the landmarks. We tuned the parameters patch size $\in \{\frac{1}{8}, \frac{2}{8}, \frac{3}{8}, \frac{4}{8}\}$ of the face, number of cells in the patch $\in \{1, 2^2, 4^2, 8^2\}$ and number of orientations $\in \{4, 8\}$. The best result for each Landmark can be seen in Tables 1 and 2.

LBP. Local Binary Pattern calculates the number of neighbor pixels that have a greater gray scale value than the center. We used the same setup as for HoG with a grid of cells for which we calculated the LBP for all pixels. For each of those cells we calculated the histogram and used that as features. We tuned the parameters patch size $\in \{\frac{1}{8}, \frac{2}{8}, \frac{3}{8}, \frac{4}{8}\}$, number of cells in the patch $\in \{1, 2^2, 4^2, 8^2\}$ and the radius $\in \{1, 2, 3, 4\}$ in which the LBP is calculated. For that we used the scikit-image library function "local_binary_pattern" with the method parameter "uniform" [15]. The best result for each Landmark can be seen in Tables 1 and 2.

SIFT. The Scale-invariant feature transform is based on a grid of HoG. It is largely invariant to illumination and local affine distortions [17]. For SIFT we used the OpenCV implementation and applied the same method with the grid as for HoG and LBP. We tuned the parameters patch size $\in \{\frac{1}{8}, \frac{2}{8}, \frac{3}{8}, \frac{4}{8}\}$ and number of cells in the patch $\in \{1, 2, 4, 8\}$. The best result for each Landmark can be seen in Tables 1 and 2.

Table 1. Validation results on AFLW for SIFT, HoG and LBP using different parameters. 'O' stands for Orientations and 'TC95' for TrueCorrect95.

Landmark	SIFT				HoG					LBP				
	Size	Cells	R^2	TC95	O	Size	Cells	R^2	TC95	Radius	Size	Cells	R^2	TC95
chinC	2.5	4^2	−0.74	0.40	4	3.8	8^2	−1.95	0.35	3	3.8	8^2	−6.6	0.30
eyeL	1.2	4^2	0.37	0.52	8	2.5	4^2	0.06	0.48	2	2.5	8^2	−2.4	**0.45**
eyeR	1.2	4^2	0.37	**0.60**	4	3.8	8^2	0.12	0.51	3	2.5	8^2	−2.1	0.43
mouthC	1.2	4^2	−0.25	0.57	8	3.8	4^2	−0.14	**0.56**	3	2.5	8^2	−4.8	0.34
mouthL	3.8	4^2	0.21	0.50	4	5.0	8^2	−0.24	0.46	4	3.8	2^2	−2.4	0.32
mouthR	2.5	4^2	0.13	0.42	8	1.2	2^2	−1.62	0.43	4	2.5	8^2	−3.5	0.41
noseC	1.2	2^2	0.20	0.55	8	2.5	1^2	−3.06	0.38	3	2.5	4^2	−2.0	0.34

Table 2. Validation results on HELEN for SIFT, HoG and LBP using different parameters. 'O' stands for Orientations and 'TC95' for TrueCorrect95.

Landmark	SIFT				HoG					LBP				
	Size	Cells	R^2	TC95	O	Size	Cells	R^2	TC95	Radius	Size	Cells	R^2	TC95
chinC	2.5	4^2	−1.08	0.37	4	5.0	8^2	−0.809	0.29	3	2.5	8^2	−4.24	0.27
eyeL	3.8	4^2	0.50	0.59	4	5.0	4^2	0.457	0.70	2	2.5	4^2	−0.54	0.40
eyeR	1.2	4^2	0.66	0.68	4	5.0	8^2	0.589	**0.77**	4	3.8	4^2	−0.83	**0.57**
mouthC	2.5	4^2	0.20	0.55	8	5.0	8^2	0.051	0.50	4	5.0	8^2	−1.86	0.31
mouthL	1.2	4^2	0.39	**0.77**	4	3.8	8^2	0.024	0.60	4	2.5	8^2	−1.89	0.27
mouthR	3.8	4^2	−0.10	0.51	8	3.8	4^2	−0.389	0.47	3	2.5	8^2	−1.63	0.27
noseC	2.5	4^2	0.67	0.74	8	5.0	4^2	0.519	0.75	3	5.0	4^2	−0.70	0.44

3.2 Individual Confidence of Landmark Detectors

In this section we describe how we predict the R^2 score that is explained in Sect. 2.4 by using the features from Sect. 3.1 for a single landmark at a time.

As training data we used the annotations from AFLW [11] and HELEN [12] as described in Sect. 2.2. In order to train a robust predictor we generate 5 different annotations per face from the data set. We introduced a Gaussian error in distance to the actual position with a standard deviation of 10% of the face size. That corresponds to 13 pixels of the 128 pixel face patch side length. The angle was chosen uniformly randomly.

We use the Support Vector Regression (SVR) as predictor for the confidence and therefore have to tune the C and ϵ parameter, as well as the kernel type. To do so we used a 5 fold cross validation. The parameters that we used in the exhaustive search are $C \in \{0.3, 0.5, 0.7\}$, $\epsilon \in \{0.01, 0.05, 0.1\}$, kernel $\in \{$Radial Basis Function, Linear$\}$. We chose the scikit-learn [15] implementation for the exhaustive search, cross validation and the SVR. The SVR is based on libsvm [18].

Feature Parameters Search. First we extracted all feature descriptors for each parameter configuration and each landmark as described in Sect. 3.1. Each of those features were then used to train the SVR in a exhaustive search for the best parameters. The results which are reported on the validation set can be seen in Tables 1 and 2. According to the TrueCorrect95 scores for the AFLW dataset SIFT is the best or almost the best feature descriptor for all landmarks. For HELEN this is the case for five of seven landmarks. The size of the training and validation set was 2300 and 480 respectively.

Feature Combination Search. Secondly we searched over all combinations of different descriptors for each landmark. For both datasets we used about 2300 samples for training. For each sample the feature extraction for the best combination of features and all landmarks takes 107 ms (Sect. 2.4). The result which is reported on the validation set can be seen in Table 3. It shows that in 86% of the cases SIFT is contained in the best feature combination.

Several parameters influence the failure detection accuracy measurement. First the standard deviation that is used to fit the confidence C (Sect. 2.4) for both the ground truth MAE and the predicted MAE. This was set to 10% of the face size. Secondly the ground truth precision threshold, which determines the ground truth correct and failures. It was set to 0.65 at which both the predicted landmarks form the Uricar [9] and Kazemi [10] Detector have an error rate that is not too unbalanced. That confidence C corresponds to a distance error of about 9.2%. Figure 5 reports the resulting error rate on a range from 0% to 40% of the face size. Thirdly the prediction precision threshold has to be set to an operating point where the rate of correctly marked as correct is above 95%. The tuning and reporting is done on disjunct subsets of the validation set. The tuning for the fixed ground truth precision threshold of 0.65 is shown in Fig. 7. The operating point that is then used to calculate the TrueCorrect95 is marked in those plots.

Table 3. Results for best two Feature Descriptor combinations for each Landmark individually on the AFLW Dataset.

Landmark	AFLW			HELEN		
	Descriptors	R^2	TrueCorrect95	Descriptors	R^2	TrueCorrect95
chinC	hog, sift	−0.54	0.35	sift	−0.60	0.26
chinC	sift	−0.48	0.33	hog	−0.89	0.25
eyeL	sift	0.43	**0.75**	hog, lbp	0.39	0.63
eyeL	hog, sift	0.44	0.73	hog, sift, lbp	0.54	0.59
eyeR	hog, sift	0.38	0.62	hog, lbp	0.54	0.75
eyeR	sift	0.36	0.55	hog, sift	0.72	0.74
mouthC	sift	−0.05	0.50	sift, lbp	−0.18	0.55
mouthC	hog, sift	−0.01	0.49	hog, sift	0.29	0.54
mouthL	hog, sift, lbp	0.27	0.61	hog, sift	0.37	0.59
mouthL	sift	0.26	0.59	sift	0.35	0.54
mouthR	sift	0.15	0.57	sift	0.06	0.45
mouthR	hog, sift	0.16	0.56	hog, sift	0.07	0.44
noseC	sift, lbp	0.27	0.59	hog, sift, lbp	0.68	**0.80**
noseC	hog, sift, lbp	0.27	0.59	hog, sift	0.68	0.79

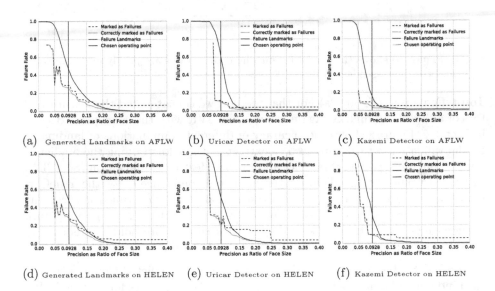

(a) Generated Landmarks on AFLW   (b) Uricar Detector on AFLW   (c) Kazemi Detector on AFLW

(d) Generated Landmarks on HELEN   (e) Uricar Detector on HELEN   (f) Kazemi Detector on HELEN

Fig. 5. The failure rate is the result of tuning the threshold to a value where the rate of correct landmarks that are marked as correct is above 95%. All plots were generated with the model that is described in this section and show the combination of nose, left mouth corner and left eye.

3.3   Joint Confidence of Landmark Detectors

In order to increase the accuracy of failure detection, we used different combinations of facial landmarks to train the predictor. The two setups that we used are shown in Fig. 6. As training data we used again the annotations from AFLW [11] and HELEN [12] as described in Sect. 2.2 and generate landmarks similarly as in Sect. 3.2. The difference is that we have to model a predicted face annotation that is both shifted overall and individually for each landmark compared to the ground truth. So we use the superposition of two errors. One is generated for each landmark with a standard deviation of 7% of the face size. The other is generated once for the whole face with a standard deviation of 10% of the face size. Both have a uniformly randomly generated angle. The total error is the superposition of both. The ground truth confidence C was calculated from the MAE of the generated landmarks with a precision of 10% of the face size.

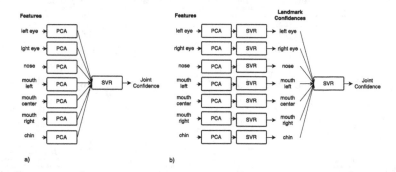

Fig. 6. Figure (a) shows the configuration where all features are jointly feeded into the SVR and (b) the Cascaded SVR.

Joint Features. (Figure 6a) For that method we concatenated the feature vectors of different landmarks. Again we used the SVR with the same properties as in Sect. 3.2. Figure 8 shows the accuracy of different combinations of landmarks on the validation set that are used to extract features for training the predictor. For training we used 10000 samples and the best feature combination (Sect. 3.2). In order to tune the threshold for the predicted confidence C for the TrueCorrect95 we use a disjunct partition of the validation set and tuned it according to Fig. 7. Figure 5 shows the result for different thresholds for the ground truth confidence C. The vertical tick shows the prior chosen threshold that is used to be able to compare Uricar [9] and Kazemi [10] Detector prediction.

Cascaded SVRs. In our second setup we used the predicted Confidence C of each individual landmark from the SVR (Sect. 3.2) as features for a second SVR (Fig. 6b). Therefore we predicted the confidence C on the test and validation set that we chose from AFLW and HELEN (Sect. 2.2). The second SVR was then trained on the validation set because the errors of the predicted confidence C from the training set would be overoptimistic. For the training the same parameters as in Sect. 3.2 and 3200 samples were used.

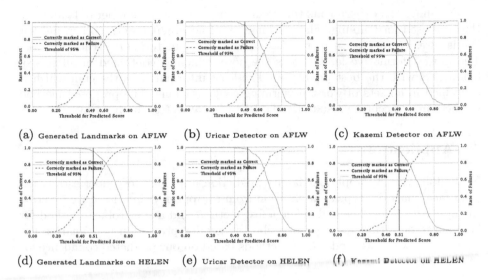

(a) Generated Landmarks on AFLW   (b) Uricar Detector on AFLW   (c) Kazemi Detector on AFLW

(d) Generated Landmarks on HELEN   (e) Uricar Detector on HELEN   (f) Kazemi Detector on HELEN

Fig. 7. The failure rate is shown at fixed ground truth precision threshold of 0.65 for different prediction thresholds. All plots were generated with the same model that is described in this section and show the combination of nose, left mouth corner and left eye.

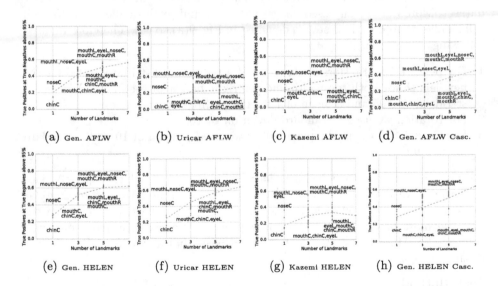

(a) Gen. AFLW   (b) Uricar AFLW   (c) Kazemi AFLW   (d) Gen. AFLW Casc.

(e) Gen. HELEN   (f) Uricar HELEN   (g) Kazemi HELEN   (h) Gen. HELEN Casc.

Fig. 8. Rate of Failures that are correctly marked as failures at an operating point where the rate of Non Failures that are correctly marked as Non Failures is above 95% for different combinations of landmarks. The dashed line represents the mean for each number of landmarks. Those with caption "Casc." were computed with the Cascaded SVR model the others with the Joint SVR. "Gen." stands for Generated Landmarks.

For almost all experiments the accuracy increases with number of landmarks that are used for training. The combination of nose, left mouth corner and left eye is already very close to the best combination in terms of TrueCorrect95. The most informative landmark is the nose, which is contained in all sets of landmarks with the best prediction. The least informative is the chin which is contained in all sets of worst performing configurations. The proposed method predicts more accurate results for the HELEN dataset than on AFLW.

4   Application

In the following we apply our failure prediction approach to gender estimation from face images, a typical application from computer vision. First we describe the application pipeline and define a fast method and a slow robust method for landmarks detection, then we describe the gender prediction and analyze the trade off between fast and robust methods based on our predicted confidence for the landmark detections.

4.1   Pipeline

Our pipeline of how we predict the gender of a person is depicted in Fig. 9. In order to make the prediction more robust we distorted the images from AFLW [11]. In this step we generated 5 new images from one in the AFLW dataset. Moreover we only used images with yaw and pitch smaller than 15° and roll smaller than 60°. This step is depicted in Fig. 9(a) to (b). Therefore we used following distortions:

1. For perspective transformation we used a Gaussian distributed ratio for the new length for both top width and the right height of the image. We used a standard of 0.05 and a mean of 1 for the scaling factor.
2. For rotation we used a standard deviation of 10.
3. For Gaussian noise in the pixel values we used a mean of 10 and a standard deviation of 5.

The next step is to detect the face. To do so we used the openCV face detector with the Haar feature-based cascade model that returns a rectangular in which the face is detected and a score for the goodness of the detection. Because most faces are already rotated properly we penalize detections with large angles. Moreover we filter detections of faces that are very small compared to the image size.

Fast Method. In order to detect the faces we first rotate each image by 45° from −90 to 90°. The runtime per image is 2.67s. Then we apply the face detection to each image and choose the highest scored.

The Uricar Detector [9] is then used to predict the landmarks of the detected face. In that step we apply our failure detection that was train for a combination for all landmarks according to Sect. 3.3. The face detection for this method takes

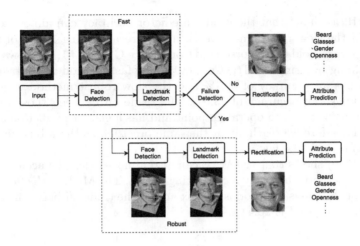

Fig. 9. Pipeline of handling failures for gender estimation.

147 ms per image. For all faces with landmarks that were predicted as being wrong we use the slow robust method to predict new landmarks.

Slow Robust Method. For the robust method we fall back to the face detection from openCV. To get a better aligned image we rotate it before the detection in steps of 5° from −90 to 90°. The runtime per image is 20.1 s. On this new aligned face detection that we got from the highest scored image, we apply the landmark detector.

4.2   Gender Prediction

From the landmarks obtained by either the fast or slow robust method we extract the features for the gender prediction. For that we calculate a grid of SIFT and LBP descriptors of the landmarks eye left, eye right, nose center, mouth left, mouth center, mouth right and chin center as described in Fig. 3(d). Moreover a Grid of SIFT descriptors over the whole face. As classifier we use the SVC from scikit-learn [15]. We use 5 fold cross-validation to determine the best parameter from a exhaustive search over $C \in \{0.1, 0.3, 0.5, 0.7\}$ and the kernels $\in$ {linear, poly (with degree $= 3$), RBF and sigmoid}.

4.3   Trade Off Between Fast and Robust

Figure 10 shows the trade off between the fast and robust method over a range of thresholds for the predicted confidence. All faces that were detected as failed detection are recalculated with the robust method. The threshold for the ground truth confidence was fixed to 0.65 and the Confidence C was calculated with a standard deviation of 10%. The plots are reported on the test set and all parameters were tuned on a disjunct set.

Figure 10(a) shows that the gender predictor accuracy can almost reach the accuracy of the slow robust method when only recalculating 15.4% of the face images. The threshold for the predicted Confidence C is at 0.51, which was chosen during training by using the TrueCorrect95. The fast method which takes 2.92 s per image has an accuracy of 69.1% and the slow robust method which takes 20.3 s per image an accuracy of 73.6%. The trade off between those two takes 6.05 s per image at chosen operating point and has a gender prediction accuracy of 72.8%. Therefore we reach about the same precision as the robust method in 70% less time or 3.36× faster.

Figure 10(b) shows that the MAE can also almost reach the accuracy of the slow robust method when only recomputing 15.4%. The MAE of the fast method is 23.9 px, of the robust method is 19.8 px and of the trade off between those two at the operating point is 20.4 px.

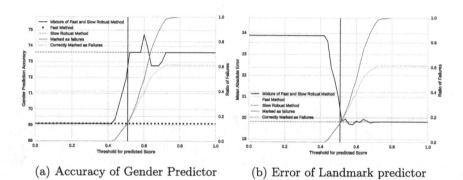

(a) Accuracy of Gender Predictor   (b) Error of Landmark predictor

Fig. 10. Trade off between fast method and slow robust method based on predicted Confidence C of each face for Uricar Detector on AFLW.

5   Conclusion

This paper proposed a novel method to predict the confidence of detected landmarks. To train the model we used the error in distance as measure for the confidence of a landmark. With this method it is possible to decide in a trade off between number of detected failures and the accuracy of detecting failures. Meaning that if we want to have less correct face detections detected as failed, we can use a higher confidence threshold to detect only those as failure that are more likely to be failed. To validate our results we used two popular face image data sets and two landmark detectors. To demonstrate the benefit of our method we showed that we can improve the speed of a gender predictor pipeline by adding a fast method. It falls back to a slow robust method if the face is detected as being failed. That gave us a speed up of ×3 while achieving almost the same accuracy as the robust method.

Acknowledgment. This work was supported by the EU Framework 7 project ReMeDi (# 610902) and by the ETH General Fund (OK).

References

1. Viola, P., Jones, M.: Rapid object detection using a boosted cascade of simple features. In: Proceedings of the 2001 IEEE Computer Society Conference on Computer Vision and Pattern Recognition, CVPR 2001, vol. 1, pp. I-511–I-518 (2001)
2. Felzenszwalb, P.F., Girshick, R.B., McAllester, D., Ramanan, D.: Object detection with discriminatively trained part-based models. IEEE Trans. Pattern Anal. Mach. Intell. **32**, 1627–1645 (2010)
3. Girshick, R., Donahue, J., Darrell, T., Malik, J.: Rich feature hierarchies for accurate object detection and semantic segmentation. In: The IEEE Conference on Computer Vision and Pattern Recognition (CVPR) (2014)
4. Benenson, R., Mathias, M., Timofte, R., Van Gool, L.: Pedestrian detection at 100 frames per second. In: 2012 IEEE Conference on Computer Vision and Pattern Recognition (CVPR), pp. 2903–2910 (2012)
5. Kalal, Z., Mikolajczyk, K., Matas, J.: Forward-backward error: automatic detection of tracking failures. In: 2010 20th International Conference on Pattern Recognition (ICPR), pp. 2756–2759 (2010)
6. Timofte, R., Kwon, J., Van Gool, L.: PICASO. pixel correspondences and soft match selection for real-time tracking. Comput. Vis. Image Underst. (CVIU) **153**, 151–162 (2016)
7. Zhu, X., Ramanan, D.: Face detection, pose estimation, and landmark localization in the wild. In: 2012 IEEE Conference on Computer Vision and Pattern Recognition (CVPR), pp. 2879–2886 (2012)
8. Mathias, M., Benenson, R., Pedersoli, M., Gool, L.: Face detection without bells and whistles. In: Fleet, D., Pajdla, T., Schiele, B., Tuytelaars, T. (eds.) ECCV 2014. LNCS, vol. 8692, pp. 720–735. Springer, Cham (2014). doi:10.1007/978-3-319-10593-2_47
9. Uricár, M., Franc, V., Thomas, D., Sugimoto, A., Hlavác, V.: Real-time multi-view facial landmark detector learned by the structured output SVM. In: 2015 11th IEEE International Conference and Workshops on Automatic Face and Gesture Recognition (FG), vol. 2, pp. 1–8. IEEE (2015)
10. Kazemi, V., Sullivan, J.: One millisecond face alignment with an ensemble of regression trees. In: Proceedings of the IEEE Conference on Computer Vision and Pattern Recognition, pp. 1867–1874 (2014)
11. Köstinger, M., Wohlhart, P., Roth, P.M., Bischof, H.: Annotated facial landmarks in the wild: a large-scale, real-world database for facial landmark localization. In: 2011 IEEE International Conference on Computer Vision Workshops (ICCV Workshops), pp. 2144–2151 (2011)
12. Le, V., Brandt, J., Lin, Z., Bourdev, L., Huang, T.S.: Interactive facial feature localization. In: Fitzgibbon, A., Lazebnik, S., Perona, P., Sato, Y., Schmid, C. (eds.) ECCV 2012. LNCS, vol. 7574, pp. 679–692. Springer, Heidelberg (2012). doi:10.1007/978-3-642-33712-3_49
13. Tsochantaridis, I., Joachims, T., Hofmann, T., Altun, Y.: Large margin methods for structured and interdependent output variables. J. Mach. Learn. Res. **6**, 1453–1484 (2005)
14. Criminisi, A., Shotton, J., Robertson, D., Konukoglu, E.: Regression forests for efficient anatomy detection and localization in CT studies. In: Menze, B., Langs, G., Tu, Z., Criminisi, A. (eds.) MCV 2010. LNCS, vol. 6533, pp. 106–117. Springer, Heidelberg (2011). doi:10.1007/978-3-642-18421-5_11

15. Pedregosa, F., Varoquaux, G., Gramfort, A., Michel, V., Thirion, B., Grisel, O., Blondel, M., Prettenhofer, P., Weiss, R., Dubourg, V., Vanderplas, J., Passos, A., Cournapeau, D., Brucher, M., Perrot, M., Duchesnay, E.: Scikit-learn: machine learning in Python. J. Mach. Learn. Res. **12**, 2825–2830 (2011)
16. Dalal, N., Triggs, B.: Histograms of oriented gradients for human detection. In: IEEE Computer Society Conference on Computer Vision and Pattern Recognition, CVPR 2005, vol. 1, pp. 886–893. IEEE (2005)
17. Lowe, D.G.: Object recognition from local scale-invariant features. In: The Proceedings of the Seventh IEEE International Conference on Computer Vision, vol. 2, pp. 1150–1157. IEEE (1999)
18. Chang, C.C., Lin, C.J.: LIBSVM: a library for support vector machines. ACM Trans. Intell. Syst. Technol. **2**, 1–27 (2011). Software available at http://www.csie.ntu.edu.tw/~cjlin/libsvm

Fitting a 3D Morphable Model to Edges: A Comparison Between Hard and Soft Correspondences

Anil Bas[1]([✉]), William A.P. Smith[1], Timo Bolkart[2], and Stefanie Wuhrer[3]

[1] Department of Computer Science, University of York, York, UK
{ab1792,william.smith}@york.ac.uk
[2] Multimodal Computing and Interaction, Saarland University,
Saarbrücken, Germany
tbolkart@mmci.uni-saarland.de
[3] Morpheo Team, Inria Grenoble Rhône-Alpes, Grenoble, France
stefanie.wuhrer@inria.fr

Abstract. In this paper we explore the problem of fitting a 3D morphable model to single face images using only sparse geometric features (edges and landmark points). Previous approaches to this problem are based on nonlinear optimisation of an edge-derived cost that can be viewed as forming soft correspondences between model and image edges. We propose a novel approach, that explicitly computes hard correspondences. The resulting objective function is non-convex but we show that a good initialisation can be obtained efficiently using alternating linear least squares in a manner similar to the iterated closest point algorithm. We present experimental results on both synthetic and real images and show that our approach outperforms methods that use soft correspondence and other recent methods that rely solely on geometric features.

1 Introduction

Estimating 3D face shape from one or more 2D images is a longstanding problem in computer vision. It has a wide range of applications from pose-invariant face recognition [1] to creation of 3D avatars from 2D images [2]. One of the most successful approaches to this problem is to use a statistical model of 3D face shape [3]. This transforms the problem of shape estimation to one of model fitting and provides a strong statistical prior to constrain the problem.

The model fitting objective can be formulated in various ways, the most obvious being an analysis-by-synthesis approach in which appearance error is directly optimised [3]. However, feature-based methods [4,5] are in general more robust and lead to optimisation problems less prone to convergence on local minima. In this paper, we focus on fitting to edge features in images.

Image edges convey important information about a face. The occluding boundary provides direct information about 3D shape, for example a profile view reveals strong information about the shape of the nose. Internal edges,

© Springer International Publishing AG 2017
C.-S. Chen et al. (Eds.): ACCV 2016 Workshops, Part II, LNCS 10117, pp. 377–391, 2017.
DOI: 10.1007/978-3-319-54427-4_28

caused by texture changes, high curvature or self occlusion, provide information about the position and shape of features such as lips, eyebrows and the nose. This information provides a cue for estimating 3D face shape from 2D images or, more generally, for fitting face models to images.

In Sect. 2 we introduce relevant background. In Sect. 3 we present a method for fitting to landmarks with known model correspondence. Our key contribution is in Sect. 4 where we present a novel, fully automatic algorithm for fitting to image edges with hard correspondence. By hard correspondence, we mean that an explicit correspondence is computed between projected model vertex and edge pixel. For comparison, in Sect. 5 we describe our variant of previous methods [4,6,7] that fit to edges using soft correspondence. By soft correspondence, we mean that an energy term that captures many possible edge correspondences is minimised. Finally, we compare the two approaches experimentally and others from the recent literature in Sect. 6.

1.1   Related Work

Landmark Fitting. 2D landmarks have long been used as a way to initialize a morphable model fit [3]. Breuer et al. [8] obtained this initialisation using a landmark detector providing a fully automatic system. More recently, landmarks have been shown to be sufficient for obtaining useful shape estimates in their own right [9]. Furthermore, noisily detected landmarks can be filtered using a model [10] and automatic landmark detection can be integrated into a fitting algorithm [11]. In a similar manner to landmarks, local features can be used to aid the fitting process [5].

Edge Fitting. An early example of using image edges for face model fitting is the Active Shape Model (ASM) [12] where a 2D boundary model is aligned to image edges. In 3D, contours have been used directly for 3D face shape estimation [13] and indirectly as a feature for fitting a 3DMM. The earliest work in this direction was due to Moghaddam et al. [14] who fitted a 3DMM to silhouettes extracted from multiple views. From a theoretical standpoint, Lüthi et al. [15] explored to what degree face shape is constrained when contours are fixed.

Romdhani et al. [4] include an edge distance cost as part of a hybrid energy function. Texture and outer (silhouette) contours are used in a similar way to LM-ICP [16] where correspondence between image edges and model contours is "soft". This is achieved by applying a distance transform to an edge image. This provides a smoothly varying cost surface whose value at a pixel indicates the distance (and its gradient, the direction) to the closest edge. This idea was extended by Amberg et al. [6] who use it in a multi-view setting and smooth the edge distance cost by averaging results with different parameters. In this way, the cost surface also encodes the saliency of an edge. Keller et al. [7] showed that such approaches lead to a cost function that is neither continuous nor differentiable. This suggests the optimisation method must be carefully chosen.

Edge features have also been used in other ways. Cashman and Fitzgibbon [17] learn a 3DMM from 2D images by fitting to silhouettes. Zhu et al. [18]

present a method that can be seen as a hybrid of landmark and edge fitting. Landmarks that define boundaries are allowed to slide over the 3D face surface during fitting. A recent alternative to optimisation-based approaches is to learn a regressor from extracted face contours to 3DMM shape parameters [19].

Fitting a 3DMM to a 2D image using only geometric features (i.e. landmarks and edges) is essentially a non-rigid alignment problem. Surprisingly, the idea of employing an iterated closest point [20] approach with hard edge correspondences (in a similar manner to ASM fitting) has been discounted in the literature [4]. In this paper, we pursue this idea and develop an iterative 3DMM fitting algorithm that is fully automatic, simple and efficient (and we make our implementation available[1]). Instead of working in a transformed distance-to-edge space and treating correspondences as "soft", we compute an explicit correspondence between model and image edges. This allows us to treat the model edge vertices as a landmark with known 2D position, for which optimal pose or shape estimates can be easily computed.

State of the Art. The most recent face shape estimation methods are able to obtain considerably higher quality results than the purely model-based approaches above. They do so by using pixel-wise shading or motion information to apply finescale refinement to an initial shape estimate. For example, Suwajanakorn et al. [21] use photo collections to build an average model of an individual which is then fitted to a video and finescale detail added by optical flow and shape-from-shading. Cao et al. [22] take a machine learning approach and train a regressor that predicts high resolution shape detail from local appearance.

Our aim in this paper is not to compete directly with these methods. Rather, we seek to understand what quality of face reconstruction it is possible to obtain using solely sparse, geometric information. The output of our method may provide a better initialisation for state of the art refinement techniques or remove the need to have a person specific model.

2   Preliminaries

Our approach is based on fitting a 3DMM to face images under the assumption of a scaled orthographic projection. Hence, we begin by introducing scaled orthographic projection and 3DMMs.

2.1   Scaled Orthographic Projection

The scaled orthographic, or weak perspective, projection model assumes that variation in depth over the object is small relative to the mean distance from camera to object. Under this assumption, the projected 2D position of a 3D point $\mathbf{v} = [u\ v\ w]^{\mathrm{T}}$ given by $\mathbf{SOP}[\mathbf{v}, \mathbf{R}, \mathbf{t}, s] \in \mathbb{R}^2$ does not depend on the distance of the point from the camera, but only on a uniform scale s given by the ratio of the focal length of the camera and the mean distance from camera to object:

---

[1] Matlab implementation: github.com/waps101/3DMM_edges.

$$\mathbf{SOP}[\mathbf{v}, \mathbf{R}, \mathbf{t}, s] = s \begin{bmatrix} 1 & 0 & 0 \\ 0 & 1 & 0 \end{bmatrix} \mathbf{R}\mathbf{v} + s\mathbf{t} \qquad (1)$$

where the pose parameters $\mathbf{R} \in \mathbb{R}^{3 \times 3}$, $\mathbf{t} \in \mathbb{R}^2$ and $s \in \mathbb{R}^+$ are a rotation matrix, 2D translation and scale respectively.

2.2 3D Morphable Model

A 3D morphable model is a deformable mesh whose shape is determined by the shape parameters $\boldsymbol{\alpha} \in \mathbb{R}^S$. Shape is described by a linear model learnt from data using Principal Components Analysis (PCA). So, the shape of any face can be approximated as:

$$\mathbf{f}(\boldsymbol{\alpha}) = \mathbf{P}\boldsymbol{\alpha} + \bar{\mathbf{f}}, \qquad (2)$$

where $\mathbf{P} \in \mathbb{R}^{3N \times S}$ contains the S principal components, $\bar{\mathbf{f}} \in \mathbb{R}^{3N}$ is the mean shape and the vector $\mathbf{f}(\boldsymbol{\alpha}) \in \mathbb{R}^{3N}$ contains the coordinates of the N vertices, stacked to form a long vector: $\mathbf{f} = [u_1 \ v_1 \ w_1 \ \dots \ u_N \ v_N \ w_N]^{\mathrm{T}}$. Hence, the ith vertex is given by: $\mathbf{v}_i = [f_{3i-2} \ f_{3i-1} \ f_{3i}]^{\mathrm{T}}$. For convenience, we denote the sub-matrix corresponding to the ith vertex as $\mathbf{P}_i \in \mathbb{R}^{3 \times S}$ and the corresponding vertex in the mean face shape as $\bar{\mathbf{f}}_i \in \mathbb{R}^3$, such that the ith vertex is given by: $\mathbf{v}_i = \mathbf{P}_i\boldsymbol{\alpha} + \bar{\mathbf{f}}_i$. Similarly, we define the row corresponding to the u component of the ith vertex as $\mathbf{P}_{iu}$ (similarly for v and w) and define the u component of the ith mean shape vertex as $\bar{f}_{iu}$ (similarly for v and w).

3 Fitting with Known Correspondence

We begin by showing how to fit a morphable model to L observed 2D positions $\mathbf{x}_i = [x_i \ y_i]^{\mathrm{T}}$ $(i = 1 \dots L)$ arising from the projection of corresponding vertices in the morphable model. We discuss in Sect. 4 how these correspondences are obtained in practice. Without loss of generality, we assume that the ith 2D position corresponds to the ith vertex in the morphable model. The objective of fitting a morphable model to these observations is to obtain the shape and pose parameters that minimise the reprojection error, E_{lmk}, between observed and predicted 2D positions:

$$E_{\mathrm{lmk}}(\boldsymbol{\alpha}, \mathbf{R}, \mathbf{t}, s) = \frac{1}{L} \sum_{i=1}^{L} \| \mathbf{x}_i - \mathbf{SOP}\left[\mathbf{P}_i\boldsymbol{\alpha} + \bar{\mathbf{f}}_i, \mathbf{R}, \mathbf{t}, s\right] \|^2. \qquad (3)$$

The scale factor in front of the summation makes the magnitude of the error invariant to the number of landmarks. This problem is multilinear in the shape parameters and the SOP transformation matrix. It is also nonlinearly constrained, since $\mathbf{R}$ must be a valid rotation matrix. Although minimising E_{lmk} is a non-convex optimisation problem, a good initialisation can be obtained using alternating linear least squares and this estimate subsequently refined using nonlinear optimisation. This is the approach that we take.

3.1   Pose Estimation

We make an initial estimate of $\mathbf{R}$, $\mathbf{t}$ and s using a simple extension of the POS algorithm [23]. Compared to POS, we additionally enforce that $\mathbf{R}$ is a valid rotation matrix. We begin by solving an unconstrained system in a least squares sense. We stack two copies of the 3D points in homogeneous coordinates, such that $\mathbf{A}_{2i-1} = [u_i \ v_i \ w_i \ 1 \ 0 \ 0 \ 0 \ 0]$ and $\mathbf{A}_{2i} = [0 \ 0 \ 0 \ 0 \ u_i \ v_i \ w_i \ 1]$ and form a long vector of the corresponding 2D points $\mathbf{d} = [x_1 \ y_1 \ \cdots \ x_L \ y_L]^{\mathrm{T}}$. We then solve for $\mathbf{k} \in \mathbb{R}^8$ in $\mathbf{A}\mathbf{k} = \mathbf{d}$ using linear least squares. We define $\mathbf{r}_1 = [k_1 \ k_2 \ k_3]$ and $\mathbf{r}_2 = [k_5 \ k_6 \ k_7]$. Scale is given by $s = (\|\mathbf{r}_1\| + \|\mathbf{r}_2\|)/2$ and the translation vector by $\mathbf{t} = [k_4/s \ k_8/s]^{\mathrm{T}}$. We perform singular value decomposition on the matrix formed from $\mathbf{r}_1$ and $\mathbf{r}_2$:

$$\mathbf{U}\mathbf{S}\mathbf{V}^{\mathrm{T}} = \begin{bmatrix} \mathbf{r}_1 \\ \mathbf{r}_2 \\ \mathbf{r}_1 \times \mathbf{r}_2 \end{bmatrix} \tag{4}$$

The rotation matrix is given by $\mathbf{R} = \mathbf{U}\mathbf{V}^{\mathrm{T}}$. If $\det(\mathbf{R}) = -1$ then we negate the third row of $\mathbf{U}$ and recompute $\mathbf{R}$. This guarantees that $\mathbf{R}$ is a valid rotation matrix. This approach gives a good initial estimate which we subsequently refine with nonlinear optimization of E_{lmk} with respect to $\mathbf{R}$, $\mathbf{t}$ and s.

3.2   Shape Estimation

With a fixed pose estimate, shape parameter estimation under scaled orthographic projection is a linear problem. The 2D position of the ith vertex as a function of the shape parameters is given by: $s\mathbf{R}_{1\ldots2}(\mathbf{P}_i\boldsymbol{\alpha} + \bar{\mathbf{f}}_i) + s\mathbf{t}$. Hence, each observed vertex adds two equations to a linear system. Concretely, for each image we form the matrix $\mathbf{C} \in \mathbb{R}^{2L \times S}$ where

$$\mathbf{C}_{2i-1} = s(\mathbf{R}_{11}\mathbf{P}_{iu}^{\mathrm{T}} + \mathbf{R}_{12}\mathbf{P}_{iv}^{\mathrm{T}} + \mathbf{R}_{13}\mathbf{P}_{iw}^{\mathrm{T}})$$

and

$$\mathbf{C}_{2i} = s(\mathbf{R}_{21}\mathbf{P}_{iu}^{\mathrm{T}} + \mathbf{R}_{22}\mathbf{P}_{iv}^{\mathrm{T}} + \mathbf{R}_{23}\mathbf{P}_{iw}^{\mathrm{T}})$$

and vector $\mathbf{h} \in \mathbb{R}^{2L}$ where

$$\mathbf{h}_{2i-1} = x_i - s(\mathbf{R}_1\bar{\mathbf{f}}_i + \mathbf{t}_1) \quad \text{and} \quad \mathbf{h}_{2i} = y_i - s(\mathbf{R}_2\bar{\mathbf{f}}_i + \mathbf{t}_2).$$

We solve $\mathbf{C}\boldsymbol{\alpha} = \mathbf{h}$ in a least squares sense subject to an additional constraint to ensure plausibility of the solution. We follow Brunton et al. [24] and use a hyperbox constraint on the shape parameters. This avoids having to choose a regularisation weight but ensures that each parameter lies within k standard deviations of the mean by introducing a linear inequality constraint on the shape parameters (we use $k = 3$ in our experiments). Hence, the problem can be solved in closed form as an inequality constrained linear least squares problem.

3.3   Nonlinear Refinement

Having alternated pose and shape estimation for a fixed number of iterations, finally we perform nonlinear optimisation of E_{lmk} over $\boldsymbol{\alpha}$, $\mathbf{R}$, $\mathbf{t}$ and s simultaneously. We represent $\mathbf{R}$ in axis-angle space to ensure that it remains a valid rotation matrix and we retain the hyperbox constraint on $\boldsymbol{\alpha}$. We minimise E_{lmk} using the trust-region-reflective algorithm [25] as implemented in the Matlab lsqnonlin function.

4   Fitting with Hard Edge Correspondence

The method in Sect. 3 enables a 3DMM to be fitted to 2D landmark positions if the correspondence between landmarks and model vertices is known. Edges, for example caused by occluding boundaries, do not have a fixed correspondence to model vertices. Hence, fitting to edges requires shape and pose estimation to happen in conjunction with establishing correspondence between image and model edges. Our proposed approach establishes these correspondences explicitly by finding the closest image edge to each model boundary vertex (subject to additional filtering to remove unreliable matches). Our method comprises the following steps:

1. Detect facial landmarks
2. Initialise shape and pose estimates by fitting to landmarks only
3. Improve initialisation using iterated closest edge fitting
4. Nonlinear optimisation of hybrid objective function containing landmark, edge and prior terms

We describe each of these steps in more detail in the rest of this section.

4.1   Landmarks

We use landmarks both for initialisation and as part of our overall objective function as one cue for shape estimation. We apply a facial landmark detector that is suitable for operating on "in the wild" images. This provides approximate positions of facial landmarks for which we know the corresponding vertices in the morphable model. We use these landmark positions to make an initial estimate of the pose and shape parameters by running the method in Sect. 3 with only these corresponding landmark positions. Note that any facial landmark detector can be used at this stage. In our experiments, we show results with a recent landmark detection algorithm [26] that achieves state-of-the-art performance and for which code is provided by the authors. In our experimental evaluation, we include the results of fitting to landmarks only.

4.2   Edge Cost

We assume that a subset of pixels have been labelled as edges and stored as the set $\mathcal{E} = \{(x,y)|(x,y) \text{ is an edge}\}$. In practice, we compute edges by applying the Canny edge detector with a fixed threshold to the input image.

Model contours are computed based on the pose and shape parameters as the occluding boundary of the 3D face. The set of occluding boundary vertices, $\mathcal{B}(\boldsymbol{\alpha}, \mathbf{R}, \mathbf{t}, s)$, are defined as those lying on a mesh edge whose adjacent faces have a change of visibility. This definition encompasses both outer (silhouette) and inner (self-occluding) contours. Since the viewing direction is aligned with the z-axis, this is tested simply by checking if the sign of the z-component of the triangle normal changes on either side of the edge. In addition, we check that potential edge vertices are not occluded by another part of the mesh (using z-buffering) and we ignore edges that lie on a mesh boundary since they introduce artificial edges. In this paper, we deal only with occluding contours (both inner and outer). If texture contours were defined on the surface of the morphable model, it would be straightforward to include these in our approach.

We define the objective function for edge fitting with hard correspondence as the sum of squared distances between each projected occluding boundary vertex and the closest edge pixel:

$$E_{edge}(\boldsymbol{\alpha}, \mathbf{R}, \mathbf{t}, s) = \tag{5}$$

$$\frac{1}{|\mathcal{B}(\boldsymbol{\alpha}, \mathbf{R}, \mathbf{t}, s)|} \sum_{i \in \mathcal{B}(\boldsymbol{\alpha}, \mathbf{R}, \mathbf{t}, s)} \min_{(x,y) \in \mathcal{E}} \| [x\ y]^T - \mathbf{SOP}\left[\mathbf{P}_i \boldsymbol{\alpha} + \bar{\mathbf{f}}_i, \mathbf{R}, \mathbf{t}, s \right] \|^2.$$

Note that the minimum operator is responsible for computing the hard correspondences. This objective is non-convex since the minimum of a set of convex functions is not convex [27]. Hence, we require a good initialisation to ensure convergence to a minimum close to the global optimum. Fitting to landmarks only does not provide a sufficiently good initialisation. So, in the next subsection we describe a method for obtaining a good initial fit to edges, before incorporating the edge cost into a hybrid objective function in Sect. 4.5.

4.3 Iterated Closest Edge Fitting

We propose to refine the landmark-only fit with an initial fit to edges that works in an iterated closest point manner. That is, for each projected model contour vertex, we find the closest image edge pixel and we treat this as a known correspondence. In conjunction with the landmark correspondences, we again run the method in Sect. 3. This leads to updated pose and shape parameters, and in turn to updated model edges and correspondences. We iterate this process for a fixed number of iterations. We refer to this process as Iterated Closest Edge Fitting (ICEF) and provide an illustration in Fig. 1. On the left we show an input image with the initial landmark detection result. In the middle we show the initial shape and pose obtained by fitting only to landmarks. On the right we show image edge pixels in blue and projected model contours in green (where nearest neighbour edge correspondence is considered reliable) and in red (where correspondence is considered unreliable). The green/blue correspondences are used for the next iteration of fitting.

Finding the image edge pixel closest to a projected contour vertex can be done efficiently by storing the image edge pixels in a kd-tree. We filter the resulting

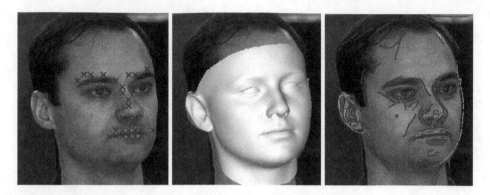

Fig. 1. Iterated closest edge fitting for initialisation of the edge fitting process. Left: input image with automatically detected landmarks. Middle: overlaid shape obtained by fitting only to landmark. Right: image edges in blue, model boundary vertices with image correspondences in green, unreliable correspondences in red. (Color figure online)

correspondences using two commonly used heuristics. First, we remove 5% of the matches for which the distance to the closest image edge pixel is largest. Second, we remove matches for which the image distance divided by s exceeds a threshold (chosen as 10 in our experiments). The division by scale factor s makes this choice invariant to changes in image resolution.

4.4   Prior

Under the assumption that the training data of the 3DMM forms a Gaussian cloud in high dimensional space, then we expect that each of the shape parameters follows a normal distribution with zero mean and variance given by the eigenvalue, λ_i, associated with the corresponding principal component. We find that including a prior term that captures this assumption significantly improves performance over using the hyperbox constraint alone. The prior penalises deviation from the mean shape as follows:

$$E_{\text{prior}}(\boldsymbol{\alpha}) = \sum_{i=1}^{S} \left(\frac{\alpha_i}{\sqrt{\lambda_i}} \right)^2. \tag{6}$$

4.5   Nonlinear Refinement

Finally, we perform nonlinear optimisation of a hybrid objective function comprising landmark, edge and prior terms:

$$E(\boldsymbol{\alpha}, \mathbf{R}, \mathbf{t}, s) = w_1 E_{\text{lmk}}(\boldsymbol{\alpha}, \mathbf{R}, \mathbf{t}, s) + w_2 E_{\text{edge}}(\boldsymbol{\alpha}, \mathbf{R}, \mathbf{t}, s) + w_3 E_{\text{prior}}(\boldsymbol{\alpha}), \tag{7}$$

where w_1, w_2 and w_3 weight the contribution of each term to the overall energy. The landmark and edge terms are invariant to the number of landmarks and edge vertices which means we do not have to tune the weights for each image (for example, for the results in Table 1 we use fixed values of: $w_1 = 0.15$, $w_2 = 0.45$ and $w_3 = 0.4$). We retain the hyperbox constraint and so the hybrid objective is a constrained nonlinear least squares problem and we again optimise using the trust-region-reflective algorithm.

For efficiency and to avoid problems of continuity and differentiability of the edge cost function, we follow [6] and keep occluding boundary vertices, $\mathcal{B}$, fixed for a number of iterations of the optimiser. After a number of iterations, we recompute the vertices lying on the occluding boundary and restart the optimiser.

5   Fitting with Soft Edge Correspondence

We compare our approach with a method based on optimising an edge cost function, in the same spirit as previous work [4,6,7]. We follow the same approach as Amberg et al. [6] to compute the edge cost function, however we further improve robustness by also integrating over scale. For our edge detector, we use gradient magnitude thresholding with non-maxima suppression. Given a set of edge detector sensitivity thresholds $\mathcal{T}$ and scales $\mathcal{S}$, we compute $n = |\mathcal{T} \times \mathcal{S}|$ edge images, $E^1, \ldots, E^n$, using each pair of image scale and threshold values. We compute the Euclidean distance transform, $D^1, \ldots, D^n$, for each edge image (i.e. the value of each pixel in D^i is the distance to the closest edge pixel in E^i). Finally, we compute the edge cost surface as:

$$S(x,y) = \frac{1}{n} \sum_{i=1}^{n} \frac{D^i(x,y)}{D^i(x,y) + \kappa}. \tag{8}$$

The parameter κ determines the influence range of an edge in an adaptive manner. Amberg et al. [6] suggest a value for κ of 1/20th the expected size of the head in pixels. We compute this parameter automatically from the scale s. An example of an edge cost surface is shown in Fig. 2. To evaluate the edge cost, we

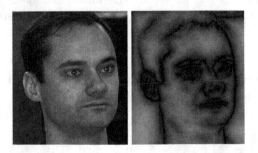

Fig. 2. Edge cost surface with soft correspondence (right) computed from input image (left)

compute model contour vertices as in Sect. 4.2, project them into the image and interpolate the edge cost function using bilinear interpolation:

$$E_{\text{softedge}}(\boldsymbol{\alpha}, \mathbf{R}, \mathbf{t}, s) = \frac{1}{|\mathcal{B}(\boldsymbol{\alpha}, \mathbf{R}, \mathbf{t}, s)|} \sum_{i \in \mathcal{B}(\boldsymbol{\alpha}, \mathbf{R}, \mathbf{t}, s)} S(\mathbf{SOP} \left[\mathbf{P}_i \boldsymbol{\alpha} + \bar{\mathbf{f}}_i, \mathbf{R}, \mathbf{t}, s \right]).$$

(9)

As with the hard edge cost, we found that the best performance was achieved by also including the landmark and prior terms in a hybrid objective function. Hence, we minimise:

$$E(\boldsymbol{\alpha}, \mathbf{R}, \mathbf{t}, s) = w_1 E_{\text{lmk}}(\boldsymbol{\alpha}, \mathbf{R}, \mathbf{t}, s) + w_2 E_{\text{softedge}}(\boldsymbol{\alpha}, \mathbf{R}, \mathbf{t}, s) + w_3 E_{\text{prior}}(\boldsymbol{\alpha}). \quad (10)$$

We again initialise by fitting to landmarks only using the method in Sect. 4.1, retain the hyperbox constraint and optimise using the trust-region-reflective algorithm. We use the same weights as for the hard correspondence method in our experiments.

6   Experimental Results

We present two sets of experimental results. First, we use synthetic images with known ground truth 3D shape in order to quantitatively evaluate our method and provide comparison to previous work. Second, we use real images to provide qualitative evidence of the performance of our method in uncontrolled conditions. For the 3DMM in both sets of experiments we use the Basel Face Model [28].

6.1   Quantitative Evaluation

We begin with a quantitative comparative evaluation on synthetic data. We use the 10 out-of-sample faces supplied with the Basel Face Model and render orthographic images of each face in 9 poses (rotations of $0°$, $\pm15°$, $\pm30°$, $\pm50°$ and $\pm70°$ about the vertical axis). We show sample input images for one subject in Fig. 3. In all experiments, we report the mean Euclidean distance between ground truth and estimated face surface in mm after Procrustes alignment.

In the first experiment, we use ground truth landmarks. Specifically, we use the 70 Farkas landmarks, project the visible subset to the image (yielding between 37 and 65 landmarks per image) and round to the nearest pixel. In Table 1 we show results averaged over pose angle and over the whole dataset. As a baseline, we show the error if we simply use the average face shape. We

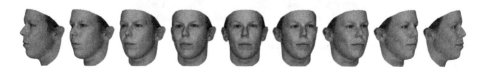

Fig. 3. Synthetic input images for one subject

Table 1. Mean Euclidean vertex distance (mm) with ground truth landmarks

Method	Rotation angle									Mean
	$-70°$	$-50°$	$-30°$	$-15°$	$0°$	$15°$	$30°$	$50°$	$70°$	
Average face	3.35	3.35	3.35	3.35	3.35	3.35	3.35	3.35	3.35	3.35
Proposed (landmarks only)	2.67	2.60	2.58	2.64	2.56	2.49	2.50	2.54	2.63	2.58
Aldrian and Smith [9]	2.64	2.60	2.55	2.54	**2.49**	2.42	2.43	2.44	2.54	2.52
Romdhani et al. [4] (soft)	2.65	2.59	2.58	2.61	2.59	2.50	2.50	2.46	2.51	2.55
Proposed (ICEF)	2.38	2.40	2.51	**2.38**	2.52	2.45	2.43	2.38	2.3	2.42
Proposed (hard)	**2.35**	**2.26**	**2.38**	2.40	2.51	**2.39**	**2.40**	**2.20**	**2.26**	**2.35**

then show the result of fitting only to landmarks, i.e. the method in Sect. 3. We include two comparison methods. The approach of Aldrian and Smith [9] uses only landmarks but with an affine camera model and a learnt model of landmark variance. The soft edge correspondence method of Romdhani et al. [4] is described in Sect. 5. The final two rows show two variants of our proposed methods: the fast Iterated Closest Edge Fitting version and the full version with nonlinear optimisation of the hard correspondence cost. Average performance over the whole dataset is best for our method and, in general, using edges over landmarks only and applying nonlinear optimisation improves performance. The performance improvement of our methods over landmark-only methods improves with pose angle. This suggest that edge information becomes more salient for non-frontal poses.

The second experiment is identical to the first except that we add Gaussian noise of varying standard deviation to the ground truth landmark positions. In Table 2 we show results averaged over all poses and subjects.

In the final experiment we use landmarks that are automatically detected using the method of Zhu and Ramanan [26]. This enables us to include comparison with the recent fitting algorithm of Zhu et al. [18]. We use the author's own implementation which only works with a fixed set of 68 landmarks. This means that the method cannot be applied to the more extreme pose angles where fewer landmarks are detected. In this more challenging scenario, our method again gives the best overall performance and is superior for all pose angles.

Table 2. Mean Euclidean vertex distance (mm) with noisy landmarks

Method	Landmark noise std. dev.					
	$\sigma = 0$	$\sigma = 1$	$\sigma = 2$	$\sigma = 3$	$\sigma = 4$	$\sigma = 5$
Proposed (landmarks only)	2.58	2.60	2.61	2.68	2.76	2.85
Aldrian and Smith [9]	2.52	2.53	2.55	2.62	2.65	2.73
Romdhani et al. [4] (soft)	2.55	2.57	2.57	2.62	2.70	2.76
Proposed (ICEF)	2.42	2.43	2.43	2.50	2.57	2.60
Proposed (hard)	**2.35**	**2.36**	**2.35**	**2.39**	**2.47**	**2.50**

Table 3. Mean Euclidean vertex distance (mm) with automatically detected landmarks

Method	Rotation angle									Mean
	−70°	−50°	−30°	−15°	0°	15°	30°	50°	70°	
Proposed (landmarks only)	6.79	6.84	5.19	5.74	5.68	6.34	6.48	7.04	7.74	6.43
Zhu et al. [18]	N/A	N/A	4.63	5.09	4.19	5.22	4.92	N/A	N/A	N/A
Romdhani et al. [4] (soft)	4.46	3.42	3.66	3.78	3.77	3.57	4.31	4.19	4.73	3.99
Proposed (ICEF)	3.70	3.32	3.26	3.23	3.37	3.50	3.43	4.07	3.52	3.49
Proposed (hard)	**3.43**	**3.20**	**3.19**	**3.09**	**3.30**	**3.36**	**3.36**	**3.84**	**3.41**	**3.35**

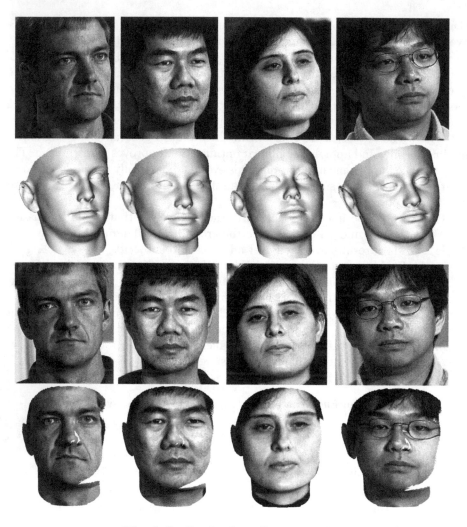

Fig. 4. Qualitative frontalisation results

6.2   Qualitative Evaluation

In Fig. 4 we show qualitative examples from the CMU PIE [29] dataset. Here, we fit to images (first row) in a non-frontal pose using automatically detected landmarks [26] and show the reconstruction in the second row. We texture map the image onto the mesh, rotate to frontal pose (bottom row) and compare to an actual frontal view (third row). Finally, we show qualitative examples from the Labelled Faces in the Wild dataset [30] in Fig. 5. Again, we texture map the image to the mesh and show a range of poses. These results show that our method is capable of robustly and fully automatically fitting to unconstrained images.

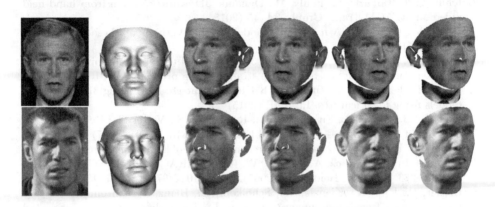

Fig. 5. Qualitative pose editing results

7   Conclusions

We have presented a fully automatic algorithm for fitting a 3DMM to single images using hard edge correspondence and compared it to existing methods using soft correspondence. In 3D-3D alignment, the soft correspondence of LM-ICP [16] is demonstrably more robust than hard ICP [20]. However, in the context of 3D-2D nonrigid alignment, a soft edge cost function is neither continuous nor differentiable since contours appear, disappear, split and merge under parameter changes [7]. This makes its optimisation challenging, unstable and highly dependent on careful choice of optimisation parameters. Although our proposed algorithm relies on potentially brittle hard correspondences, solving for shape and pose separately requires only solution of a linear problem and, together, optimisation of a multilinear problem. This makes iterated closest edge fitting very fast and it provides an initialisation that allows the subsequent nonlinear optimisation to converge to a better optimum. We believe that this explains the improved performance over edge fitting with soft correspondence.

There are many ways this work can be extended. First, we could explore other ways in which the notion of soft correspondence is formulated. For example, we could borrow from SoftPOSIT [31] or Blind PnP [32] which both estimate pose with unknown 3D-2D correspondence. Second, we could incorporate any of the refinements to standard ICP [33]. Third, we currently use only geometric information and do not fit texture. Finally, we would like to extend the method to video using a model that captures expression variation and incorporating temporal smoothness constraints.

References

1. Blanz, V., Vetter, T.: Face recognition based on fitting a 3D morphable model. IEEE Trans. Pattern Anal. Mach. Intell. **25**, 1063–1074 (2003)
2. Ichim, A.E., Bouaziz, S., Pauly, M.: Dynamic 3D avatar creation from hand-held video input. ACM Trans. Graph. **34**, 45 (2015)
3. Blanz, V., Vetter, T.: A morphable model for the synthesis of 3D faces. In: SIG-GRAPH (1999)
4. Romdhani, S., Vetter, T.: Estimating 3D shape and texture using pixel intensity, edges, specular highlights, texture constraints and a prior. In: CVPR (2005)
5. Huber, P., Feng, Z., Christmas, W., Kittler, J., Rätsch, M.: Fitting 3D morphable models using local features. In: ICIP (2015)
6. Amberg, B., Blake, A., Fitzgibbon, A., Romdhani, S., Vetter, T.: Reconstructing high quality face-surfaces using model based stereo. In: ICCV (2007)
7. Keller, M., Knothe, R., Vetter, T.: 3D reconstruction of human faces from occluding contours. In: Gagalowicz, A., Philips, W. (eds.) MIRAGE 2007. LNCS, vol. 4418, pp. 261–273. Springer, Heidelberg (2007). doi:10.1007/978-3-540-71457-6_24
8. Breuer, P., Kim, K., Kienzle, W., Schölkopf, B., Blanz, V.: Automatic 3D face reconstruction from single images or video. In: Proceedings of the FG, pp. 1–8 (2008)
9. Aldrian, O., Smith, W.A.P.: Inverse rendering of faces with a 3D morphable model. IEEE Trans. Pattern Anal. Mach. Intell. **35**, 1080–1093 (2013)
10. Amberg, B., Vetter, T.: Optimal landmark detection using shape models and branch and bound. In: Proceedings of the ICCV (2011)
11. Schönborn, S., Forster, A., Egger, B., Vetter, T.: A monte carlo strategy to integrate detection and model-based face analysis. In: Weickert, J., Hein, M., Schiele, B. (eds.) GCPR 2013. LNCS, vol. 8142, pp. 101–110. Springer, Heidelberg (2013). doi:10.1007/978-3-642-40602-7_11
12. Cootes, T.F., Taylor, C.J., Cooper, D., Graham, J.: Active shape models - their training and application. Comput. Vis. Image Underst. **61**, 38–59 (1995)
13. Atkinson, G.A., Smith, M.L., Smith, L.N., Farooq, A.R.: Facial geometry estimation using photometric stereo and profile views. In: Tistarelli, M., Nixon, M.S. (eds.) ICB 2009. LNCS, vol. 5558, pp. 1–11. Springer, Heidelberg (2009). doi:10.1007/978-3-642-01793-3_1
14. Moghaddam, B., Lee, J., Pfister, H., Machiraju, R.: Model-based 3D face capture with shape-from-silhouettes. In: Proceedings of the FG (2003)
15. Lüthi, M., Albrecht, T., Vetter, T.: Probabilistic modeling and visualization of the flexibility in morphable models. In: Hancock, E.R., Martin, R.R., Sabin, M.A. (eds.) Mathematics of Surfaces 2009. LNCS, vol. 5654, pp. 251–264. Springer, Heidelberg (2009). doi:10.1007/978-3-642-03596-8_14
16. Fitzgibbon, A.W.: Robust registration of 2D and 3D point sets. Image Vis. Comput. **21**, 1145–1153 (2003)

17. Cashman, T.J., Fitzgibbon, A.W.: What shape are dolphins? Building 3D morphable models from 2D images. IEEE Trans. Pattern Anal. Mach. Intell. **35**, 232–244 (2013)

18. Zhu, X., Lei, Z., Yan, J., Yi, D., Li, S.Z.: High-fidelity pose and expression normalization for face recognition in the wild. In: Proceedings of the CVPR, pp. 787–796 (2015)

19. Sánchez-Escobedo, D., Castelán, M., Smith, W.: Statistical 3D face shape estimation from occluding contours. Comput. Vis. Image Underst. **142**, 111–124 (2016)

20. Besl, P.J., McKay, N.D.: A method for registration of 3-D shapes. IEEE Trans. Pattern Anal. Mach. Intell. **14**, 239–256 (1992)

21. Suwajanakorn, S., Kemelmacher-Shlizerman, I., Seitz, S.M.: Total moving face reconstruction. In: Fleet, D., Pajdla, T., Schiele, B., Tuytelaars, T. (eds.) ECCV 2014. LNCS, vol. 8692, pp. 796–812. Springer, Heidelberg (2014). doi:10.1007/978-3-319-10593-2_52

22. Cao, C., Bradley, D., Zhou, K., Beeler, T.: Real-time high-fidelity facial performance capture. ACM Trans. Graph. **34**, 46 (2015)

23. Dementhon, D.F., Davis, L.S.: Model-based object pose in 25 lines of code. Int. J. Comput. Vis. **15**, 123–141 (1995)

24. Brunton, A., Salazar, A., Bolkart, T., Wuhrer, S.: Review of statistical shape spaces for 3D data with comparative analysis for human faces. Comput. Vis. Image Underst. **128**, 1–17 (2014)

25. Coleman, T., Li, Y.: An interior, trust region approach for nonlinear minimization subject to bounds. SIAM J. Optim. **6**, 418–445 (1996)

26. Zhu, X., Ramanan, D.: Face detection, pose estimation, and landmark localization in the wild. In: Proceedings of the CVPR (2012)

27. Grant, M., Boyd, S., Ye, Y.: Disciplined convex programming. In: Liberti, L., Maculan, N. (eds.) Global Optimization: From Theory to Implementation, pp. 155–210. Springer, Heidelberg (2006)

28. Paysan, P., Knothe, R., Amberg, B., Romdhani, S., Vetter, T.: A 3D face model for pose and illumination invariant face recognition. In: Proceedings of the AVSS (2009)

29. Sim, T., Baker, S., Bsat, M.: The CMU pose, illumination, and expression database. IEEE Trans. Pattern Anal. Mach. Intell. **25**, 1615–1618 (2003)

30. Huang, G.B., Ramesh, M., Berg, T., Learned-Miller, E.: Labeled faces in the wild: a database for studying face recognition in unconstrained environments. Technical report 07–49, University of Massachusetts, Amherst (2007)

31. David, P., DeMenthon, D., Duraiswami, R., Samet, H.: SoftPOSIT: simultaneous pose and correspondence determination. In: Heyden, A., Sparr, G., Nielsen, M., Johansen, P. (eds.) ECCV 2002. LNCS, vol. 2352, pp. 698–714. Springer, Heidelberg (2002). doi:10.1007/3-540-47977-5_46

32. Moreno-Noguer, F., Lepetit, V., Fua, P.: Pose priors for simultaneously solving alignment and correspondence. In: Forsyth, D., Torr, P., Zisserman, A. (eds.) ECCV 2008. LNCS, vol. 5303, pp. 405–418. Springer, Heidelberg (2008). doi:10.1007/978-3-540-88688-4_30

33. Rusinkiewicz, S., Levoy, M.: Efficient variants of the ICP algorithm. In: Proceedings of the 3DIM (2001)

Multiple Facial Attributes Estimation Based on Weighted Heterogeneous Learning

Hiroshi Fukui[1], Takayoshi Yamashita[1][✉], Yuu Kato[1], Ryo Matsui[1],
T. Ogata[2], Yuji Yamauchi[1], and Hironobu Fujiyoshi[1]

[1] Chubu University, 1200, Matuoto-cho, Kasugai, Aichi, Japan
yamashita@cs.chubu.ac.jp
[2] Abeja Inc., 4-1-20, Toranomon, Minato-ku, Tokyo, Japan

Abstract. To estimate multiple face attributes, independent classifier for each attribute are trained such as facial point detection, gender recognition, and age estimation in the conventional approach. It is inefficient because the computational cost of training and testing increases with the number of tasks. To address this problem, heterogeneous learning is able to train a single classifier to perform multiple tasks. Heterogeneous learning is simultaneously train regression and recognition tasks, thereby reducing both training and testing time. However, it is difficult to obtain equivalent performance for set of single task classifiers due to variance of training error of each task. In this paper, we propose weighted heterogeneous learning of a convolutional neural network with a weighted error function. Our method outperformed the conventional method in terms of facial attribute recognition, especially for regression tasks such as facial point detection, age estimation, and smile ratio estimation.

1 Introduction

Facial attributes estimation such as facial point, gender, and age has been used in marketing strategies and social networking services. Marketing strategies recommend specific items that are matched to the client requirement. Various social networking services based on facial recognition techniques have recently been developed that can estimate age from a facial image with a high accuracy.

To estimate multiple face attributes, independent classifier for each attributes are trained such as facial point detection, gender recognition, and age estimation. Active appearance model (AAM) [1] and conditional regression forest (CRF) [2] are common approaches for facial point detection. Age estimation and gender recognition are classified by a support vector machine (SVM) or a decision tree using facial points or local binary pattern (LBP) features [3,4]. With the increase of deep learning, the deep convolutional neural network (CNN) [5] has become a common classifier for facial point detection [6–11], age estimation [12–15], and gender recognition [16–18]. The conventional approach must prepare multiple classifiers for each task. This approach is inefficient because the computational cost of training and testing increases with the number of tasks. To address this problem, heterogeneous learning [19] trains a single classifier to

© Springer International Publishing AG 2017
C.-S. Chen et al. (Eds.): ACCV 2016 Workshops, Part II, LNCS 10117, pp. 392–406, 2017.
DOI: 10.1007/978-3-319-54427-4_29

perform multiple recognition tasks. A CNN trained using heterogeneous learning has output units that correspond to each task. Thus, a single network classifies multiple tasks simultaneously and the computational cost does not vary with the number of tasks. In this paper, we use a heterogeneous learning CNN for facial point detection, gender recognition, age estimation, race recognition, and smile rate estimation.

Conventional heterogeneous learning has used the mean squared error function for regression tasks and the cross entropy error function for recognition tasks during the training process. The error ranges of mean squared error function and cross entropy error function are noticeably different. Therefore, we integrate the error range from 0 to 1 by exchanging cross entropy error function for mean squared error function for recognition tasks. However, if we integrate the training error functions, difference of training error is occured, as shown in Fig. 1(a). This difference of training error is occured by difference between label value of regression task and recognition task. The label value of regression tasks is a continuous value from 0 to 1, whereas the label value for classification tasks is a discrete value of 0 or 1. Consequently, facial point detection performance suffers, as shown in Fig. 1(b). Therefore, differences between training errors negatively affect the training process for heterogeneous learning for a CNN.

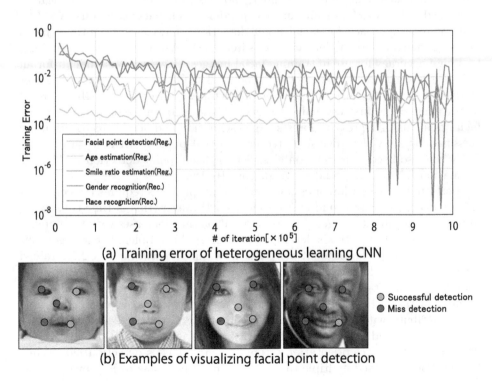

(a) Training error of heterogeneous learning CNN

(b) Examples of visualizing facial point detection

Fig. 1. Training error and example results by heterogeneous learning for a CNN

In this paper, we propose weighted heterogeneous learning for a CNN. First, we select a basis task from all tasks. Additionally, we define subtasks, not including the basis task. We weight the error function for the subtasks. Our method suppresses the training error and dispersion training errors by weighting the cost function for the subtasks. Weighted heterogeneous learning for a CNN improves the recognition performance by stable training when introducing the proposed method.

2 Facial Image Analysis Using Heterogeneous Learning CNN

We categorize related work into facial image analysis and heterogeneous learning. First, we describe the related publications in these categories and then further discuss problems with existing heterogeneous learning for a CNN method as applied to facial image analysis.

2.1 Related Work

Marketing strategies and social networking services have used facial attribute information, such as facial point, gender, and age. In particular, facial point have been used as features for estimating age, gender, and facial expressions. AAM [1] is a common approach for facial point detection. AAM detects optimal facial point by changing face model parameters iteratively. AAM can detect facial point to a high accuracy for use in training facial images. However, it is difficult for an unknown testing sample to detect facial point. The CRF proposed by Dantone et al. detects facial point using regression forests for each face pose [2]. CRF consists of two stages: the first estimates the facial pose, and the second regresses the facial point using regression forests. Age estimation and gender recognition are classified by a SVM or decision tree using facial point or LBP features [3,4]. CNN has also become a common classifier for facial point detection [6–11], age estimation [12–15], and gender recognition [16–18].

Performing recognition or estimation for multiple tasks requires the construction of classifiers corresponding to each task. However, this is time-consuming during training and testing, and the computation time increases with the number of tasks. One of the methods developed to address this problem is heterogeneous learning, which performs multiple tasks in a single network. A CNN trained for heterogeneous learning has units that output the recognition results corresponding to each task. The computational cost does not directly depend on the number of tasks. Heterogeneous learning can estimate and recognize multiple facial attribute with high accuracy by combining CNN [20–24]. Zhang et al. proposed a method to perform multiple tasks such as facial point detection, gender classification, face orientation estimation, and glasses detection [20]. While the method estimated multiple tasks, its main purpose was to improve the performance of the primary task, such as facial point detection. It thus assigned weighted error functions to each task. When the error decreased sufficiently, the training of the task was terminated earlier to avoid over-fitting to a specific task.

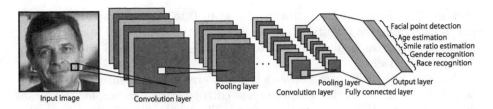

Fig. 2. Heterogeneous learning for a CNN

2.2 Heterogeneous Learning

Figure 2 shows the structure of a heterogeneous learning for a CNN. First, M training samples are chosen randomly to form a mini-batch. We used mini-batch training when updating CNN parameters. During mini-batch training, the error E is calculated and backpropagated to update the parameters θ of the network. For each backpropagation [25] iteration, the samples in the mini-batch are selected randomly from the dataset. When the CNN is trained using heterogeneous learning, the recognition and regression tasks are combined in a single network and each task has an independent error function. The mean squared error in Eq. (1) and the cross entropy in Eq. (2) are employed as the error functions of the recognition and regression tasks, respectively.

$$E_t^{Regression} = \frac{1}{M} \sum_{m=1}^{M} \| \mathbf{y} - \mathbf{o} \|_2^2 \tag{1}$$

$$E_t^{Recognition} = \frac{1}{M} \sum_{m=1}^{M} -\mathbf{y} \log \mathbf{o} \tag{2}$$

The errors E_t of the sample m for all tasks $\{t|1,\ldots,1\}$ are accumulated and propagated once per iteration.

$$\begin{aligned} \theta &\leftarrow \theta + \Delta\theta \\ &= \theta - \eta \frac{\partial \sum_{t=1}^{T} E_t}{\partial \theta} \end{aligned} \tag{3}$$

The parameters θ of the CNN are updated using the differential of the accumulated error with the training coefficient η.

2.3 CNN Based on Heterogeneous Learning

Figure 1(a) shows the training errors of five tasks using heterogeneous learning for a CNN. There are differences between the training errors for all tasks.

The differences between training errors occur because of the error function for regression tasks and recognition tasks. There are noticeable differences between the error ranges of the mean squared error function and the cross entropy error function. The mean squared error ranges from 0 to 1, and the cross entropy error ranges from 0 to infinity. Thus, we integrate the error range from 0 to 1 by exchanging the cross entropy error function for the mean squared error function for recognition tasks.

However, the differences between training error occur if integration errors range from 0 to 1 by exchanging the cross entropy error function for the mean squared error function for recognition tasks. The label value of regression tasks is a continuous value from 0 to 1, whereas the label value of recognition tasks is a discrete value of 0 or 1. Thus, recognition tasks develop more differences between the training errors than regression tasks. These causes negatively affect heterogeneous learning during the training process. Thus, facial point detection performance suffers due to the lowest training error, as shown in Fig. 1(b).

3   Proposed Method

Conventional heterogeneous learning calculates the training error under Eq. (1) evenly. Hence, differences between training errors occur because of differences between label values for regression tasks and recognition tasks. The proposed method weights the error function of the training error E_t for subtasks. The proposed method stabilizes the training error by weighting each task and improves the heterogeneous learning performance.

3.1   Training of Single Task CNN

First, we obtain training error by training CNN of a single task for each task. Unlike in training error of heterogeneous learning, the training error of a single task CNN is not interference training error between other tasks. Therefore, we will be able to obtain a stable basis value using training error of single task CNN, when computing basis values for each task. In this paper, this CNN training is repeated until the training criterion condition is satisfied.

3.2   Computing the Weights of Error Functions

We compute basis value N_t for each task that gave weight to training error functions using training error of single task CNN, as shown in Fig. 3. However, these training errors are varied by each iteration. Therefore, we calculate the basis value N_t using the normal distribution of the training error for each task. If it reflects training error for normal distribution, the normal distribution of the training error for task t connotes 99.7% in the interval that sums the average μ and 3-fold vertical 3σ. The other interval is the dispersion of training error, and we can calculate the basis value N_t that is negatively affected by ignoring the interval. Thus, we use the basis value N_t that sums the average μ and 3-fold vertical 3σ.

$$N_t = \mu + 3\sigma \tag{4}$$

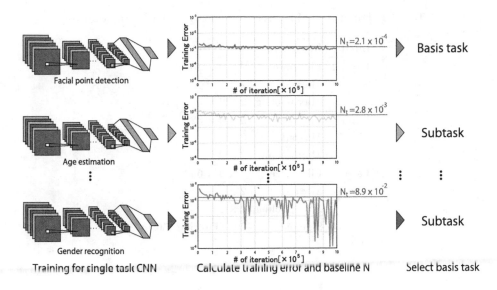

Fig. 3. Selection basis task

After calculating the basis value, we select a basis task. In this paper, we select the basis task for the lowest basis value N_t. Thus, the facial point detection task is the basis task and the other tasks are subtasks. After selecting the basis task and subtasks, we calculate the weight w_t for each subtask. The basis value N_f of the facial point detection task and basis value N_t of the other tasks are used in Eq. (4).

$$w_t = \frac{N_f}{N_t} \tag{5}$$

3.3 Training of Weighted Heterogeneous Learning

We give weight to error function for each subtask, as shown in Eq. (6). The first term in Eq. (6) is an error function of the main task. The second term in Eq. (6) is an error function of subtasks.

$$E = \frac{1}{M} \sum_{m=1}^{M} \left(||\boldsymbol{y}_{f,m} - \boldsymbol{o}_{f,m}||_2^2 + \sum_{t=1, t\neq f}^{T-1} w_t ||\boldsymbol{y}_{t,m} - \boldsymbol{o}_{t,m}||_2^2 \right) \tag{6}$$

Note that $\boldsymbol{y}_{f,m}$ and $\boldsymbol{o}_{f,m}$ are the label value and output of the facial point detection task, respectively. Additionally, the weight w_t is constant for each iteration. We update the CNN parameters $\boldsymbol{\theta}$, such as weight filter and connection weight, using backpropagation in Eq. (3).

4 Experiments

We evaluate the proposed method by comparing its performance with those of the CNN for a single task and conventional heterogeneous learning. In these

Table 1. Parameters of heterogeneous learning CNN structure

Input	Image size	100×100
Layer 1	Filter size	$9 \times 9 \times 16$
	Maxout	2
	Max pooling	2×2
Layer 2	Filter size	$9 \times 9 \times 32$
	Maxout	2
	Max pooling	2×2
Layer 3	Filter size	$9 \times 9 \times 64$
	Maxout	2
	Max pooling	2×2
Layer 4	Sigmoid	200 (Dropout:50%)
Output		17

experiments, we perform facial point detection, gender recognition, race recognition, age estimation and smile ratio estimation. For facial point detection, we use regression estimation to detect five facial points: the left eye, right eye, nose, left mouth, and right mouth, Smile ratio estimation is identified as regression of the value between 0 and 99. Note that, smile ratio label is the average of some smile ratios that some people are given as labels. Age label is identified as regression of the value between 0 and 66. Race recognition is identified as Asian, White, or Black.

We employ a CNN that consists of three convolutional layers and three fully connectied layers, as shown in Table 1. In training of single task CNN, convolution layers and fully connected layers are have the same structure. In contrast, number of units in the output layer is equal to the number of classes for each facial attribute task. The total number of iterations to update the parameters is 1,000,000, the training coefficient η is set to 0.001, and the mini-batch size is 10. The comparison dataset consists of 53,663 facial images that were captured by aggregating face images from the Web. However, almost no published dataset has been given any facial attribute labels, because we created a facial attribute dataset that has been given five facial attribute. Note that the training sample consists of 42,663 images and the test sample consists of 11,000 images. The input images are 100×100 grayscale. We will publish this facial attribute dataset as soon as it is ready. The evaluation method of facial point detection is the same as that of Dantone et al. [2]. In age and smile ratio estimation, we judge estimation to be successful if the difference between output and label is connoted by the threshold, which are ± 5 years and 10%.

4.1 Comparison of Training Errors

Figure 4 shows the training error for each task for the proposed method. The training error for conventional heterogeneous learning is different from the

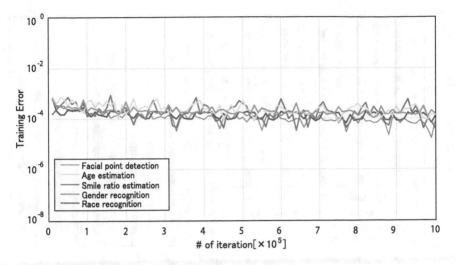

Fig. 4. Comparison of the training error for proposed method

convergence value for each task, and the training error varies suddenly for the recognition task, as shown in Fig. 1(a). Additionally, training errors of the proposed method for each task are lower overall than those of conventional heterogeneous learning. The proposed method has a unified training error for each task, and suppresses the dispersion training error variation. To achieve this result, the proposed method is stably trained by weighting the error function.

4.2 Comparison of Performance for the Proposed Method

In Fig. 5, we compare the performances of single task learning conventional heterogeneous learning and the proposed method. The accuracy of the regression tasks is lower for single task learning than conventional heterogeneous learning, especially for the facial point detection task. This is because facial point detection most negatively affects other tasks training errors, such as the difference between training error variation. Compared with conventional heterogeneous learning and the proposed method, we improved performance by approximately 5% and the accuracy of the facial point detection task by approximately 14%. This means that the proposed method is stably able to train by extracting available facial features.

Figure 6 shows an example of facial image analysis using conventional heterogeneous learning and the proposed method. The first and third columns show result of examples of conventional heterogeneous learning, and the second and fourth columns show results of the proposed method. Additionally, the left images are the input image and result of facial point detection by the conventional heterogeneous learning or proposed method, and the text on their right is results of subtasks such as gender, age, race, and smile ratio. The green points are facial points detected by conventional heterogeneous learning or proposed

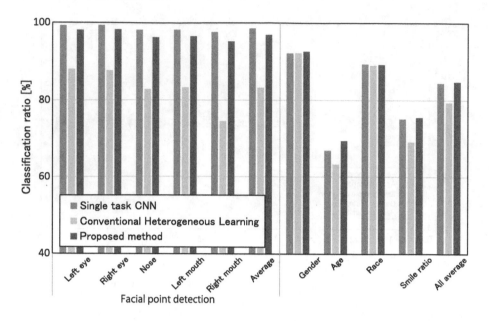

Fig. 5. Comparison of the proposed method and other method

method. The red text is inaccurate recognition or estimation. As shown in Fig. 6, we observe that the proposed method is robust to faces with large pose variation, lighting, and severe occlusion. Additionally, the processing time of our method is approximately 22 ms to analyze one image on an Intel Core i7-4790 (3.4 GHz) with 8 GB of memory, and the processing time is approximately 1.8 ms to analyze one image on GeForce GTX980.

5   Disscusion

In this section, we define the effectiveness of the proposed method by comparing various viewpoints of conventional heterogeneous learning and the proposed method.

5.1   Performance of Integrating Training Error Functions

In conventional heterogeneous learning, mean square error function and cross entropy error function are employed as error functions of the regression and recognition tasks, respectively. However, there are noticeable differences between the error ranges of the mean squared error function and the cross entropy error function. Thus, we integrate the error range from 0 to 1 by exchanging the cross entropy error function for the mean squared error function for recognition tasks. In this section, we evaluate integrating the error range by changing training error functions.

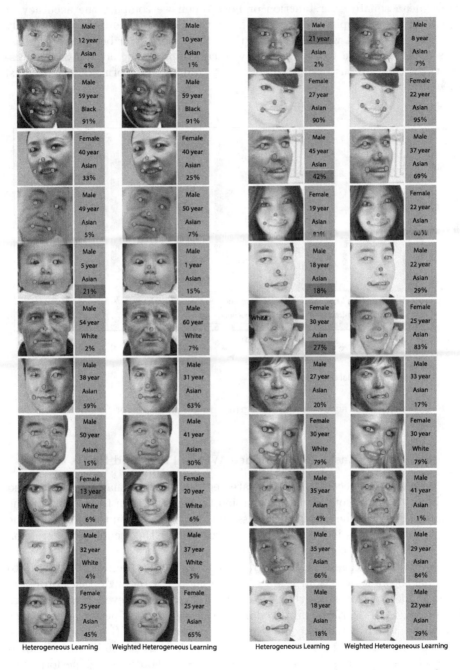

Fig. 6. Comparison of examples of facial image analysis (Color figure online)

Figure 7 shows the experimental results of classification accuracy that integrates mean square error function or not. When we compare the accuracy of regression tasks, proposed method is improved performance by approximately 20%. This mean that we can suppress the difference of error between regression tasks and recognition tasks, and this way can improve the accuracy of regression tasks that are susceptible to affect the training error of recognition tasks, especially.

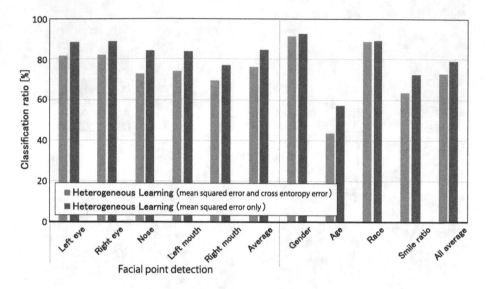

Fig. 7. Results of accuracy that integrates mean square error function or not

5.2 Regression Tasks Performance When Threshold Shifts

In experiments at Sect. 4.2, we evaluate regression tasks that set to be fixed threshold, and if the output of a regression task is over than the threshold, the output is correct. If the output of regression task is under than the threshold, the output is missing classification. Therefore, we evaluate the accuracy of regression tasks by shifting the threshold for each method.

Figure 8 shows classification accuracy that shifts the threshold between 5 to 20 for facial point detection. If we compare the CNN of single task, proposed method is less performance than CNN of single task. However, If we compare the heterogeneous learning, proposed method is significantly better performance than conventional heterogeneous learning. Figure 9(a) and (b) show classification accuracy that shifts the threshold between 5 to 20 for smile ratio estimation and age estimation, respectively. Proposed method is better performance than conventional heterogeneous learning and CNN of single task in age estimation. Improving the performance of age estimations was caused by improving

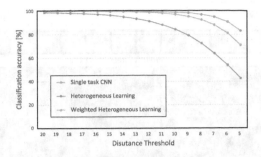

Fig. 8. Classification accuracy of facial point detection

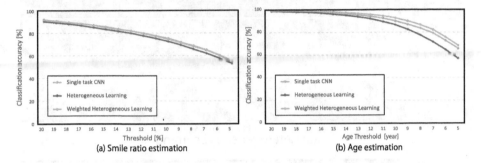

Fig. 9. Classification accuracy of smile ratio and age estimations

the facial point detection indebted proposed method. When CNN is trained with facial position by using heterogeneous learning, CNN easily focuses on facial part, and improving the performance by getting features that effectual estimate.

5.3   Visualization Weight Filters and Feature Maps

Weight filters and feature maps of CNN with heterogeneous learning are visualized in Fig. 10. Note that, we visualize them in the first layer. Figure 10(a) shows visualization of weight filters and feature maps of conventional heterogeneous learning, and Fig. 10(b) shows visualization of weight filters and feature maps of the proposed method. Weight filters of conventional heterogeneous learning are shown a clear contrast between light and shape, as shown in Fig. 10(a). However, conventional heterogeneous learning was outputted weak response at facial part such as eye and mouth in feature maps. On the other hand, weight filters of proposed method were noisy, nevertheless, proposed method was outputted strong response at facial part such as eye and mouth in feature maps, as shown in Fig. 10(b).

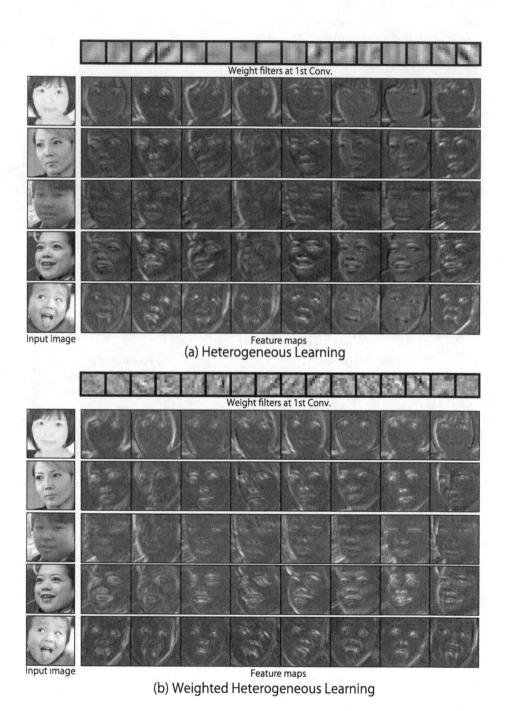

Weight filters at 1st Conv.

Input image   Feature maps
(a) Heterogeneous Learning

Weight filters at 1st Conv.

Input image   Feature maps
(b) Weighted Heterogeneous Learning

Fig. 10. Visualization weight filters and feature maps

6   Conclusion

In this paper, we proposed a method to improve the performance of heterogeneous learning for facial image analysis. As a result, compared with conventional heterogeneous learning, the proposed method improved performance by approximately 5% and the accuracy of the facial point detection task by approximately 14%.

References

1. Cootes, T.F., Edwards, G.J., Taylor, C.J.: Active appearance models. In: Burkhardt, H., Neumann, B. (eds.) ECCV 1998. LNCS, vol. 1407, pp. 484–498. Springer, Heidelberg (1998). doi:10.1007/BFb0054760
2. Dantone, M., Gall, J., Fanelli, G., Gool, L.V.: Real-time facial feature detection using conditional regression forests. In: Computer Vision and Pattern Recognition (2012)
3. Lian, H.-C., Lu, B.-L.: Multi-view gender classification using local binary patterns and support vector machines. In: Wang, J., Yi, Z., Zurada, J.M., Lu, B.-L., Yin, H. (eds.) ISNN 2006. LNCS, vol. 3972, pp. 202–209. Springer, Heidelberg (2006). doi:10.1007/11760023_30
4. Guo, G., Fu, Y., Dyer, C.R., Huang, T.S.: Image-based human age estimation by manifold learning and locally adjusted robust regression. IEEE Trans. Image Process. **17**, 1178–1188 (2008)
5. Krizhevsky, A., Ilva, S., Hinton, G.E.: ImageNet classification with deep convolutional neural network. In: Advances in Neural Information Processing System, vol. 25, pp. 1097–1105 (2012)
6. Sun, Y., Wang, X., Tang, X.: Deep convolutional network cascade for facial point detection. In: Computer Vision and Pattern Recognition (2013)
7. Zhou, E., Fan, H., Cao, Z., Jiang, Y., Yin, Q.: Extensive facial landmark localization with coarse-to-fine convolutional network cascade. In: IEEE International Conference on Computer Vision Workshops (2013)
8. Yamashita, T., Watasue, T., Yamauchi, Y., Fujiyoshi, H.: Facial point detection using convolutional neural network transferred from a heterogeneous task. In: International Conference on Image Processing (2015)
9. Kimura, M., Yamashita, T., Yamauchi, Y., Fujiyoshi, H.: Facial point detection based on a convolutional neural network with optimal mini-batch procedure. In: IEEE International Conference on Image Processing (2015)
10. Wu, Y., Ji, Q.: Discriminative deep face shape model for facial point detection. Int. J. Comput. Vis. **113**, 37–53 (2015)
11. Jourabloo, A., Liu, X.: Large-pose face alignment via CNN-based dense 3D model fitting. In: Computer Vision and Pattern Recognition (2016)
12. Yan, C., Lang, C., Wang, T., Du, X., Zhang, C.: Age estimation based on convolutional neural network. In: Ooi, W.T., Snoek, C.G.M., Tan, H.K., Ho, C.-K., Huet, B., Ngo, C.-W. (eds.) PCM 2014. LNCS, vol. 8879, pp. 211–220. Springer, Heidelberg (2014). doi:10.1007/978-3-319-13168-9_22
13. Eidinger, E., Enbar, R., Hassner, T.: Age and gender estimation of unfiltered faces. In: IEEE Press Transactions on Information Forensics and Security (2014)
14. Zhu, Y., Yan, L., Guowang, M., Guodong, G.: A study on apparent age estimation. In: IEEE International Conference on Computer Vision Workshops (2015)

15. Kuang, Z., Huang, C., Zhang, W.: Deeply learned rich coding for cross-dataset facial age estimation. In: IEEE International Conference on Computer Vision Workshops, pp. 96–101 (2015)
16. Tivive, F.H.C., Bouzerdoum, A.: A gender recognition system using shunting inhibitory convolutional neural networks. In: Neural Networks, pp. 5336–5341 (2006)
17. Antipov, G., Berrani, S.A., Dugelay, J.-L.: Minimalistic CNN-based ensemble model for gender prediction from face images. Pattern Recogn. Lett. **70**, 59–65 (2015)
18. Levi, G., Hassner, T.: Age and gender classification using convolutional neural networks. In: Computer Vision and Pattern Recognition (2015)
19. Argyriou, A., Evgeniou, T., Pontil, M.: Convex multi-task feature learning. Mach. Learn. **73**(3), 243–272 (2008). Kluwer Academic Publishers
20. Zhang, Z., Luo, P., Loy, C.C., Tang, X.: Facial landmark detection by deep multi-task learning. In: Fleet, D., Pajdla, T., Schiele, B., Tuytelaars, T. (eds.) ECCV 2014. LNCS, vol. 8694, pp. 94–108. Springer, Heidelberg (2014). doi:10.1007/978-3-319-10599-4_7
21. Devries, T., Biswaranjan, K., Taylor, G.W.: Multi-task learning of facial landmarks and expression. In: Computer and Robot Vision (2014)
22. Zhang, Z., Luo, P., Loy, C.C., Tang, X.: Learning deep representation for face alignment with auxiliary attributes. arXiv preprint arXiv:1408.3967 (2015)
23. Yim, J., Jung, H., Yoo, B., Choi, C., Park, D., Kim, J.: Rotating your face using multi-task deep neural network. In: Computer Vision and Pattern Recognition (2015)
24. Ranjan, R., Vishal, M.P., Chellappa, R.: HyperFace: a deep multi-task learning framework for face detection, landmark localization, pose estimation, and gender recognition. arXiv preprint arXiv:1603.01249 (2016)
25. Rumelhart, D.E., Hinton, G.E., Williams, R.J.: Learning representations by back-propagating errors. In: Neurocomputing, pp. 696–699 (1988)

Reliable Age Estimation Based on Apt Gabor Features Selection and SVM

ArulMurugan Ambikapathi[1(✉)], Yi-Tseng Cheng[2], Gee-Sern(Jison) Hsu[2], and Cheng-Hua Hsieh[2]

[1] Utechzone Co. Ltd., New Taipei City, Taiwan
aareul@ieee.org
[2] Artificial Vision Laboratory, Mechanical Engineering,
National Taiwan University of Science and Technology, Taipei, Taiwan
jison@mail.ntust.edu.tw

Abstract. Automatic and reliable facial image-based age estimation is becoming an intriguing research in computer vision and other related applications. The degrees of apparent visibility of the ageing process on a person's face significantly differ from person to person, and thereby making facial image analysis based age estimation, a great challenge. In this paper, we propose a robust method for age estimation based on detailed feature analysis of the frontal facial image of a person. We employ the Gabor transformation based filters for feature extraction and implement exhaustive search to find the most appropriate orientations, scales and sizes of Gabor filters, for accurate age estimation using multi-class SVM classifiers. The deployment of conventional Adaboost classifier is also discussed to show the significance of our approach that involves exhaustive search. The proposed method is evaluated and compared on a standard FG-NET database according to the Leave-One-Person-Out (LOPO) protocol, and Mean Absolute Error (MAE), along with the Cumulative Score are adopted as quantification performance measures. The experimental results demonstrates the superior efficacy of the proposed method over many of the state-of-the-art methods under test.

1 Introduction

In recent years, age estimation has become an important research of concern in the field of face recognition. The applications of facial image based age estimation ranges from a simple facial attribute recognition to highly sensitive criminal investigations. Apart from common imaging problems such as illumination, pose variation and expression change, the reflection of human ageing on a person's face could differ drastically due to varying life styles and working environments. In addition, challenges get enhanced with the usage of cosmetics, wearing eye glasses, and ageing pattern differences among people from different ethnicity.

The major existing categories of facial image based age estimation procedures include the following [1]: Anthropometric models (AM), Active appearance models (AAM), Ageing pattern subspace (AGES), Age manifold (AMF), and

© Springer International Publishing AG 2017
C.-S. Chen et al. (Eds.): ACCV 2016 Workshops, Part II, LNCS 10117, pp. 407–416, 2017.
DOI: 10.1007/978-3-319-54427-4_30

Appearance models (APM). The AM describes the cranio-facial growth, which is observed to be the greatest change in the early age, i.e. from birth to adulthood [2]. From adulthood to old age, cranio-facial shape changes slightly and the most apparent change becomes the skin texture. Kwon and Lobo [3] combined AMs and wrinkle information to classify human age into three categories, i.e. child, young adult and senior adult. The AAM is a statistical model generated by combining a model of shape variation with a model of intensity variation in a sample of shape-normalized images, based on Principal Component Analysis (PCA). For instance, [4] is an AAM-based approach for age estimation, in which an ageing function has been proposed and performance comparisons between different classifiers are discussed. On the other hand, the AGES consists of a group of algorithms to model ageing pattern that is defined as a sequence of personal images sorted in temporal order, by constructing a representative subspace. Geng et al. [5] proposed an AGES method to model and represent the ageing pattern. The age of a test image is determined according to the projection score to the respective age-group representative subspace which yields the minimum reconstruction error. AMF uses supervised manifold embedding technique to learn the age image distribution on an intrinsic low-dimensional manifold. Some typical AMF approaches such as Orthogonal Locality Preserving Projections (OLPP), Conformal Embedding Analysis (CEA), Locally Adjusted Robust Regression (LARR) and Synchronized Sub-manifold Embedding (SSE) can be found in [6, 7].

Finally, APM aims to extract age-related features based on the facial appearance, more specifically, facial features. The proposed work belongs to this category. It is worth to recall the existing works that belongs to this group. Gunay et al. utilized the texture descriptor Local Binary Pattern (LBP) [8] for extracting the facial features. Gao et al. employed Gabor features [9], while Yan et al. proposed to use Spatially Flexible Patch (SFP) as local feature descriptor [10]. Suo et al. used four types of features: topology, geometry, photometry and configuration, and devised a hierarchical face model for age estimation [11]. The system of Guo et al. used Biologically Inspired Features (BIF), which is based on a feed-forward model of the primate visual object recognition pathway [7].

AM might be useful to distinguish minors from adults, but not appropriate for classifying different adult ages. Besides, AM is sensitive to the head pose and ignores useful texture information. AAM considers both shape and texture information, but requires precise feature points location for shape fitting, which is difficult to achieve in natural environment. To implement AGES, a complete sequence of personal images of different ages is necessary, which is not available in most of existing age databases. Applying PCA method iteratively is a way to handle this problem but this procedure increases the complexity of this algorithm. AMF requires large amount of images labeled with the corresponding age to learn the embedded manifold with statistical sufficiency.

Owing to the above mentioned limitations associated with the AM, AAM, AGES, and AMF, in this work we endeavor to develop a reliable AMF based approach. Again, the most sensitive and important aspect for AMF based approach will be the precise definition and extraction of the age-related features. The proposed method applies Gabor transformation based filters (Gabor filters) to extract appearance features, and implement exhaustive search to identify the set of Gabor filters with particular orientation, scale and filter size, that can generate the most discriminative features for age estimation. Features generated by each and every Gabor filter are labeled with the corresponding age-group and trained to build a multi-class support vector machine (SVM) classifier, where the number of classes corresponds to the number of age-groups that need to be classified. After the training, an exhaustive search is performed, and the n classifiers with final class scores that closely matches with the age-groups are chosen and their corresponding Gabor parameters will be the ones considered for feature extraction. Thus, there are n Gabor filters chosen, which results in n multi-class SVM classifiers. The age of a test image is then determined by averaging the n classification results. The conventional Adaboost classifier is also considered for comparison with the proposed strategy that involves exhaustive search. Though the exhaustive search approach is naive, it is highly effective in most of the scenarios (as will be demonstrated in Sect. 3). Unlike traditional methods that concatenates features generated by a bank of Gabor filters together to build one classifier, the proposed classification design can deal with the huge dimensionality issue by appropriate selection and usage of the Gabor features.

The rest of the paper is organized as follows. The proposed age estimation method is introduced in Sect. 2, along with Gabor feature extraction and the classification design. Experiment results and comparisons are presented in Sect. 3. Section 4 summarizes the work, and some conclusions and future directions are discussed.

2   The Proposed Age Estimation Methodology

In this section, we will present the proposed age estimation methodology. Specifically, we will discuss the details pertaining to the utilization of Gabor filters to extract facial appearance features, and the employment of exhaustive search to identify the most suitable Gabor filters with specific orientation, scale and filter size, for age estimation. A simple, but effective exhaustive search based multi-class SVM classification design to solve the high dimensionality problem in the Gabor features extraction is described alongside. The typical Adaboost classification method is briefly presented so as to facilitate a later discussion on the merit of the exhaustive search based method when compared to the Adaboost feature selection method, in Sect. 3.

2.1   Gabor Feature Extraction and Classification Design

It is apparent that convolution with a detailed bank of Gabor filters with different orientation and scales provides a more complete description of a given image. The Gabor filter can be formulated as the following equation:

$$\psi_{\mu,\nu}(z) = \frac{\|k_{\mu,\nu}\|^2}{\sigma^2} e^{-(\|k_{\mu,\nu}\|^2 \|z\|^2 / 2\sigma^2)} \left[e^{ik_{\mu,\nu}z} - e^{-(\sigma^2/2)} \right] \tag{1}$$

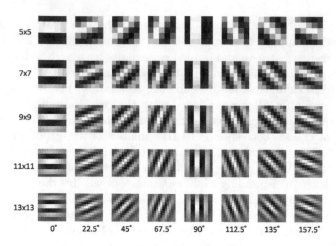

Fig. 1. Displaying discrete Gabor filters with different filter sizes and orientations for a fixed scale $\nu = 0$

where μ and ν determine the orientation and scale of the Gabor filter, $z = (x, y)$ represents the pixel position, $\|\cdot\|$ is the norm operator, and wave vector $k_{\mu,\nu}$ is defined as $k_{\mu,\nu} = k_\nu e^{i\phi\mu}$. Where $k_\nu = k_{\max}/f^v$ and $\phi_\mu = \pi\mu/8$. f is the frequency spacing factor. In this work, we have set $\sigma = 2\pi$, $k_{\max} = \pi/2$ and $f = \sqrt{2}$.

To utilize Gabor wavelets transformation, it is common to adopt five different scales and eight orientations, i.e. $\nu \in \{0, \cdots, 4\}$ and $\mu \in \{0, \cdots, 7\}$. Different filter sizes are jointly taken into consideration in this work, which is seldom considered in other researches. Figure 1 shows discrete Gabor filters with different orientations and filter sizes for the same scale $\nu = 0$. Though Gabor features extraction results in stronger feature description for a given image, this method is computationally expensive due to the large feature dimensions involved (owing to varying orientation values, scaling values, and kernel sizes). To identify and extract the most representative Gabor features, we have performed an investigation based on exhaustive search of the Gabor filters under different combinations of orientation, scale and filter size. The detailed procedure is described below.

For each image in the training data set, both the real and imaginary components of a Gabor-filtered image are recorded and concatenated into a single feature vector. Then, features generated by each Gabor filter are labeled with the corresponding age-group and trained to build a multi-class support vector machine (SVM) classifier pertaining to that particular Gabor filter. Here, the number of classes correspond to the number of age-groups that need to be classified. Thus, in the training phase, totally $5 \times 8 \times 5 = 200$ multi-class SVM classifiers are generated as 5 scales, 8 orientations and 5 filter sizes are considered in this work, and each test image will then result in 200 classification scores. Following which, the Gabor filters that owns the most discriminative power for age estimation are identified as the ones that has the classification scores close to the true age-group values. Precisely, the mean absolute error (MAE, defined in Sect. 3) of SVM classification results is calculated and the best combination out of the 200 SVM classifiers is identified as the set of classifiers that generates the minimum MAE. Once the set of apt Gabor filters are identified, they are then used to extract the corresponding Gabor features from the image during the testing phase, and the class scores obtained by those corresponding classifiers are noted. The age of the person in a given test image is then finally determined by averaging the classification scores.

2.2 Adaboost Feature Selection

For sake of comparison and evaluation, we also briefly discuss and employ the conventional Adaboost classifier. The Adaboost is a machine learning, classification algorithm that constructs a strong classifier by linear combination of weak classifiers with corresponding weighting factors produced in the learning process. The weighting factors are adaptively learned according to the mis-classified samples. The string classifier $H(\times)$ is generated with Adaboost algorithm and can be formulated as $H(\times) = sign\left(\sum_{t=1}^{T} \alpha_t h_t(\times)\right)$, where h_t is the weak classifier produced in iteration t during learning process, α_t is the weight of each weak classifier, and T is the total number of iterations. In the ensuing section, we will also experimentally compare the exhaustive search based classification accuracy with that of Adaboost classifier.

3   Experimental Analysis and Discussions

To validate the feasibility and effectiveness of our method, it is evaluated on the standard FG-NET ageing database [10]. FG-NET database contains 1002 gray-level or color facial images, formed by 82 subjects, aged from 0 to 69. Each subject has 6 to 18 images. This database contains images under challenging factors including illumination variation, pose variation, expression change, and wearing accessories. Figure 2 shows some of the samples from this database.

To most efficiently use this database, Leave-One-Person-Out (LOPO) protocol is adopted here. The LOPO protocol refers to the partition of data into N-1

Fig. 2. Samples from FG-NET ageing database

samples (here person) for training and 1 for testing, when N is not sufficiently large. Note that there are several images available for each person. This protocol is a common practice when dealing with limited data available for an statistical study. Mean Absolute Error (MAE) and Cumulative Scores (CS) are computed to measure the performance of the proposed method and other contemporary state-of-the-art methods under test. The MAE is defined as

$$MAE = \frac{1}{N} \sum_{K=1}^{N} \left| \hat{l}_k - l_k \right|, \tag{2}$$

where $\hat{l}_k$ and l_k represent the estimated age and the ground truth of the kth image, respectively, and N is the number of test samples. The CS is defined as

$$CS(j) = N_{e \leq j}/N \times 100\%, \tag{3}$$

where $N_{e \leq j}$ is the number of test images that results in an absolute error no higher than j years.

Face and eye detector developed by Viola and Jones [12] are applied on each image, to identify the ROIs. Image pre-processing performed in this experiment includes size normalization, in-plane rotation elimination, face cropping, and histogram equalization. The former three is based on the detected eyes location and the final cropped face image consists of 64 × 64 pixels. The investigation on the number of Gabor filters shows that the performance advances with combining more Gabor filters. The minimum MAE obtained is 4.97 by the combination of 13 Gabor filters (identified by exhaustive search, from the 200 Gabor filter based SVM classifiers). The finalized 13 Gabor filters are shown in Fig. 3.

For comparison, the following state-of-the-art methods are considered in addition to the methods [5,7,10], mentioned above in Sect. 1. While [13] considered the automatic design of a regressor from the training samples to identify the age labels, [14] proposed a method to address the problem of ordinal/rank label prediction. Two AMF based approaches proposed in [15,16] are noteworthy, and hence considered in this experiment. In addition, a Probabilistic Fusion

Filter size	scale(v) · orientation(μ)	Gabor wavelets	Filter size	scale(v) · orientation(μ)	Gabor wavelets
5 X 5	v = 0, μ = 5		11 X 11	v = 0, μ = 3	
5 X 5	v = 2, μ = 4		11 X 11	v = 0, μ = 4	
7 X 7	v = 0, μ = 4		13 X 13	v = 2, μ = 5	
7 X 7	v = 1, μ = 0		13 X 13	v = 3, μ = 3	
9 X 9	v = 1, μ = 5		13 X 13	v = 3, μ = 5	
9 X 9	v = 2, μ = 0		13 X 13	v = 4, μ = 6	
9 X 9	v = 3, μ = 6				

Fig. 3. Most discriminative Gabor filters for age estimation, as selected by the exhaustive search.

Table 1. MAE of the proposed method for various age ranges using FG-NET Database

FG-NET		
Range	#img	Ours
0–9	371	2.96
10–19	339	3.29
20–29	144	5.38
30–39	79	10.46
40–49	46	13.81
50–59	15	19.69
60–69	8	29.31
Total	1002	4.97

Table 2. MAE comparison of all the methods under test for the images in the FG-NET Database

Method	FG-NET
AGES [5]	6.77
RUN [13]	5.78
BM [14]	5.33
SSE [15]	5.21
LARR [16]	5.07
PFA [17]	4.97
RPK [10]	4.95
BIF [7]	4.77
Ours	4.97

Approach (PFA) [17] is also evaluated. The Table 1 presents the MAE obtained by the proposed method, for various age-group ranges, and Table 2 is the MAE (averaged over all the age ranges) comparisons for all the methods under test. It is evident that the performance of the proposed method is better than many of the state-of-the-art methods and comparable to other methods under test. On the other hand, Fig. 4, shows the performance of different approaches in terms of CS, for the age range of 0 to 10 years. Here again, the performance of the proposed method is on par with the state-of-the-art methods, under test. It is interesting to note that, almost 90% of the estimated errors occurs for age less than or equal to 10 years, for most methods, which indicates that the age estimation in this age group (0 to 10) is the most challenging scenario. Figure 4 further reveals that there is 10% errors for ages greater than 10 years.

The next experiment is regarding the comparison of the performance of the proposed method with those of Adaboost classifier. For computational efficiency, we randomly separated the database into two sets 80% for training and 20%

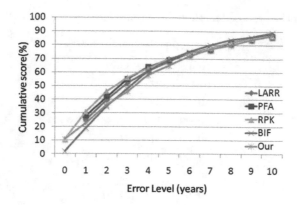

Fig. 4. Cumulative scores for various age estimation methods for error levels from 0 to 10 years.

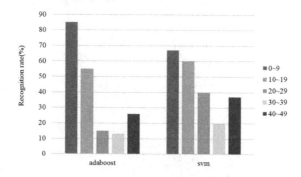

Fig. 5. Comparison of Recognition rates of Adaboost method and the proposed method.

for testing. Facial images are convolved by Gabor filters and the features are recorded, such that each feature vector comprises $64 \times 64 \times 5 \times 8 \times 5$ elements (5 scales, 8 orientations and 5 filter sizes). Figure 5 is the comparison of the recognition rates with Adaboost classifier and the proposed exhaustive search based method. It can be inferred that Adaboost classification accuracy is superior to the proposed method only in age range from 0 to 9; in all the other age ranges, the proposed method significantly fairs well. Finally, the minimum, maximum and average MAE while using Adaboost algorithm is 6.21, 7.72 and 6.93 respectively; as compared to 5.51, 6.84 and 6.02, obtained by using exhaustive search based SVM classifier. This further validates that for age estimation, the proposed method also outperforms the Adaboost classifier based age estimation.

4   Conclusion and Future Directions

In this work, we have proposed an effective age estimation methodology based on the extraction and selection of apt Gabor features, thereby smartly handling the issue of high dimensionality of the Gabor features. The resulting Gabor features that are selected based on the exhaustive search approach are then used to classify the age-group of the give test image according to the scores from the respective trained multi-class SVM classifiers. Comparisons to other approaches illustrated the efficacy of the proposed method. Combining multiple age-related features, along with the Gabor features is currently under investigation. Some of the potential future directions in the line of this work include more robust age estimation particularly for the age group 0–10 years, analysis of more robust classifiers for multi-class age classification, and building a strong bench-mark data base for more rigorous validation of the age estimation methodologies.

References

1. Fu, Y., Guo, G., Huang, T.S.: Age synthesis and estimation via faces: a survey. IEEE PAMI **32**, 1955–1976 (2010)
2. Todd, J.T., Mark, L.S., Shaw, R.E., Pittenger, J.B.: The perception of human growth. Sci. Am. **242**, 1–32 (1980)
3. Kwon, Y.H., Lobo, D.V.: Age classification from facial images. Comput. Vis. Image Underst. **74**, 1–21 (1999)
4. Lanitis, A., Draganova, C., Christodoulou, C.: Comparing different classifiers for automatic age estimation. IEEE Trans. Syst. Man Cybern. Part B (Cybern.) **34**, 621–628 (2004)
5. Geng, X., Zhou, Z.H., Zhang, Y., Li, G., Dai, H.: Learning from facial aging patterns for automatic age estimation. In: Proceedings of the 14th ACM International Conference on Multimedia, pp. 307–316. ACM (2006)
6. Fu, Y., Huang, T.S.: Human age estimation with regression on discriminative aging manifold. IEEE Trans. Multimedia **10**, 578–584 (2008)
7. Mu, G., Guo, G., Fu, Y., Huang, T.S.: Human age estimation using bio-inspired features. In: IEEE Conference on Computer Vision and Pattern Recognition, CVPR 2009, pp. 112–119 (2009)
8. Gunay, A., Nabiyev, V.V.: Automatic age classification with LBP. In: 23rd International Symposium on Computer and Information Sciences, ISCIS 2008, pp. 1–4. IEEE (2008)
9. Gao, F., Ai, H.: Face age classification on consumer images with Gabor feature and fuzzy LDA method. In: Tistarelli, M., Nixon, M.S. (eds.) ICB 2009. LNCS, vol. 5558, pp. 132–141. Springer, Heidelberg (2009). doi:10.1007/978-3-642-01793-3_14
10. Yan, S., Zhou, X., Liu, M., Hasegawa-Johnson, M., Huang, T.S.: Regression from patch-kernel. In: IEEE Conference on Computer Vision and Pattern Recognition, CVPR 2008, pp. 1–8 (2008)
11. Suo, J., Wu, T., Zhu, S., Shan, S., Chen, X., Gao, W.: Design sparse features for age estimation using hierarchical face model. In: 8th IEEE International Conference on Automatic Face Gesture Recognition, FG 2008, pp. 1–6 (2008)
12. Viola, P., Jones, M.: Robust real-time object detection. Int. J. Comput. Vis. **4**, 34–47 (2001)
13. Yan, S., Wang, H., Tang, X., Huang, T.S.: Learning auto-structured regressor from uncertain nonnegative labels. In: 2007 IEEE 11th International Conference on Computer Vision, pp. 1–8 (2007)
14. Yan, S., Wang, H., Huang, T.S., Yang, Q., Tang, X.: Ranking with uncertain labels. In: 2007 IEEE International Conference on Multimedia and Expo, pp. 96–99 (2007)

15. Yan, S., Wang, H., Fu, Y., Yan, J., Tang, X., Huang, T.S.: Synchronized submanifold embedding for person-independent pose estimation and beyond. IEEE Trans. Image Process. **18**, 202–210 (2009)
16. Guo, G., Fu, Y., Dyer, C.R., Huang, T.S.: Image-based human age estimation by manifold learning and locally adjusted robust regression. IEEE Trans. Image Process. **17**, 1178–1188 (2008)
17. Guo, G., Fu, Y., Dyer, C.R., Huang, T.S.: A probabilistic fusion approach to human age prediction. In: IEEE Computer Society Conference on Computer Vision and Pattern Recognition Workshops, CVPRW 2008, pp. 1–6 (2008)

VFSC: A Very Fast Sparse Clustering to Cluster Faces from Videos

Dinh-Luan Nguyen$^{(\boxtimes)}$ and Minh-Triet Tran

University of Science, VNU-HCMC, Ho Chi Minh City, Vietnam
1212223@student.hcmus.edu.vn, tmtriet@fit.hcmus.edu.vn

Abstract. Face clustering is a task to partition facial images into disjoint clusters. In this paper, we investigate a specific problem of face clustering in videos. Unlike traditional face clustering problem with a given collection of images from multiple sources, our task deals with set of face tracks with information about frame ID. Thus, we can exploit two kinds of prior knowledge about the temporal and spatial information from face tracks: sequence of faces in the same track and contemporary faces in the same frame. We utilize this forehand lore and characteristic of low rank representation to introduce a new light weight but effective method entitled Very Fast Sparse Clustering (VFSC). Since the superior speed of VFSC, the method can be adapted into large scale real-time applications. Experimental results with two public datasets (BF0502 and Notting-Hill), on which our proposed method significantly breaks the limits of not only speed but also accuracy clustering of state-of-the-art algorithms (up to 250 times faster and 10% higher in accuracy), reveal the imminent power of our approach.

1 Introduction

Object clustering is a work of organizing/grouping objects based on their characteristics. The characteristics here can be either low-level features, like raw input pixels, or high-level features extracted by a specific descriptor. Since clustering is one of the essential parts leading to success of other areas such as segmentation [1], tracking [2], recognition [3] and still challenged by noisy input data, it is really an allured field to be fully exploited.

Face clustering in video is a specific area of object clustering but it has special interests because it possesses wide range of applications in daily life, including auto tagging/naming character, video organization, video summarization, content retrieval, etc. Although several efforts [4,5] have been proposed to solve this task, it is still challenging and inspiring by the difficulties of real-life captured video input. For instance, the uncontrolled environments surrounding faces lead to the huge face appearances, erratic illumination, head position, pose and facial expression. Furthermore, occlusion and blur may occur during capturing faces by character's gestures or actions. Thus, these are the main reasons for face clustering not being regarded as a saturated area.

© Springer International Publishing AG 2017
C.-S. Chen et al. (Eds.): ACCV 2016 Workshops, Part II, LNCS 10117, pp. 417–433, 2017.
DOI: 10.1007/978-3-319-54427-4_31

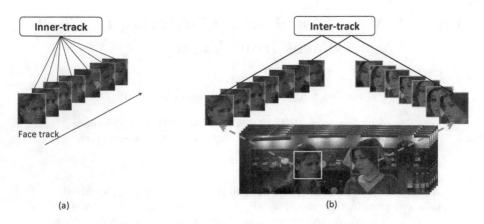

Fig. 1. Prior knowledge of face tracks in video extracted from "Buffy the Vampire Slayer" TV series. (a) Facial images in the same track must belongs to one person (Inner-track relation); (b) Two faces are concurrent in the same frame or two tracks overlap at least one frame are the possession of different people (Inter-track relation).

There is an existing misconception between face clustering and face recognition that we need to clarify before moving on. Face clustering, as mentioned above, is an unsupervised problem in which no labeled faces are given before processing. Whereas face recognition utilizes labeled images to train a classifier to return person face ID in the test set. This procedure is supervised problem.

In clustering faces retrieved from video, there is some information we can exploit. First, face images are not discrete. They usually come in sequences called face tracks, obtained by detection or tracking stage. In this paper, the input tracks for processing do not contain any contamination. In specific, all images from one track belong to one person and exclude any images from other. This leads to two other following characteristics [6]: *Inner-track*: images in one track belongs to the same person; and *Inter-track*: tracks having concurrent images are usually related to different people. These characteristics are useful to boost up the accuracy of the clustering process described in our proposed system. Figure 1 visualizes these two properties in clustering facial images in video scenario.

To deal with clustering problem, several works [7–9] have been proposed to split data into several clusters corresponding to relevant subspaces. The idea of representing one data point as the linear combination of others in the identical subspace [8,10] seems to get promising results with low-rank representation [8,9, 11]. However, these approaches neglect information from face tracks and require high computation cost. Besides, the data-driven method with prior knowledge [6], called probabilistic constraint, successfully exploits the valuable characteristic of faces in video but still gets low accuracy in comparison with sparse methods.

Main Contribution. Inspirited by these two main directions, in this paper, we present a new light weight approach inherited the idea of constraint and sparse representation methods to foster the accuracy and make our method applicable in real time. In specific, we create a sparse low-rank data representation integrated with prior knowledge from face tracks. This representation is a coefficient matrix which pulls images in one cluster closer and push images from different clusters far away. Sparsity appears for the linear performance of faces within tracks. In short, there are three main contributions in our work.

- Create a coalescence between sparse clustering and constraint knowledge characteristic.
- Explore an adaptive light weight method to significantly reduce computational complexity and boost up running time.
- Address the bias of evaluation protocol in previous works and proposed a new justified one.

Experiments in two public datasets show that our proposed method is superior than previous works and becomes a new state-of-the-art in this area in not only accuracy but also speed.

The rest of our paper is organized as follows: Sect. 2 reviews some related works on sparse subspace and constraint clustering methods. Our primary contributions for proposing a new effectively light weight clustering method for video faces in the wild are carefully discussed in Sect. 3. Section 4 shows experimental results and comparisons to other state-of-the-art techniques on two face datasets from real-world videos. Finally, conclusion and our discussion are given in Sect. 5.

2   Related Works

A lot of existing works [12–14] utilize data-driven methods, which focus on creating favorable distance metrics or transforming given data into new spaces, to foster inter-track differences. The first endeavor is clustering faces in videos by using affine invariant distance measurement [12]. Manifold-Manifold Distance (MMD) method proposed by Wang *et al.* [15] divides a nonlinear manifold into local linear subspaces. Specifically, from one subspace in the involved manifolds, it amalgamates distance between subspace pairs. Work of Arandjelovic and Cipolla [16] exploits the coherence of disparities between manifolds by grouping faces appearance in an anisotropic space. Another similar approach is to combine clustering with Bayesian method [17] to count dissimilar people in face images. In addition, Wolf *et al.* [18] propose Matched Background Similarity (MBGS) measurement to point out the distinction between face images having similar background. However, all mentioned works heavily depend on the data quality. Thus, they are unsteady in real-world videos. There are works [19,20] that make use of information from face tracks by modifying the distance matrix so that faces with inner-track relation come closer whereas faces in inter-track are thrust far away. Based on the Hidden Markov Random Fields (HMRF) model, HMRF-com [6], a probabilistic constrained clustering method, gets competitive results by exploiting prior knowledge of faces in video. However, this HMRF method and other related works [20,21] require high computation, which is inapplicable for real time systems.

Another kind of approach is to use subspace clustering methods. Several works [7–9] propose solutions for face clustering problem. While the problem in this paper is face images from videos in the wild, those works use the ideal or rectified images to process. Furthermore, they do not utilize the prior knowledge in video track effectively. Shijie *et al.* [22] propose Weighted Block-Sparse Low Rank Representation (WBSLRR) and claim that they exploit useful information from face images by integrating this information into their self-expressive matrix. This strategy seems to get promising results in comparison with other methods in spectral subspace clustering. However, all of these works require heavy parameter tuning for optimization. Thus, it takes huge time to process, depends on input data and is far from being applicable into real world. Based on these two trends of clustering mentioned above, it is essential to have a novel approach that does not rely on major parameter tuning but still produce good performance in real time. The following system proposed in Sect. 3 will resolve the inquiry.

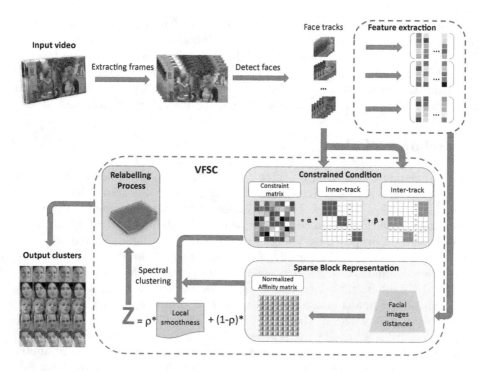

Fig. 2. An end-to-end system for face clustering in video. From given input video, we pre-process each frames to get face tracks and image features. Information from face tracks is an input for Constrained Condition module while image features are the input for Sparse Block Representation one. Constraint and affinity matrix returned from these module are used to create final data representation matrix Z. As a high accuracy face detector requires long processing time to complete its task, we do not apply it on every frame but only selected frames. Each detected face from interval frames is used as the input for the tracking process.

3 Proposed Method

3.1 Problem Specification

Given a set of unlabeled data

$$X = \{X_i\}_{i=1}^m$$

where $X_i \in R^{d \times n_i}$ denotes the ith track in the total m face tracks, d is the dimension of each image vector. The goal is to divide X into K separate groups, where each group comprises tracks of the same subject. The number of faces in the dataset is $n = \sum_{i=1}^m n_i$, where n_i is the number of face images per track. The information of prior knowledge can be formulated into $n \times n$ symmetric matrices $W^{(in)} = \left\{ W_{ij}^{(in)} \right\}$ and $W^{(it)} = \left\{ W_{ij}^{(it)} \right\}; i, j = \overline{1 \dots n}$, where $W_{ij}^{(in)} = 1$, $W_{ij}^{(it)} = -1$ if the ith and j^{th} images belong to inner-track or inter-track relations, respectively.

3.2 System Architecture

The modules of proposed method in an end-to-end clustering video face system is illustrated in Fig. 2. Our system has 3 parts: sparse block representation, constrained condition, and relabelling process. To demonstrate the ability to integrate into end-to-end real-time application, from the given video, we apply face detector [23,24] to extract faces from interval frames. Since the high accuracy face detector take times to complete its task, we use these detected faces to stimulate for tracking process. An online tracking [25] is used to accelerate the extracting frame procedure.

Because the tracks'length returned by detection or tracking phase is different from others, we normalize it by subsampling $n_i = N; i = \overline{1 \dots m}$ images from each track. For instance, given a L-length track, we choose N images at position $1 + k * (L/N); k = \overline{0 \dots N-1}$ to represent this track. This way of sampling has the advantage of reserving the characteristics of the track in comparison with other methods such as choosing the first-N or last-N images in each track.

3.3 Sparse Block Representation

We inherit the idea of sparse subspace clustering [10], the problem can be regarded as finding the optimal $n \times n$ face relationship representation matrix Z such that $X = ZX$. Then, by applying spectral clustering method [7,11], we get the final face clustering result. As X is collected from tracks of face images, it can be seen as a union of linear subspaces [8]. Thus to solve the problem, we must solve the convex approximation of $rank(Z)$ [8,26]:

$$\min_Z \|Z\|_* \quad \text{s.t.} \, X = ZX$$

where $\|Z\|_*$ denotes the nuclear norm of Z. Traditional methods requires objective functions to adapt to the optimization of Z like $\|Z\|_1$ [10], $\|Z\|_F^2$ [7], or even adding new regularization terms of matrix itself [11,22]. However, in the paper, we propose to manage matrix Z based on input data X without heavy parameters for optimization. Thus, this technique has the advantage of being light weight and applicable in real time.

We apply PCA [27] to reduce the dimension of input data and lessen the computation cost. Based on the distances between data points, we construct a list of k-nearest neighbors for each point. Thus, we get n lists corresponding to n points in X. Specifically, let $\mathcal{N}_p(x_i)$ denotes the p nearest neighbors of x_i, matrix $D = \{d_{ij}\}; i, j = \overline{1 \dots n}$, where

$$d_{ij} = \begin{cases} 0, & x_j \notin \mathcal{N}_p(x_i) \text{ or } i = j \\ exp(-d(x_i, x_j)/\sigma_i \sigma_j), & \text{otherwise} \end{cases} \tag{1}$$

is created to represent the affinity between data points. We construct $\sigma_i = d(x_i, x_p)$, where x_p is the pth nearest neighbor of x_i and the distance is measured as L_2 norm. We use $L2$ for the default metric because it is robust with the range of input value X in general. Thus, we get the normalized affinity matrix $L = G^{-\frac{1}{2}} D G^{-\frac{1}{2}}$ where G is a diagonal matrix and $G_{ii} = \sum_{j=1}^{m} D_{ij}^2$. This matrix is integrated with constrain matrix in Sect. 3.4 to get final data representation matrix Z.

3.4   Constrained Condition

As in Sect. 3.1, by exploiting prior knowledge, we get $n \times n$ symmetric matrix $W^{(in)}$ and $W^{(it)}$ corresponding to inner-track and inter-track relation respectively. To improve the non-adaptive constrained condition in [6], which treats these matrices equally, we create new relation matrix W as follows:

$$W = \alpha W^{(in)} + \beta W^{(it)} \tag{2}$$

where α, $\beta \in \mathbb{R}$ are the adaptive weight of inner-track and inter-track relation respectively, which are dependent on the input dataset. The Eq. 2 has the advantage that it does not treat $W^{(in)}$ and $W^{(it)}$ as the same importance. These weights are adaptive to the character of the input data. Since $W^{(in)}$ and $W^{(it)}$ are sparse, constrained matrix W also be sparse. Let $\mu = \arg\min_W W$ and $\nu = \arg\max_W W$, we get $\mu \leq W_{ij} \leq \nu$ where $w_{ij} \rightarrow \mu$ means the relationship between face image i^{th} and j^{th} is close to inter-track and $W_{ij} \rightarrow \nu$ means it in the vicinity of inner-track relation. Matrix W can be explained that it represents for the cost of prior knowledge.

Since $W^{(in)}$ and $W^{(it)}$ are sparse, matrix W is also sparse and heavily depends on the input dataset, it is not sufficient to be reliable to grant information for clustering scheme. Thus, we assuage this limitation by converting the constraint between pair wise images into soft one. Following the idea of the local smoothness method [28], we define the smooth constraint S as follows:

$$S = \left(1 - \frac{\gamma}{2}\right)^2 (I - \gamma L)^{-1} W (I - \gamma L)^{-1} \tag{3}$$

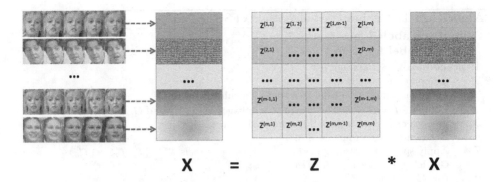

Fig. 3. Block sparse representation. This illustration depicts the relationship between input data X and self-data representation Z. Different color strips denote the different face tracks. Each block $Z^{(i,j)}$ denotes the coefficient of track i represented by track j. (Color figure online)

where $\gamma = \frac{1}{\exp\left(\frac{\alpha+\beta}{\exp(\alpha+\beta)}\right)}$ and I is an eye matrix. Equation 3 can be interpreted as the suppression of weight value α and β. Specifically, if α and β are too big or small, we get $\gamma \to 1$ so that S is a normalized smoothness. Otherwise, S is varied and depends on each weight value.

We convert the sparse low rank constraint, from prior knowledge, into the constrained matrix S. Since the magnitude of both sparse matrix L and constraint matrix S are too small, we get a natural combination of these matrices into final relationship matrix

$$Z = \exp\left(\frac{1}{2}(\rho S + (1-\rho)L)\right) \tag{4}$$

where $\rho \in (0,1)$ is the user-defined trade-off parameter.

3.5 Relabelling Process

Since Z is the data representation matrix and includes the coefficients to exhibit the relationship between faces in track X^i and those in track X^j, it can be regarded as the matrix of sub-matrix $Z^{(i,j)} \in R^{n_i \times n_j}$: $Z = \left\{Z^{(i,j)}\right\}; i,j = \overline{1\ldots m}$. The ideal of Z is that for the inner-track relation, $Z^{(i,j)}$ has the similar coefficient values if track i and j are from the same subject, otherwise, it is approximately zero; for the inter-track relation, the differences between coefficients of dissimilar clusters are must significant so that it can be easy for spectral clustering process. Figure 3 visualizes the ideal matrix Z as well as the clustering problem.

The spectral clustering [28,29] is widely used to obtain cluster labels from Z, so we inherit this approach as other works [9,11]. However, in those works, labels returning from spectral clustering process are taken by the most common label mode or even the raw ones. This has the disadvantage of neglecting the

---

Algorithm 1. The proposed algorithm VFSC

Input: unlabeled face tracks $X = \{X_i\}_{i=1}^m$
Output: label for each track $T = \{T_1, T_2, \ldots, T_m\}$

1: **procedure** VFSC ALGORITHM
2:     Compute distance matrix D as in 1
3:     Based on information from face tracks, calculate W from $W^{(in)}$ and $W^{(it)}$ as in 2
4:     Compute $L = G^{-\frac{1}{2}}DG^{-\frac{1}{2}}$ where G is a diagonal matrix and $G_{ii} = \sum_{j=1}^m D_{ij}^2$
5:     Smooth constraint matrix S is computed as in 3
6:     Get representation matrix Z in 4
7:     Apply spectral clustering method to return raw labels $T' = \{T'_1, T'_2, \ldots, T'_m\}$
8:     Get label distribution M_i in each track l'_i
9:     **for** $T'_i \in T'$ **do**
10:         Get $M_i^{(1)}, M_i^{(2)}, M_i^{(3)}$ are percentages of the three most common labels in track T'_i
11:         **if** $M_i^{(1)} - M_i^{(r)} \leq \phi; r \in \{2, 3\}$ **then**
12:             Rerun VFSC with label of track i is the r-th most common label in T'_i
13:         **end if**
14:         $T = Normalize(T')$
15:     **end for**
16: **end procedure**

---

possibility of other labels if the distribution of those labels are roughly similar. Thus, we propose to consider the second and third mode of the label returned. We verify these modes because the correct labels do not always belong to the most frequent one. Specifically, if the proportion between the first mode and the others smaller than threshold value $\phi = 10\%$, we rerun our method with this corresponding mode for current track. Otherwise, we use the first mode. The details of our proposed system as well as relabelling process are summarized in Algorithm 1.

4   Experiments

4.1   Datasets

We verify the merit of proposed system by using two public face datasets comprising Notting-Hill [6] and BF0502 [30]. The comparison here is not only the accuracy but also running time of all candidates. The Notting-Hill dataset is retrieved from "Notting Hill" movie, having 5 main characters comprising 76 tracks of 4660 faces. Whereas the BF0502 one has 17337 faces in 229 tracks belonging to 6 main casts in "Buffy the Vampire Slayer" TV series. Although it might be useful to extract high-level features by applying deep learning techniques [31,32], we still use raw RGB pixel as the input in system for fair comparison with other works.

To estimate the difficulty of each dataset, we create a distinction value based on face images in inner-track and inter-track relation. Specifically, we calculate the $L2$-norm distance between face images in the same track and those in dissimilar tracks. The distinction value is the ratio between image distance of inter-track over inner-track. This value can be interpreted that the smaller the value is, the more difficulty to cluster face tracks. Table 1 describes the information of both datasets. Not only has the dataset Notting-Hill smaller number of people as well as number of faces images and tracks but also larger distinction value in comparison with dataset BF0502, Notting-Hill is really easier for clustering. Therefore it gets high accuracy of clustering by state-of-the-art techniques and optimized clustering results by the proposed system described in Sects. 4.2 and 4.3.

Table 1. Comparison between two face datasets in the real world. Symbol "#" denotes the phrase "the number of".

Dataset	#People	#Face	#Dimension	#Tracks	#Overlap	#Distinction value
BF0502	6	17737	1937	229	20	0.96
Notting-Hill	5	4660	18000	76	0	1.62

Evaluation Protocol. In many existing works, the evaluation is still biased because they use the ratio between the number of correct clustered face images over the number of images (denoted as **"Acc-1"** in our experiments). Since the length of each track is vary and mainly depends on the dataset. If one dataset has a long track while the number of track is small, this track will dominate the whole accuracy of this dataset.

Thus, we propose our metric that to compute the accuracy of clustering process that is the ratio of correct labeled track over the number of track (**"Acc-2"**). This evaluation will get rid of the unbalance between tracks' lengths. However, we also compare the accuracy between our proposed method with all previous works on both evaluation metrics.

Besides, since the output of methods having optimization process is varied, we repeat each works 30 times and report the mean accuracy and its standard deviation. To demonstrate the ability being integrated into real-life application, running time is also compared. All methods are executed on the same environment for fair comparison.

Baseline Clustering Methods. We compare our proposed method with state-of-the-arts in various kind of approaches in clustering problem to verify the superiority of our VFSC system. Specifically, the following are the types of algorithms that we consider in the comparison with our proposed method in experiments:

- *Traditional clustering:* We utilize K-means [21] as a baseline with two approaches given in [6]. No useful information of inner and inter-track is exploited in these methods. Firstly, using K-means to cluster processed dataset by PCA is called "K-means 1". Secondly, "K-means 2" indicates the using K-means in Stage 2 of Algorithm 2 in [6].

- *Constrained clustering:* Gaussian mixture models combines with image constraints is the idea of the Penalized Probabilistic Clustering (PPC) [19] method we use in comparison. Besides, following the advice of [6], we set the mixture coefficient parameter π in [19] to $\frac{1}{m}$ and size of all constraints in PPC is equal to 1.
- *Metric learning based:* In this type of approach, since the work of [33] is too slow to apply in the real-life application, we compare with the unsupervised logistic discriminant metric learning (ULDML) method [20]. There are also two settings called "ULDML-cl", a complete link hierarchical clustering method based on distance matrix between face tracks, and "ULDML-km", a K-means approach to utilize learned metric.
- *Hidden Markov Random Fields based:* A probabilistic constrained clustering approach called HMRF-com [6] is taken into consideration and gives out promising result from prior knowledge.
- *Compressed sensing based subspace clustering:* These methods have the same goal that optimize sparse matrix Z such as Sparse Subspace Clustering (SSC) [10], Least Squares Regression (LSR) [7], Correlation Adaptive Subspace Segmentation (CASS) [9], Low Rank Representation (LRR) [8], Low Rank Sparse Subspace Clustering (LRSSC) [11] and Weighted Block-Sparse Low Rank Representation (WBSLRR) [22]. However, the differences between those methods are the approaches to regularize on matrix Z in their objective function.

4.2 Comparison with Non Sparse Method

The results of the comparison between state-of-the-art techniques in non sparse subspace clustering and our proposed method are shown in Table 2. There are three configurations in our comparison: using inter-track relation only, inner-track only and both. From the table, we can observe that the proposed method surpasses all existing techniques in both datasets in all configurations. Specifically, in comparison with the second best method HMRF-com, we significantly increase the clustering accuracy: about 17.43% increases in inner-track (from 47.77% to 65.20%), 17.54% (from 48.83% to 66.37%) increases in inter-track and 16.67% (from 50.30% to 67.07%) increases in the whole setting on BF0502 dataset. ULDML-cl method reveals the similar characteristic like HMRF-com, which is the combination of both inner and inter-track help boost up the accuracy overall. Results of K-means 2 are better than K-means 1 in both datasets. However, they are still far from the satisfaction in comparison of accuracy.

In Notting-Hill dataset, the overall accuracy of all techniques seem to be high because it is regarded as an "easy" one based on the comparison of two datasets in Table 1 and analysis in Sect. 4.1. That is the reason why the proposed method approaches the optimal value (98.68%) clustering on this dataset while only got 66.37% on BF0502 dataset. However, there is still one track in Notting-Hill dataset is wrongly labeled. As showed in Fig. 4, there are wrongly clustered samples in the forth row. Those are blurrier and noisier than others. This hazy characteristic leads to wrong clustered in spectral clustering process. Thus, it is

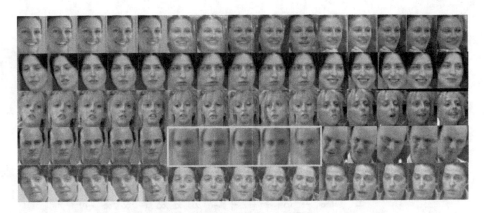

Fig. 4. Difficult track in Notting-Hill dataset. This track is blur and captured in low condition so that it is the big challenge to completely cluster Notting-Hill dataset.

very difficult to get 100% right label unless tuning parameters. Besides, the accuracy evaluation metric Acc-2 is usually lower than Acc-1 in both datasets. This phenomenon can be explained that easily clustered facial images are belonged to long track. Specifically, in 5 clusters of Notting-Hill dataset, HMRF-com can achieve up to 90.66%, 70.72%, 100%, 93.95% and 43.47% respectively in accuracy. However, since the number of images of the second and the fifth cluster has only nearly 30% of the whole dataset, the overall accuracy of HMRF-com is approximately 84.39%. In Acc-2 evaluation, long tracks and short tracks are equaly in computing accuracy so that it is unbiased by input datasets. Although all previous works significantly degrade the clustering results while applying Acc-2 evaluation, our proposed method still reserve the high accuracy in both datasets.

Furthermore, the proposed method is superior in not only clustering accuracy but also speed. Our performance on BF0502 dataset only takes 2.37 s to finish clustering while the others take up to 119.26 (HMRF-com) or even 437.32 (ULDML-km) seconds to complete this task. Since we are based on sparse representation and computation and do not need any optimization parts, the running time is fast enough to be integrated in the real-life application. Further discussion on the running time is presented in Sect. 4.3 when compared with other state-of-the-art in sparse clustering methods. In general, the experiment results clearly reveal that VFSC has better usage of prior knowledge (i.e. the inner-track and inter-track relation) of facial images in video than other existing works.

4.3 Comparison with Sparse Low-Rank Method

The aim of this comparison is to prove superior robustness and speed of proposed method. Table 3 reports the comparison of mean accuracy in two evaluation metrics as well as standard deviations and running times of six subspace clustering works and the proposed method, where $\|Z\|_F = \left(\sum_{i,j} Z_{i,j}^2 \right)^{\frac{1}{2}}$ denotes

Table 2. Comparison with non sparse clustering methods

Setting			Methods						HMRF com [6]	VFSC (Proposed)
			Kmeans		PPC [19]	ULDML				
			1	2		km	cl			
BF0502	Acc-1	Inner	39.31 ± 4.51	42.05 ± 5.45	43.64 ± 4.61	29.05 ± 2.84	39.01 ± 0.00		47.77 ± 3.31	65.20 ± 0.00
		Inter	39.31 ± 4.51	42.05 ± 5.45	38.22 ± 3.02	39.61 ± 1.42	47.97 ± 0.00		48.83 ± 4.05	66.37 ± 0.00
		Both	39.31 ± 4.51	42.05 ± 5.45	42.54 ± 3.98	41.62 ± 0.00	49.29 ± 0.00		50.30 ± 2.73	67.06 ± 0.00
	Acc-2	Inner	37.12 ± 2.62	39.30 ± 3.49	39.77 ± 2.94	31.75 ± 3.39	38.27 ± 0.00		45.62 ± 2.01	58.43 ± 0.00
		Inter	37.12 ± 2.62	39.30 ± 3.49	37.12 ± 3.57	37.82 ± 0.00	47.01 ± 0.00		46.61 ± 1.94	62.44 ± 0.00
		Both	37.12 ± 2.62	39.30 ± 3.49	40.11 ± 4.03	39.94 ± 0.79	48.28 ± 0.00		48.52 ± 3.49	66.37 ± 0.00
	Time(s)		72.06	87.93	61.46	437.32	410.56		119.26	2.37
Notting-Hill	Acc-1	Inner	69.16 ± 3.22	73.43 ± 8.12	79.71 ± 2.14	72.66 ± 12.78	51.72 ± 0.00		81.33 ± 0.43	97.83 ± 0.00
		Inter	69.16 ± 3.22	73.43 ± 8.12	77.05 ± 3.12	73.87 ± 5.98	35.91 ± 0.00		82.36 ± 2.67	97.59 ± 0.00
		Both	69.16 ± 3.22	73.43 ± 8.12	78.88 ± 5.15	73.18 ± 8.66	36.87 ± 0.00		84.39 ± 1.47	98.73 ± 0.00
	Acc-2	Inner	69.73 ± 5.26	71.05 ± 2.63	77.63 ± 1.31	70.13 ± 3.57	48.68 ± 0.00		79.26 ± 3.59	97.36 ± 0.00
		Inter	69.73 ± 5.26	71.05 ± 2.63	71.85 ± 6.57	70.94 ± 3.48	32.89 ± 0.00		80.01 ± 2.83	97.36 ± 0.00
		Both	69.73 ± 5.26	71.05 ± 2.63	72.37 ± 7.89	71.03 ± 2.11	42.10 ± 0.00		81.94 ± 1.16	98.68 ± 0.00
	Time(s)		14.77	15.08	10.03	84.24	89.57		19.58	0.83

the Frobenius norm of Z. In general, subspace clustering is better than the non sparse clustering methods given in Sect. 4.2 in accuracy but it is too slow because of heavy computation. We can clearly notice that the the proposed VFSC system significantly outperform all subspace clustering state-of-the-arts. Like the explanation in Sect. 4.2, accuracy achieved in Acc-2 evaluation metric is lower than one in Acc-1 for all works except the proposed VFSC. Specifically, VFSC gets up to 4.30% and 2.44% in Acc-1 and 11.35% and 5.26% in Acc-2 higher than the WBSLRR, which is regarded as the best method for face clustering, in comparison on BF0502 and Notting-Hill dataset respectively. Since WBSLRR has $\Omega(Z)$ for their regularization part to encourage the block-sparsity of Z, WBSLRR's result is better than LRR. Besides, the general sparsity neglecting the prior information between face tracks leads LRSSC get only 58.08% in BF0502 and 94.03% in Notting-Hill dataset. LSR and SSC method are just the special cases of LRR and LRSSC so that they has poor performance, however, they are still better than some non sparse methods described in Sect. 4.2.

Besides the accuracy in clustering, system's robustness is also important. This characteristic can be reflected from the standard deviation. For instance, a small standard deviation reveals good robustness whereas a big one means the corresponding method is easily fluctuated and fragile, depending on the input dataset. As can be observed from Tables 2 and 3, sparse subspace clustering has the smaller standard deviation than the non sparse ones in Notting-Hill dataset. However, they are nearly equal in BF0502 dataset because this dataset has more faces and face tracks with small distinction value, which is not easy to solve completely. On the contrary to all of existing works, the proposed VFSC system presents zero standard deviation in both datasets with different configurations. One explanation is that we do not have regularization terms in constructing data representation matrix Z. Although LRSSC method gets zero standard deviation in Notting-Hill dataset, it get up to 5.37% in variation in BF0502 dataset. It can be explained that BF0502 is more difficult than the other. However, SSC method shows the contradiction. It achieves only 1.74% (Acc-2) standard deviation in BF0502 while jumping up to 6.57% in Notting-Hill dataset. Hence, SSC is easily fluctuate with the given data. Thus, our method, whose standard deviations are zero in all setting, is more robust and does not depend on the difficulty of input dataset.

Furthermore, our proposed VFSC method also shows the superior in speed in comparison with state-of-the-art of sparse techniques. In specific, we boost up running time from 129.61/8.27 s (LSR) to 2.37/0.83 s in BF0502 and Notting-Hill dataset respectively. Although LSR is the fastest among existing works, its accuracy is one of the worsts. Furthermore, while collating with WBSLRR on these two datasets, our proposed method still surpass this method about approximately 10% and 5% in accuracy (Acc-2) and 250 times faster in speed.

When comparing sparse and non sparse clustering method, we address the following comment: Non sparse methods prioritize the speed rather than clustering accuracy while sparse subspace clustering ones focus on the accuracy by adjusting parameters to satisfy regularization constraints but neglect the

Table 3. Comparison with sparse clustering method

| Methods | Regularization on Z | BF0502 | | | Notting-Hill | | |
		Acc-1	Acc-2	Time	Acc-1	Acc-2	Time
LSR [7]	$\|Z\|_F^2$	50.19 ± 1.93	46.72 ± 2.62	129.61	89.89 ± 0.00	85.52 ± 0.00	8.27
SSC [10]	$\|Z\|_1$	36.52 ± 0.91	33.18 ± 1.74	24558.27	75.50 ± 7.90	65.78 ± 6.57	2892.00
LRR [8]	$\|Z\|_*$	51.17 ± 2.94	48.03 ± 3.05	1201.08	93.11 ± 0.00	88.15 ± 0.00	34.07
CASS [9]	$\sum_{i=1}^{n} \|X \, diag(Ze_i)\|$	N/A	N/A	N/A	93.18 ± 0.00	90.78 ± 0.00	29672.40
LRSSC [11]	$\|Z\|_* + \gamma \|Z\|_1$	58.08 ± 5.37	54.14 ± 1.31	8218.24	94.03 ± 0.00	92.13 ± 0.00	538.51
WBSLRR [22]	$\|Z\|_* + \gamma \Omega(Z)$	62.76 ± 1.10	55.02 ± 3.49	692.66	96.29 ± 0.00	93.42 ± 0.00	201.93
VFSC (Proposed)	$\exp\left(\frac{1}{2}(\rho S + (1-\rho)L)\right)$	$\mathbf{67.06 \pm 0.00}$	$\mathbf{66.37 \pm 0.00}$	**2.37**	$\mathbf{98.73 \pm 0.00}$	$\mathbf{98.68 \pm 0.00}$	**0.83**

running time. The proposed VFSC system joins the advantages and assuage the defects of both trends. Therefore VFSC achieves superior results in accuracy and speed of clustering faces in video problems.

5 Conclusion

This paper presents VFSC system, a light weight novel approach to handle face clustering in video by utilizing the prior useful knowledge from face track and characteristic of sparse data representation. An end-to-end system for adapting into real-life application is also discussed. New evaluation metric is also given to solve the bias of existing works which depends largely on the input dataset. The experimental results reveals the truth that our method outperforms all state-of-the-art techniques in not only sparse but also non sparse clustering trends and become a new state-of-the-art method in clustering area. Since the swiftness, stability and high accuracy of our method while tackling face clustering problem, it is applicable to be adapted into very large scale tasks. Finally, for the future work, the accuracy of our system can be further improved by applying deep learning techniques in the post-processing stage.

References

1. Yao, H., Duan, Q., Li, D., Wang, J.: An improved k-means clustering algorithm for fish image segmentation. Math. Comput. Model. **58**, 790–798 (2013)
2. Kang, Z., Landry, S.J.: An eye movement analysis algorithm for a multielement target tracking task: maximum transition-based agglomerative hierarchical clustering. IEEE Trans. Hum.-Mach. Syst. **45**, 13–24 (2015)
3. Huang, Z., Wang, R., Shan, S., Chen, X.: Projection metric learning on Grassmann manifold with application to video based face recognition. In: Proceedings of the IEEE Conference on Computer Vision and Pattern Recognition, pp. 140–149 (2015)
4. Aggarwal, C.C., Reddy, C.K.: Data Clustering: Algorithms and Applications. CRC Press, Boca Raton (2013)
5. Sang, J., Xu, C.: Robust face-name graph matching for movie character identification. IEEE Trans. Multimedia **14**, 586–596 (2012)
6. Wu, B., Zhang, Y., Hu, B.G., Ji, Q.: Constrained clustering and its application to face clustering in videos. In: Proceedings of the IEEE Conference on Computer Vision and Pattern Recognition, pp. 3507–3514 (2013)
7. Lu, C.-Y., Min, H., Zhao, Z.-Q., Zhu, L., Huang, D.-S., Yan, S.: Robust and efficient subspace segmentation via least squares regression. In: Fitzgibbon, A., Lazebnik, S., Perona, P., Sato, Y., Schmid, C. (eds.) ECCV 2012. LNCS, vol. 7578, pp. 347–360. Springer, Heidelberg (2012). doi:10.1007/978-3-642-33786-4_26
8. Liu, G., Lin, Z., Yan, S., Sun, J., Yu, Y., Ma, Y.: Robust recovery of subspace structures by low-rank representation. IEEE Trans. Pattern Anal. Mach. Intell. **35**, 171–184 (2013)
9. Lu, C., Feng, J., Lin, Z., Yan, S.: Correlation adaptive subspace segmentation by trace lasso. In: Proceedings of the IEEE International Conference on Computer Vision, pp. 1345–1352 (2013)

10. Elhamifar, E., Vidal, R.: Sparse subspace clustering. In: IEEE Conference on Computer Vision and Pattern Recognition, CVPR 2009, pp. 2790–2797. IEEE (2009)
11. Wang, Y.X., Xu, H., Leng, C.: Provable subspace clustering: when LRR meets SSC. In: Advances in Neural Information Processing Systems, pp. 64–72 (2013)
12. Fitzgibbon, A., Zisserman, A.: On affine invariant clustering and automatic cast listing in movies. In: Heyden, A., Sparr, G., Nielsen, M., Johansen, P. (eds.) ECCV 2002. LNCS, vol. 2352, pp. 304–320. Springer, Heidelberg (2002). doi:10.1007/3-540-47977-5_20
13. Fitzgibbon, A.W., Zisserman, A.: Joint manifold distance: a new approach to appearance based clustering. In: Proceeding of the IEEE Computer Society Conference on Computer Vision and Pattern Recognition, vol. 1, p. I-26. IEEE (2003)
14. Hu, Y., Mian, A.S., Owens, R.: Sparse approximated nearest points for image set classification. In: 2011 IEEE Conference on Computer Vision and Pattern Recognition (CVPR), pp. 121–128 (2011)
15. Wang, R., Shan, S., Chen, X., Gao, W.: Manifold-manifold distance with application to face recognition based on image set. In: IEEE Conference on Computer Vision and Pattern Recognition, CVPR 2008, pp. 1–8. IEEE (2008)
16. Arandjelović, O., Cipolla, R.: Automatic cast listing in feature-length films with anisotropic manifold space. In: 2006 IEEE Computer Society Conference on Computer Vision and Pattern Recognition, vol. 2, pp. 1513–1520. IEEE (2006)
17. Prince, S.J., Elder, J.H.: Bayesian identity clustering. In: 2010 Canadian Conference on Computer and Robot Vision (CRV), pp. 32–39. IEEE (2010)
18. Wolf, L., Hassner, T., Maoz, I.: Face recognition in unconstrained videos with matched background similarity. In: 2011 IEEE Conference on Computer Vision and Pattern Recognition (CVPR), pp. 529–534. IEEE (2011)
19. Lu, Z.L., Leen, T.K.: Penalized probabilistic clustering. Neural Comput. **19**, 1528–1567 (2007)
20. Cinbis, R.G., Verbeek, J., Schmid, C.: Unsupervised metric learning for face identification in TV video. In: 2011 IEEE International Conference on Computer Vision (ICCV), pp. 1559–1566. IEEE (2011)
21. Bishop, C.M.: Pattern recognition. Mach. Learn. **128**, 1–58 (2006)
22. Xiao, S., Tan, M., Xu, D.: Weighted block-sparse low rank representation for face clustering in videos. In: Fleet, D., Pajdla, T., Schiele, B., Tuytelaars, T. (eds.) ECCV 2014. LNCS, vol. 8694, pp. 123–138. Springer, Cham (2014). doi:10.1007/978-3-319-10599-4_9
23. Nguyen, D.L., Nguyen, V.T., Tran, M.T., Yoshitaka, A.: Adaptive wildnet face network for detecting face in the wild. In: Eighth International Conference on Machine Vision, International Society for Optics and Photonics, p. 98750S (2015)
24. Nguyen, D.L., Nguyen, V.T., Tran, M.T., Yoshitaka, A.: Boosting speed and accuracy in deformable part models for face image in the wild. In: 2015 International Conference on Advanced Computing and Applications (ACOMP), pp. 134–141. IEEE (2015)
25. Zhang, K., Zhang, L., Yang, M.H.: Fast compressive tracking. IEEE Trans. Pattern Anal. Mach. Intell. **36**, 2002–2015 (2014)
26. Zeng, Z., Chan, T.-H., Jia, K., Xu, D.: Finding correspondence from multiple images via sparse and low-rank decomposition. In: Fitzgibbon, A., Lazebnik, S., Perona, P., Sato, Y., Schmid, C. (eds.) ECCV 2012. LNCS, vol. 7576, pp. 325–339. Springer, Heidelberg (2012). doi:10.1007/978-3-642-33715-4_24
27. Jolliffe, I.: Principal Component Analysis. Wiley, Hoboken (2002)

28. Lu, Z., Ip, H.H.S.: Constrained spectral clustering via exhaustive and efficient constraint propagation. In: Daniilidis, K., Maragos, P., Paragios, N. (eds.) ECCV 2010. LNCS, vol. 6316, pp. 1–14. Springer, Heidelberg (2010). doi:10.1007/978-3-642-15567-3_1

29. Zelnik-Manor, L., Perona, P.: Self-tuning spectral clustering. In: Advances in Neural Information Processing Systems, pp. 1601–1608 (2004)

30. Everingham, M., Sivic, J., Zisserman, A.: Hello! My name is.. Buffy"-automatic naming of characters in TV video. In: BMVC, vol. 2, p. 6 (2006)

31. Nguyen, D.-L., Nguyen, V.-T., Tran, M.-T., Yoshitaka, A.: Deep convolutional neural network in deformable part models for face detection. In: Bräunl, T., McCane, B., Rivera, M., Yu, X. (eds.) PSIVT 2015. LNCS, vol. 9431, pp. 669–681. Springer, Heidelberg (2016). doi:10.1007/978-3-319-29451-3_53

32. Girshick, R., Iandola, F., Darrell, T., Malik, J.: Deformable part models are convolutional neural networks. In: Proceedings of the IEEE Conference on Computer Vision and Pattern Recognition, pp. 437–446 (2015)

33. Vretos, N., Solachidis, V., Pitas, I.: A mutual information based face clustering algorithm for movie content analysis. Image Vis. Comput. **29**, 693–705 (2011)

Deep or Shallow Facial Descriptors? A Case for Facial Attribute Classification and Face Retrieval

Rasoul Banaeeyan[1(⊠)], Mohd Haris Lye[1], Mohammad Faizal Ahmad Fauzi[1], Hezerul Abdul Karim[1], and John See[2]

[1] Faculty of Engineering, Multimedia University, Cyberjaya, Malaysia
banaeeyan@gmail.com, {haris.lye,faizal1,hezerul}@mmu.edu.my
[2] Faculty of Computing and Informatics, Multimedia University, Cyberjaya, Malaysia
johnsee@mmu.edu.my

Abstract. With the largely growing quantity of face images in the social networks and media, different face analyzing systems are developed to be employed in real-world situations such as face recognition, facial expression detection, or automated face tagging. Two demanding face-related applications are studied in this paper: facial attribute classification and face image retrieval. The main common issue with most of the attribute classifiers and face retrieval systems is that they fail to perform well under various facial expressions, pose variations, geometrical deformation, and photometric alterations. On one hand, the emerging role of deep CNNs (convolutional neural networks) has shown superior results in tasks like object recognition, face recognition, etc. On the other hand, their applications are yet to be more investigated in facial attribute classification and face retrieval. In this study, we compare the performance of shallow and deep facial descriptors in the two mentioned applications by proposing to exploit distinctive facial features from a very deep pre-trained CNN for attribute classification as well as constructing deep attribute-driven feature vectors for face retrieval. According to the results, the higher accuracy of the attribute classifiers and superior performance of the face retrieval system is demonstrated.

1 Introduction

Recent developments in social networks and medias have heightened the need for analyzing images because of the important applications they have in real-world contexts such as large-scale image retrieval [1,2], automated image annotation [3,4], and scene understanding [5]. Millions of images are uploaded to social networks every day, Facebook, for instance, process approximately 350 million faces daily [6]. Of all these images, 45% are estimated to encompasses faces [7] highlighting the importance of processing faces for purposes such as face recognition [8], facial expression detection [9], or face annotation [10]. However, retrieving similar face images using a single query face image remains a challenging problem in many real-world applications, particularly in law enforcement investigations

C.-S. Chen et al. (Eds.): ACCV 2016 Workshops, Part II, LNCS 10117, pp. 434–448, 2017.
DOI: 10.1007/978-3-319-54427-4_32

Fig. 1. Sample faces of one person under different expressions, head poses, facial occlusions, geometrical deformations, and photometric alterations.

where it assists authorities with narrowing leads and, subsequently, confirming a suspects identity. The goal of such content-based Face Image Retrieval (FIR) systems is to retrieve top most similar faces from a large database faces (distracting or gallery). Such databases are challenging due to the presence of diverse face images from multiple persons. There are also large intra-class variations due to facial expressions, pose variations, geometrical deformation, and photometric alterations (Fig. 1). Hence, FIR has been an active field of re-search among the scholars over the recent years [11–18].

In order to foster a new attribute-driven FIR system to fill in the gap of the existing research, this study will examine the emerging role of very deep pre-trained convolutional neural net-work (CNN) [8] in the context of supervised semantic facial attribute classification and, more importantly, its impact on performance, mean average precision (mAP), of FIR systems whilst using attribute-driven feature vectors. Our research contributions are listed as:

- Semantic facial attribute classifiers are trained by exploiting salient facial features from the last fully connected layer of a very deep pre-defined CNN (VGG-Face) [8] which is designed and trained for the task of face recognition. Although the quantity of training faces is not large, the accuracy is higher than shallow-based attribute classifiers.
- We compared the accuracy of the proposed deep descriptor classifiers with those of the shallow descriptors (D-SIFT, G-LBP, and G-HOG) to investigate their effectiveness regarding prediction of each facial attributes (eight attributes). This is the only comparative study, to the best of our knowledge, com-paring deep and shallow features in the context of face attribute classification.
- The mean average precision of (mAP) shallow attribute-driven face image retrieval is increased by more than double by exploiting deep attribute-driven

facial features resulted from the trained binary classifiers. We compared the performance (mAP) of proposed deep attribute-driven FIR with the performance of shallow attribute-driven FIRs with respect to different gallery (distracting) face database sizes (2k, 4k, 6k, 8k, and 10k) constructed using faces from CelebA database [19].

This paper is organized as follows. The next section briefly discusses recent related works in face attribute classification and face image retrieval. Section 3 presents the methodology of our proposed system while Sect. 4 and 5 discusses the experimental results and discussion. It is followed by Sect. 6 which is conclusion of the paper.

2   Related Work

This research is tightly related to two research areas, comprising facial attribute classification, and face image retrieval (FIR). The conventional facial attribute classifiers extract low-level features (SIFT, LBP, HOG, etc.) within different localized face components (two eyes, nose, and two mouth corners) from a given face and feed them to a supervised classifier such as linear SVM [20]. Following this approach, along with extracting features from overlapped components, they tolerate some degrees of error in delocalized face components [15]. Recently, the studies in [19,21] extracted features from deep CNNs aiming for training more accurate face attribute classifiers. In [21] the CNN architecture in [8] was adopted and the cost function was replaced with a new proposed one. The new architecture was trained using 160k face images from CelebA [19] database and it took 2,500 iterations to a complete one epoch on the collection of training faces. The second work presented in [19] used two pre-trained CNNs, known as LNet, and ANet, to fine-tune them for tasks of face localization and attribute classification, respectively. To train LNet and ANet, 1.2 million images of 1k distinct objects, and 160k face images of 8k different identities were employed, respectively. However, fine-tuning a pre-trained deep CNN or training a newly designed one is very resource-demanding (a large corpus of training images is required) and computationally heavy, as with the works in [19,21]. These two studies, to date, are the only ones exploiting deep CNNs for the task of supervised facial attribute classification.

To increase the performance of FIR systems several methods have been proposed recently. In [16], to give distinctiveness to the facial features, soft biometrics along with facial marks (moles, scars, etc.) were taken into account to increase the FIR precision. An automated facial mark detection was proposed for this purpose too. The semi-supervised FIR method in [22] was the first to introduce the application of sparse-coded features constructed from locally extracted features. This work achieved scalability by proposing a novel coding scheme and combining inverted indexing and sparse features. The work in [15] was the first introducing a multi-reference re-ranking and identity quantization model to achieve scalability and precision while retrieving face from a database of 1 million images. An unsupervised FIR method based on voting scheme and classification was introduced in [11]. The faces are represented using local features and then, to speed computations, an approximate nearest neighbor (ANN) search is used to find matching features. Another work in [12] attempted to take advantage of a deep neural net (NN) to pre-filter the database faces using extracted features from the learned NN. A k-NN search was employed to find the top-k most similar faces while retrieving.

The state-of-the-art FIR systems achieve scalability and precision by exploiting both high-level (semantic) and low-level facial attributes [17,18]. They first retrieve the N top similar faces using semantic attribute-driven features, then the re-rank them, to get the top $M(M < N)$ relevant faces (e g top 10), by their low-level facial descriptors extracted from sub-regions (patches) within different localized face regions (components) [15]. To date, the only attribute-driven FIR techniques are those proposed in [17,18,23].

3   Method

An overall pipeline of our FIR approach is presented in Fig. 2. Given a face image as a query, a database of gallery faces (distracting), and a database of training faces, the system begins by extracting the deep facial features using a very deep pre-trained CNN, followed by training eight supervised binary classifiers for eight facial attributes (Fig. 3). Then, extracted deep facial features from the query face and gallery sets are fed to the eight trained classifiers in the previous stage in order to compute the Fuzzy attribute scores of each unique attribute. Finally, the attribute scores calculated for each face image are concatenated to construct a deep attribute-driven facial feature vector for that face. The procedure for the last step is illustrated in Fig. 4.

3.1   VGG-Face – A Very Deep Pre-trained CNN

In our proposed method, we extract facial features from the very deep pre-trained CNN (VGG-Face) [8] which is designed and trained for the task of face recognition. This architecture is referred to as very deep as it comprises a long sequence of convolutional layers. The VGG-Face has 37 layers and is was trained by a triplet-loss method using 2.6 million faces (2,633 identities) collected from the wild. To solve the problem of recognizing $N = 2,633$ distinctive identities, the deep CNN architectures were bootstrapped (N-ways classification). Then, each training image was associated with a training score vector considering the last fully-connected layer including N linear predictors. Finally, all scores were compared to the ground-truth by calculating the statistical soft-max log-loss and their scores were further improved by triplet-loss tuning.

3.2   Facial Feature Extraction

For each face image, four 224×224 patches are cropped (from the center with horizontal flip and four corners) and an averaged feature vector is constructed

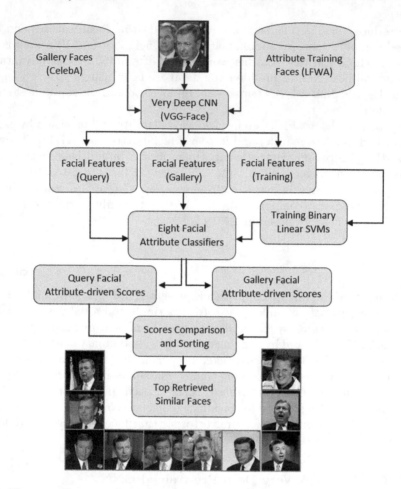

Fig. 2. The overall system level diagram of the proposed FIR system: faces in green boxes are true positives, and those in red boxes indicate false positives. (Color figure online)

based on them. In order to take advantage of multi-scale testing, the given face is scaled to different sizes of 512, 384, and 256 pixels followed by replicating the cropping procedure for all of them. To extract the final facial features, the 4,096-dimensional vector is extracted from the last fully-connected layer in the VGG-Face which is the results of averaging all feature vectors.

3.3 Binary Linear SVMs

For the attribute classification task, features are extracted from both training and test images. Both set of images contain positive and negative samples from

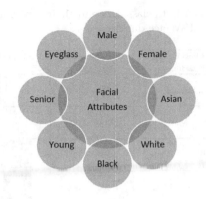

Fig. 3. List of eight facial attributes employed in this research.

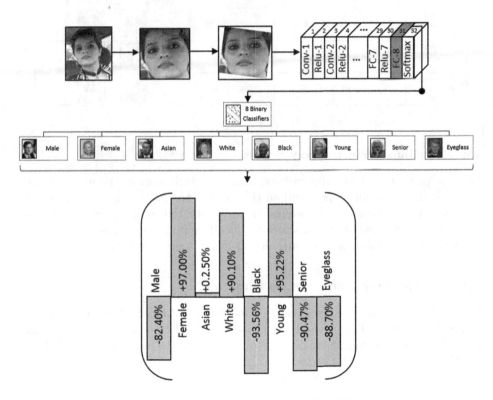

Fig. 4. Construction of deep attribute-driven facial feature vectors

the corresponding attribute. Eight linear binary SVM classifiers are trained and tested on the eight facial attributes classification tasks.

The choice and the number of positive and negative samples from LFW [24] dataset are listed in Table 1.

Table 1. Types of facial attribute classifiers and their choice and number of training face images selected from LFW face dataset.

Type of Classifiers	Selection of Positive and Negative Face Images	# Training and Testing Faces	Sample Faces used for Training
Male Classifier	Male vs. Female	2,000 – 2,000	
Female Classifier	Female vs. Male	2,000 – 2,000	
Asian Classifier	Asian vs. (Black + White)	1,000 – 1,000	
White Classifier	White vs. (Asian + Black)	1,000 – 1,000	
Black Classifier	Black vs.(Asian + White)	500 - 500	
Young Classifier	Young vs. Senior	2,000 – 2,000	
Senior Classifier	Senior vs. Young	2,000 – 2,000	
Eyeglass Classifier	Eyeglass vs. No Eyeglass	2,000 – 2,000	

3.4   Attribute-Driven Fuzzy Scores

In this part, the 4,096 dimensional facial feature vectors (from Sect. 3.1) extracted from the query face and all the faces in the gallery database are used as inputs to the eight attribute classifiers constructed in the previous section. Then, all the resulted attribute scores for each face image are L2 normalized and concatenated to build the attribute-driven Fuzzy scores for that face (Fig. 4).

3.5   Scores Comparison and Sorting

Finally, by comparing the Euclidean distance between the attribute-driven Fuzzy scores of the query face and all the faces in the gallery database, we sort the faces in the gallery from the minimum to maximum. The first top N similar faces are the first N in the final sorted list.

4   Experiments

In this section, two sets of experiments are conducted to evaluate the performance of the attribute classifiers using deep descriptor and deep attribute-driven FIR system with respect to different database sizes ranging from 2k to 10k face images.

4.1 Databases

The LFW (labeled faces in the wild) [24] is a face image dataset developed for studying the problem of unconstrained face recognition. The database includes over 13,000 images of faces collected from the WWW. From LFW database, the training faces are selected as detailed in Table 1. Following the experiments performed in [17,18], we select 12 unique individuals from LFW database with more than 50 face images per person (totally 1,560 faces), from these collection, we selected 10 face images for each individual (12 individuals) to be served as our query faces (totally 120 faces). Samples and numbers of selected faces for both quay and gallery sets are shown in Table 2. Using Viola-Jones face detector [25], all the faces are detected from the entire collection of face images and then they are resized to 224 × 224 using bilinear interpolation method.

CelebA [19] dataset (Celebrity Faces Attributes) is a scalable face image dataset including over 200k faces of celebrities (over 10k distinct people). From this dataset, we select five subsets of 2k, 4k, 6k, 8k and 10k faces to be served as our gallery databases while performing face retrieval. There is no overlap between the faces in the training set and those relevant faces in the gallery databases.

Table 2. Names of the individuals in the LFW used as query set and their associated numbers of faces used in query and gallery sets.

No	Person Name	Sample Face	# of Query Faces	# of Faces in the gallery	Total number of faces
1	Ariel Sharon		10	67	77
2	Colin Powell		10	226	236
3	Donald Rumsfeld		10	111	121
4	George W Bush		10	520	530
5	Gerhard Schroeder		10	99	109
6	Hugo Chavez		10	61	71
7	Jacques Chirac		10	42	52
8	Jean Chretien		10	45	55
9	John Ashcroft		10	43	53
10	Junichiro Koizumi		10	50	60
11	Serena Williams		10	42	52
12	Tony Blair		10	134	144

4.2   Performance Evaluation and Comparison

For face retrieval, mAP at 100% recall is used as performance measurement. The mAP is defined as the mean of all average precisions resulted from inquiring each unique face query image.

In order to compare the performance of deep facial descriptors with the shallow descriptors in the contexts of attribute classification and face retrievals, three feature descriptors are selected as Histogram of Oriented Gradients (HOG) [26] which its successfully used in face recognition high accuracy [27], Local Binary Pattern (LBP) [28] which has proved to work well in face recognition tasks as reported in [29], and Scale Invariant Feature Transform (SIFT) [30].

4.3   Parameter Settings

To facilitate the procedure of re-implementing the proposed methodology, here we report the parameter settings of the binary linear SVM classifiers and shallow feature extractors. For the SVMs a linear kernel function was used and SMO (Sequential Minimal Optimization) method was selected to find the separating hyperplanes. Two input parameters for the kernel MLP (Multilayer Perceptron) are set as 1 and -1. To extract the HOG, LBP, and SIFT features we used the grid version (G-HOG, G-LBP, and G-SIFT) with the grid step of 8×8 and block width of 32×32.

4.4   Software and Hardware

For G-SIFT extraction, VLFeat Toolbox (0.9.20) [31] was used, and other programming parts are done using Matlab R2016a. Our experiments are performed on a desktop computer with Windows 7 (64 bit), 16 GB RAM, Intel Core i7-2600 CPU @ 3.40GHz.

5   Results and Discussion

In this section, the accuracy of eight trained binary attribute classifiers, as well as mAP of the proposed FIR system, are presented. Furthermore, recommendations on future works are founded on analysis of the results.

5.1   Facial Attribute Classification Performance

Based on the accuracies reported for each attribute in the Fig. 5, performance of deep facial descriptors is always better than the shallow descriptors, but of the shallow ones, G-SIFT is comparable with the highest (achieved by deep descriptors). Regarding the young and senior classification accuracies, their difference between G-SIFT and Deep CNN is inconsiderable, it is 0.45% and 1.05% for the young and senior classification (age category), respectively. On the other hand,

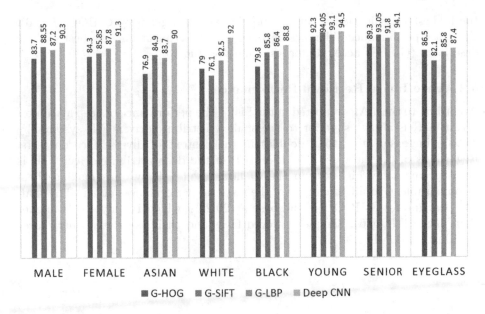

Fig. 5. Accuracy of various facial attribute classifiers. Accuracies using three types of shallow descriptors (G-HOG, G-LBP, and D-SIFT) and one deep descriptor.

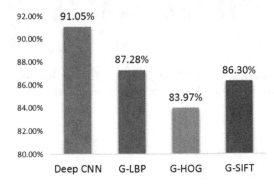

Fig. 6. Overall performance of deep and shallow facial descriptors in the task of attribute classification.

the biggest difference can be seen in the Asian, white, and black classifications (race category). The performance of G-LBP is comparable with deep CNN in the eyeglass classification while the difference is 1.6%. Finally, the accuracy of deep descriptors is higher than the shallow descriptors in the gender classification task (male and female). Overall, of the shallow facial descriptors, G-LBP has the highest classification accuracy and the lowest is given by G-HOG (Fig. 6). The total performance difference between deep descriptors and the best shallow descriptor (G-LBP) accounts for less than 4% (Fig. 6).

5.2   Face Image Retrieval Performance

According to the mAPs reported in the Fig. 7, the performance of deep attribute-driven FIR is always superior to the shallow attribute-driven FIR systems by 24.63% increase (more than double) in the context of diverse face datasets. Regardless of the database size, the mAP of the deep attribute-driven FIR is always remarkably higher than the shallow-based FIR systems making them incomparable. Overall, among all of the retrieval performances attained by shallow descriptors, G-HOG has the best at 20.20% which is only 1.02% higher than G-SIFT (Fig. 6). Sample retrieved faces and their associated query faces are illustrated in Fig. 9.

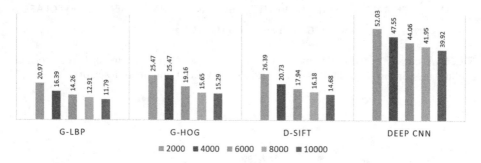

Fig. 7. Mean average precision (mAP) of different FIR systems using shallow and deep attribute-driven feature vectors. Results are based on different database sizes.

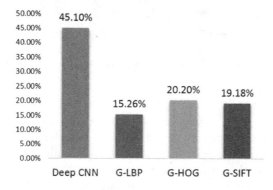

Fig. 8. Overall performance of deep and shallow facial descriptors for face image retrieval task.

Fig. 9. Examples of the top-10 retrieved face images in the proposed FIR system. Query images are given 3 in blue boxes. The faces are ranked from left to right, and top to bottom. True positives are indicated by green boxes while false positives by red boxes. (Color figure online)

5.3 Discussion and Recommendation

Overall, it can be inferred that shallow facial descriptors used in this study (G-HOG, G-LBP, and G-SIFT) are still capable of classifying eight facial attributes (Fig. 3) with a slight difference compared to the deep facial descriptors, but their performances largely degraded in the task of attribute-driven FIR as opposed to the deep facial descriptors. Regarding shallow descriptors, while the G-LBP achieved the best performance for attribute classification, the worst retrieval performance was attained by the same descriptor. Conversely, G-HOG has the best performance for the face retrieval task but it also performed the worst in attribute classification. This highlights the fact that handcrafted shallow facial descriptors are often tailored to perform well in certain applications only, as it does not learn the best possible set of features in a natural way.

Despite the superior performance of deep facial descriptors in both face retrieval and facial attribute classification, they are computationally intensive and time-consuming. In the case of G-SIFT (as stated in Fig. 10), the speed of both feature extraction and classification are more than double in contrast with the deep CNN descriptors. The speed is remarkably higher for G-LBP and G-HOG with more than two orders of magnitude (10^2) faster than generating the deep descriptors.

Therefore, we recommended that future research in the field of facial attribute classification should attempt to focus more on designing CNNs that are able to achieve higher accuracy while decreasing the execution time for both feature

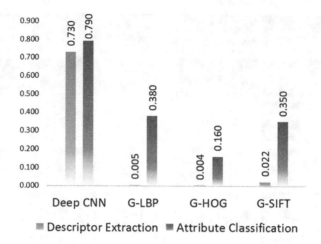

Fig. 10. Average execution time for the both descriptor extraction and attribute classification according to deep and shallow descriptors.

extraction and classification. As most state-of-the-art research in this field exploit a large number of training (positive and negative) faces to achieve good results, CNNs are most suited to provide the necessary generalization.

With respect to an attribute-driven face retrieval system, as the retrieval performance is correlated with the performance of facial attribute classification. Thus, it is essential to increase the number of facial attributes used to describe faces while utilizing more accurate attribute classifiers. In future, we intend to design an attribute-specific CNN architecture for face retrieval that is both efficient (small number of network parameters) and effective (maximizing the accuracy).

6   Conclusion

In this paper, we exploited deeply learned facial descriptors to increase the accuracy of face attribute classifiers compared to conventional shallow descriptors (HOG, LBP, and SIFT). We also utilized deep attribute-driven facial feature vectors to enhance the performance of an FIR system. Although the accuracy of the deep attribute classifier is comparable to that of shallow attribute classifiers, its performance in the face retrieval task increased by more than double. This demonstrates the capability of deeply learned facial descriptors in improving the performance of an FIR system.

In this preliminary research, we have used a reasonable amount of face images (10k maximum) to train a reduced set of attribute classifiers, owing to computational limitations. Hence in future, we intend to further validate this work on larger number of faces from the selected datasets (particularly the large-scale CelebA dataset). Besides, a larger number of attributes will also be used to provide more discrimination power to the deep attribute-driven feature vectors while scoring and retrieving similar-appearing faces. Also, we intend to use features from localized face components and to further investigate their effect on the overall performance of both tasks.

Acknowledgment. This research was fully funded by the Ministry of Science, Technology, and Innovation (MOSTI), Malaysia, Project Number 01-02-01-SF0232. We also gratefully acknowledge the feedback from anonymous reviewers.

References

1. Lv, Y., Ng, W.W., Zeng, Z., Yeung, D.S., Chan, P.P.: Asymmetric cyclical hashing for large scale image retrieval. IEEE Trans. Multimedia **17**, 1225–1235 (2015)
2. Tang, J., Li, Z., Wang, M., Zhao, R.: Neighborhood discriminant hashing for large-scale image retrieval. IEEE Trans. Image Process. **24**, 2827–2840 (2015)
3. Cao, X., Zhang, H., Guo, X., Liu, S., Meng, D.: SLED: semantic label embedding dictionary representation for multilabel image annotation. IEEE Trans. Image Process. **24**, 2746–2759 (2015)
4. Murthy, V.N., Maji, S., Manmatha, R.: Automatic image annotation using deep learning representations. In: Proceedings of the 5th ACM on International Conference on Multimedia Retrieval, pp. 603–606 (2015)
5. Zitnick, C.L., Vedantam, R., Parikh, D.: Adopting abstract images for semantic scene understanding. IEEE Trans. Pattern Anal. Mach. Intell. **38**, 627–638 (2016)
6. Kotenko, J.: Facebook reveals we upload a whopping 350 million photos to the network daily. DigitalTrends.com (2013)
7. Kumar, N., Belhumeur, P., Nayar, S.: FaceTracer: a search engine for large collections of images with faces. In: Forsyth, D., Torr, P., Zisserman, A. (eds.) ECCV 2008. LNCS, vol. 5305, pp. 340–353. Springer, Heidelberg (2008). doi:10.1007/978-3-540-88693-8_25
8. Parkhi, O.M., Vedaldi, A., Zisserman, A.: Deep face recognition. In: British Machine Vision Conference, vol. 1, p. 6 (2015)
9. Agrawal, S., Khatri, P.: Facial expression detection techniques: based on viola and jones algorithm and principal component analysis. In: 2015 Fifth International Conference on Advanced Computing and Communication Technologies, pp. 108–112 (2015)
10. Zhang, Y., Tang, Z., Zhang, C., Liu, J., Lu, H.: Automatic face annotation in tv series by video/script alignment. Neurocomputing **152**, 316–321 (2015)
11. Utsumi, Y., Sakano, Y., Maekawa, K., Iwamura, M., Kise, K.: Scalable face retrieval by simple classifiers and voting scheme. In: Ji, Q., Moeslund, T.B., Hua, G., Nasrollahi, K. (eds.) FFER 2014. LNCS, vol. 8912, pp. 99–108. Springer, Heidelberg (2015). doi:10.1007/978-3-319-13737-7_9
12. Wang, D., Jain, A.K.: Face retriever: pre-filtering the gallery via deep neural net. In: 2015 International Conference on Biometrics (ICB), pp. 473–480 (2015)
13. Wang, D., Hoi, S.C., He, Y., Zhu, J., Mei, T., Luo, J.: Retrieval-based face annotation by weak label regularized local coordinate coding. IEEE Trans. Pattern Anal. Mach. Intell. **36**, 550–563 (2014)
14. Smith, B.M., Zhu, S., Zhang, L.: Face image retrieval by shape manipulation. In: 2011 IEEE Conference on Computer Vision and Pattern Recognition (CVPR), pp. 769–776 (2011)

15. Wu, Z., Ke, Q., Sun, J., Shum, H.Y.: Scalable face image retrieval with identity-based quantization and multireference reranking. IEEE Trans. Pattern Anal. Mach. Intell. **33**, 1991–2001 (2011)
16. Park, U., Jain, A.K.: Face matching and retrieval using soft biometrics. IEEE Trans. Inf. Forensics Secur. **5**, 406–415 (2010)
17. An, L., Zou, C., Zhang, L., Denney, B.: Scalable attribute-driven face image retrieval. Neurocomputing **172**, 215–224 (2016)
18. Chen, B.C., Chen, Y.Y., Kuo, Y.H., Hsu, W.H.: Scalable face image retrieval using attribute-enhanced sparse codewords. IEEE Trans. Multimedia **15**, 1163–1173 (2013)
19. Liu, Z., Luo, P., Wang, X., Tang, X.: Deep learning face attributes in the wild. In: Proceedings of the IEEE International Conference on Computer Vision, pp. 3730–3738 (2015)
20. Klare, B.F., Klum, S., Klontz, J.C., Taborsky, E., Akgul, T., Jain, A.K.: Suspect identification based on descriptive facial attributes. In: 2014 IEEE International Joint Conference on Biometrics (IJCB), pp. 1–8 (2014)
21. Rudd, E., Günther, M., Boult, T.: Moon: a mixed objective optimization network for the recognition of facial attributes. arXiv preprint arXiv:1603.07027 (2016)
22. Chen, B.C., Kuo, Y.H., Chen, Y.Y., Chu, K.Y., Hsu, W.: Semi-supervised face image retrieval using sparse coding with identity constraint. In: Proceedings of the 19th ACM International Conference on Multimedia, pp. 1369–1372 (2011)
23. Li, Y., Wang, R., Liu, H., Jiang, H., Shan, S., Chen, X.: Two birds, one stone: jointly learning binary code for large-scale face image retrieval and attributes prediction. In: Proceedings of the IEEE International Conference on Computer Vision, pp. 3819–3827 (2015)
24. Huang, G.B., Ramesh, M., Berg, T., Learned-Miller, E.: Labeled faces in the wild: a database for studying face recognition in unconstrained environments. Technical report 07–49, University of Massachusetts, Amherst (2007)
25. Viola, P., Jones, M.J.: Robust real-time face detection. Int. J. Comput. Vis. **57**, 137–154 (2004)
26. Dalal, N., Triggs, B.: Histograms of oriented gradients for human detection. In: 2005 IEEE Computer Society Conference on Computer Vision and Pattern Recognition (CVPR 2005), vol. 1, pp. 886–893 (2005)
27. Déniz, O., Bueno, G., Salido, J., De la Torre, F.: Face recognition using histograms of oriented gradients. Pattern Recogn. Lett. **32**, 1598–1603 (2011)
28. Ojala, T., Pietikainen, M., Maenpaa, T.: Multiresolution gray-scale and rotation invariant texture classification with local binary patterns. IEEE Trans. Pattern Anal. Mach. Intell. **24**, 971–987 (2002)
29. Yang, B., Chen, S.: A comparative study on local binary pattern (LBP) based face recognition: LBP histogram versus LBP image. Neurocomputing **120**, 365–379 (2013)
30. Lowe, D.G.: Distinctive image features from scale-invariant keypoints. Int. J. Comput. Vis. **60**, 91–110 (2004)
31. Vedaldi, A., Fulkerson, B.: VLFeat: an open and portable library of computer vision algorithms. In: Proceedings of the 18th ACM International Conference on Multimedia, pp. 1469–1472 (2010)

A Main Directional Maximal Difference Analysis for Spotting Micro-expressions

Su-Jing Wang[1]([✉]), Shuhang Wu[2], and Xiaolan Fu[1]

[1] CAS Key Laboratory of Behavioral Science, Institute of Psychology,
Beijing 100101, China
wangsujing@psych.ac.cn
[2] College of Information Science and Engineering, Northeastern University,
Shenyang, China

Abstract. Micro-expressions are facial expressions that have a short duration (generally less than 0.5 s), involuntary appearance and low intensity of movement. They are regarded as unique cues revealing the hidden emotions of an individual. Although methods for the spotting and recognition of general facial expressions have been investigated, little progress has been made in the automatic spotting and recognition of micro-expressions. In this paper, we proposed the Main Directional Maximal Difference (MDMD) analysis for micro-expression spotting. MDMD uses the magnitude of maximal difference in the main direction of optical flow as a feature to spot facial movements, including micro-expressions. Based on block-structured facial regions, MDMD obtains more accurate features of the movement of expressions for automatically spotting micro-expressions and macro-expressions from videos. This method obtains both the temporal and spatial locations of facial movements. The evaluation was performed on two spontaneous databases (CAS(ME)^2 and CASME) containing micro-expressions and macro-expressions.

1 Introduction

The telling of a lie exists everywhere in human social intercourse. Lies are extremely difficult to detect, and although everyone experiences deceiving others or being deceived, even specialists cannot detect them. The polygraph is widely employed in the traditional lie-detection systems, which monitor uncontrolled changes in heart rate and electro-dermal responses when the subject is telling a lie. However, the polygraph makes incursions into the private space of the subject, and the subject can take steps to conceal their genuine emotions [1]. Recently, a type of expression called a micro-expression, which is involuntary, of short duration and low intensity, has aroused the broad concern of affective computational psychologists and researchers. Micro-expressions may appear when individuals are likely to hide their real emotions, and they are very rapid and minute, especially in high-stakes situations [2,3]. Ekman professed that micro-expressions may be the most reliable clue for detecting lies [3].

Micro-expressions can be detected using a concealed camera during a conversation or interview; therefore, the person will not realize that he is being judged

© Springer International Publishing AG 2017
C.-S. Chen et al. (Eds.): ACCV 2016 Workshops, Part II, LNCS 10117, pp. 449–461, 2017.
DOI: 10.1007/978-3-319-54427-4_33

on whether he is lying. Spotting micro-expressions automatically from videos in a trial may considerably contribute to judicial officials detecting clues of deception by culprits. The automatic spotting of micro-expressions in long videos thus offers great potential.

Regarding micro-expression recognition, a number of papers have been published in recent years. Polikovsky et al. [4] recognized micro-expressions based on the 3D-Gradients orientation histogram descriptor. Pfister et al. [5] developed the Temporal Interpolation Model (TIM), handling dynamic features by spatiotemporal local texture descriptors (SLTD), and then used Support Vector Machine (SVM), Multiple Kernel Learning (MKL) and Random Forest (RF) classifiers to recognize spontaneous facial micro-expressions. They [6] also proposed a new spatiotemporal local texture descriptor (CLBP-TOP) to differentiate spontaneous vs. posed (SVP) facial expressions. Wang et al. [7] utilized Discriminant Tensor Subspace Analysis (DTSA), which treats a gray facial image as a third-order tensor and Extreme Learning Machine (ELM). However, subtle movements of micro-expressions may be lost in this method. Wang et al. [8,9] set up a novel color space model, Tensor Independent Color Space (TICS), because color could provide useful information for expression recognition. Then, they [10] used the sparse part of Robust PCA (RPCA) to extract the subtle motion information of micro-expressions and Local Spatiotemporal Directional Features (LSTD) to extract the local texture features.

The number of papers on micro-expression spotting is smaller than that on micro-expression recognition. Shreve et al. [11,12] primarily used a robust optical flow method [13] for computing the strain from the measured displacement (motion) observed in the video sequence to differentiate macro-expressions and micro-expressions. Polikovsky et al. [4,14] calculated the durations of the three phases of micro-expressions by the 3D-Gradients orientation histogram descriptor. Moilanen et al. [15] proposed a method based on Local Binary Pattern (LBP) histogram features to obtain both the temporal locations and spatial locations for micro-expression spotting. Current research on micro-expression spotting is constrained mostly by micro-expression databases, which employ cropped micro-expression samples or short videos. One exception to this is Shreve et al. [12], but the database used in Shreves study is not yet publicly available.

2   Main Directional Maximal Difference Analysis

2.1   Face Alignment, Face Cropping and Block-Structure

The inner eye corners were calibrated manually in the first frame of videos to align faces using nonreflective similarity transformation. Nonreflective similarity transformation supports translation, rotation, and isotropic scaling. It has four degrees of freedom and requires two pairs of points, which is similar to affine transformation that needs three pairs of non-collinear points. The inner eye corners are relatively steady [16] and are used in the rest of the video. The original image is shown in Fig. 1(left), and the aligned image is shown in Fig. 1(right).

Fig. 1. An example of face alignment.

The method, called Discriminative Response Map Fitting (DRMF) [17], can obtain the outline points of a face and was employed to carry out face cropping. The cropped face image was divided into blocks. We utilized a 6 × 6 block structure [15] that comprises all the crucial parts of a face and guarantees a relatively low computational complexity. The block structure was based on the horizontal distance between the inner eye corners, the vertical distance between the nasal spine point and the horizontal line connecting the inner eye corners. It is adaptable to faces of different sizes and maintained for each video in this paper because of the measure of face cropping, which is shown in Fig. 2.

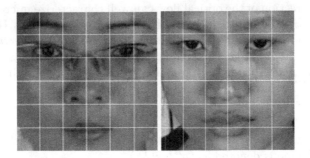

Fig. 2. Examples of the facial 6 × 6 block structure.

2.2   Main Directional Maximal Difference Analysis

Given a video with n frames, the current frame is denoted as F_i. F_{i-k} is the k-th frame before F_i, and F_{i+k} is the k-th frame after F_i. The optical flow between the F_{i-k} frame (Head Frame) and the F_i frame (Current Frame) after alignment is denoted (u^{HC}, v^{HC})[1]. Similarly, the optical flow between the F_{i-k} frame (Head Frame), and the F_{i+k} frame (Tail Frame) is denoted (u^{HT}, v^{HT}).

---

[1] For convenience, (u^{HC}, v^{HC}) means the displacement of any point.

The (u^{HC}, v^{HC}) and (u^{HT}, v^{HT}) are converted from Euclidean coordinates to polar coordinates (ρ^{HC}, θ^{HC}) and (ρ^{HT}, θ^{HT}), where ρ and θ represent the magnitude and direction, respectively.

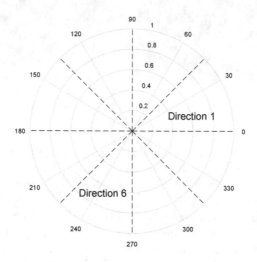

Fig. 3. 8 directions in polar coordinates.

The main direction of the optical flow can well characterize micro-expressions [18]. Based on the directions $\{\theta^{HC}\}$, all the optical flow vectors $\{(\rho^{HC}, \theta^{HC})\}$ are separated into eight directions (see Fig. 3). The *Main Direction* Θ is the direction that has the largest number of optical flow vectors among the eight directions. The main directional optical vector $(\rho_M^{HC}, \theta_M^{HC})$ is the optical flow vector (ρ^{HC}, θ^{HC}) that falls in the *Main Direction* Θ.

$$\{(\rho_M^{HC}, \theta_M^{HC})\} = \{(\rho^{HC}, \theta^{HC}) | \theta^{HC} \in \Theta\} \tag{1}$$

The optical flow vector corresponding to $(\rho_M^{HC}, \theta_M^{HC})$ between the F_{i-k} frame and the F_{i+k} frame is denoted as $(\rho_M^{HT}, \theta_M^{HT})$.

$$\{(\rho_M^{HT}, \theta_M^{HT})\} = \{(\rho^{HT}, \theta^{HT}) | (\rho^{HT}, \theta^{HT}) \text{ and } (\rho_M^{HC}, \theta_M^{HC})$$
$$\text{are two different vectors of the same point in } F_{i-k}\} \tag{2}$$

After sorting the differences $\rho_M^{HC} - \rho_M^{HT}$ in descending order, the maximal difference d^i is the mean difference value of the first $\frac{1}{3}$ of the differences $\rho_M^{HC} - \rho_M^{HT}$ to characterize the frame F_i as the formulation

$$d = \frac{3}{g} \sum \max_{\frac{g}{3}} \{\rho^{HC} - \rho^{HT}\} \tag{3}$$

where $g = |\{(\rho^{HC}, \theta^{HC})\}|$ is the number of elements in the subset $\{(\rho^{HC}, \theta^{HC})\}$, and $\max_m S$ denotes a set that comprises the first m maximal elements in subset S.

In practice, we employed the 6×6 block structure introduced in Sect. 2.1. We will calculate the maximal difference d_b^i $(b = 1, 2, \ldots, 36)$ for each block in F_i frame.

It was discovered that picking out approximately one-third of the difference values can obtain a better distinction [15]. For the frame F_i, there are 36 maximal differences d_b^i owing to the 6×6 block structure. Similarly, we arranged the 36 maximal differences d_b^i in descending order. $\bar{d}^i$ is the maximal differences of the first $\frac{1}{3}$, i.e., 12, of the 36 mean values to characterize the frame F_i feature:

$$\bar{d}^i = \frac{1}{12} \sum \max_{12} \{d_b^i\} \quad b = 1, 2, \ldots, 36 \tag{1}$$

If a person maintains a neutral expression at F_{i-k}, an emotional expression such as disgust will appear at the onset frame between F_{i-k} and F_i and be repressed at the offset frame between F_i and F_{i+k}, and her facial expression will recover to a neutral expression at F_{i+k}, as is presented in Fig. 4(a). Under this circumstance, the movement between F_i and F_{i-k} is deservedly more intense than that between F_{i+k} and F_{i-k} because the expressions are neutral at both F_{i+k} and F_{i-k}. Therefore, the $\bar{d}^i$ value will be large. In another situation, if a person maintains a neutral expression from F_{i-k} to F_{i+k}, the movement between F_i and F_{i-k} will be similar to that between F_{i+k} and F_{i-k}, so the $\bar{d}^i$ value will be small. In a long video, an emotional expression sometimes appears at the onset frame before F_{i-k} and is repressed at the offset frame after F_{i+k} (see Fig. 4(b)), so the $\bar{d}^i$ value will be small if k is set to a small value. However, k cannot take too large a value, as this could influence the accuracy of computing the optical flow.

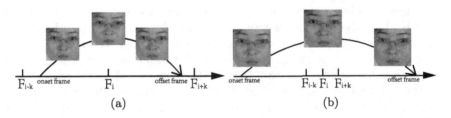

(a)   (b)

Fig. 4. (a) An emotional expression starts at the onset frame between F_{i-k} and F_i, is repressed at the offset frame between F_i and F_{i+k} and recovers a neutral expression at F_{i+k}; (b) An emotional expression of a person starts at the onset frame before F_{i-k} and is repressed at the offset frame after F_{i+k}.

2.3   Expression Spotting

We employed the relative difference vector for eliminating background noise which was computed by

$$r^i = \bar{d}^i - \frac{1}{2}\left(\bar{d}^{i-k+1} + \bar{d}^{i+k-1}\right) \quad i = k+1, k+2, \ldots, n-k \tag{5}$$

This is shown in Fig. 5(a), except for the first and the last k frames of a video. The negative difference values indicate that the movement between F_i and F_{i-k} is subtler than the movement between F_{i+k} and F_{i-k}. Accordingly, all negative difference values were set to zero (see Fig. 5(b)).

Thresholding was used to ascertain the frames that have the highest intensity of the facial movements in a video,

$$
\begin{aligned}
&\textbf{if } r_{max} - r_{mean} < a: \\
&threshold = r_{mean} + b \times (r_{max} - r_{mean}) + c \\
&\textbf{else}: \\
&threshold = r_{mean} + b \times (r_{max} - r_{mean})
\end{aligned}
\tag{6}
$$

where $r_{mean} = \frac{1}{n-2k} \sum_{i=k+1}^{n-k} r^i$ and $r_{max} = \max_{i=k+1}^{n-k} r^i$ are the average and the maximum of all r^i for the entire video. The parameter a is employed due to the condition that the expressions are so subtle that the difference between the emotional expressions and neutral expression in long videos is small, and b is a variable parameter in the range $[0, 1]$. The threshold is more adaptive to improve the robustness of micro-expression detection in long videos, and it is shown in Fig. 5 as the red dashed line. The difference values above the red dashed line are the frames that show expressions.

3   Experiments

3.1   Evaluation on CAS(ME)^2

The Chinese Academy of Sciences Macro-Expressions and Micro-Expressions (CAS(ME)^2) database is the first publicly available database comprising both spontaneous macro-expressions and micro-expressions in long videos (Part A) and separate samples (Part B). Macro-expressions and micro-expressions were collected from the same participants under the same experimental conditions.

In the CAS(ME)^2 database, Part A has 87 long videos that include spontaneous macro-expressions and micro-expressions collected from 22 participants and Part B contains 300 spontaneous macro-expression samples and 57 micro-expression samples. To our knowledge, there are no publicly available databases that contain macro- and micro-expressions in long videos that can be used for expression detection. The CAS(ME)^2 database used a Logitech Pro C920 camera with 30 frames per second and a resolution of 640×480 pixels, which satisfied the constraint of brightness constancy. The expression samples were selected from more than 600 elicited facial movements and were coded with the onset, apex, and offset frames, with the AUs marked, emotions labeled, and a self-report for each expression.

In the experiments, we use 59^2 videos which include 152 macro-expressions and 38 micro-expressions. The maximum duration of the macro-expressions in this database is more than 500 ms and less than 4 s, and the maximum duration of the micro-expressions is no greater than 500 ms. The average durations of the macro-expressions and micro-expressions are approximately 1305 ms and 419 ms, respectively. According to the average durations of macro-expressions and micro-expressions, k is set to 12, with $a = 6.6$, $b = 0.15$, $c = 0.15$. The spotting results of a video named 'disgust02' of subject no. 2 are presented in Fig. 5. The green areas denote the durations of expressions or blinking.

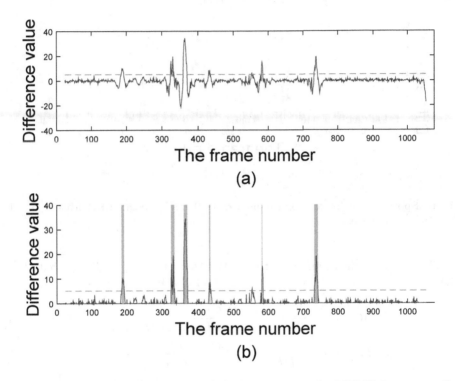

Fig. 5. Spotting results of a video named 16_0102 using the MDMD feature on the CAS(ME)^2 database. (Color figure online)

The LBP feature was used [15] as a comparison by computing the LBP histogram in the (8, 1) neighborhood with $k = 12$ and $p = 0.25$. The spotting results of the same video are presented in Fig. 6. Apparently, the difference by using the MDMD feature is more notable than that of using the LBP feature on the same video.

---

^{2} 28 videos were removed because of relatively large movements of the head.

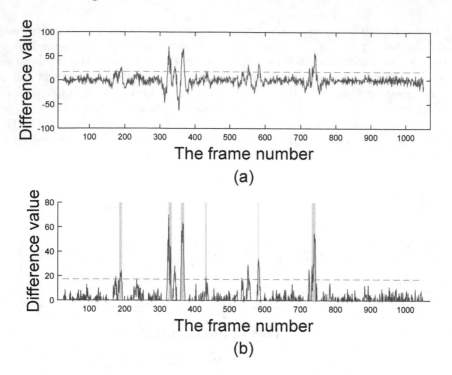

Fig. 6. Spotting results of a video named 16_0102 by using LBP feature on the CAS(ME)^2 database.

The results of the accuracy of the MDMD feature compared with that of the LBP feature are shown in a Table 1 with the following equations:

$$ACC = \frac{spotted\ frames\ which\ are\ expressions}{all\ the\ numbers\ of\ spotted\ frames} \tag{7}$$

$$TOTAL = \frac{spotted\ expressions}{all\ the\ numbers\ of\ expressions} \tag{8}$$

$$MIC = \frac{spotted\ micro - expressions}{all\ the\ numbers\ of\ micro - expressions} \tag{9}$$

$$MAC = \frac{spotted\ macro - expressions}{all\ the\ numbers\ of\ macro - expressions} \tag{10}$$

We regarded the spotted frames as true results if they fell within the span of $k/2$ before or after the onset or offset of the truth frames provided by the labels of the database. The distance between spotted peaks was not considered, owing to the fact that there are several overlapping micro-expressions in this database. Eye blinks were treated as true results as well because eye blinks can express emotions, e.g., the squinting of the eyes (nervousness or disagreement) or rolling of the eyes (contempt) [3,19]. They will cause rapid movements in some regions around the eyes [15] and exist in several micro-expressions. An evaluation

Table 1. Evaluation by using MDMD and LBP on CAS(ME)^2 database

	ACC	TOTAL	MIC	MAC
MDMD	61.3%	74.6%	55.3%	79.5%
LBP	42.7%	70.4%	47.3%	76.2%

Table 2. Evaluation according to emotional category by using MDMD and LBP on CAS(ME)^2 database

Category	Number	TOTAL		MIC		MAC	
		LBP	MDMD	LBP	MDMD	LBP	MDMD
Positive	54	70.3%	**77.8%**	0%	**25%**	76%	**82%**
Negative	68	72.1%	**75%**	20%	**30%**	81%	**82.8%**
Surprise	14	**71.4%**	64.3%	66.7%	66.7%	**75%**	62.5%
Others	54	69.8%	**73.6%**	66.7%	**72.2%**	71.4%	**74.3%**

according to the emotional category on the CAS(ME)^2 database is shown in the Table 2. The MDMD feature performed better than the LBP feature under the conditions of Positive, Negative and Others.

3.2 Evaluation on CASME

The Chinese Academy of Sciences Micro-Expression (CASME) database [20] includes 195 spontaneous facial micro-expressions recorded by two 60 fps cameras. The samples were selected from more than 1,500 facial expressions. The CASME database is divided into two classes: Set A and Set B. The samples in Set A were recorded with a BenQ M31 consumer camera with 60 fps, with the resolution set to 1280 × 720 pixels. The participants were recorded in natural light. The samples in Set B were recorded with a Point Grey GRAS-03K2C industrial camera with 60 fps, with the resolution set to 640 × 480 pixels. The participants were recorded in a room with two LED lights. This experiment was performed on Set B because of the fixed light source. The average duration of the micro-expression samples in Set B was approximately 299 ms. All of the samples in Set B were utilized for evaluation in this experiment. According to the average duration of the micro-expressions, k is set to 5 with $a = 1.69$, $b = 0.6$, and $c = 0.3$. The spotting results of a video named EP12_2_5 from subject 8 with 77 frames is presented in Fig. 7, and a micro-expression of disgust is successfully spotted from approximately frame 11 to frame 13.

The LBP feature was used [15] as a means of comparison by computing the LBP histogram in the (8, 1) neighborhood with $k = 5$ and $p = 0.85$. The spotting results of the same video are presented in Fig. 8, and there is no micro-expression at frame 35.

The accuracy of the MDMD feature compared with that of the LBP feature is shown in the Table 3. We regarded the spotted frames as true results if they

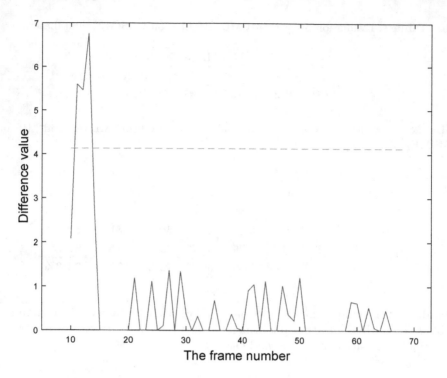

Fig. 7. Spotting results of a video named EP12_2_5 from subject 8 by using MDMD feature on the CASME Set B.

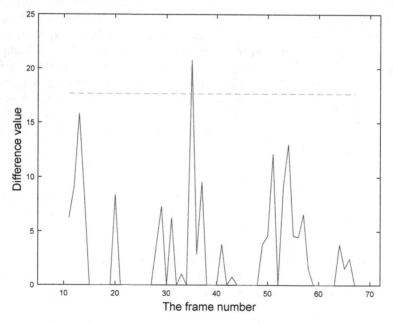

Fig. 8. Spotting results of a video named EP12_2_5 from subject 8 by using LBP feature on the CASME Set B.

fall between the onset and offset of the truth frames provided by the labels of the database. The constraint in Set B was stricter than that of CAS(ME)^2 because all the micro-expression in CASME-B are short videos. Eye blinks were considered as true results. The value for k is set to 2 in a video called EP12_4_4 from subject 8 of Set B due to it being a short video with only 13 frames. The flickering light in the video of the CASME-B database influenced the accuracy of

Table 3. Evaluation by using MDMD and LBP on CASME Set B

	ACC	MIC
MDND	67.8%	53.5%
LBP	57.2%	58.4%

Table 4. Evaluation according to emotional category by using MDMD and LBP on CASME-B database

Category	Number	MIC	
		LBP	MDMD
Disgust	42	64.3%	59.5%
Sadness	6	50%	**50%**
Surprise	14	57.1%	**71.4%**
Tense	23	60.9%	39.1%
Repression	10	50%	**50%**
Fear	1	100%	**100%**
Happiness	5	20%	**20%**

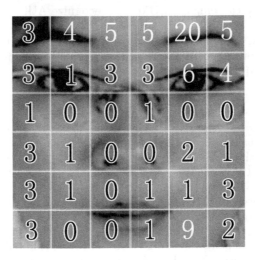

Fig. 9. An example of difference values in the space when a person frowned with a curl of her lip.

the results when using the MDMD feature. The capability of the MDMD feature may be higher when it is employed in long videos. Evaluation according to the emotional category on the Set B database is presented in Table 4. The MDMD feature performed better than the LBP feature under the condition of Surprise, while it performed nearly the same under the conditions of Sadness, Repression, Fear, and Happiness.

4 Conclusions

In this paper, we proposed the Main Directional Maximal Difference (MDMD) Analysis for micro-expression spotting. We pre-processed databases, including facial alignment, cropping and division, primarily by nonreflective similarity transformation. Based on block-structured facial regions, we calculated a robust local optical flow. We proposed that MDMD will obtain more accurate features of the movement of expressions. The MDMD features were used to spot micro-expressions.

Evaluation was performed on MDMD using two spontaneous databases (CAS(ME)^2 and CASME) by comparison with LBP [15]. For the CAS(ME)^2 database, the efficiency and accuracy of the MDMD feature were superior to those of LBP. For the CASME database, the flickering light in the video influenced the accuracy of MDMD. The capability of MDMD feature may be higher in long videos than in short videos.

MDMD achieves not only temporal detection but also the spatial location of facial movements. An example of the 8 largest difference values colored as white, of a person who frowned with a curl of her lip from the CAS(ME)^2 database is presented in Fig. 9. The expression of the frown was labeled AU 4 (brow lower) according to the FACS to convey a negative emotion.

In the future, we can employ MDMD to recognize AU.

Acknowledgments. This work was supported by grants from the National Natural Science Foundation of China (61379095, 61375009), and the Beijing Natural Science Foundation (4152055).

References

1. Michael, N., Dilsizian, M., Metaxas, D., Burgoon, J.K.: Motion profiles for deception detection using visual cues. In: Daniilidis, K., Maragos, P., Paragios, N. (eds.) ECCV 2010. LNCS, vol. 6316, pp. 462–475. Springer, Heidelberg (2010). doi:10.1007/978-3-642-15567-3_34
2. Ekman, P., Friesen, W.V.: Nonverbal leakage and clues to deception. Psychiatry **32**, 88–106 (1969)
3. Ekman, P.: Telling Lies: Clues to Deceit in the Marketplace, Politics, and Marriage. WW Norton & Company, New York (2009). (Revised Edition)
4. Polikovsky, S., Kameda, Y., Ohta, Y.: Facial micro-expressions recognition using high speed camera and 3D-gradient descriptor. In: 3rd International Conference on Imaging for Crime Detection and Prevention (ICDP 2009)

5. Pfister, T., Li, X., Zhao, G., Pietikäinen, M.: Recognising spontaneous facial micro-expressions. In: 2011 IEEE International Conference on Computer Vision (ICCV), pp. 1449–1456. IEEE (2011)
6. Pfister, T., Li, X., Zhao, G.: Differentiating spontaneous from posed facial expressions within a generic facial expression recognition framework. In: 2011 IEEE International Conference on Computer Vision Workshops (ICCV Workshops), pp. 868–875. IEEE (2011)
7. Wang, S.J., Chen, H.L., Yan, W.J., Chen, Y.H., Fu, X.: Face recognition and micro-expression recognition based on discriminant tensor subspace analysis plus extreme learning machine. Neural Process. Lett. **39**, 25–43 (2014)
8. Wang, S.J., Yan, W.J., Li, X., Zhao, G., Fu, X.: Micro-expression recognition using dynamic textures on tensor independent color space. In: 2014 22nd International Conference on Pattern Recognition (ICPR), pp. 4678–4683. IEEE (2014)
9. Wang, S.J., Yan, W.J., Li, X., Zhao, G., Zhou, C.G., Fu, X., Yang, M., Tao, J.: Micro-expression recognition using color spaces. IEEE Trans. Image Process. **24**, 6034–6047 (2015)
10. Wang, S.-J., Yan, W.-J., Zhao, G., Fu, X., Zhou, C.-G.: Micro-expression recognition using robust principal component analysis and local spatiotemporal directional features. In: Agapito, L., Bronstein, M.M., Rother, C. (eds.) ECCV 2014. LNCS, vol. 8925, pp. 325–338. Springer, Cham (2015). doi:10.1007/978-3-319-16178-5_23
11. Shreve, M., Godavarthy, S., Manohar, V., Goldgof, D., Sarkar, S.: Towards macro- and micro-expression spotting in video using strain patterns. In: 2009 Workshop on Applications of Computer Vision (WACV), pp. 1–6. IEEE (2009)
12. Shreve, M., Godavarthy, S., Goldgof, D., Sarkar, S.: Macro- and micro-expression spotting in long videos using spatio-temporal strain. In: 2011 IEEE International Conference on Automatic Face & Gesture Recognition and Workshops (FG 2011), pp. 51–56. IEEE (2011)
13. Black, M.J., Anandan, P.: The robust estimation of multiple motions: parametric and piecewise-smooth flow fields. Comput. Vis. Image Underst. **63**, 75–104 (1996)
14. Polikovsky, S., Kameda, Y., Ohta, Y.: Detection and measurement of facial micro-expression characteristics for psychological analysis. Kameda's Publ. **110**, 57–64 (2010)
15. Moilanen, A., Zhao, G., Pietikainen, M.: Spotting rapid facial movements from videos using appearance-based feature difference analysis. In: 2014 22nd International Conference on Pattern Recognition (ICPR), pp. 1722–1727. IEEE (2014)
16. Valstar, M.F., Pantic, M.: Fully automatic recognition of the temporal phases of facial actions. IEEE Trans. Syst. Man Cybern. Part B: Cybern. **42**, 28–43 (2012)
17. Asthana, A., Zafeiriou, S., Cheng, S., Pantic, M.: Robust discriminative response map fitting with constrained local models. In: 2013 IEEE Conference on Computer Vision and Pattern Recognition (CVPR), pp. 3444–3451. IEEE (2013)
18. Liu, Y.J., Zhang, J.K., Yan, W.J., Wang, S.J., Zhao, G., Fu, X.: A main directional mean optical flow feature for spontaneous micro-expression recognition. IEEE Trans. Affect. Comput. **7**(4), 299–310 (2015)
19. Ekman, P.: Lie catching and microexpressions. In: The Philosophy of Deception, pp. 118–133 (2009)
20. Yan, W.J., Wu, Q., Liu, Y.J., Wang, S.J., Fu, X.: CASME database: a dataset of spontaneous micro-expressions collected from neutralized faces. In: 2013 10th IEEE International Conference and Workshops on Automatic Face and Gesture Recognition (FG), pp. 1–7. IEEE (2013)

Aesthetic Evaluation of Facial Portraits Using Compositional Augmentation for Deep CNNs

Magzhan Kairanbay[✉], John See, and Lai-Kuan Wong

Center for Visual Computing, Faculty of Computing and Informatics,
Multimedia University, 63100 Cyberjaya, Malaysia
magzhan.kairanbay@gmail.com, {johnsee,lkwong}@mmu.edu.my

Abstract. Digital facial portrait photographs make up a massive portion of photos in the web. A number of methods for evaluating the aesthetics of photographs have been proposed recently. However, there have been a little work in the research community to address the aesthetics of targeted image domain, such as portraits. This paper introduces a new compositional-based augmentation scheme for aesthetic evaluation of portraits by well-known deep convolutional neural network (DCNN) models. We present a set of feature augmentation methods that take into account compositional photographic rules to ensure that the aesthetic in portraits are not hindered by standard transformations used for DCNN models. On a portrait subset of the large-scale AVA dataset, the proposed approach demonstrated a reasonable improvement in classification performance over the baseline and vanilla deep learning approaches.

1 Introduction

The ubiquity of smartphones and popularity of social media networks have contributed to an increasing number of photographs, particularly self-portraits or "selfies". A recent study by Bakhshi et al. [1] showed that photos with faces are more popular than other types of pictures. Nevertheless, casual photographers tend to produce many unprofessionally taken photos that are less aesthetically pleasing, and proceed to share them with other people in their circles. As such, an automatic assessment of the photo aesthetics would be useful to provide advice as to which of these photos should be retaken. This is also helpful for designers to select aesthetically pleasing photos for magazine covers, books, and news articles. However, different type of photos needs different type of photographic rules. Even, the most applicable rules, such as rule of thirds [2] could be applied differently for portraits and other type of photos [3].

Deep learning have been widely used in computer vision and natural language processing fields in recent years. In particular, deep convolutional neural network (DCNN) models have made revolutionary steps in object detection and recognition tasks [4–6]. The usage of graphic processing units (GPUs) have made it possible to handle large amounts of data, which directly provides better generalization and discrimination of models. Despite these technological advances and a growing research interest in computational aesthetics ever since the work

© Springer International Publishing AG 2017
C.-S. Chen et al. (Eds.): ACCV 2016 Workshops, Part II, LNCS 10117, pp. 462–474, 2017.
DOI: 10.1007/978-3-319-54427-4_34

Fig. 1. Compositional-based cropping for DCNN inputs: (a) A portrait can be cropped aesthetically from power points defined by rule of thirds composition principle. (b) Full image warping *(top)* and center-cropping *(bottom)* can typically distort composition in portraits. (c) The four sub-images based on the power points.

of Datta et al. [7], very little work has been done in the specific domain of facial or portrait aesthetics, with a number of works treating the aesthetic problem generically [8–10] or categorically depending on its content [11] or query [12].

Motivated by these developments, we are inspired to improve existing deep learning techniques to better evaluate facial aesthetics. In this paper, we present a new compositional-based augmentation scheme for deep convolutional neural networks (DCNN) that can boost the aesthetic evaluation of facial portraits. Conventional sampling of sub-images (i.e. warping, center-crop, random crop) as receptive fields of a DCNN is rigid and can potentially distort the composition in portraits (Fig. 1). As such, we exploit the rule of thirds composition rule to generate a feasible set of inputs to the network, which are in turn aggregated to form a deeply-learned augmented feature set. On a portrait-based dataset derived from the large-scale AVA dataset, our approach is able to surpass the performance of handcrafted and compositional-based features [3,13] which have been a staple for defining beauty in portraits. The objective of this paper is to provide the methods of evaluating the portraits aesthetics using deep neural networks.

The following sections are organized as follows: Sect. 2 reviews the current works in literature, Sect. 3 describes the dataset that was used, Sect. 4 elaborates on the proposed method in detail, Sect. 5 reports the experimental results with further discussion on them, while Sect. 6 concludes the paper.

2 Related Work

The task of evaluating the aesthetic quality of photographs is becoming increasingly popular in recent years. As the notion of aesthetics is a subjective one, substantial human effort to rate photos is needed. Nonetheless, the creation of various datasets such as CUHKPQ [11] and AVA [9], from online photo communities such as Photo.net and DPChallenge [14], have made it possible to evaluate solutions with ground truth labels. Works from the early efforts of Datta et al. [7] to the recent use of popular descriptors [8] have concentrated on the extraction of hand-crafted visual features for aesthetic evaluation of photos. In [7], as many as 66 pre-determined features, depicting colourfulness, composition, textures and statistical properties were crafted while [8] went for generic representations such as SIFT and Color descriptors, supplemented by Bag-of-Words (BoW) and Fisher Vector (FV) encodings and spatial pyramids (SP).

Different forms of convolutional neural networks [10,15,16] have been applied to solve this task, with the intent of learning features in a more natural and unguided manner. We take particular interest in the work of Lu et al. [10], which generated two heterogeneous inputs to represent global and local cues respectively. Global views are represented by normalized inputs such as center-crop, warp and padding, while local views are represented by randomly-cropped inputs of smaller parts of the image. These views were used generically for any type of photo, and may not be suitable for photos with facial portraits. However, a majority of DCNN-based works showed scant attempts to consider compositional impact relating to photographic rules.

However, not many works focused on the aesthetics of photos from a specific domain, particularly in facial portraits. The works in [3,13,17] try to rate the aesthetic quality of face portraits and understand the visual features that define the beauty of digital portraits, while another recent work [18] attempted to answer the question of "how to take a good selfie?" from the perspective of image popularity.

Khan and Vogel [3] made a comparative study on the use of traditional features [7] versus their choice of features that are focused on spatial and color composition. Firstly, template based spatial composition features were constructed from three templates for power points, power lines and their combination which were modified from [19]. Secondly, highlight and shadow composition features were obtained by extracting the face illumination, background contrast, brightness and size of the face as features. Their results show that the proposed features are more informative, more concise and better equipped to rank portraiture composition. The proposed features are more accurate by 6% and smaller 9 times in feature length compared to traditional features. More importantly, the authors postulated that spatial composition is more important than highlight and shadow composition when identifying aesthetics.

Redi et al. [13] introduced their own features to describe the portrait aesthetics. These features have been derived from the following photographic dimensions: compositional rules (lighting, sharpness, traditional compositional features), semantics and scene content (object bank feature), basic quality metrics (noise, exposure quality, JPEG quality), portrait specific features (facial landmark sharpness, landmark statistics and face/background contrasts) and fuzzy properties (emotion, originality, memorability and uniqueness). Their feature analysis showed that race, gender and age of a person are uncorrelated with photographic beauty, while soft and hard image properties such as sharpness of facial landmarks, image contrast, exposure, uniqueness and originality are directly related to portrait aesthetics. The proposed method (after shrinking to a selected set of 12 positively-correlated features) outperformed the baseline method of [3] by 14.66% on the same data. However, using only face-related features without any selection, it performed better by only ~1%. Another work by Lienhart et al. [17] applied similar types of features on three small datasets, while also performing aesthetic score prediction by support vector regression. But their work does not have the similar rigor and extensiveness as that of [13].

3 Dataset

The Aesthetic Visual Analysis (AVA) [9] dataset is the largest known dataset that is widely used to evaluate the aesthetic value of the images. The dataset contains more than 250,000 images from DPChallenge [14] with ground truth data such as human ratings ranging from 1 to 10 (with average 210 ratings per photo), with further meta-annotations from 66 textual semantics and 14 photographic styles from 72 different photo challenges. Following the work in [13], we select from the entire AVA dataset photos that are tagged with "Portraits" and also those whose challenge title contained the words Portrait, Portraiture or Portraits. Then, we filter further by detecting for photos that contained the presence of one or more faces with Face++ API [20]. A total of 10,141 images were obtained to form the portrait subset dataset. For better clarity, we refer to this subset as "AVA-Portraits" from this point onwards.

The same evaluation methodology as [13] was used. The dataset is randomly divided into five equal partitions. To ensure fair experimentation, we obtained the partition information from the authors. One partition will be used as test set and the rest as training set. Using 5-fold cross validation, the final accuracy is determined by averaging the minimum and maximum result of these five different cases. A portrait image is classified as aesthetically high if its aesthetic score is more than 5.5 (mean user score) [13]. Otherwise, the image is classified as aesthetically low.

4 Proposed Approach

Convolutional neural networks (CNN) are biologically-inspired variants of the multilayer perceptron, dating back to the work of LeCun et al. [21]. However, it started to be increasingly popular only in 2012, after the work of Krizhevsky et al. [4] introduced an eight-layer deep CNN network, winning the ImageNet Large Scale Visual Recognition Challenge (ILSVRC) 2012. From then on, convolutional neural network models have shown unrivaled performance in ILSVRC each year. In ILSVRC, models are trained and tested on the ImageNet dataset [22], which consists of more than 14 million of images with 1,000 classes. The classes comprise of a large variety of objects such as car, motocycle, person, cats, etc. Realistically, an immense amount of computational power is necessary to be able to train weights in these deep networks at a reasonable amount of

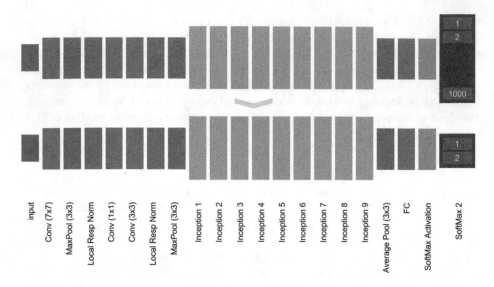

Fig. 2. Changing the number of nodes in the last layer of GoogleNet.

time. Thus, pre-trained models are commonly available for application to different related classification problems. However, in view of the different number of classes in other problems, the number of nodes in the last softmax layer should be replaced with that equivalent to the number of classes. In our case, the softmax layers are 2×1 as it is a binary classification problem (good aesthetics, bad aesthetics) (See Fig. 2).

There are a number of popular pre-trained models available for public use. In this work, we use the Overfeat [5] (winner of the localization task in ILSVRC 2013) and GoogleNet [6] (winner of the classification task in ILSVRC 2014). For each of these models, the number of nodes in last (softmax) layer is reduced from 1000 to 2, and the weights are fine-tuned with the new training data. The second last layer of the network is then taken as the feature vector for classification by a softmax classifier (or multi-class binary logistic regression). We use the Sklearn-theano [23] library, an open source library on the Theano framework for the implementation of this research.

4.1 Feature Augmentation Using Photographic Rules

Feature augmentation is a process intended to increase the feature vector length at a certain layer by aggregating features obtained from various transformed versions of the input image. We distinguished this from *data augmentation*, which is a well-known technique [4,24] used especially for learning deep representations by generating additional number of training samples. By augmentation we mean perturbing an image by transformations (e.g. cropping, flipping, or a combination), leaving the underlying class unchanged.

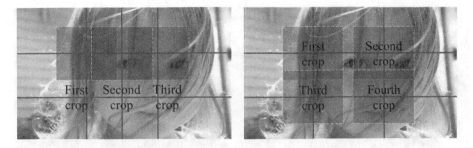

Fig. 3. (a) Three-crop by rule of thirds. (b) Four-crop by rule of thirds.

Fig. 4. (a) Five-crop: four crops from four different corners + center crop. (b) Five-crop: four crops from four corners + face crop.

In our work, we propose a feature augmentation scheme based on compositional photographic rules. This is motivated by the fact that typical image transformations such as cropping and flipping do not make aesthetic sense when it comes to face portraits. They may aggravate the actual composition of the portrait since a majority of cropping are done randomly or without applying compositional principles, while warping can potentially distort the face aspect ratio. The rule of thirds [2,19,25] is one of the popular photographic rules which divides the image to nine parts by 3 × 3 (see Fig. 3). The four intersections of the rule of thirds lines are called power points. According to this popular rule, a photo is considered aesthetically appealing if the object of interest is located near one of these four power points (Fig. 4).

We design a number of compositional cropping arrangements that are motivated by photographic rules associated to rule of thirds. Firstly, the baseline transformation involves warping the image to a constant width and height (corresponding to the model) and passing it as a single input to the DCNN model. We denote this in our experimental reporting as '**warp**'.

According to [3], the face, which is the main subject of interest in the portrait, should be located at the top power line by keeping the eyes in the upper third of the image area. Meanwhile, the symmetrical composition rule states that it is aesthetically pleasing to center the face in the image frame. Inspired by these

Fig. 5. Sample cropped faces detected by Face++ API

photographic composition rules, we modify the rule of thirds for portraits by taking two crops centered on the top two power points (left and right) and a center crop between these two crops (see Fig. 3a). All three crops are passed to the DCNN model as individual inputs to generate feature vectors that are then concatenated. We denote this technique as '**c3**'. The third technique (denoted as '**c4**') uses four different sub-images of constant height and width, center cropped from four power points based on the rule of thirds (see Fig. 3b). The augmented feature vector is formed by concatenating the four feature vectors generated from each crop.

Extending on to five-crop method, we propose two different variants. The first (denoted as '**c5**') takes four sub-images from the four corners of the image, and the center crop as the fifth. Since the fifth crop provides information from the middle of the photo, the use of four crops from the corners yields better coverage than using crops centered at all four power points. The second variant (denoted as '**c5f**') crops the face region as the fifth crop. Face++ [20] API was used to detect the location of the center coordinate, and a bounding box that defines the sub-image is cropped and squashed to a constant width and height. Figure 5 shows some sample face regions detected by Face++ API and used as the fifth crop for method '**c5f**'. Similar to previous schemes, DCNN features resulting from all five crops are concatenated. Due to slight unreliability of the Face++ API, only about 77.24% of all faces from the AVA-Portraits dataset can be detected. As such, the affected images fall back on the center crop as the fifth crop. Alternatively, we also manually annotated the faces in the remaining affected images. We denote this as '**c5f-m**'.

For all compositional cropping methods, we crop the sub-images based on a fixed width and height of 256 pixels, which are subsequently resized to receptive field sizes of their respective networks i.e. 234 for Overfeat, 224 for GoogleNet, both in RGB color space with zero mean.

Table 1. Aesthetic classification accuracy and dispersion error of the proposed methods compared against baseline methods.

Methods	Accuracy (%)	Error ± (%)
Baseline A [2]	54.585	1.25
Overfeat-warp	63.211	1.48
GoogleNet-warp	63.751	0.85
Overfeat-c3	63.861	0.44
Overfeat - c4	64.133	0.62
Baseline B [10]	64.240	1.76
GoogleNet-c3	64.244	0.64
GoogleNet-c4	64.519	1.06
Overfeat-c5	64.536	1.11
GoogleNet-c5	64.962	1.20
GoogleNet-c5f	**65.216**	1.20
GoogleNet-c5f-m	**65.330**	1.01

5 Results and Discussions

5.1 Overall Results

We define two baselines for the purpose of comparisons – one by [3] which relies on traditional spatial and color composition features ('**Baseline A**'), and another [13] which uses a discriminative set of visual features ('**Baseline B**'). Two DCNN models were utilized in our experiments: the 8-layer **Overfeat** model and the 22-layer very deep **GoogleNet** model.

Table 1 shows the proposed and baseline approaches on the AVA-Portraits dataset. We achieve the best result reported thus far for this dataset, of 65.33%. From the feature augmentation results, we can see that the aesthetic classification accuracy generally increases as more crops were added to augment the DCNN-learned features (see Fig. 7a). The best performing scheme uses five crops, with the variant that takes a face as fifth crop marginally higher than the one with the center crop. Manual annotation of the faces (that were not detected by Face++) resulted in further improvements. This feature augmentation scheme showed that having the face as foreground and the four other crops as background provides sufficient representation to mimic the foreground and background contrast rule raised in [3]. Figure 6 shows some sample images that were correctly classified (true positive, true negative) and incorrectly classified (false positive, false negative).

Meanwhile, we observe that the GoogleNet model outperforms the Overfeat model for each feature augmentation approach (see Fig. 7b). This is expected of a more contemporary DCNN architecture that is deeper and had previously yielded better results at the ILSVRC. Of course, this comes at the expense of heavier computational cost. Figure 8 shows the breakdown of classification performance into

Fig. 6. Classification result of some images from the AVA-Portraits dataset using the GoogleNet-c5f-m method. The four quadrants show sample images based on their actual ground truth aesthetic class and predicted aesthetic class (positive side indicates high aesthetic quality, negative side indicates low aesthetic quality).

the the five parts as specified in [13]. There is no monopoly by any of the competing methods in the proposed scheme while the warping technique is noticeably poor.

Conventional single image warping produced quite poor results, much worser than Baseline B [13] which relies on an eclectic bunch of features that are both handcrafted and learned. It is important to point out that DCNN based methods do not required careful selection of features [3], or needing to eliminate features that are poorly correlated [13]. This itself gives an immense advantage towards methods that are able to learn good features on it own with minimal intervention.

5.2 Influence of Small Faces

We made an interesting observation that the AVA-Portraits dataset contains so-called "portrait" images with small faces inside (see Fig. 10 for some samples). This may be caused by the fact that photo challenges in DPChallenge [14] are based on themes that can be subjectively interpreted by participants (as they deemed fit), or that these photos could exemplify certain artistic qualities associated to portraiture. Hence, this is the reason why some photos are not captured closed-up, frontal, or from portrait-like angles.

We identified images with "small faces" by considering faces that have a bounding box area of smaller or equal to 49 pixels[1]. We investigated if small faces

---

[1] This size was roughly determined by observation. Most of the other images appear to have substantially larger faces.

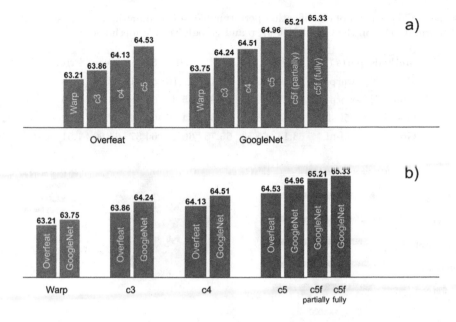

Fig. 7. Aesthetic classification accuracy grouped by (a) deep convolutional neural network models, and (b) type of feature augmentation scheme.

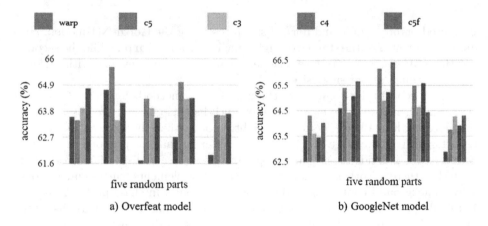

Fig. 8. Classification accuracy of the five random parts of the AVA-Portraits dataset using Overfeat and GoogleNet models.

can possibly influence the overall aesthetics of these portrait images. The whole dataset, and likewise, each of the five parts, contain approximately 17% of small face portrait images (Fig. 9). So that we can conclude that small face portrait images, distributed almost evenly across the five parts. A total of 1,388 small face portrait images were found while the remaining 6,445 images contain normal size faces (those not marked by Face++ were ignored). Two sets of experiments were

Table 2. Comparison of classification performance between methods using small faces and normal faces on the GoogleNet-warp and GoogleNet-c5f methods.

Methods/parts (%)	1st	2nd	3rd	4th	5th	Mean	Error
GoogleNet-warp-sf	65.10	62.96	63.12	62.16	59.91	62.50	2.59
GoogleNet-warp-nf	61.77	65.37	62.85	65.35	63.58	63.57	1.80
GoogleNet-c5f-sf	62.08	65.31	66.66	64.86	64.04	64.37	2.29
GoogleNet-c5f-nf	63.14	65.53	65.75	66.18	64.37	64.66	1.51

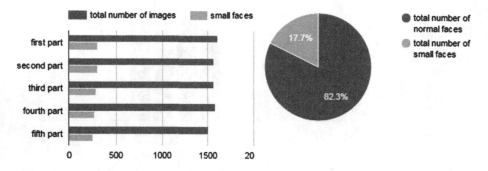

Fig. 9. Proportion of small faces in AVA-Portraits dataset

conducted on both small and normal size faces using the GoogleNet model. The first experiment uses the standard single image warping '**warp**', while the second uses the '**c5f**' five crop with face scheme. We indicate the faces sizes with '**nf**' for normal size and '**sf**' for small size.

Table 2 reports the accuracy of these evaluated methods for each of the five parts, and their respective mean accuracy and dispersion error. We observe that the photos with small faces performed slightly poorer than their normal face counterparts, with a larger spread of results. Considering that the proportion of small faces are evenly divided in all parts (Fig. 9), the increase in dispersion error could mean that there are other non-face related aesthetic elements (such as image condition, background context, artistic styles, etc.) found in these portrait images that could potentially influence their actual ratings. Small faces are also likely to not contribute any meaningful impression of beauty to the viewer, in contrast to the way how close-up portraits do.

6 Conclusion

In conclusion, we introduce a new compositional-based augmentation scheme for deep convolutional neural network (DCNN) models to evaluate the aesthetic quality of facial portraits. Photographic rules relating to composition principles are adapted for portraits with the purpose of enriching inputs to the network with more visually pleasing crops. This way, we are able to generate a

Fig. 10. Samples where small faces are found in "portrait" photos

deeply-learned augmented feature set that can improve classification. On the AVA-Portraits dataset (a subset of the AVA dataset first used in [13]) consisting of more than 10,000 portrait-like images, our approach demonstrated an improvement in aesthetic classification over handcrafted baselines and standard DCNN models, achieving a top accuracy of 65.33%. In future, we intend to extend this idea by learning the compositional principles in facial portraits directly from the deep learning model itself. Also, we aim to explore other photo composition concepts besides the rule of thirds, such as visual balance and diagonal dominance.

References

1. Bakhshi, S., Shamma, D.A., Gilbert, E.: Faces engage us: photos with faces attract more likes and comments on instagram. In: Proceedings of the 32nd Annual ACM Conference on Human factors in Computing Systems, pp. 965–974. ACM (2014)
2. Krages, B.: Photography: The Art of Composition. Skyhorse Publishing Inc., New York (2012)
3. Khan, S.S., Vogel, D.: Evaluating visual aesthetics in photographic portraiture. In: Proceedings of the Eighth Annual Symposium on Computational Aesthetics in Graphics, Visualization, and Imaging, pp. 55–62. Eurographics Association (2012)
4. Krizhevsky, A., Sutskever, I., Hinton, G.E.: Imagenet classification with deep convolutional neural networks. In: Advances in Neural Information Processing Systems, pp. 1097–1105 (2012)
5. Sermanet, P., Eigen, D., Zhang, X., Mathieu, M., Fergus, R., LeCun, Y.: Overfeat: integrated recognition, localization and detection using convolutional networks. arXiv preprint (2013). arXiv:1312.6229

6. Szegedy, C., Liu, W., Jia, Y., Sermanet, P., Reed, S., Anguelov, D., Erhan, D., Vanhoucke, V., Rabinovich, A.: Going deeper with convolutions. In: Proceedings of the IEEE Conference on Computer Vision and Pattern Recognition, pp. 1–9 (2015)
7. Datta, R., Joshi, D., Li, J., Wang, J.Z.: Studying aesthetics in photographic images using a computational approach. In: Leonardis, A., Bischof, H., Pinz, A. (eds.) ECCV 2006. LNCS, vol. 3953, pp. 288–301. Springer, Heidelberg (2006). doi:10.1007/11744078_23
8. Marchesotti, L., Perronnin, F., Larlus, D., Csurka, G.: Assessing the aesthetic quality of photographs using generic image descriptors. In: 2011 International Conference on Computer Vision, pp. 1784–1791. IEEE (2011)
9. Murray, N., Marchesotti, L., Perronnin, F.: AVA: A large-scale database for aesthetic visual analysis. In: 2012 IEEE Conference on Computer Vision and Pattern Recognition (CVPR), pp. 2408–2415. IEEE (2012)
10. Lu, X., Lin, Z., Jin, H., Yang, J., Wang, J.Z.: RAPID: rating pictorial aesthetics using deep learning. In: Proceedings of the 22nd ACM International Conference on Multimedia, pp. 457–466. ACM (2014)
11. Tang, X., Luo, W., Wang, X.: Content-based photo quality assessment. IEEE Trans. Multimed. **15**, 1930–1943 (2013)
12. Tian, X., Dong, Z., Yang, K., Mei, T.: Query-dependent aesthetic model with deep learning for photo quality assessment. IEEE Trans. Multimed. **17**, 2035–2048 (2015)
13. Redi, M., Rasiwasia, N., Aggarwal, G., Jaimes, A.: The beauty of capturing faces: rating the quality of digital portraits. In: 11th IEEE International Conference and Workshops on Automatic Face and Gesture Recognition (FG), vol. 1, pp. 1–8 (2015)
14. Dpchallenge - a digital photography contest. http://www.dpchallenge.com/. Accessed 26 Aug 2016
15. Wang, Z., Dolcos, F., Beck, D., Chang, S., Huang, T.S.: Brain-inspired deep networks for image aesthetics assessment. arXiv preprint (2016). arXiv:1601.04155
16. Lihua, G., Fudi, L.: Image aesthetic evaluation using paralleled deep convolution neural network. arXiv preprint (2015). arXiv:1505.05225
17. Lienhard, A., Ladret, P., Caplier, A.: Low level features for quality assessment of facial images. In: 10th International Conference on computer Vision Theory and Applications, VISAPP, pp. 545–552 (2015)
18. Kalayeh, M.M., Seifu, M., LaLanne, W., Shah, M.: How to take a good selfie? In: Proceedings of the 23rd ACM International Conference on Multimedia, pp. 923–926 (2015)
19. Obrador, P., Schmidt-Hackenberg, L., Oliver, N.: The role of image composition in image aesthetics. In: IEEE International Conference on Image Processing, pp. 3185–3188 (2010)
20. Face++: Leading Face Recognition on Cloud. http://www.faceplusplus.com/. Accessed 26
21. LeCun, Y., Bottou, L., Bengio, Y., Haffner, P.: Gradient-based learning applied to document recognition. Proc. IEEE **86**, 2278–2324 (1998)
22. ImageNet Database and Challenge. http://www.image-net.org/. Accessed 26
23. Sklearn-theano library. https://sklearn-theano.github.io. Acceseed 26
24. Chatfield, K., Simonyan, K., Vedaldi, A., Zisserman, A.: Return of the devil in the details: delving deep into convolutional nets. arXiv preprint (2014). arXiv:1405.3531
25. Child, J.: Studio Photography: Essential Skills. CRC Press, Boca Raton (2013)

Discrete Geometry and Mathematical Morphology for Computer Vision

Discrete Polynomial Curve Fitting Guaranteeing Inclusion-Wise Maximality of Inlier Set

Fumiki Sekiya[1]($\boxtimes$) and Akihiro Sugimoto[2]

[1] Department of Informatics,
SOKENDAI (The Graduate University for Advanced Studies),
Tokyo, Japan
sekiya@nii.ac.jp
[2] National Institute of Informatics, Tokyo, Japan
sugimoto@nii.ac.jp

Abstract. This paper deals with the problem of fitting a discrete polynomial curve to 2D noisy data. We use a discrete polynomial curve model achieving connectivity in the discrete space. We formulate the fitting as the problem to find parameters of this model maximizing the number of inliers i.e., data points contained in the discrete polynomial curve. We propose a method guaranteeing inclusion-wise maximality of its obtained inlier set.

1 Introduction

Curve fitting to noisy data (i.e., containing outliers) is an essential task in many applications such as object recognition, image segmentation and shape approximation. Continuous curve models have been used for fitting in most cases even though data dealt with in a computer are discrete.

The method most commonly used for continuous curve fitting in the presence of noise is RANdom SAmple Consensus (RANSAC) [1], which uses random sampling to estimate model parameters, and then choose the ones having the largest number of inliers, i.e., data points explained by the parameters. For its robustness and simplicity, RANSAC is used in a wide range of problems in computer vision. The main drawback of RANSAC (and most of its variants) is however that it does not guarantee any deterministic properties on its output. It also requires an empirical error threshold to define an inlier, which affects the output. Another popular approach for the task is to use the Hough transform [2,3], which allows to find model parameters consistent with many data points in the space of the model parameters. This method however requires to manually set the resolution to discretize the parameter space, which affects the output.

As long as a continuous curve model is fitted to discrete data, an error threshold is required to determine if a data point is explained by the model. By using a discrete curve model, on the other hand, we can define an inlier without an empirical error threshold. A discrete curve model is defined as a set of discrete points to represent a discretized curve. For curve discretization, it has been considered to be important to preserve the topological properties (e.g., connectivity)

© Springer International Publishing AG 2017
C.-S. Chen et al. (Eds.): ACCV 2016 Workshops, Part II, LNCS 10117, pp. 477–492, 2017.
DOI: 10.1007/978-3-319-54427-4_35

of an original curve [4–9]. For example, a Jordan curve (i.e., simple closed curve) in a 2D image allows to partition the image into two connected regions. Such a property is useful in computer graphics, computer vision and image processing (see [10]), and therefore should be preserved in discretization. Based on the idea, several discrete curve models have been developed [11–16] to achieve some consistent topological properties in the discrete space. It is therefore preferable to use such a model for curve fitting.

Curve fitting to noisy data in 2D has been studied for discrete lines [17–20], discrete circles [21–24] and discrete polynomial curves [25]. For lines and circles, models having connectivity have been used. For polynomial curves, on the other hand, only one type of discrete polynomial curve model without guaranteeing any topological property has been used. Discrete polynomial curve fitting therefore has yet to be studied for a model having consistent topological properties.

In this paper we deal with the problem of fitting a discrete polynomial curve to 2D noisy data. We use the discrete curve model introduced by Toutant et al. [16] to define our discrete polynomial curve. This is because this model guarantees connectivity in the discrete space [26], and is closely related to the morphological discretization [27–29]. To be precise, this model corresponds to the morphological discretization with a structuring element called the k-adjacency flake [15], which is defined for $k = 0, 1$ in 2D and achieves different topological properties depending on k. Note that in this paper we limit ourselves to define our discrete polynomial curve only for $k = 0$. We formulate our problem as to find parameters of this model that maximize the number of inliers, where an inlier is defined as a point contained in the discrete polynomial curve.

We propose for this problem a method guaranteeing inclusion-wise maximality of its obtained inlier set (i.e., there exists no larger inlier set in the sense of set inclusion). Note that an inclusion-wise maximal inlier set does not necessarily have the maximum cardinality. Our method runs in the space of parameters (coefficients) of the discrete polynomial curve model. In the parameter space a discrete polynomial curve is represented by a point, while a data point or a set of data points gives a feasible region shaped like a polytope where any discrete polynomial curve represented by a point in the region contains the data point(s). Given any initial inlier set, the method adds new data points to the inlier set one by one with tracking its feasible region in the parameter space, until inclusion-wise maximality is achieved. The feasible region is generally an infinite set so that it is impossible to store all its points in a computer. We solve this problem by focusing only on a finite number of points corresponding to the notion of the vertices of a polytope. Our method thus does not require any discretization of the parameter space, which is a major difference from the Hough transform.

2   Problem Formulation

A continuous polynomial curve of degree d in the xy-plane is represented by $y - \sum_{l=0}^{d} a_l x^l = 0$ with coefficients $a_0, \ldots, a_{d-1} \in \mathbb{R}$ and $a_d \in \mathbb{R} \setminus \{0\}$. Toutant et al. [16] introduced its discretized form in $\mathbb{Z}^2$, i.e., a *discrete polynomial curve* $D(a_0, \ldots a_d)$ by

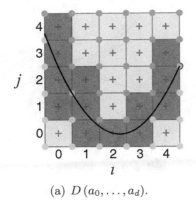

(a) $D(a_0, \ldots, a_d)$.

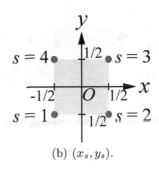

(b) (x_s, y_s).

Fig. 1. Discrete polynomial curve. (a) shows $D(a_0, \ldots, a_d)$ (red integer points) for $d = 2$ and $(a_0, a_1, a_2) = (2.5, -2.25, 0.5)$ with its continuous counterpart (depicted in black). For $(i, j) \in \mathbb{Z}^2$ and $s \in \{1, \ldots, 4\}$, $(i + x_s, j + y_s)$ is depicted in green, purple or yellow: green if $(j + y_s) - \sum_{l=0}^{d} a_l (i + x_s)^l > 0$; purple if $(j + y_s) - \sum_{l=0}^{d} a_l (i + x_s)^l < 0$; yellow if $(j + y_s) - \sum_{l=0}^{d} a_l (i + x_s)^l = 0$. $(i, j) \in \mathbb{Z}^2$ is in $D(a_0, \ldots, a_d)$ if $\{(i + x_s, j + y_s) \mid s = 1, \ldots, 4\}$ contains green and purple points, or an yellow point. In (b), (x_s, y_s) is depicted in red for $s = 1, \ldots, 4$. (Color figure online)

$$
D(a_0, \ldots, a_d) = \left\{ (i, j) \subset \mathbb{Z}^2 \ \middle| \ \begin{array}{c} \min\limits_{s \in \{1, \ldots, 4\}} \left[(j + y_s) - \sum\limits_{l=0}^{d} a_l (i + x_s)^l \right] \\ \leq 0 \leq \\ \max\limits_{s \in \{1, \ldots, 4\}} \left[(j + y_s) - \sum\limits_{l=0}^{d} a_l (i + x_s)^l \right] \end{array} \right\}, \quad (1)
$$

where $(x_1, y_1) = \left(-\frac{1}{2}, -\frac{1}{2}\right)$, $(x_2, y_2) = \left(\frac{1}{2}, -\frac{1}{2}\right)$, $(x_3, y_3) = \left(\frac{1}{2}, \frac{1}{2}\right)$ and $(x_4, y_4) = \left(-\frac{1}{2}, \frac{1}{2}\right)$. See Fig. 1 for an illustration of $D(a_0, \ldots, a_d)$.

Let $P = \{ (i_p, j_p) \in \mathbb{Z}^2 \mid p = 1, \ldots, n \}$ be a finite set (i.e., $n < \infty$) of integer points (data). For a discrete polynomial curve $D(a_0, \ldots, a_d)$, a point (i_p, j_p) $(p = 1, \ldots, n)$ is called an *inlier* if $(i_p, j_p) \in D(a_0, \ldots, a_d)$, while otherwise it is called an *outlier*. Our goal is to find $D(a_0, \ldots, a_d)$ that maximizes the number of inliers for given data P and a degree d, where we permit $a_d = 0$ so that discrete polynomial curves of degree less than d are covered as well.

When d is fixed, $D(a_0, \ldots, a_d)$ is determined only by $a_0, \ldots, a_d$. We therefore consider the problem in the *parameter space* $\{(a_0, \ldots, a_d)\} = \mathbb{R}^{d+1}$, instead of the *data space* $\mathbb{Z}^2$ where P resides. A discrete polynomial curve in the data space is represented as a point in the parameter space. A data point in P, on the other hand, is represented as a region in the parameter space, which is defined as follows.

For $p = 1, \ldots, n$, we define the *feasible region* R_p for the pth data (i_p, j_p) by

$$
R_p = \left\{ (a_0, \ldots, a_d) \in \mathbb{R}^{d+1} \ \middle| \ \begin{array}{c} \min\limits_{s \in \{1, \ldots, 4\}} h_{(p,s)}(a_0, \ldots, a_d) \\ \leq 0 \leq \\ \max\limits_{s \in \{1, \ldots, 4\}} h_{(p,s)}(a_0, \ldots, a_d) \end{array} \right\}, \quad (2)
$$

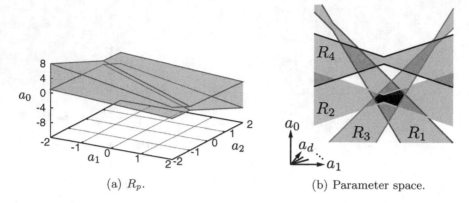

(a) R_p.   (b) Parameter space.

Fig. 2. Feasible regions in the parameter space. (a) shows R_p for $d = 2$ and $(i_p, j_p) = (1, 0)$. Each point in the data space is represented in the parameter space by an unbounded concave polytope like this. (b) shows intersections among the feasible regions for four individual data points, which are indexed from 1 to 4. $(a_0, \ldots, a_d)$ in a darker region has a larger number of inliers in the data space.

where

$$h_{(p,s)}(a_0, \ldots, a_d) = (j_p + y_s) - \sum_{l=0}^{d} (i_p + x_s)^l a_l. \tag{3}$$

See Fig. 2(a) for an illustration of R_p. We remark that $(i_p, j_p) \in D(a_0, \ldots, a_d)$ iff $(a_0, \ldots, a_d) \in R_p$.

We also define a feasible region for a set of data. For $\Pi \subset \{1, \ldots, n\}$, we define the *feasible region* R_Π for the set $\{(i_p, j_p) \mid p \in \Pi\}$ of data by $R_\Pi = \bigcap_{p \in \Pi} R_p$, which is also written as

$$R_\Pi = \left\{ (a_0, \ldots, a_d) \in \mathbb{R}^{d+1} \; \middle| \; \begin{array}{c} \max\limits_{p \in \Pi} \min\limits_{s \in \{1, \ldots, 4\}} h_{(p,s)}(a_0, \ldots, a_d) \\ \leq 0 \leq \\ \min\limits_{p \in \Pi} \max\limits_{s \in \{1, \ldots, 4\}} h_{(p,s)}(a_0, \ldots, a_d) \end{array} \right\}. \tag{4}$$

We remark that R_Π may be bounded or unbounded, convex or concave, and connected or disconnected as can be seen in Fig. 2(b) (e.g., $R_{\{1,2,3\}}$ is bounded and convex, while $R_{\{2,4\}}$ is unbounded and disconnected).

$R_\Pi = \emptyset$ if no $(a_0, \ldots, a_d) \in \mathbb{R}^{d+1}$ satisfies $(i_p, j_p) \in D(a_0, \ldots, a_d)$ for $\forall p \in \Pi$. Our problem is therefore formulated as follows.

Problem 1. Given P and d, find $\Pi \subset \{1, \ldots, n\}$ that has the maximum cardinality providing $R_\Pi \neq \emptyset$, and $(a_0, \ldots, a_d) \in \mathbb{R}^{d+1}$ satisfying $(a_0, \ldots, a_d) \in R_\Pi$ for that Π.

We remark that the solution Π is not necessarily unique. In the example in Fig. 2(b), the data index set Π that we would like to find is $\{1, 2, 3\}$.

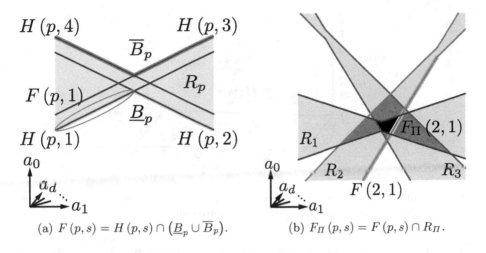

(a) $F(p,s) = H(p,s) \cap \left(\underline{B}_p \cup \overline{B}_p\right)$.

(b) $F_\Pi(p,s) = F(p,s) \cap R_\Pi$.

Fig. 3. Concepts to represent a flat part on the surface of a feasible region. In (b), $\Pi = \{1, 2, 3\}$.

3 Properties of Feasible Regions

3.1 Concepts and Notations

Our approach to find an inclusion-wise maximal inlier set is as follows: starting from an arbitrary $\Pi \subset \{1, \ldots, n\}$ satisfying $R_\Pi \neq \emptyset$, we iteratively search $p \in \{1, \ldots, n\} \setminus \Pi$ such that $R_{\Pi \cup \{p\}}$ $(= R_\Pi \cap R_p) \neq \emptyset$ and add it to Π, where every time we update Π ($\Pi := \Pi \cup \{p\}$) we compute its corresponding R_Π. By repeating this procedure until there is no such p, we can ensure an inclusion-wise maximal inlier set. It is however impossible to store all points in R_Π in a computer, since R_Π is generally an infinite set when $R_\Pi \neq \emptyset$. We therefore focus only a finite number of points in R_Π that correspond to the notion of the vertices of a polytope (Fig. 4). As a vertex of a polytope is defined as an intersection point of flat parts on the surface of the polytope called facets, we first need a notion for R_Π corresponding to a facet of a polytope.

Let $p \in \{1, \ldots, n\}$ be any data index. For $s = 1, \ldots, 4$, we first define $H(p,s) = \{(a_0, \ldots, a_d) \in \mathbb{R}^{d+1} \mid h_{(p,s)}(a_0, \ldots, a_d) = 0\}$. $H(p,s)$ is a hyperplane included in R_p (Fig. 3(a)), which determines a flat part on the surface of R_p. To represent the surface of R_p, we then define $\underline{B}_p = \{(a_0, \ldots, a_d) \in \mathbb{R}^{d+1} \mid \min_{s \in \{1, \ldots, 4\}} h_{(p,s)}(a_0, \ldots, a_d) = 0\}$ and $\overline{B}_p = \{(a_0, \ldots, a_d) \in \mathbb{R}^{d+1} \mid \max_{s \in \{1, \ldots, 4\}} h_{(p,s)}(a_0, \ldots, a_d) = 0\}$. $\underline{B}_p$ and $\overline{B}_p$ are the "lower" and "upper" boundaries (with a_0 considered as the height) of R_p (cf. Fig. 3(a)). We remark that $\underline{B}_p$ is determined only by $s = 1, 2$ while $\overline{B}_p$ is determined only by $s = 3, 4$, and that $\underline{B}_p \cap \overline{B}_p = \emptyset$. The flat part of $\underline{B}_p \cup \overline{B}_p$ (i.e., facet of R_p) determined by $H(p,s)$, for $s = 1, \ldots, 4$, is then represented by $F(p,s) = H(p,s) \cap \left(\underline{B}_p \cup \overline{B}_p\right)$ (Fig. 3(a)). Note that $\underline{B}_p \cup \overline{B}_p = \bigcup_{s \in \{1, \ldots, 4\}} F(p,s)$.

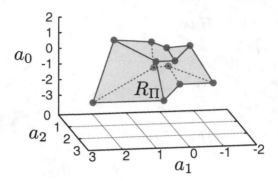

Fig. 4. C_Π for $d = 2$ and $\Pi = \{1, 2, 3\}$ with $(i_1, j_1) = (-2, 3)$, $(i_2, j_2) = (0, 0)$, $(i_3, j_3) = (2, 5)$. Points in C_Π are depicted in red. (Color figure online)

Let $\Pi \subset \{1, \ldots, n\}$ be any set of data indices. Since $R_\Pi = \bigcap_{p \in \Pi} R_p$, the flat part of the boundary of R_Π (i.e., facet of R_Π) determined by $H(p, s)$, for $(p, s) \in \Pi \times \{1, \ldots, 4\}$, is obtained as a subset of $F(p, s)$. Namely, it is represented by $F_\Pi(p, s) = F(p, s) \cap R_\Pi$. See Fig. 3(b) for an illustration of $F_\Pi(p, s)$. We remark that $F_\Pi(p, s)$ can be empty for some $(p, s) \in \Pi \times \{1, \ldots, 4\}$.

We now define a subset of R_Π corresponding to the vertices of a polytope. In $\mathbb{R}^{d+1}$, $d + 1$ hyperplanes intersect at one point when their normal vectors are linearly independent. We therefore can specify a finite subset of R_Π by enumerating $(a_0, \ldots, a_d) \in \bigcap_{\lambda=1}^{d+1} F_\Pi(p_\lambda, s_\lambda)$ for all $(p_1, s_1), \ldots, (p_{d+1}, s_{d+1}) \in \Pi \times \{1, \ldots, 4\}$ such that $H(p_1, s_1), \ldots, H(p_{d+1}, s_{d+1})$ are linearly independent. Namely, C_Π identifies a subset (i.e., vertices) of R_Π:

$$
C_\Pi = \left\{ (a_0, \ldots, a_d) \in \mathbb{R}^{d+1} \;\middle|\; \begin{array}{l} \text{there exist} \\ (p_1, s_1), \ldots, (p_{d+1}, s_{d+1}) \in \Pi \times \{1, \ldots, 4\} \\ \text{such that } (a_0, \ldots, a_d) \in \bigcap_{\lambda=1}^{d+1} F_\Pi(p_\lambda, s_\lambda) \\ \text{and } H(p_1, s_1), \ldots, H(p_{d+1}, s_{d+1}) \text{ are} \\ \text{linearly independent} \end{array} \right\}. \tag{5}
$$

See Fig. 4 for an illustration of C_Π. We remark that C_Π is a finite set since $\Pi \times \{1, \ldots, 4\}$ has only a finite number of elements. We also define the family of sets of $d + 1$ elements of $\Pi \times \{1, \ldots, 4\}$ determining elements of C_Π:

$$
\Psi_\Pi = \left\{ \begin{array}{c} \{(p_1, s_1), \ldots, (p_{d+1}, s_{d+1})\} \\ \subset \Pi \times \{1, \ldots, 4\} \end{array} \;\middle|\; \begin{array}{l} H(p_1, s_1), \ldots, H(p_{d+1}, s_{d+1}) \\ \text{are linearly independent and} \\ \text{their intersection point is in} \\ \bigcap_{\lambda=1}^{d+1} F_\Pi(p_\lambda, s_\lambda) \end{array} \right\}. \tag{6}
$$

We remark that different sets in Ψ_Π may determine the same element of C_Π.

3.2   Updated Feasible Region by an Additional Inlier

Here we give four properties of R_Π (Theorems 1–4) which are important for enabling the approach described in the beginning of Sect. 3.1, where R_Π is represented by C_Π. We start with the following lemma required to prove three of those properties, which states the condition for m ($m = 1,\ldots,d$) flat parts $F_\Pi(p_1,s_1),\ldots,F_\Pi(p_m,s_m)$ on the surface of R_Π to contribute to determining a point in C_Π. The proof of this lemma is provided in Appendix A.

Lemma 1. *Let $\Pi \subset \{1,\ldots,n\}$ be a data index set such that $R_\Pi \neq \emptyset$ and R_Π is bounded, and let $(p_1,s_1),\ldots,(p_m,s_m)$ be m ($m = 1,\ldots,d$) elements of $\Pi \times \{1,\ldots,4\}$. There exists a set in Ψ_Π containing $(p_1,s_1),\ldots,(p_m,s_m)$ if (i) $\bigcap_{\lambda=1}^m F_\Pi(p_\lambda,s_\lambda) \neq \emptyset$ and (ii) $H(p_1,s_1),\ldots,H(p_m,s_m)$ are linearly independent.*

Since R_Π is represented by C_Π in our approach, it is important that $C_\Pi \neq \emptyset$ whenever $R_\Pi \neq \emptyset$. This can be proven under the condition that R_Π is bounded. Note that $\Psi_\Pi \neq \emptyset$ is equivalent with $C_\Pi \neq \emptyset$.

Theorem 1. *For $\Pi \subset \{1,\ldots,n\}$ such that $R_\Pi \neq \emptyset$ and R_Π is bounded, $\Psi_\Pi \neq \emptyset$.*

Proof. It is obvious that there exists $(p,s) \in \Pi \times \{1,\ldots,4\}$ satisfying $F_\Pi(p,s) \neq \emptyset$. From Lemma 1, then, there exists a set in Ψ_Π containing (p,s). $\Psi_\Pi \neq \emptyset$, accordingly. $\square$

From Theorem 1, for any $\Pi \subset \{1,\ldots,n\}$ such that R_Π is bounded, we can always obtain a point in R_Π by computing C_Π. There are however $\binom{4|\Pi|}{d+1}$ ways to pick $d+1$ different elements from $\Pi \times \{1,\ldots,4\}$, so that checking all those combinations to compute C_Π is not practical for Π with a large number of elements. In the following we give a relationship between Ψ_Π and $\Psi_{\Pi\cup\{p\}}$, for $p \in \{1,\ldots,n\}\setminus\Pi$, with which we can reduce the computational cost for obtaining $\Psi_{\Pi\cup\{p\}}$ when we have Ψ_Π.

For $\Pi \subsetneq \{1,\ldots,n\}$ and $p \in \{1,\ldots,n\} \setminus \Pi$, we define $\Phi^1_{\Pi,p}$ to be the family of sets each of whose set is obtained by replacing an element of a set in Ψ_Π with (p,s) for $s \in \{1,\ldots,4\}$:

$$\Phi^1_{\Pi,p} = \left\{ \begin{array}{c} \{(p_1,s_1),\ldots,(p_d,s_d)\} \\ \cup\{(p,s)\} \end{array} \middle| \begin{array}{l} \{(p_1,s_1),\ldots,(p_d,s_d)\} \text{ is a subset} \\ \text{of a set in } \Psi_\Pi \text{ and } s = 1,\ldots,4 \end{array} \right\}. \quad (7)$$

We also define $\Phi^2_{\Pi,p}$ to be the family of sets each of whose set is obtained by replacing two elements of a set in Ψ_Π with $(p,1)$ and $(p,2)$, or $(p,3)$ and $(p,4)$:

$$\Phi^2_{\Pi,p} = \left\{ \begin{array}{c} \{(p_1,s_1),\ldots,(p_{d-1},s_{d-1})\} \\ \cup\{(p,s),(p,s')\} \end{array} \middle| \begin{array}{l} \{(p_1,s_1),\ldots,(p_{d-1},s_{d-1})\} \text{ is} \\ \text{a subset of a set in } \Psi_\Pi \\ \text{and } (s,s') = (1,2),(3,4) \end{array} \right\}. \quad (8)$$

We then have the following relationship among Ψ_Π, $\Psi_{\Pi\cup\{p\}}$, $\Phi^1_{\Pi,p}$ and $\Phi^2_{\Pi,p}$.

Theorem 2. *For Π such that R_Π is bounded, $\Psi_{\Pi\cup\{p\}} \subset \Psi_\Pi \cup \Phi^1_{\Pi,p} \cup \Phi^2_{\Pi,p}$.*

Proof. We assume $\Psi_{\Pi\cup\{p\}} \neq \emptyset$; otherwise the statement is obviously true. Let $\psi = \{(p_1,s_1),\ldots,(p_{d+1},s_{d+1})\}$ be a set in $\Psi_{\Pi\cup\{p\}}$. We show $\psi \in \Psi_\Pi \cup \Phi^1_{\Pi,p} \cup$

$\Phi_{\Pi,p}^2$. We assume that m of $p_1, \ldots, p_{d+1} \in \Pi \cup \{p\}$ are equal to p and the others belong to Π. If ψ contains $(p, \underline{s})$ and $(p, \overline{s})$ for any $\underline{s} \in \{1, 2\}$ and $\overline{s} \in \{3, 4\}$, then $\bigcap_{\lambda=1}^{d+1} F_{\Pi \cup \{p\}} (p_\lambda, s_\lambda) = \emptyset$ from $F(p, \underline{s}) \cap F(p, \overline{s}) = \emptyset$ (recall that $F(p, \underline{s}) \subset \underline{B}_p$, $F(p, \overline{s}) \subset \overline{B}_p$ and $\underline{B}_p \cap \overline{B}_p = \emptyset$), contradicting $\psi \in \Psi_{\Pi \cup \{p\}}$. $m \geq 3$ is therefore impossible, because in that case $(p, \underline{s})$ and $(p, \overline{s})$ are necessarily contained in ψ. We therefore have $m \leq 2$, where when $m = 2$ the two elements of ψ corresponding to p are either $(p, 1)$ and $(p, 2)$, or $(p, 3)$ and $(p, 4)$.

We first consider the case of $m = 0$. We then have $(p_1, s_1), \ldots, (p_{d+1}, s_{d+1}) \in \Pi \times \{1, \ldots, 4\}$. From $\psi \in \Psi_{\Pi \cup \{p\}}$, $\bigcap_{\lambda=1}^{d+1} F_{\Pi \cup \{p\}} (p_\lambda, s_\lambda) \neq \emptyset$ and $H(p_1, s_1), \ldots,$ $H(p_{d+1}, s_{d+1})$ are linearly independent. Since $F_\Pi (p_\lambda, s_\lambda) \supset F_{\Pi \cup \{p\}} (p_\lambda, s_\lambda)$ $(\lambda = 1, \ldots, d+1)$, we have $\bigcap_{\lambda=1}^{d+1} F_\Pi (p_\lambda, s_\lambda) \neq \emptyset$. $\psi \in \Psi_\Pi$, consequently.

We next consider the case of $m = 1, 2$. Without loss of generality, we assume $p_1, \ldots, p_{d+1-m} \in \Pi$. We prove $\psi \in \Phi_{\Pi,p}^m$ by showing that assuming otherwise leads to a contradiction. Namely, we assume that there exists no such set in Ψ_Π that contains $(p_1, s_1), \ldots, (p_{d+1-m}, s_{d+1-m})$. Lemma 1 then suggests that $\bigcap_{\lambda=1}^{d+1-m} F_\Pi (p_\lambda, s_\lambda) = \emptyset$ (which implies $\bigcap_{\lambda=1}^{d+1-m} F_{\Pi \cup \{p\}} (p_\lambda, s_\lambda) = \emptyset$) or $H(p_1, s_1), \ldots, H(p_{d+1-m}, s_{d+1-m})$ are not linearly independent. This contradicts $\psi \in \Psi_{\Pi \cup \{p\}}$.   □

With Theorem 2, we do not have to evaluate all the sets of $d + 1$ elements of $(\Pi \cup \{p\}) \times \{1, \ldots, 4\}$ to compute $\Psi_{\Pi \cup \{p\}}$, but only those in $\Psi_\Pi \cup \Phi_{\Pi,p}^1 \cup \Phi_{\Pi,p}^2$. For $\{(p_1, s_1), \ldots, (p_{d+1}, s_{d+1})\} \in \Psi_\Pi$, $\{(p_1, s_1), \ldots, (p_{d+1}, s_{d+1})\} \in \Psi_{\Pi \cup \{p\}}$ is verified as soon as the corresponding $(a_0, \ldots, a_d) \in C_\Pi$ satisfies $(a_0, \ldots, a_d) \in R_p$. This is because that we have $(a_0, \ldots, a_d) \in \bigcap_{\lambda=1}^{d+1} F_\Pi (p_\lambda, s_\lambda)$ from $(a_0, \ldots, a_d) \in C_\Pi$, and $F_\Pi (p_\lambda, s_\lambda) \cap R_p = F_{\Pi \cup \{p\}} (p_\lambda, s_\lambda)$ $(\lambda = 1, \ldots, d+1)$. For $\{(p_1, s_1), \ldots, (p_{d+1}, s_{d+1})\} \in \Phi_{\Pi,p}^m$ $(m = 1, 2)$, on the other hand, we have to check if $\bigcap_{\lambda=1}^{d+1} H(p_\lambda, s_\lambda)$ has the unique element $(a_0, \ldots, a_d)$, and $(a_0, \ldots, a_d) \in \bigcap_{\lambda=1}^{d+1} F_{\Pi \cup \{p\}} (p_\lambda, s_\lambda)$. We remark that here it suffices to evaluate $(a_0, \ldots, a_d) \in \bigcap_{\lambda=1}^{d+1} F_\Pi (p_\lambda, s_\lambda)$ (i.e., we do not have to verify $(a_0, \ldots, a_d) \in R_p$) because $p_\lambda = p$ for some $\lambda \in \{1, \ldots, d+1\}$ so that we have $(a_0, \ldots, a_d) \in H(p, s_\lambda) \subset R_p$. The computational cost for evaluating $(a_0, \ldots, a_d) \in R_\Pi$, which is required for checking if $(a_0, \ldots, a_d) \in F_\Pi (p_\lambda, s_\lambda)$, can be reduced by using the following property of R_Π.

Let $\bigcup \Psi_\Pi$ denote the union of all sets in Ψ_Π. For $\Pi \subset \{1, \ldots, n\}$, then, we define

$$R_\Pi^* = \left\{ (a_0, \ldots, a_d) \in \mathbb{R}^{d+1} \,\middle|\, \begin{array}{c} \max\limits_{p \in \Pi^*} \min\limits_{s \in \Sigma_\Pi(p)} h_{(p,s)} (a_0, \ldots, a_d) \\ \leq 0 \leq \\ \min\limits_{p \in \Pi^*} \max\limits_{s \in \Sigma_\Pi(p)} h_{(p,s)} (a_0, \ldots, a_d) \end{array} \right\}, \qquad (9)$$

where

$$\Pi^* = \left\{ p \in \Pi \,\middle|\, (p, s) \in \bigcup \Psi_\Pi \text{ for some } s \in \{1, \ldots, 4\} \right\} \qquad (10)$$

and

$$\Sigma_\Pi (p) = \left\{ s \in \{1, \ldots, 4\} \,\middle|\, (p, s) \in \bigcup \Psi_\Pi \right\}. \qquad (11)$$

Note that $\Sigma_\Pi(p) \neq \emptyset$ for $p \in \Pi^*$. R_Π^* is equivalent with Eq. (4) where only $h_{(p,s)}(a_0, \ldots, a_d)$ for (p, s) contained in $\bigcup \Psi_\Pi$ are involved. We now show that Eq. (9) serves as a simpler form of Eq. (4) when R_Π is bounded.

Theorem 3. *For Π such that R_Π is bounded, $R_\Pi^* = R_\Pi$.*

Proof. Suppose that $(p, s) \in \Pi \times \{1, \ldots, 4\}$ is not in $\bigcup \Psi_\Pi$. Lemma 1 (in the case of $m = 1$) then implies $F_\Pi(p, s) = \emptyset$, which means that (p, s) does not contribute to determining the boundary of R_Π. $\square$

Since the theorems above hold true only when R_Π is bounded, it is important to know when R_Π is bounded. We conclude this section by giving a sufficient condition for which R_Π is bounded. Recall that the coordinates of the pth data $(p = 1, \ldots, n)$ is denoted by (i_p, j_p).

Theorem 4. *Let $\Pi = \{p_1, \ldots, p_{d+1}\} \subset \{1, \ldots, n\}$ be a set of $d+1$ data indices such that $\left| i_{p_\lambda} - i_{p_\mu} \right| > 1$ for $\forall \lambda \neq \mu$. R_Π is bounded. For any $\Pi' \subset \{1, \ldots, n\}$ such that $\Pi' \supset \Pi$, therefore, $R_{\Pi'}$ is bounded.*

Proof. We show that a superset $R_\Pi' \subset \mathbb{R}^{d+1}$ of R_Π defined in the following is bounded.

For $\lambda = 1, \ldots, d+1$, we first define

$$
R_{p_\lambda}' = \left\{ (a_0, \ldots, a_d) \in \mathbb{R}^{d+1} \; \middle| \; \begin{array}{c} \min\limits_{(x', y') \in S} \left[(j_{p_\lambda} + y') - \sum\limits_{l=0}^{m} (i_{p_\lambda} + x')^l a_l \right] \\ \leq 0 \leq \\ \max\limits_{(x', y') \in S} \left[(j_{p_\lambda} + y') - \sum\limits_{l=0}^{m} (i_{p_\lambda} + x')^l a_l \right] \end{array} \right\},
$$
(12)

where $S = \left\{ (x', y') \in \mathbb{R}^2 \; \middle| \; \max \{|x'|, |y'|\} \leq \frac{1}{2} \right\}$. S is the square having (x_1, y_1), $\ldots$, (x_4, y_4) as its vertices. We therefore have $R_{p_\lambda}' \supset R_{p_\lambda}$. Since S is connected in $\mathbb{R}^2$, the intermediate value theorem allows to rewrite Eq. (12) as

$$
R_{p_\lambda}' = \left\{ (a_0, \ldots, a_d) \in \mathbb{R}^{d+1} \; \middle| \; \begin{array}{l} (j_{p_\lambda} + y') = \sum\limits_{l=0}^{d+1} (i_{p_\lambda} + x')^l a_l \\ \text{for some } (x', y') \in S \end{array} \right\}.
$$
(13)

We then define R_Π' by $R_\Pi' = \bigcap_{\lambda=1}^{d+1} R_{p_\lambda}'$, which is written as

$$
R_\Pi' = \left\{ (a_0, \ldots, a_d) \in \mathbb{R}^{d+1} \; \middle| \; \begin{array}{l} (j_{p_\lambda} + y_\lambda') = \sum\limits_{l=0}^{d+1} (i_{p_\lambda} + x_\lambda')^l a_l \\ \text{for some } (x_1', y_1'), \ldots, (x_{d+1}', y_{d+1}') \in S \end{array} \right\}.
$$
(14)

R_Π' is therefore obtained by collecting $(a_0, \ldots, a_d) \in \mathbb{R}^{d+1}$ satisfying

---

Algorithm 1. Discrete polynomial curve fitting.

Require: P, d and $I \subset \{1, \ldots, n\}$ such that $R_I \neq \emptyset$ and R_I is guaranteed by Theorem 4 to be bounded.

Ensure: $\Pi \subset \{1, \ldots, n\}$ and C_Π.

1: Initialize $\Pi :=$ any set of $d + 1$ elements of I satisfying the property in Theorem 4.
2: Initialize $\Pi^C := \emptyset$.
3: Compute Ψ_Π and C_Π.
4: **while** $I \setminus \Pi \neq \emptyset$ **do**
5:     $p :=$ any data index in $I \setminus \Pi$.
6:     Compute $\Psi_{\Pi \cup \{p\}}$ and $C_{\Pi \cup \{p\}}$ by Algorithm 2
7:     $\Pi := \Pi \cup \{p\}$.
8: **end while**
9: **while** $\{1, \ldots, n\} \setminus \left(\Pi \cup \Pi^C \right) \neq \emptyset$ **do**
10:     $p :=$ any data index in $\{1, \ldots, n\} \setminus \left(\Pi \cup \Pi^C \right)$.
11:     Compute $\Psi_{\Pi \cup \{p\}}$ and $C_{\Pi \cup \{p\}}$ by Algorithm 2
12:     **if** $\Psi_{I \cup \{p\}} \neq \emptyset$ **then**
13:         $\Pi := \Pi \cup \{p\}$.
14:     **else**
15:         $\Pi^C := \Pi^C \cup \{p\}$.
16:     **end if**
17: **end while**
18: **return** Π and C_Π.

$$
\begin{pmatrix} j_{p_1} + y'_1 \\ j_{p_2} + y'_2 \\ \vdots \\ j_{p_{d+1}} + y'_{d+1} \end{pmatrix} = \begin{pmatrix} 1 & i_{p_1} + x'_1 & (i_{p_1} + x'_1)^2 & \cdots & (i_{p_1} + x'_1)^d \\ 1 & i_{p_2} + x'_2 & (i_{p_2} + x'_2)^2 & \cdots & (i_{p_2} + x'_2)^d \\ \vdots & \vdots & \vdots & \ddots & \vdots \\ 1 & i_{p_{d+1}} + x'_{d+1} & (i_{p_{d+1}} + x'_{d+1})^2 & \cdots & (i_{p_{d+1}} + x'_{d+1})^d \end{pmatrix} \begin{pmatrix} a_0 \\ a_1 \\ \vdots \\ a_d \end{pmatrix}
$$
(15)

for all combinations of $(x'_1, y'_1), \ldots, (x'_{d+1}, y'_{d+1}) \in S$. Since the $(d+1) \times (d+1)$ matrix in Eq. (15) is a Vandermonde matrix, its determinant is given by

$$
\prod_{1 \leq \lambda < \mu \leq d+1} \left((i_{p_\mu} + x'_\mu) - (i_{p_\lambda} + x'_\lambda) \right),
$$
(16)

none of whose factors can be zero from $\left| i_{p_\mu} - i_{p_\lambda} \right| > 1$ for $\forall \lambda \neq \mu$. For any fixed $(x'_1, y'_1), \ldots, (x'_{d+1}, y'_{d+1}) \in S$, therefore, $(a_0, \ldots, a_d)$ is uniquely determined by Eq. (15) to be a point with the coordinates of finite values. It follows from this that R'_Π is bounded.   □

4   Algorithm

4.1   Algorithm Ensuring Inclusion-Wise Maximal Inlier Set

Our method for discrete polynomial curve fitting, described in Algorithm 1, requires an initial inlier set and ensures an inclusion-wise maximal inlier set containing the initial set. The initial inlier set is represented by its corresponding

---

Algorithm 2. Update of Ψ_Π and C_Π for an additional inlier.

Require: $P, d, \Pi \subset \{1, \ldots, n\}, p \in \{1, \ldots, n\} \setminus \Pi, \Psi_\Pi$ and C_Π.

Ensure: $\Psi_{\Pi \cup \{p\}}$ and $C_{\Pi \cup \{p\}}$.

```
 1: Initialize Ψ := ∅ and C := ∅.
 2: for all ψ ∈ Ψ_Π do
 3:     (a_0, ..., a_d) := the point in C_Π corresponding to ψ.
 4:     if (a_0, ..., a_d) ∈ R_p then
 5:         Ψ := Ψ ∪ {ψ} and C := C ∪ {(a_0, ..., a_d)}.
 6:     end if
 7: end for
 8: Compute Φ¹_{Π,p} and Φ²_{Π,p} (Eqs. (7) and (8)).
 9: for all ψ = {(p_1, s_1), ..., (p_{d+1}, s_{d+1})} ∈ Φ¹_{Π,p} ∪ Φ²_{Π,p} do
10:     if ∩^{d+1}_{λ=1} H(p_λ, s_λ) has the unique point then
11:         (a_0, ..., a_d) := the unique point in ∩^{d+1}_{λ=1}(p_λ, s_λ)
12:         if (a_0, ..., a_d) ∈ R*_Π ∩ ∩^{d+1}_{λ=1} F(p_λ, s_λ) then
13:             Ψ := Ψ ∪ {ψ} and C := C ∪ {(a_0, ..., a_d)}.
14:         end if
15:     end if
16: end for
17: return Ψ = Ψ_{Π∪{p}} and C = C_{Π∪{p}}.
```

---

index set I in Algorithm 1. The algorithm divides the data indices $1, \ldots, n$ into two classes Π and Π^{C}: those for inliers are sorted into Π, while those for outliers into Π^{C}. Π is first initialized to be a set of $d + 1$ indices in I, for which Ψ_Π and C_Π are computed at low cost using Eq. (6). In the two while-loops, then, we add new data indices to Π one by one accordingly computing corresponding Ψ_Π and C_Π using Algorithm 2. The first while-loop in Algorithm 1 is to obtain Ψ_I and C_I, where $\Psi_{\Pi \cup \{p\}}$ in each iteration cannot be empty from $R_I \neq \emptyset$ (Theorem 1). Note that R_I is nonempty and bounded. The second while-loop is to obtain Ψ_Π and C_Π for Π such that $\Pi \supset I$, where a data index p is sorted into Π^{C} if $\Psi_{\Pi \cup \{p\}} = \emptyset$, i.e., $R_{\Pi \cup \{p\}} = \emptyset$ (Theorem 1).

Algorithm 2 shows how to compute $\Psi_{\Pi \cup \{p\}}$ and $C_{\Pi \cup \{p\}}$ for $\Pi \subset \{1, \ldots, n\}$ and $p \in \{1, \ldots, n\} \setminus \Pi$ when Ψ_Π and C_Π are known. The algorithm evaluates each set in $\Psi_\Pi \cup \Phi^1_{\Pi,p} \cup \Phi^2_{\Pi,p}$ to check if it is in $\Psi_{\Pi \cup \{p\}}$ (Theorem 2). The first for-loop is to evaluate the sets in Ψ_Π, while the second for-loop is to evaluate the sets in $\Phi^1_{\Pi,p} \cup \Phi^2_{\Pi,p}$. Why a set in $\Psi_\Pi \cup \Phi^1_{\Pi,p} \cup \Phi^2_{\Pi,p}$ is verified to be in $\Psi_{\Pi \cup \{p\}}$ in this way is explained in Sect. 3.2 (after the proof of Theorem 2). In the second loop we use Theorem 3 ($R^*_\Pi = R_\Pi$) to reduce the computational cost for checking if $(a_0, \ldots, a_d) \in F_\Pi(p_\lambda, s_\lambda)$ ($= R_\Pi \cap F(p_\lambda, s_\lambda)$) for $\lambda = 1, \ldots, d + 1$.

Since $\Psi_{\Pi \cup \{p\}} \neq \emptyset$ if $R_{\Pi \cup \{p\}} \neq \emptyset$ from Theorem 1, after the second loop in Algorithm 1 we obtain an inclusion-wise maximal inlier set, which is equivalently stated in the following theorem.

Theorem 5. *Let $\Pi \subset \{1, \ldots, n\}$ be a data index set obtained by Algorithm 1. There exists no $\Pi' \subset \{1, \ldots, n\}$ satisfying $\Pi' \supset \Pi$ and $R_{\Pi'} \neq \emptyset$.*

The output of Algorithm 1 depends on the initial inlier set (i.e., I), and therefore how to determine I is an important issue. Most straightforwardly we can just set I to be random $d + 1$ data indices for which R_I is bounded according to Theorem 4, where $R_I \neq \emptyset$ can be evaluated by computing C_I (Theorem 1). Since our objective is to maximize the number of inliers, however, it is better to give I that is as large as possible. Note that Algorithm 1 outputs $\Pi \subset \{1, \ldots, n\}$ such that $\Pi \supset I$. To reduce the possibility of being trapped in a local optimum with a small number of inliers, it is also important for I not to be contaminated with noise. For the acquisition of such I, we can use a robust estimation algorithms such as RANSAC [1].

The output of Algorithm 1 also depends on the order in which data are added to the initial inlier set, i.e., how to choose p in Line 10. Choosing p corresponding to an outlier for the optimal solution here may make it impossible for many data to be added. The performance of the algorithm therefore might be improved by incorporating a procedure to select a "good" p, which is out of the scope of this paper.

4.2   Computational Complexity

We give the computational cost required for Algorithm 1. We remark that here we discuss the computational cost depending the number n of data where the degree d is treated as a constant. The computational cost for each iteration in the two while-loops (i.e., the computational cost for Algorithm 2), depending on $|\Pi|$, is written as $\left|\Psi_\Pi \cup \Phi^1_{\Pi,p} \cup \Phi^2_{\Pi,p}\right|$ multiplied by the computational cost required for checking if a set in $\Psi_\Pi \cup \Phi^1_{\Pi,p} \cup \Phi^2_{\Pi,p}$ is in $\Psi_{\Pi \cup \{p\}}$.

We first consider the order of $|\Psi_\Pi|$. We remark that $\left|\Phi^1_{\Pi,p}\right|$ and $\left|\Phi^2_{\Pi,p}\right|$ depend on $|\Psi_\Pi|$: in fact, we generally have $\left|\Phi^1_{\Pi,p}\right| = 4(d+1)|\Psi_\Pi|$ and $\left|\Phi^2_{\Pi,p}\right| = 2\binom{d+1}{2}|\Psi_\Pi| = d(d+1)|\Psi_\Pi|$. Since a set in Ψ_Π is composed of $d + 1$ elements of $\Pi \times \{1, \ldots, 4\}$, $|\Psi_\Pi|$ is bounded by the number of ways to pick $d + 1$ elements of $\Pi \times \{1, \ldots, 4\}$, i.e., $\binom{4|\Pi|}{d+1} = \mathcal{O}\left(|\Pi|^{d+1}\right)$. This upper bound is reduced by removing sets of $d + 1$ elements of $\Pi \times \{1, \ldots, 4\}$ containing $(p, \underline{s})$ and $(p, \overline{s})$ for any $p \in \Pi$, $\underline{s} \in \{1, 2\}$ and $\overline{s} \in \{3, 4\}$ (such sets cannot be in $\Psi_{\Pi \cup \{p\}}$ since $F(p, \underline{s}) \cap F(p, \overline{s}) = \emptyset$), which however does not change the order $\mathcal{O}\left(|\Pi|^{d+1}\right)$.

We next consider the computational cost for evaluating $\psi \in \Psi_{\Pi \cup \{p\}}$ for $\psi = \{(p_1, s_1), \ldots, (p_{d+1}, s_{d+1})\} \in \Psi_\Pi \cup \Phi^1_{\Pi,p} \cup \Phi^2_{\Pi,p}$. For $\psi \in \Psi_\Pi$, we only have to check if the corresponding $(a_0, \ldots, a_d) \in C_\Pi$ satisfies $(a_0, \ldots, a_d) \in R_p$, which takes a constant cost $\mathcal{O}(1)$. For $\psi \in \Phi^1_{\Pi,p} \cup \Phi^2_{\Pi,p}$, on the other hand, the computational cost is $\mathcal{O}(|\Pi|)$: we first have to check if $(a_0, \ldots, a_d) \in \bigcap_{\lambda=1}^{d+1} H(p_\lambda, s_\lambda)$ uniquely exists (computational cost: $\mathcal{O}(1)$), and if so, we then have to check if $(a_0, \ldots, a_d) \in \bigcap_{\lambda=1}^{d+1} F_\Pi(p_\lambda, s_\lambda)$ (computational cost: $\mathcal{O}(|\Pi|)$).

Since $\mathcal{O}(|\Psi_\Pi|) = \mathcal{O}\left(|\Phi^1_{\Pi,p} \cup \Phi^2_{\Pi,p}|\right) = \mathcal{O}\left(|\Pi|^{d+1}\right)$, the computational cost for each iteration in the two while-loops is therefore obtained as $\mathcal{O}\left(|\Pi|^{d+1}\right) \times$

$\mathcal{O}\left(|\Pi|\right) = \mathcal{O}\left(|\Pi|^{d+2}\right)$. In the first iteration $|\Pi| = d + 1$, and in the last iteration $|\Pi| = n - 1$ at most. The theoretical computational cost for Algorithm 1 is therefore $\sum_{m=d+1}^{n-1} \mathcal{O}\left(m^{d+2}\right) = \mathcal{O}\left(n^{d+2}\right)$.

5   Conclusions

We dealt with the problem of fitting a discrete polynomial curve to 2D noisy data, for which we proposed a method guaranteeing inclusion-wise maximality of its obtained inlier set. The method is constructed based on our investigation on the properties of the feasible regions in the parameter space corresponding to input data points. Evaluation of the practical performance of the proposed method is left for future work. This work may be extended to implicit functions ($f\left(x,y\right) = 0$) and surface fitting in 3D.

A   Appendix: Proof of Lemma 1

Proof. For $m = 1, \ldots, d$, let $(p_1, s_1), \ldots, (p_m, s_m) \in \Pi \times \{1, \ldots, 4\}$ satisfy (i) and (ii) in Lemma 1. It suffices to show that there always exists $(p_{m+1}, s_{m+1}) \in \Pi \times \{1, \ldots, 4\}$ such that $\bigcap_{\lambda=1}^{m+1} F_\Pi\left(p_\lambda, s_\lambda\right) \neq \emptyset$ and $H\left(p_1, s_1\right), \ldots, H\left(p_{m+1}, s_{m+1}\right)$ are linearly independent. See Fig. 5 for an illustration of this proof.

Let $(a'_0, \ldots, a'_d) \in \mathbb{R}^{d+1}$ be a point in $\bigcap_{\lambda=1}^{m} F_\Pi\left(p_\lambda, s_\lambda\right)$. No proof is required for the case where there exists $(p_{m+1}, s_{m+1}) \in \Pi \times \{1, \ldots, 4\}$ such that $(a'_0, \ldots, a'_d) \in F_\Pi\left(p_{m+1}, s_{m+1}\right)$ and $H\left(p_1, s_1\right), \ldots, H\left(p_{m+1}, s_{m+1}\right)$ are linearly independent. We therefore assume otherwise. Since $F_\Pi\left(p, s\right) \subset H\left(p, s\right)$ for any (p, s), we have $(a'_0, \ldots, a'_d) \in \bigcap_{\lambda=1}^{m} H\left(p_\lambda, s_\lambda\right)$. $\bigcap_{\lambda=1}^{m} H\left(p_\lambda, s_\lambda\right)$ is a $(d+1-m)$-dimensional flat ($d + 1 - m \geq 1$), and therefore we may consider a half-line in $\bigcap_{\lambda=1}^{m} H\left(p_\lambda, s_\lambda\right)$ running from $(a'_0, \ldots, a'_d)$. A point in the half-line is represented by $(a''_0(r), \ldots, a''_d(r))$ where $a''_l(r) = a'_l + r v_l$ ($l = 0, \ldots, d$) with some non-zero vector $(v_0, \ldots, v_d) \in \mathbb{R}^{d+1}$ and a non-negative parameter $r \in \mathbb{R}_{\geq 0}$: $(a''_0(r), \ldots, a''_d(r)) = (a'_0, \ldots, a'_d)$ for $r = 0$, and as we increase the value of r, the point traces the half-line in the direction of the vector $(v_0, \ldots, v_d)$.

Since $F_\Pi\left(p_\lambda, s_\lambda\right)$ ($\subset R_\Pi$) is bounded for $\lambda = 1, \ldots, m$, a large enough r satisfies $(a''_0(r), \ldots, a''_d(r)) \notin \bigcap_{\lambda=1}^{m} F_\Pi\left(p_\lambda, s_\lambda\right)$. Let r'_1 be the maximum value of r such that any $r \leq r'_1$ satisfies $(a''_0(r), \ldots, a''_d(r)) \in R_\Pi$ (note that this may be satisfied for some $r > r'_1$ when R_Π is concave). Let r'_2, on the other hand, be the maximum value of r satisfying $(a''_0(r), \ldots, a''_d(r)) \in \bigcap_{\lambda=1}^{m} F\left(p_\lambda, s_\lambda\right)$ (note that this is satisfied for any $r < r'_2$ since $F\left(p_\lambda, s_\lambda\right)$ is convex for $\lambda = 1, \ldots, m$), where we put $r'_2 = \infty$ if it is satisfied for any $r > 0$. Then, $r' = \min\{r'_1, r'_2\}$ is the maximum value of r such that any $r \leq r'$ satisfies $(a''_0(r), \ldots, a''_d(r)) \in \bigcap_{\lambda=1}^{m} F_\Pi\left(p_\lambda, s_\lambda\right)$ (recall that $F_\Pi\left(p, s\right) = F\left(p, s\right) \cap R_\Pi$). We now show that $(a''_0(r'), \ldots, a''_d(r')) \in F_\Pi\left(p_{m+1}, s_{m+1}\right)$ for some $(p_{m+1}, s_{m+1}) \in \Pi \times \{1, \ldots, 4\}$ where $(p_{m+1}, s_{m+1}) \neq (p_\lambda, s_\lambda)$ for $\lambda = 1, \ldots, m$. We remark that, for such (p_{m+1}, s_{m+1}), $H\left(p_1, s_1\right), \ldots, H\left(p_{m+1}, s_{m+1}\right)$ are linearly independent, since otherwise it is impossible to have $(a''_0(r'), \ldots, a''_d(r')) \in H\left(p_{m+1}, s_{m+1}\right)$ whereas $(a''_0(0), \ldots, a''_d(0)) \notin H\left(p_{m+1}, s_{m+1}\right)$.

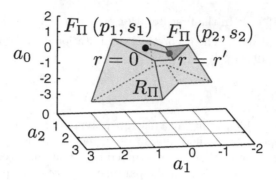

Fig. 5. Illustration for the proof of Lemma 1. We assume $m = 1$ here. Let the black point depict $(a'_0, \ldots, a'_d) = (a''_0(0), \ldots, a''_d(0))$. As we increase the value of r from zero, $(a''_0(r), \ldots, a''_d(r))$ traces a half-line in $H(p_1, s_1)$ running from $(a'_0, \ldots, a'_d)$. For $r = r'$ (intuitively speaking, just before $(a''_0(r), \ldots, a''_d(r))$ comes off $F_\Pi(p_1, s_1)$), then, $(a''_0(r), \ldots, a''_d(r))$ is in $F_\Pi(p_1, s_1) \cap F_\Pi(p_2, s_2)$ for some $(p_2, s_2) \in \Pi \times \{1, \ldots, 4\}$, so that we have $m = 2$ with $F_\Pi(p_1, s_1)$ and $F_\Pi(p_2, s_2)$. $(a''_0(r'), \ldots, a''_d(r'))$ is depicted in a red point. (Color figure online)

We first assume the case of $r' = r'_1 < r'_2$: as we increase the value of r past r', $(a''_0(r), \ldots, a''_d(r))$ gets out of R_Π, i.e., $R_{p_{m+1}}$ for some $p_{m+1} \in \Pi$ (recall that $R_\Pi = \bigcap_{p \in \Pi} R_p$). We remark that $p_{m+1} \neq p_1, \ldots, p_m$ for $\lambda = 1, \ldots, m$ since any r satisfies $(a''_0(r), \ldots, a''_d(r)) \in H(p_\lambda, s_\lambda) \subset R_{p_\lambda}$. For this p_{m+1}, therefore, we have $(a''_0(r'), \ldots, a''_d(r')) \in B_{p_{m+1}}$, i.e., $(a''_0(r'), \ldots, a''_d(r')) \in F(p_{m+1}, s_{m+1})$ for some $s_{m+1} \in \{1, \ldots, 4\}$. $(a''_0(r'), \ldots, a''_d(r')) \in F(p_{m+1}, s_{m+1}) \cap R_\Pi = F_\Pi(p_{m+1}, s_{m+1})$, accordingly.

We next assume the case of $r' = r'_2 \leq r'_1$: as we increase the value of r past r', $(a''_0(r), \ldots, a''_d(r))$ gets out of $F(p_\lambda, s_\lambda)$ for some $\lambda \in \{1, \ldots, m\}$. Without loss of generality, we assume that $\lambda = 1$, and $(a''_0(r), \ldots, a''_d(r)) \in \overline{B}_{p_1}$ for $r \leq r'$. We then have

$$h_{(p_1, s_1)}(a''_0(r), \ldots, a''_d(r)) = \max_{s \in \{1, \ldots, 4\}} h_{(p_1, s)}(a''_0(r), \ldots, a''_d(r)) \text{ for } r \leq r',$$
(17)

while

$$h_{(p_1, s_1)}(a''_0(r), \ldots, a''_d(r)) < \max_{s \in \{1, \ldots, 4\}} h_{p_1, s}(a''_0(r), \ldots, a''_d(r)) \text{ for } r > r'. \quad (18)$$

This suggests that for some $s'_1 \in \{1, \ldots, 4\} \setminus \{s_1\}$ we have

$$0 = h_{(p_1, s_1)}(a''_0(r'), \ldots, a''_d(r')) = h_{(p_1, s'_1)}(a''_0(r'), \ldots, a''_d(r'))$$
$$= \max_{s \in \{1, \ldots, 4\}} h_{(p_1, s)}(a''_0(r'), \ldots, a''_d(r')), \quad (19)$$

Note that $h_{(p_1, s_1)}(a''_0(r), \ldots, a''_d(r)) = 0$ (i.e., $(a''_0(r), \ldots, a''_d(r)) \in H(p_1, s_1)$) is satisfied for any r. $(a''_0(r'), \ldots, a''_d(r')) \in F(p_1, s'_1) \cap R_\Pi = F_\Pi(p_1, s'_1)$, consequently. □

References

1. Fischler, M.A., Bolles, R.C.: Random sample consensus: a paradigm for model fitting with applications to image analysis and automated cartography. Commun. ACM **24**, 381–395 (1981)
2. Duda, R.O., Hart, P.E.: Use of the Hough transformation to detect lines and curves in pictures. Commun. ACM **15**, 11–15 (1972)
3. Ballard, D.H.: Generalizing the Hough transform to detect arbitrary shapes. Pattern Recogn. **13**, 111–122 (1981)
4. Cohen-Or, D., Kaufman, A.: Fundamentals of surface voxelization. Graph. Models Image Process. **57**, 453–461 (1995)
5. Brimkov, V.E., Andres, E., Barneva, R.P.: Object discretization in higher dimensions. In: Borgefors, G., Nyström, I., Baja, G.S. (eds.) DGCI 2000. LNCS, vol. 1953, pp. 210–221. Springer, Heidelberg (2000). doi:10.1007/3-540-44438-6_18
6. Tajine, M., Ronse, C.: Topological properties of Hausdorff discretization, and comparison to other discretization schemes. Theoret. Comput. Sci. **283**, 243–268 (2002)
7. Lincke, C., Wüthrich, C.A.: Surface digitizations by dilations which are tunnel-free. Discret. Appl. Math. **125**, 81–91 (2003)
8. Brimkov, V.E., Barneva, R.P., Brimkov, B.: Minimal offsets that guarantee maximal or minimal connectivity of digital curves in nD. In: Brlek, S., Reutenauer, C., Provençal, X. (eds.) DGCI 2009. LNCS, vol. 5810, pp. 337–349. Springer, Heidelberg (2009). doi:10.1007/978-3-642-04397-0_29
9. Brimkov, V.E.: On connectedness of discretized objects. In: Bebis, G., Boyle, R., Parvin, B., Koracin, D., Li, B., Porikli, F., Zordan, V., Klosowski, J., Coquillart, S., Luo, X., Chen, M., Gotz, D. (eds.) ISVC 2013. LNCS, vol. 8033, pp. 246–254. Springer, Heidelberg (2013). doi:10.1007/978-3-642-41914-0_25
10. Khalimsky, E., Kopperman, R., Meyer, P.R.: Computer graphics and connected topologies on finite ordered sets. Topol. Its Appl. **36**, 1–17 (1990)
11. Andres, E., Acharya, R., Sibata, C.: Discrete analytical hyperplanes. Graph. Models Image Process. **59**, 302–309 (1997)
12. Andres, E., Jacob, M.A.: The discrete analytical hyperspheres. IEEE Trans. Vis. Comput. Graph. **3**, 75–86 (1997)
13. Andres, E.: Discrete linear objects in dimension n: the standard model. Graph. Models **65**, 92–111 (2003)
14. Andres, E., Roussillon, T.: Analytical description of digital circles. In: Debled-Rennesson, I., Domenjoud, E., Kerautret, B., Even, P. (eds.) DGCI 2011. LNCS, vol. 6607, pp. 235–246. Springer, Heidelberg (2011). doi:10.1007/978-3-642-19867-0_20
15. Toutant, J.L., Andres, E., Roussillon, T.: Digital circles, spheres and hyperspheres: From morphological models to analytical characterizations and topological properties. Discret. Appl. Math. **161**, 2662–2677 (2013)
16. Toutant, J.-L., Andres, E., Largeteau-Skapin, G., Zrour, R.: Implicit digital surfaces in arbitrary dimensions. In: Barcucci, E., Frosini, A., Rinaldi, S. (eds.) DGCI 2014. LNCS, vol. 8668, pp. 332–343. Springer, Cham (2014). doi:10.1007/978-3-319-09955-2_28
17. Zrour, R., Kenmochi, Y., Talbot, H., Shimizu, I., Sugimoto, A.: Combinatorial optimization for fitting of digital line and plane. In: International Workshop on Computer Vision and Its Application to Image Media Processing, Satellite Workshop of the 3rd Pacific-Rim Symposium on Image and Video Technology, pp. 35–41 (2009)
18. Kenmochi, Y., Buzer, L., Talbot, H.: Efficiently computing optimal consensus of digital line fitting. In: International Conference on Pattern Recognition (ICPR2010), pp. 1064–1067. IEEE (2010)

19. Zrour, R., Kenmochi, Y., Talbot, H., Buzer, L., Hamam, Y., Shimizu, I., Sugimoto, A.: Optimal consensus set for digital line and plane fitting. Int. J. Imaging Syst. Technol. **21**, 45–57 (2011)

20. Sere, A., Sie, O., Andres, E.: Extended standard Hough transform for analytical line recognition. In: Proceedings of International Conference on Sciences of Electronics, Technologies of Information and Telecommunications (SETIT2012), pp. 412–422. IEEE (2012)

21. Zrour, R., Largeteau-Skapin, G., Andres, E.: Optimal consensus set for annulus fitting. In: Debled-Rennesson, I., Domenjoud, E., Kerautret, B., Even, P. (eds.) DGCI 2011. LNCS, vol. 6607, pp. 358–368. Springer, Heidelberg (2011). doi:10.1007/978-3-642-19867-0_30

22. Largeteau-Skapin, G., Zrour, R., Andres, E., Sugimoto, A., Kenmochi, Y.: Optimal consensus set and preimage of 4-connected circles in a noisy environment. In: Proceedings of the International Conference on Pattern Recognition (ICPR2012), pp. 3774–3777. IEEE (2012)

23. Largeteau-Skapin, G., Zrour, R., Andres, E.: $O(n^3 \log n)$ time complexity for the optimal consensus set computation for 4-connected digital circles. In: Gonzalez-Diaz, R., Jimenez, M.-J., Medrano, B. (eds.) DGCI 2013. LNCS, vol. 7749, pp. 241–252. Springer, Heidelberg (2013). doi:10.1007/978-3-642-37067-0_21

24. Zrour, R., Largeteau-Skapin, G., Andres, E.: Optimal consensus set for nD fixed width annulus fitting. In: Barneva, R.P., Bhattacharya, B.B., Brimkov, V.E. (eds.) IWCIA 2015. LNCS, vol. 9448, pp. 101–114. Springer, Heidelberg (2015). doi:10.1007/978-3-319-26145-4_8

25. Sekiya, F., Sugimoto, A.: Fitting discrete polynomial curve and surface to noisy data. Ann. Math. Artif. Intell. **75**, 135–162 (2015)

26. Sekiya, F., Sugimoto, A.: On connectivity of discretized 2D explicit curve. In: Ochiai, H., Anjyo, K. (eds.) Mathematical Progress in Expressive Image Synthesis II. MI, vol. 18, pp. 33–44. Springer, Heidelberg (2015). doi:10.1007/978-4-431-55483-7_4

27. Heijmans, H.: Morphological discretization. In: Geometrical Problems of Image Processing, pp. 99–106 (1991)

28. Heijmans, H.J.A.M., Toet, A.: Morphological sampling. CVGIP: Image Underst. **54**, 384–400 (1991)

29. Heijmans, H.J.A.M.: Discretization of morphological operators. J. Vis. Commun. Image Represent. **3**, 182–193 (1992)

A Discrete Approach for Decomposing Noisy Digital Contours into Arcs and Segments

Phuc Ngo[1,2(✉)], Hayat Nasser[1,2], and Isabelle Debled-Rennesson[1,2]

[1] Université de Lorraine, LORIA, UMR 7503, Vandoeuvre-lès-Nancy 54506, France
{hoai-diem-phuc.ngo,hayat.nasser,isabelle.debled-rennesson}@loria.fr
[2] CNRS, LORIA, UMR 7503, Vandoeuvre-lès-Nancy 54506, France

Abstract. In the paper, we present a method for decomposing a discrete noisy curve into arcs and segments which are the frequent primitives in digital images. This method is based on two tools: dominant point detection using adaptive tangential cover and tangent space representation of the polygon issued from detected dominant points. The experiments demonstrate the robustness of the method w.r.t. noise.

Keywords: Adaptive tangential cover · Dominant point detection · Tangent space · Curve reconstruction · Vectorization

1 Introduction

The extraction of meaningful features from image contour is an important problem in computer vision. Many existing methods used critical points or straight segments as meaningful features to construct high level descriptors of images. Among the primitives (straight lines, circles, circular arcs, ellipses, parabolas, etc.,) arc and straight segment are those appearing often in images, especially in graphic document images. Furthermore, a combination of circular arcs and straight line segments is an interesting solution to avoid an approximation of arcs by many straight segments or critical points. Many methods have been proposed to decompose a planar curve into arcs and segments. Surveys of the different approaches can be found in [1,2]. Several of these methods work in a transformed domain [3,4].

Inspired by the works about dominant point detection using adaptive tangential cover (ATC) [5] and tangent space representation of the polygon issued from dominant points detected [1,6], we present in this paper a method for decomposing digital curves into arcs and segments. More precisely, the algorithm first detects the dominant points of the input curve. Then, it transforms the polygon of the detected dominant points into the *tangent space* [4]. In [1,6], it is shown that a sequence of chords of an arc corresponds to a sequence of collinear points in the tangent space. Thus, the problem of digital arc detection becomes digital straight line recognition in the tangent space. In other words, one can use the tangent space to identify the segments and arcs of a curve. However, the tangent space is very sensitive to noise. By taking the advantage of the discrete structure

© Springer International Publishing AG 2017
C.-S. Chen et al. (Eds.): ACCV 2016 Workshops, Part II, LNCS 10117, pp. 493–505, 2017.
DOI: 10.1007/978-3-319-54427-4_36

of ATC to noisy curves, the proposed method uses the ATC to detect dominant points and thus makes it more robust to noise.

The paper is organized as follows: Sect. 2 consists of definitions and results used in this paper. The main results are presented in Sect. 3 with the proposed decomposition algorithm. Section 4 shows the experimental results and comparisons for segmenting digital curve into arcs and segments. Finally, in Sect. 5, we conclude and discuss the perspectives.

2   Discrete Curve Structure

2.1   Maximal Blurred Segments

Maximal blurred segment has been introduced by Debled-Rennesson *et al.* in [7] as an extension of arithmetical discrete line [8] with a width parameter for noisy or disconnected digital contours.

Definition 1. *An **arithmetical discrete line** $\mathcal{D}(a, b, \mu, \omega)$, with a direction vector (b, a), a lower bound μ and an arithmetic thickness ω (with $a, b, \mu, \omega \in \mathbb{Z}$ and $\gcd(a, b) = 1$) is the set of integer points (x, y) verifying $\mu \leq ax - by < \mu + \omega$.*

Definition 2. *A set S_f is a **blurred segment of width** ν if there exists a discrete line $\mathcal{D}(a, b, \mu, \omega)$ containing S_f has the vertical (or horizontal) distance $d = \frac{\omega - 1}{\max(|a|, |b|)}$ equal to the vertical (or horizontal) thickness of the convex hull of S_f, and $d \leq \nu$.*

Let C be a discrete curve and $C_{i,j}$ a sequence of points of C indexed from i to j. Let denote the predicate "$C_{i,j}$ is a blurred segment of width ν" as $BS(i, j, \nu)$.

Definition 3. *$C_{i,j}$ is called a **maximal blurred segment (MBS) of width** ν and noted $MBS(i, j, \nu)$ iff $BS(i, j, \nu)$, $\neg BS(i, j + 1, \nu)$ and $\neg BS(i - 1, j, \nu)$.*

These notions are illustrated in Figs. 1 and 2. The sequence of maximal segments along a digital contour is called a *tangential cover* and used in numerous

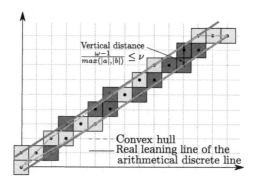

Fig. 1. Example of arithmetical discrete line $\mathcal{D}(2, 3, -3, 5)$ (grey and blue points) and a blurred segment of width $\nu = 1.4$ (grey points) bounded by D. (Color figure online)

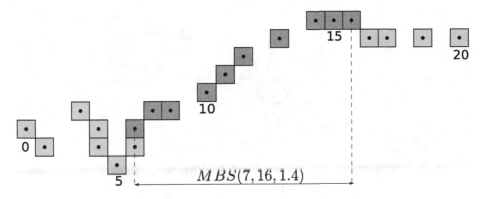

Fig. 2. Maximal blurred segment of width $\nu = 1.4$ (green points). (Color figure online)

discrete geometric estimators (see [9] for a state of the art): length, tangent, curvature estimators, detection of convex or concave parts, *etc.* In [7], a linear algorithm is proposed for computing the tangential cover of a given curve. Inspired by this algorithm, in [1,10] a quasi-linear algorithm is developed for tangential cover using MBS of a priori fixed width ν, namely **width ν tangential cover**.

Still in [1], the width ν tangential cover is proposed to deal with the noisy contours. However, this fixed width value ν needs to be manually adjusted and the method is inadequate to local amount of noise which can appear on real contours.

2.2 Adaptive Tangential Cover

Recently, a notion of **adaptive tangential cover (ATC)** was introduced in [5] which is composed of MBS with appropriated widths. More precisely, the ATC is composed of MBS of different width values varying in function of the noise perturbations presented on the contour. In particular, it uses the local noisy estimator, namely *meaningful thickness* [11], to determine the significant width locally at each point of the contour to analyze the considering curve. This meaningful thickness is used as an input parameter of the width ν tangential cover framework to compose the MBS of ATC with appropriate widths w.r.t. noise. A quasi-linear algorithm is developed in [5] to compute the ATC. Example of ATC is given in Fig. 3.

In the ATC, the obtained MBS decomposition of various widths transmits the noise levels and the geometrical structure of the given discrete curve. In [5], a non-parametric algorithm is proposed to build the ATC of a given discrete curve.

In the next section, we discuss about the method for decomposing noisy digital curves into arcs and segments. This method is based on: (1) the detection of key points, namely *dominant points*, on a curve using the ATC [5], and (2) the *tangent space* [3,4] –a tangent angle versus length segment plot– of polygon issued from dominant points detected to recognize points of arcs or segments on the considered polygon.

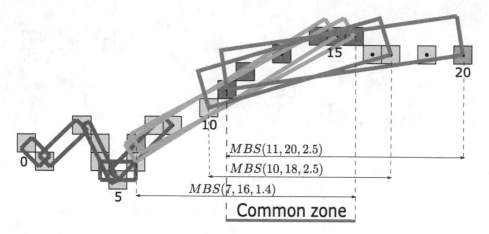

Fig. 3. Adaptive tangential cover with three width values $\nu = 1, 1.4$ and 2.5 (in blue, green and red respectively) deduced from the local noise level estimation. In pink are points in the common zone determined by the last three MBS of the ATC, and in orange are its two endpoints. (Color figure online)

3   Arcs and Segments Decomposition Method

Nguyen *et al.* in [1,6] assumed that junctions between two different primitives are dominant points of a curve, and proposed an algorithm for decomposing curves into arcs and segments using tangential cover with a fixed width value of MBS and tangent space representation of the approximated polygon of dominant points. Such an approach is not adapted to noisy curves due to the fixed width, in particular when the noise is not uniformly distributed. Furthermore, the width value is manually adjusted. Inspired by this study, we propose a suitable approach for noisy curves by using the ATC which is a free-parameter algorithm.

3.1   Dominant Point Detection

Dominant points are significant points on a curve with local maximum curvature. Such points contain a rich information which allows to characterize and describe the curve.

Issued from the dominant point detection proposed in [12,13] and the notion of ATC, a sequential algorithm is developed in [5] to determine the dominant points of a given noisy curve C. The main idea is that candidate dominant points are localized in the common zones of successive MBS of the ATC of C. In particular, this step can be done by simply verifying the beginning and ending indexes of the MBS constituting the common zone (see Fig. 3). Once the candidates are found, an angle measure is used to determine the dominant points with local extreme curvature in the common zone. More precisely, this measure is estimated as the angle between the considered point and the two left and right

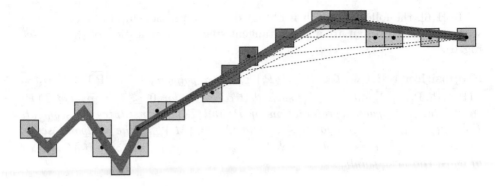

Fig. 4. Dominant points (in red) with the ATC in Fig. 3 are detected as point having the smallest angle measure in the common zone. (Color figure online)

endpoints of the left and right MBS involved in the studied common zone, and dominant point is identified as the point having a local minimum measure of angle. This is illustrated in Fig. 4.

3.2   Tangent Space Representation (or (l, α) Plane [4])

Let $P = \{P_i\}_{i=0}^m$ be a polygon, l_i length of segment P_iP_{i+1} and $\alpha_i = \angle(\overrightarrow{P_{i-1}P_i}, \overrightarrow{P_iP_{i+1}})$ such that $\alpha_i > 0$ if P_{i+1} is on the right side of $\overrightarrow{P_{i-1}P_i}$ and $\alpha_i < 0$ otherwise. A tangent space representation $T(P)$ of P is a step function which is constituted of segments $T_{i2}T_{(i+1)1}$ and $T_{(i+1)1}T_{(i+1)2}$ for $0 \le i < m$ with

$$T_{02} = (0,0),$$
$$T_{i1} = (T_{(i-1)2}.x + l_{i-1}, T_{(i-1)2}.y) \text{ for } 1 \le i \le m,$$
$$T_{i2} = (T_{i1}.x, T_{i1}.y + \alpha_i), \ 1 \le i \le (m-1).$$

In other words, the difference between two ordinates of an adjacent step in the tangent space represents the turn angle α_i of the corresponding pair of line segments $\overrightarrow{P_{i-1}P_i}$ and $\overrightarrow{P_iP_{i+1}}$. The difference between two abscissas of two endpoints of a step in the tangent space represents the length l_i of the corresponding line segments P_iP_{i+1} (see Fig. 5).

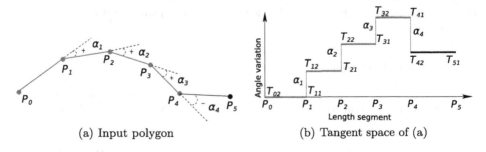

(a) Input polygon   (b) Tangent space of (a)

Fig. 5. Tangent space representation of a straight segment succession.

In [1,6], the following result is obtained for the relationship between a set of sequential chords of an arc and the tangent space representation of the set (see Fig. 6).

Proposition 1 [1,6]. *Let* $P = \{P_i\}_{i=0}^m$ *be a polygon,* $l_i = |\overrightarrow{P_iP_{i+1}}|$, $\alpha_i = \angle(\overrightarrow{P_{i-1}P_i}, \overrightarrow{P_iP_{i+1}})$ *and* $\alpha \leqslant \frac{\pi}{4}$ *such that* $\alpha_i \leqslant \alpha$ *for* $0 \leq i < n$. *Let* $T(P)$ *be the tangent space representation of* P *and* $T(P)$ *constitutes of segments* $T_{i2}T_{(i+1)1}, T_{(i+1)1}T_{(i+1)2}$ *for* $0 \leq i < m$, $M = \{M_i\}_{i=0}^{m-1}$ *the midpoint set of* $\{T_{i2}T_{(i+1)1}\}_{i=0}^{m-1}$. *P is a polygon whose vertices are on an arc only if* M *is a set of quasi collinear points.*[1]

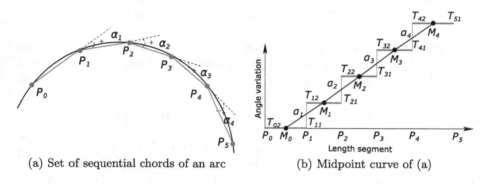

(a) Set of sequential chords of an arc   (b) Midpoint curve of (a)

Fig. 6. Tangent space and the curve of midpoints of an arc. If the midpoints is quasi collinear, then the initial points belong to a circular arc [1,6].

In other words, the arc detection of P becomes the problem of verifying the quasi collinearity of midpoints in tangent space representation of P. In particular, this quasi collinearity can be done with the algorithm of MBS of width ν with midpoints in the tangent space (see [1,6] for the details).

Particularly, in the midpoint curve M of $T(P)$, a midpoint is said an **isolated point** if the difference of ordinate values between it and one of the two neighboring midpoints is higher than a threshold α, the point corresponds to a junction of two primitives. If this condition of isolated point is satisfied with all two neighbors, the point is called a **full isolated point** and corresponds a segment in P (see Fig. 7).

3.3   Proposed Algorithm

We now describe the method to decompose a discrete curve into arcs and segments. The method is divided into three steps (see Fig. 8) and given in Algorithm 1:

---

[1] The points are said *quasi collinear* if they belong to a small width strip bounded by two real parallel lines.

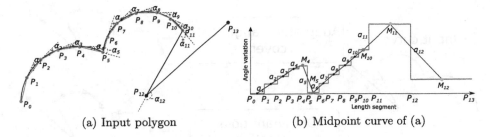

(a) Input polygon   (b) Midpoint curve of (a)

Fig. 7. Classification of midpoints in tangent space. Pink (resp. green) points are iso-lated points (resp. full isolated points) and correspond to junctions of two primitives (resp. segments), while black points are points of arcs. (Color figure online)

1. Detection of dominant points with ATC decomposition of a noisy discrete curve.
2. Tangent space representation of polygon issued from dominant points detected.
3. Computation of arcs and segments of the considered curve from the tangent space analysis and the discrete points of the curve.

Algorithm 1. Curve Decomposition into Arcs and Segments

Require: $C = (C_i)_{0 \leq i \leq n-1}$ a discrete curve of n points
$\quad\quad\quad\quad \nu$ width of MBS for collinear test in the tangent space
$\quad\quad\quad\quad \alpha$ threshold of admissible angle in the tangent space
Ensure: $ARCs$ and $SEGs$ sets of arcs and segments of C
1: **Begin**
2: $ARCs \leftarrow \emptyset, SEGs \leftarrow \emptyset, pARCs \leftarrow \emptyset, MBS_\nu \leftarrow \emptyset$
3: Detect the dominant point D of C (see Sec. 3.1)
4: Transform D into the tangent space $T(D)$ (see Sec. 3.2)
5: Construct the midpoint curve $\{M_i\}_{i=0}^{m-1}$ of $T(D)$ (see Sec. 3.2)
6: **for** $i \leftarrow 1$ to $m - 2$ **do**
7: $\quad C_{b_i} C_{e_i}$ is the part of C corresponding to M_i
8: $\quad$ **if** $(\mid M_i.y - M_{i-1}.y \mid > \alpha) \& (\mid M_i.y - M_{i+1}.y \mid > \alpha)$ **then**
9: $\quad\quad SEGs \leftarrow SEGs \cup \{\text{find a segment from } C_{b_i} C_{e_i}\}$
10: $\quad\quad MBS_\nu \leftarrow \emptyset$
11: $\quad$ **else**
12: $\quad\quad$ **if** $MBS_\nu \cup \{M_i\}$ is a MBS of width ν **then**
13: $\quad\quad\quad MBS_\nu \leftarrow MBS_\nu \cup \{M_i\}$
14: $\quad\quad\quad pARC \leftarrow pARC \cup \{C_{b_i} C_{e_i}\}$
15: $\quad\quad$ **else**
16: $\quad\quad\quad ARCs \leftarrow ARCs \cup \{\text{find an arc from } pARC\}$
17: $\quad\quad\quad pARC \leftarrow \emptyset$
18: $\quad\quad$ **end if**
19: $\quad$ **end if**
20: **end for**
21: **End**

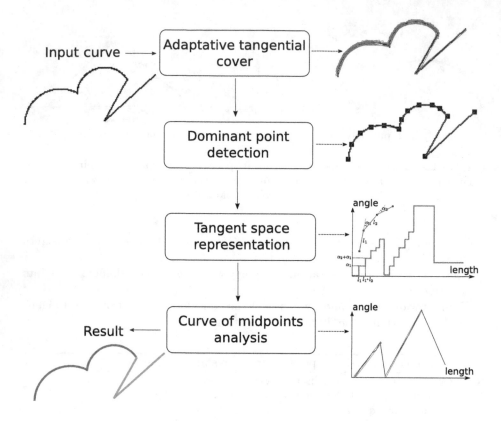

Fig. 8. Flowchart of the proposed algorithm.

In short, the decomposition process is realized by analyzing the tangent space representation of the polygon constructed from detected dominant points of the input curve. Then it outputs the sets of segments and arcs decomposed of the curve.

In Algorithm 1, Line 9 (resp. 16) is to calculate the real segments (resp. arcs) corresponding to points in the set SEGs (resp. ARCs). For the continuity of the decomposition between the primitives, we consider the endpoints as dominant points. Thus, each segment in SEGs is simply the line segment passing through the corresponding endpoints C_{b_i} and C_{e_i} –which are also the dominant points. While, each arc in ARCs is computed as the arc passing through the two endpoints of $pARC$ and best-fitting to points in $pARC$ when the squared distance is used. Due to a high angular deviation near the endpoints of an arc, we verify the best-fitting arc with one point in the central one-third portion of $pARC$ and the two endpoints of $pARC$. Note that this is improved w.r.t. the original algorithm proposed in [1,6] which simply pick-up a middle point without any optimization of arcs detected.

Furthermore, it is obvious that any two midpoints in the tangent space belong to a segment, thus a MBS. As a consequence, we can always determine an arc associated to these midpoints. In order to obtain an accurate segmentation and a better description of curve with arcs and segments in such cases, we consider an error criterion, namely integral sum of square errors, for the considered arc approximation. More precisely, we compute the error value for the part of curve approximated by an arc and this approximated by segments, if the error by the arc is smaller than this by segments then the arc is kept in the decomposition, and the segments otherwise.

Algorithm 1 has a $O(n \log n + nm)$ complexity where n is the number of points of C and m is the number of midpoints. Indeed, the ATC and dominant point detection can be performed in $O(n \log n)$ [5] and the tangent space representation of dominant points as well as the recognition of MBS of the midpoint curve are done in $O(m)$ [14]. The loop iterates over the midpoints to find the corresponding segments and arcs (Lines 6–20) is performed in $O(nm)$. More precisely, Line 9, for finding a segment, is in $O(1)$ since the segment is determined by the extremities $C_{b_i} C_{e_i}$. For finding a best-fitting arc associated to $pArc$ of the curve in Line 16, we have $O(1)$ to compute an arc passing through three points and $O(|pArc|)$ to compute the fitting error, and $|pArc| = n/3$ in the worst case. Finally, the complexity of the algorithm is $O(n \log n + nm)$.

4   Experimental Results

We now present some experiments to illustrate the efficiency of Algorithm 1 to decompose discrete curves into arcs and segments. To this end, we first show the

Fig. 9. Arcs and segments reconstruction of noisy curves. Left: input with Gaussian noise, middle left: extracted curves, middle right: ATC, right: decomposition results.

decomposed results in case of noisy data. Then, we provide some comparisons to Rosin [15] and Nguyen [1] methods. Overall the experiments, the arcs and segments are colored in red and green respectively.

From Proposition 1, we can fix the admissible angle in the tangent space $\alpha = \frac{\pi}{4}$. While the width ν for the quasi collinear test of MBS, it is mentioned in [1] that this value relies on the approximation error $sin\alpha_i \simeq \alpha_i$ in the tangent space and it is bounded by 0.1. Therefore, we have $\nu = 0.1$.

4.1   Experimentation on Noisy Curves

In order to test the robustness of the proposed algorithm towards noise, we consider Gaussian noise and a statistical noise model similar to the Kanungo noise [16], in which the probability P_d of changing the pixel located at a distance d from the shape boundary is defined as $P_d = \beta^d$ with $0 < \beta < 1$. Figure 10 shows curves obtained with different noise levels for $\beta = 0.3, 0.5, 0.7$.

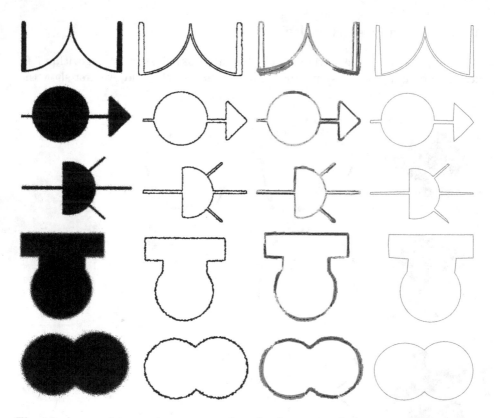

Fig. 10. Arcs and segments reconstruction of noisy curves. Left: input with Kanungo noise for $\beta = 0.3$ (first image), 0.5 (second and third images) and 0.5 (fourth and fifth images), middle left: extracted curves, middle right: ATC, right: decomposition results.

Comparing to the algorithm proposed in [1], tangential covers with fixed width value are applied for arcs and segments detection. Since the noise can appear randomly along the contour, this method is inadequate when contours present different noise levels. We observe in Figs. 9 and 10 that the ATC allows a better model of tangent cover and thus more relevant to noise. Still in Figs. 9 and 10, the experimental results of the proposed algorithm are shown. It can be seen that good reconstructions with arcs and segments are obtained even with an important and non-uniform noise on digital contours.

4.2   Comparison Results with Other Methods

We now present some comparisons of our algorithm with two algorithms proposed by Rosin [15] and Nguyen [1]. The experiments are carried out on noisy curves, technical and real images and shown in Figs. 11, 12 and 13. The modifications of the decomposition algorithm allows to improve the quality of the curves reconstructed with arcs and segments w.r.t. Nguyen's algorithm [1,6] and particularly for noisy curves.

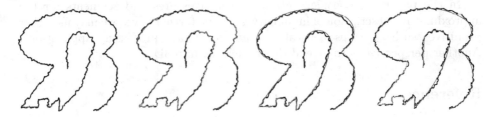

Fig. 11. Comparison result on noisy curve. Left: input curve, middle left: results with [15], middle right: results with [1], right: results with our method.

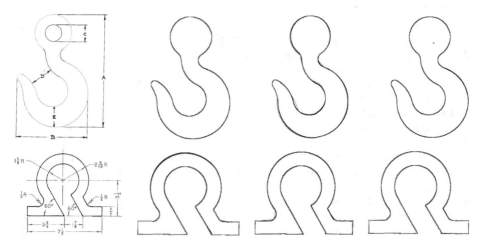

Fig. 12. Comparison result on technical images. Left: input images, middle left: results with [15], middle right: results with [1], right: results with our method.

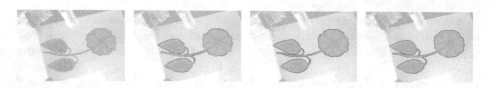

Fig. 13. Comparison result on real image. Left: input image, middle left: results with [15], middle right: results with [1], right: results with our method.

5 Conclusion

We present in this paper a discrete approach for decomposing a noisy discrete curve into arcs and segments using the notion of adaptive tangential cover deduced from the meaningful thickness. The experiments demonstrate the efficiency of the proposed approach even with an important and non-uniform noise level on input contours. An online demonstration based on the DGtal [17] and ImaGene [18] library, is available at the following website: http://ipol-geometry. loria.fr/~phuc/ipol_demo/ATC_ArcSegDecom_IPOLDemo

In the paper, we consider only two primitives arcs and segments for the approximating/description of input curves. In the further work, we may use other primitives such as eclipses, parabolas, *etc.* for image processing applications. Another perspective is to extend the framework into 3D.

References

1. Nguyen, T.P., Debled-Rennesson, I.: Decomposition of a curve into arcs and line segments based on dominant point detection. In: Heyden, A., Kahl, F. (eds.) SCIA 2011. LNCS, vol. 6688, pp. 794–805. Springer, Heidelberg (2011). doi:10.1007/978-3-642-21227-7_74
2. Akinlar, C., Topal, C.: EDCircles: a real-time circle detector with a false detection control. Pattern Recogn. **46**, 725–740 (2013)
3. Arkin, E.M., Chew, L.P., Huttenlocher, D.P., Kedem, K., Mitchell, J.S.B.: An efficiently computable metric for comparing polygonal shapes. In: Proceedings of the First Annual ACM-SIAM Symposium on Discrete Algorithms, SODA 1990, pp. 129–137 (1990)
4. Latecki, L.J., Lakämper, R.: Shape similarity measure based on correspondence of visual parts. IEEE Trans. Pattern Anal. Mach. Intell. **22**, 1185–1190 (2000)
5. Ngo, P., Nasser, H., Debled-Rennesson, I., Kerautret, B.: Adaptive tangential cover for noisy digital contours. In: Normand, N., Guédon, J., Autrusseau, F. (eds.) DGCI 2016. LNCS, vol. 9647, pp. 439–451. Springer, Cham (2016). doi:10.1007/978-3-319-32360-2_34
6. Nguyen, T.P., Debled-Rennesson, I.: Arc segmentation in linear time. In: Real, P., Diaz-Pernil, D., Molina-Abril, H., Berciano, A., Kropatsch, W. (eds.) CAIP 2011. LNCS, vol. 6854, pp. 84–92. Springer, Heidelberg (2011). doi:10.1007/978-3-642-23672-3_11
7. Debled-Rennesson, I., Feschet, F., Rouyer-Degli, J.: Optimal blurred segments decomposition of noisy shapes in linear time. Comput. Graph. **30**, 30–36 (2006)

8. Reveillès, J.P.: Géométrie discrète, calculs en nombre entiersgorithmique. Thèse d'état, Université Louis Pasteur, Strasbourg (1991)
9. Lachaud, J.-O.: Digital shape analysis with maximal segments. In: Köthe, U., Montanvert, A., Soille, P. (eds.) WADGMM 2010. LNCS, vol. 7346, pp. 14–27. Springer, Heidelberg (2012). doi:10.1007/978-3-642-32313-3_2
10. Faure, A., Buzer, L., Feschet, F.: Tangential cover for thick digital curves. Pattern Recogn. **42**, 2279–2287 (2009)
11. Kerautret, B., Lachaud, J.O.: Meaningful scales detection: an unsupervised noise detection algorithm for digital contours. Image Proc. Line **4**, 98–115 (2014)
12. Nguyen, T.P., Debled-Rennesson, I.: A discrete geometry approach for dominant point detection. Pattern Recogn. **44**, 32–44 (2011)
13. Ngo, P., Nasser, H., Debled-Rennesson, I.: Efficient dominant point detection based on discrete curve structure. In: Barneva, R.P., Bhattacharya, B.B., Brimkov, V.E. (eds.) IWCIA 2015. LNCS, vol. 9448, pp. 143–156. Springer, Cham (2015). doi:10.1007/978-3-319-26145-4_11
14. Feschet, F., Tougne, L.: Optimal time computation of the tangent of a discrete curve: application to the curvature. In: Bertrand, G., Couprie, M., Perroton, L. (eds.) DGCI 1999. LNCS, vol. 1568, pp. 31–40. Springer, Heidelberg (1999). doi:10.1007/3-540-49126-0_3
15. Rosin, P.L., Wesst, G.A.W.: Segmentation of edges into lines and arcs. Image Vis. Comput. **7**, 109–114 (1989)
16. Kanungo, T., Haralick, R.M., Stuezle, W., Baird, H.S., Madigan, D.: A statistical, nonparametric methodology for document degradation model validation. IEEE Trans. Pattern Anal. Mach. Intell. **22**, 1209–1223 (2000)
17. DGtal: Digital Geometry tools and algorithms library. http://libdgtal.org
18. Imagene, generic digital image library. http://gforge.liris.cnrs.frs/projects/imagene

Mathematical Morphology on Irregularly Sampled Signals

Teo Asplund[1(✉)], Cris L. Luengo Hendriks[2], Matthew J. Thurley[3],
and Robin Strand[1]

[1] Centre for Image Analysis, Uppsala University, Uppsala, Sweden
teo.asplund@it.uu.se
[2] Flagship Biosciences Inc., Westminster, CO, USA
[3] Luleå University of Technology, Luleå, Sweden

Abstract. This paper introduces a new operator that can be used to approximate continuous-domain mathematical morphology on irregularly sampled surfaces. We define a new way of approximating the continuous domain dilation by duplicating and shifting samples according to a flat continuous structuring element. We show that the proposed algorithm can better approximate continuous dilation, and that dilations may be sampled irregularly to achieve a smaller sampling without greatly compromising the accuracy of the result.

1 Introduction

In this paper, we will look at two-dimensional continuous domain signals and morphological operators that operate upon regularly, or irregularly, sampled signals of this kind. Regular, discrete morphological operators [13] are limited in how well they approximate their continuous-domain counterparts. This is because the continuous-domain morphological operators may introduce unbounded frequencies when applied to a band-limited signal. Thus, representing the continuous transformed signal with samples restricted to a grid is problematic, since cusps in the continuous signal cannot be represented using traditional sampling theory [9,14]. Moreover, the operators depend on local extrema, but sample points will generally not fall on the maxima and minima of the continuous signal, even for sampling frequencies high enough to accurately represent the signal. Finally, the shape of the structuring element (SE) is constrained by the sampling grid in regular, discrete morphology, i.e., the SE must be discretized according to the sampling grid.

Therefore, we introduce a way of dilating sampled 2D signals that allow for a variable sampling density in the input and output. This paper extends previous work for one-dimensional signals to two-dimensional images [1].

1.1 Previous Work

Some attempts to approach the issue of better approximating the continuous domain morphological operators have previously been made. In a paper from

© Springer International Publishing AG 2017
C.-S. Chen et al. (Eds.): ACCV 2016 Workshops, Part II, LNCS 10117, pp. 506–520, 2017.
DOI: 10.1007/978-3-319-54427-4_37

1992, Brockett and Maragos [3] develop partial differential equations that can be evolved to compute continuous morphology in the discrete domain. Many authors have improved on this work [2, 4, 11], however, these methods represent the result of the operation on a regular grid, and therefore suffer from the problems discussed above. Moreover, these methods are very slow [17].

In his thesis [16], Thurley develops a morphology on irregularly sampled data using continuous structuring elements. These operators only shift sample points vertically (i.e. changing the depth of samples), thus the morphologically transformed signal is sampled at the same points as the original signal. This means that these operators have some problems similar to the ones of the regular morphological operators mentioned above, since the sampling rate is unchanged after the transformation.

In a paper by Luengo Hendriks et al. [7], the authors introduce operators that use interpolation to better approximate continuous-domain operators, even though they lose strict adherence to certain defining properties.

1.2   Irregular Sampling

To address the issues that limit the possibility of approximating the continuous operators using regular, discrete mathematical morphology (MM), we propose applying MM to irregularly sampled signals. In this case, non-smooth parts of the signal can get a higher sampling density, thereby alleviating the issue of introducing unbounded frequencies. Moreover, one might add samples to the input signal at extrema without having to increase the sampling density everywhere. Finally, the SE will not be restricted to a regular sampling grid, and one may therefore use SEs with boundaries of subpixel precision.

To this end, we propose a new operator that is used to approximate dilation, erosion, opening, and closing in the continuous domain. The main idea is to take an irregularly sampled signal as input, duplicate and shift samples, and then suppress samples based on a continuous structuring element. The remaining samples are output as an approximation of the result of applying the *continuous* domain morphological operator to the band-limited signal represented by the input.

1.3   Background

The basic operators of mathematical morphology (MM) are dilations, erosions, openings, and closings [13]. In the continuous case, these four operators take an input signal $f : \mathbb{R}^2 \to \bar{\mathbb{R}}$ and a structuring function $g : \mathbb{R}^2 \to \bar{\mathbb{R}}$, where $\bar{\mathbb{R}} = \mathbb{R} \cup \{-\infty, \infty\}$ and output a transformed signal based on these in the following manner:

1. Dilation: $\delta_g(f)(\mathbf{x}) = \bigvee_{\mathbf{y} \in \mathbb{R}^2} f(\mathbf{y}) + \mathbf{g}(\mathbf{x} - \mathbf{y})$
2. Erosion: $\varepsilon_g(f)(\mathbf{x}) = \bigwedge_{\mathbf{y} \in \mathbb{R}^2} f(\mathbf{y}) - \mathbf{g}(\mathbf{y} - \mathbf{x})$
3. Opening: $\gamma_g(f) = \delta_g(\varepsilon_g(f))(\mathbf{x})$
4. Closing: $\psi_g(f) = \varepsilon_g(\delta_g(f))(\mathbf{x})$

An operator is called *flat* if the structuring function g only takes the values 0 and $-\infty$. The definitions are essentially the same for the discrete, grayscale case (simply change the domain of the functions to the regular grid).

These operators are non-linear, therefore, applying any of them to a band-limited signal may yield a transformed signal that is *not* band-limited. This is a problem, since a band-limited function can no longer be accurately represented by regular samples after being transformed by one of the morphological operators. Moreover, the discrete operators require a predefined sampling grid on which the input signal is defined (the domain of the sampled function), however, there is a multitude of data that is *not* regularly sampled, for example, data obtained from range cameras (e.g. using structured light, time of flight, or stereo cameras), which is often used in computer vision applications. One might also sample data irregularly on purpose, in order to increase resolution [12].

Since range cameras often collect data in an irregular grid, the data can contain gaps at surface discontinuities. These data are typically resampled onto a regular grid. The MM tools defined in this paper will be directly applicable to this type of data, yielding more efficient and precise analyses. As an example, 3D range sensors are currently applied for particle size measurement in mining and mineral processing based on morphological image processing [10].

2 Morphological Operators on Irregularly Sampled Data

To introduce the proposed operators corresponding to the four regular MM operators, we focus initially on dilations. Conceptually, the proposed operator takes the input samples, puts a structuring element at each sample (with the origin of the SE coinciding with the sample), then, looking at the collections of structuring elements from above, the SEs are clipped, such that the parts of structuring elements that are obscured from above are removed. The sample at the origin of each SE is then duplicated a number of times and each copy is shifted out to the edge of the clipped SE. The samples (including the duplicates) are the output of the dilation. This process is analogous to sliding the structuring element along the surface of the signal, tracing out the volume shaded by the SE as it moves, and finally taking the maximum of this shaded area (i.e., the top of the umbra [6]).

2.1 Naive Algorithm

In the following, we will describe how an approximation of the continuous dilation can be computed as outlined above, using a square structuring element of arbitrary size. A similar process could be used to create, for example, a disk shaped SE.

For a set of n input samples $X = \{(x_i, y_i, z_i) \mid i \in \{1, 2, 3, \ldots, n\}\}$, where z_i describes the height of a sample at position (x_i, y_i), and $x_i, y_i, z_i \in \mathbb{R}$, and a flat structuring element B_r, which is a square with sides of length $2r$ and origin in the middle. Let X be sorted in descending order on z_i. The dilation is then performed by selecting each sample $(x', y', z') \in X$ in order and dropping a copy of B_r on top of it. If any previously laid structuring elements are hit by the falling B_r, the overlapping portion is cut out from B_r. In practice, this is implemented by representing B_r as a collection of nodes along the edges of the square and removing those nodes that are part of the overlapped segment (for a disk shaped SE, for example, one could instead put these nodes along the edge of the disk at an even spacing). The nodes that are left are, essentially, the duplicated and shifted samples mentioned earlier and these become part of the output. One might instead use polygons and some polygon clipping algorithm [15, 18], however the described approach is simple and, as we shall see, works well. Importantly, the samples will not generally butt up against the edges of the neighboring SEs, which avoids creating a "staircase" effect where the output is just a collection of plateaus, instead there is room between neighboring plateaus which enables connecting them smoothly via interpolation.

2.2   Implementation

The dilation is implemented using C++ with an interface to MATLAB. The samples are processed one by one in the following manner:

1. Create 4 copies of the sample and shift them to the corners of the square;
2. For each side, create n copies, put these at equal distance from each other and the endpoints of the side;
3. For each of the $4n + 4$ copies, examine its neighborhood to see if there are any input samples at a higher position such that an SE at that sample would cover the copy. If so, suppress it;
4. For each input sample $x \in X$, examine its neighborhood to see if there are any input samples at a higher position such that an SE at that sample would cover x. If so, suppress x.
5. Output all unsuppressed samples, i.e., input samples and duplicates.

In step 3 and 5, we need to find nearby neighbors of a sample. To do this, one should store the input samples in some suitable data structure, e.g., a kd-tree. Then, in order to check if a node should be suppressed when using a square structuring element with a side of length $2r$, simply examine all neighbors within a distance of r using a range search, then check if a square centered at such a neighbor covers the sample, by checking whether the sample lies between the left and right side of the square and the upper and lower side, while having a smaller value than the neighboring sample. In the rest of this paper, we choose $n = 1$. Algorithm 1 shows pseudocode for the proposed approximate dilation.

Figure 1 shows the result of applying the naive algorithm described above to an image of size 256×256 pixels using a square structuring element B_r, for increasing values of r, and resampling the output onto the regular grid using linear interpolation. For small SEs, the result is very similar, however, for large values of r, the output of the naive algorithm has very fuzzy edges. This is because there are very few samples along the edges of the squares (relative to

Algorithm 1. This algorithm takes a set of input samples X and outputs a list of output samples Y approximating the dilation of the input by a square SE, as described in Section 2.1.

1: **function** NAIVE–DILATION(X, r)
2:     let Y be an empty array
3:     **for** each sample $(x, y, z) \in X$ **do**
4:         make eight shifted copies of (x, y, z):
5:             $c_1 = (x - r, y - r, z)$, $c_2 = (x + r, y - r, z)$,
6:             $c_3 = (x + r, y + r, z)$, $c_4 = (x - r, y + r, z)$,
7:             $s_1 = (x, y - r, z)$, $s_2 = (x + r, y, z)$,
8:             $s_3 = (x, y + r, z)$, $s_4 = (x - r, y, z)$
9:         let X' be an array containing the sample (x, y, z) and the shifted copies c_1,
10:         $c_2, c_3, c_4, s_1, s_2, s_3$, and s_4
11:         **for** each sample $s = (x', y', z') \in X'$ **do**
12:             find all input samples $\{(x_i'', y_i'', z_i'')\}$ within a distance such that a square
13:             with sides of length $2r$ centered at (x_i'', y_i'') could cover s.
14:             **if** $z_i'' > z'$ for any i **then**
15:                 remove s from X'
16:             **end if**
17:         **end for**
18:         insert all elements of X' into Y
19:     **end for**
20:     **return** Y
21: **end function**

their size), and linear interpolation is problematic in this case (in fact there are only 2159 samples in the output for the largest SE used, compared to the 65536 of the regular, discrete dilation). We shall see that this issue can be dealt with by increasing the density of samples in the output.

2.3   Increasing Sampling Density

To improve the result of the naive implementation, a simple approach would be to increase the sampling density by iteratively applying the naive algorithm using smaller SEs, until the desired size has been reached. Figure 1 shows the result of this approach. The fuzzy borders are gone, and the results look very similar to the regular, discrete dilation. This stems from the fact that the number of samples is an order of magnitude greater for the iteratively applied dilations with large SEs compared to the results for a single application (62519 samples in the output for the largest SE used, compared to 2159 for the same size SE using a single iteration).

2.4   Octagonal and Hexadecagonal Structuring Elements

By structuring element decomposition [13] we may create new SEs from the square SE we have been using so far. For example, by dilating an image with a

Fig. 1. Upper row: Regular discrete dilation. Middle row: Result of naive dilation algorithm after resampling onto the regular grid using linear interpolation. Lower row: Result of iterating the naive dilation algorithm followed by resampling onto the regular grid. From left to right, the size of the SE used was 3, 7, 11, 21, 41 (length of the sides) in the regular case and equivalent values for the naive, and iterated algorithms.

Fig. 2. Regular, discrete dilation (a) with a disk (every pixel at a distance less than the SE size). Hexadecagonal proposed dilation (b) after resampling onto the regular grid using linear interpolation. The size of the SE used was 7, and 17 in the regular case and equivalent values for the proposed algorithm (with 4 iterations for the smaller SE and 12 for the larger).

square SE, rotating the resulting samples by $\pi/4$, applying the dilation again, and rotating the samples back by $-\pi/4$, we get a dilation by an octagonal SE.

From an octagonal SE it is easy to create a hexadecagonal SE by dilating once by the octagonal SE, rotating the output by $\pi/8$, dilating again by the octagonal SE, and rotating back by $-\pi/8$. Figure 2 shows some example results of this. In this case, the number of samples for the larger SE is 143302. In similar ways one may construct a large number of SEs. It is also easy to modify the algorithm in Algorithm 1 such that the SE used is a rectangle. This allows for an even greater variety of shapes.

2.5   Erosions, Openings, and Closings

At this point, we may easily define erosions, openings, and closings by duality and composition [13], i.e., to erode a sampled signal, invert the z values, apply the dilation, and invert the z values of the output. To perform an opening, or

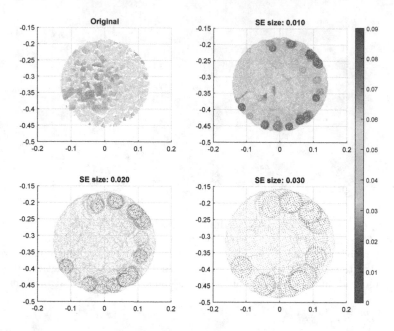

Fig. 3. Erosion of an irregularly sampled pile of rocks with a regular hexadecagonal SE of increasing size. Depending on the number of iterations used for the erosion, a denser or sparser sampling of the output can be obtained.

a closing, simply apply an erosion followed by a dilation or vice versa. Figures 3 and 4 show some examples of irregularly sampled input (a sampled surface of a pile of rocks), and the result of applying erosions and openings to these data.

2.6   Adaptive Density

As can be seen from the output for the iterated dilations, the number of output samples tends to increase significantly with the number of iterations. One may control the number of outputs by decreasing the number of iterations. Indeed, for large SEs, and few iterations, the number of output samples is a lot smaller than the number of input samples. For example, 65536 input samples becomes 2159 output samples in the initial naive algorithm described in Sects. 2.1 and 2.2, using a large SE. However, this results in a fuzzy output when the number of iterations becomes too small. This is the motivation behind adaptively increasing and decreasing the density of output samples, based on the input signal.

For a sample in a mostly flat neighborhood, only a few samples should be necessary, however, for samples in neighborhoods which vary quickly, there is a need for an increased sampling density. In order to facilitate this, two things are needed:

– Some measure of how flat the input is near any sample s, denoted by $K(s)$; and

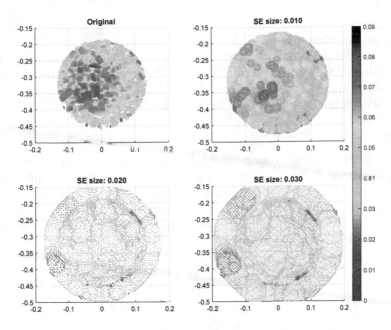

Fig. 4. Opening of an irregularly sampled pile of rocks with a regular octagon SE of increasing size. Depending on the number of iterations used for the erosion, a denser or sparser sampling of the output can be obtained. The behaviour on the border of the input samples can be changed by enforcing some border condition (e.g., by adding some samples at the lowest height level possible near the border).

– A modified algorithm that allows different numbers of duplications based on the "flatness" of a sample.

Regarding the first point, we initially select a normalized, smoothed gradient magnitude as a measure of flatness, for the sake of simplicity, i.e., for a set of n input samples $I = \{(x_i, y_i, z_i) \mid i \in \{1, 2, 3, \ldots, n\}\}$, we choose

$$K(s) = \frac{F(s) - \min_{i \in I} F(i)}{\max_{i \in I} F(i) - \min_{i \in I} F(i)}, \tag{1}$$

for any $s \in I$. Here F is defined as:

$$F(s) = (|\nabla I| * G_\sigma)[s], \tag{2}$$

where G is a Gaussian kernel with standard deviation σ. The purpose of smoothing with a Gaussian, is to increase the range of the influence of the gradient magnitude, as well as to smooth out noise. In the experiments in this paper, we fix $\sigma = 1.0$. Apart from the gradient, there are other measures that are of interest (e.g. higher order measures), however, we will only be using the K of Eq. (1) in this paper.

As for the second point, the algorithm outlined in Algorithm 1 may be modified as follows:

1. Calculate $K(s)$ at each input sample, s;
2. For each sample create the eight shifted copies;
3. Based on the value of K at the sample under consideration choose some k, apply the unmodified NAIVE-DILATION to the eight shifted copies iteratively k times (taking the output as input to the next iteration);
4. For each sample in the result of this iterated dilation, check if it should be suppressed;
5. Output all unsuppressed samples.

Pseudocode for this algorithm is shown in Algorithm 2 (note that the usage of NAIVE- DILATION in the for-loop is not ideal from a computational performance perspective, instead one could use a function that generates, in one go, a sampling of the desired continuous SE with varying density based on k).

For all adaptive operations we choose, in this paper, k using the density function ρ in the following manner:

$$k = \rho(s) = \begin{cases} 1, & \text{if } K(s) \le 0.005 \\ 2, & \text{if } 0.005 < K(s) \le 0.55 \\ 4, & \text{if } 0.55 < K(s) \le 0.95 \\ 8, & \text{otherwise} \end{cases} \tag{3}$$

The thresholds were selected empirically by testing a small set of thresholds. The function should be chosen such that the number of output points are minimized without destroying the result. Eight is chosen as the maximal number of dilations since this is the size of the SE used, although nothing precludes using more iterations than the size of the SE (it may, in fact, sometimes be desirable).

In the adaptive case the number of output samples is generally smaller than the number of input samples. Representing the image with fewer samples after the dilation makes some sense, since the dilation creates many plateaus, which can be represented using only a few samples. On the other hand, the dilation will generally (in the continuous domain) introduce unbounded frequencies, which would require infinitely many samples to correctly represent in the framework of traditional sampling theory, using sinc interpolation! However, increasing the sampling density near points whose neighborhood includes non-smooth parts of the signal, which is the purpose of ρ, should lead to a better approximation of the continuous dilation. Figure 6 shows naive, adaptive density, and dense-everywhere dilations and erosions respectively, on two different images (see Fig. 5) for a diamond shaped SE (i.e. a square SE rotated by $\pi/4$) of size 8. By dense-everywhere dilation, we mean that $\rho(s) \equiv c$ for some constant $c \in \mathbb{Z}$, such that $c > 1$. The mean square error between the interpolated result in the naive case and the dense-everywhere case ($c = 8$) is 43.4 for Fig. 5a, and 7.2 for Fig. 5b. Comparing the dense-everywhere case and the adaptive case, the error decreases to 14.8, and 3.3 respectively for Fig. 5a and b.

Irregular Data: Equation (2) can easily be computed for regularly sampled data, let us now consider how to extend this to *irregularly* sampled data. We can

Algorithm 2. This algorithm takes a set of input samples X and outputs a list of output samples Y based on the gradient magnitude at each sample of X, as described in Section 2.6, which approximates the dilation of the input samples by a square SE. The density near each input point is defined by the function ρ, which takes a sample as input, and outputs a number of iterations.

```
 1: function ADAPTIVE- DENSITY–DILATION(X, r, ρ)
 2:     let Y be an empty array
 3:     for each sample a ⊂ X do
 4:         let k = ρ(a)
 5:         let SE = {a}
 6:         for each i in {1, 2, . . . , k} do
 7:             let SE = NAIVE–DILATION(SE, r/k)
 8:         end for
 9:         for each sample s = (x′, y′, z′) ∈ SE do
10:             find all input samples {(x″ᵢ, y″ᵢ, z″ᵢ)} within a distance such that a square
11:             with sides of length 2r centered at (x″ᵢ, y″ᵢ) could cover s.
12:             if z″ᵢ > z′ for any i then
13:                 remove s from SE
14:             end if
15:         end for
16:         insert all elements of SE into Y
17:     end for
18:     return Y
19: end function
```

Fig. 5. Left: original 8-bit image used in Fig. 6. Right: regular, discrete dilation (a) and erosion (b) using diamond-shaped SE of size 8.

approximate the gradient magnitude at each sample by first creating a graph out of the input samples (for example using Delaunay triangulation). I.e., the samples V are inserted into a graph $G = (V, E)$, such that nearby samples are connected by an edge in E. Then, at each sample $s \in V$, compute the vector $\nabla_s = (\nabla_s^1, \nabla_s^2, \ldots, \nabla_s^n)$, where n is the number of neighbors of s as follows:

$$\nabla_s^i = \frac{s_z - t_z^i}{\sqrt{(s_x - t_x^i)^2 + (s_y - t_y^i)^2}}, \tag{4}$$

where t^i is the i:th neighbor of s, and the position of a sample s is denoted by s_x and s_y, and its value by s_z. This is similar to what is suggested in the

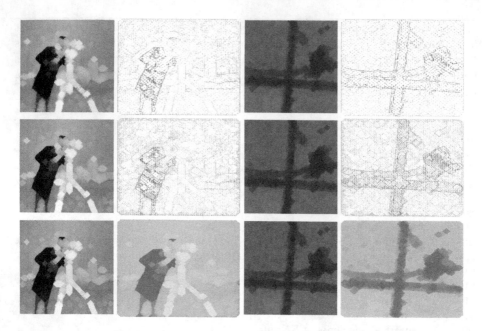

Fig. 6. Dilation and erosion of size 8 (from center to edge measured vertically) with diamond-shaped SE. Top: Naive dilation (left) and erosion (right), Mid: Adaptive (up to 8 iterations), Bottom: dense-everywhere dilation and erosion (8 iterations). Columns one and three show the result of linear interpolation of the points from the output of dilating Fig. 5a and eroding Fig. 5b, which is displayed in columns two and four. Number of points from top to bottom is 13383, 24149, and 147169 for column two and 10906, 22872, and 147169 for column four. Original contains 65536 points.

paper by Gilboa and Osher [5]. We then choose $K(s) = |\nabla_s|$ and apply Eq. (3). The range of influence of a gradient in the signal can be controlled by changing the connections in the graph (i.e. allowing connections between more distant samples will spread the influence out, while restricting the connections to only very nearby samples will lead to the value of K being more locally dependent).

3 Experiments and Results

3.1 Filtering Sparsely Sampled Signal

Filtering a sparsely sampled signal using discrete morphological operators yields a very poor approximation of the ideal, continuous-domain filtering results. The difficulty lies in the unbounded frequencies generated by the morphological operators, which are not representable using traditional sampling theory. It is necessary to introduce irregular sampling to better approximate such filtering results. Additionally, since the size and shape of the structuring element is restricted by the sampling grid, the size of the SE cannot be chosen very precisely in regular, discrete morphology.

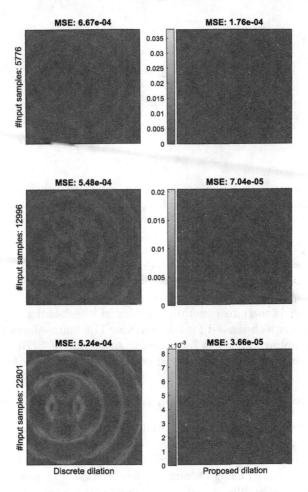

Fig. 7. Squared error when comparing the regular discrete dilation of the densely sampled function $\sin(\sqrt{x^2 + y^2}) \cdot \sin(\sqrt{(x + 2\pi)^2 + y^2})$ using a diamond shaped SE of size $\pi/3$, against dilating the same function, but sparsely sampled. Note that the colormap changes each row. Left column is regular discrete dilation, right is Naive–Dilation. The number of input samples increases each row.

Both these limitations are circumvented by allowing for irregular samplings (pre and post transformation). Since the output of the proposed operators can vary in sampling density over the output signal, the approximation of the continuous signal should improve. Moreover, the proposed operators enable freely choosing the size of the structuring element, since samples in the output can be placed at positions between input samples.

In Fig. 7 a simple function is densely sampled and dilated using regular discrete dilation with a diamond shaped SE of a size corresponding to $\pi/3$. The function is then *sparsely* sampled (less than 1/100 of the number of samples

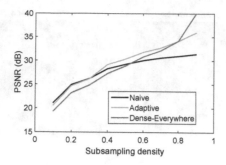

Fig. 8. Peak signal-to-noise ratio for different subsampling densities of the 8-bit image in Fig. 5a using the proposed naive, adaptive, and dense-everywhere dilation (8 iterations) compared with regular dilation of all the samples in the original image.

in the densely sampled case) and dilated using regular dilation as well as the Naive–Dilation algorithm. The figure shows the squared error between the samples in the sparsely sampled dilations and the densely sampled dilation. Linear interpolation is used to generate samples at the locations of the densely sampled dilation, which is used for comparison. The figure shows the change in error as the sampling density is increased from a total of 5776 samples at the top, to 12996 samples, and finally 22801 samples at the bottom.

3.2 Adaptive Density Dilation on Irregularly Sampled Signal

In this section we dilate Fig. 5a using a diamond SE of size 8. However, the input signal is subsampled, discarding samples randomly before dilation. The adaptive dilation described in Sect. 2.6 for irregularly sampled data is then used, where the graph is constructed using Delaunay triangulation. Figure 8 shows the peak signal-to-noise ratio for different subsampling densities when compared with the regular dilation of the original samples. The adaptive dilation gives the highest PSNR for most subsampling densities. Only at the extremes (i.e. very few samples, or almost all samples left) do naive or dense-everywhere dilation give a slightly better result. The dense-everywhere dilation creates many samples in sparsely sampled areas of the signal, based on only nearby samples (those that fall under the SE). In contrast to this, the adaptive implementation does not create as dense a sampling. This means that when we interpolate samples onto the regular grid, the adaptive and naive implementations has left sampling positions empty, which can then be interpolated, thereby using information from farther away than the size of the SE. This means that for sparsely sampled signals, one might get a worse result when using the dense-everywhere implementation, since this can create SE-shaped artifacts (essentially plateaus) where there should not be any, because of a lack of information.

Important to remember is also the fact that the number of nodes varies greatly for the different implementations. For example, at subsampling density

0.9 (i.e. 58982 input samples out of the original 65536), the naive dilation yielded 12821 samples, the adaptive 25044, and the dense-everywhere 147074.

4   Conclusions

In this paper we have presented a new way of constructing the four basic operators of mathematical morphology: dilation, erosion, opening, and closing for flat SEs. These operators, in contrast to the regular, discrete counterparts, work on irregularly sampled data and can produce output that varies in sampling density based, for example, on signal smoothness.

Morphological operators can be thought of as filters that remove superfluous detail in a controlled manner. Thus, it is reasonable to expect that the sampling density could locally be reduced for the transformed signal. Indeed, the proposed operators may reduce the number of sample points by reducing the sampling density where the operators produce plateaus, without greatly impacting the accuracy of the approximation of the continuous domain operators.

On the other hand, the continuous domain morphological operators generally produce unbounded frequencies, even if the input signal was band-limited. Therefore, regular, discrete MM operators will approximate the continuous operators poorly, since the sampling density determines the maximal frequency that can be represented, and the sampling density is determined by the fixed sampling grid. However, the variable sampling density that the proposed operators enable can mitigate such problems by increasing the density at points of the function whose neighborhood behaves non-smoothly, thereby allowing the representation of higher frequencies that are produced by the operators.

The irregular sampling, additionally, frees the structuring elements from the constraints of the regular sampling grid. In other words, one may construct SEs of any size, regardless of the sampling density. This should enable more precise measurements.

The proposed morphological operators were implemented in C++ with an interface to MATLAB. These operators are slower than their regular counterparts by a large margin, however, this is to be expected, since the lack of a regular grid makes it computationally expensive to find neighbors. Still, if the input samples are stored in a suitable data structure, reasonable performance can be achieved. Using a 2D kd-tree (we make use of nanoflann [8]), for example, dilating an input signal with more than a million samples with a diamond shaped SE of size 8 takes approximately 55 s using the NAIVE–DILATION algorithm on an Intel Xeon 3.40 GHz processor. Note that the number of samples in the output can decrease greatly, therefore, composing n operators will not, in general, lead to an n-fold increase in computation time.

Finally, we would like to note that it is possible to implement non-flat morphology using a scheme similar to the one presented, however, there is not enough space to go into detail in this paper.

Acknowledgement. Teo Asplund was funded through grant 2014-5983 from the Swedish Research Council.

References

1. Asplund, T., Luengo Hendriks, C.L., Thurley, M.J., Strand, R.: A new approach to mathematical morphology on one dimensional sampled signals. In: Proceedings of the 23rd ICPR 2016 (2016, to be published)
2. Breuß, M., Burgeth, B., Weickert, J.: Anisotropic continuous-scale morphology. In: Martí, J., Benedí, J.M., Mendonça, A.M., Serrat, J. (eds.) IbPRIA 2007. LNCS, vol. 4478, pp. 515–522. Springer, Heidelberg (2007). doi:10.1007/978-3-540-72849-8_65
3. Brockett, R.W., Maragos, P.: Evolution equations for continuous-scale morphology. In: 1992 IEEE International Conference on Acoustics, Speech, and Signal Processing, ICASSP 1992, vol. 3, pp. 125–128. IEEE (1992)
4. Brockett, R.W., Maragos, P.: Evolution equations for continuous-scale morphological filtering. IEEE Trans. Signal Process. **42**, 3377–3386 (1994)
5. Gilboa, G., Osher, S.: Nonlocal operators with applications to image processing. Multiscale Model. Simul. **7**, 1005–1028 (2008)
6. Haralick, R.M., Sternberg, S.R., Zhuang, X.: Image analysis using mathematical morphology. IEEE Trans. Pattern Anal. Mach. Intell. **9**, 532–550 (1987)
7. Luengo Hendriks, C.L., van Kempen, G.M.P., van Vliet, L.J.: Improving the accuracy of isotropic granulometries. Pattern Recogn. Lett. **28**, 865–872 (2007)
8. Nanoflann: C++ header-only fork of the FLANN library for approximate nearest neighbors. https://github.com/jlblancoc/nanoflann. Accessed 21 June 2016
9. Nyquist, H.: Certain topics in telegraph transmission theory. Trans. AIEE, 617–644 (1928). Reprinted in: Proc. IEEE **90**(2), 280–305 (2002)
10. Onederra, I., Thurley, M., Catalan, A.: Measuring blast fragmentation at Esperanza mine using high resolution 3D laser scanning. Inst. Mater. Miner. Min. Trans. Sect. A: Mining Technol. **124**, 34–46 (2015)
11. Sapiro, G., Kimmel, R., Shaked, D., Kimia, B.B., Bruckstein, A.M.: Implementing continuous-scale morphology via curve evolution. Pattern Recogn. **26**, 1363–1372 (1993)
12. Schöberl, M., Seller, J., Foessel, S., Kaup, A.: Increasing imaging resolution by covering your sensor. In: 2011 18th IEEE International Conference on Image Processing (ICIP), pp. 1897–1900. IEEE (2011)
13. Serra, J.: Image Analysis and Mathematical Morphology. Academic Press, Inc., Cambridge (1983)
14. Shannon, C.E.: Communication in the presence of noise. Proc. IRE **37**, 10–21 (1949). Reprinted in: Proc. IEEE **86**(2), 447–457 (1998)
15. Sutherland, I.E., Hodgman, G.W.: Reentrant polygon clipping. Commun. ACM **17**, 32–42 (1974)
16. Thurley, M.J.: Three dimensional data analysis for the separation and sizing of rock piles in mining. Ph.D. thesis, Monash University (2002)
17. Weickert, J.: Anisotropic Diffusion in Image Processing. Teubner, Stuttgart (1998)
18. Weiler, K., Atherton, P.: Hidden surface removal using polygon area sorting. In: ACM SIGGRAPH Computer Graphics, vol. 11, pp. 214–222. ACM (1977)

Adaptive Moving Shadows Detection Using Local Neighboring Information

Bingshu Wang[1], Yule Yuan[1], Yong Zhao[1($\boxtimes$)], and Wenbin Zou[2]

[1] School of Electronic and Computer Engineering,
Shenzhen Graduate School of Peking University, Shenzhen, China
wangbingshu@sz.pku.edu.cn, lemmas@foxmail.com, zhaoyong@pkusz.edu.cn
[2] College of Information Engineering, Shenzhen University, Shenzhen, China
wzou@szu.edu.cn

Abstract. In this paper, we propose an adaptive approach to the detection of moving shadows by employing local neighboring information. The process of the proposed approach is mainly operated by three steps: the first step is to detect the candidate shadows by RGB ratio; the second step is to extract partial accurate shadows in order to estimate accurate threshold parameters of shadow detectors; the final step is to utilize three detectors to detect real shadows from candidate shadows. The main contributions of this paper include two parts: an effective method of candidate shadows detection is presented; an adaptive mechanism of estimating threshold parameters is designed. Moreover, three detectors that consist of color, texture and gradient features are jointly utilized to detect shadows at pixel-level. Experimental results on a benchmark suit of indoor and outdoor video sequences demonstrated the proposed approach's effectiveness.

1 Introduction

The detection of moving objects has always been a stepping stone in the applications of intelligent video surveillance. A large number of excellent approaches were proposed in the area to obtain accurate and complete objects as much as possible. But there are still many challenges for object detection such as illumination changes, dynamic backgrounds, camouflage, shadows, etc. Among these challenges, shadow detection is a significant factor and has been a research field in itself as an inevitable task of computer vision applications. In real world scenes, a shadow is generated when the direct light is partially or totally occluded by objects. Generally, there are three independent components that create shadows: (a) the light source, (b) the occluding object and (c) the surface or other object onto which the shadow is cast [1]. All three can in principle move independently, which can create large number of situations of occurring shadows. In realistic scenes, two categories of shadows exist frequently: static shadows and moving shadows. Static shadows are casted by static objects such as buildings, trees, parking cars and other static objects. In general, static shadows have little influence on the object detection and are always regarded as background

© Springer International Publishing AG 2017
C.-S. Chen et al. (Eds.): ACCV 2016 Workshops, Part II, LNCS 10117, pp. 521–535, 2017.
DOI: 10.1007/978-3-319-54427-4_38

Input frame   Foreground   Chr   Geo   Phy   SR   LR   Ours

Fig. 1. An example of our method for tackling objects merging compared with five methods: Chr [3], Geo [4], Phy [5], SR [6], LR [7] and the proposed method. The blue areas represent the detected objects and the green areas represent shadows. (Color figure online)

because they change a bit or even keep unchanged in short period. By contrast, moving shadows are mainly generated by light source occlusion of opaque moving objects such as vehicles and pedestrians. It should be noted that this paper mainly focuses on moving shadows, so the detection of static shadows goes beyond the scope of this paper.

Essentially, moving shadows always share the similar motion property and obvious discrimination from the corresponding background as that of foreground objects. Shadows are always misclassified as parts of objects. Such erroneous classification easily results in distorted object shapes, the merging of objects and interferes with features description of connected objects. Therefore, moving shadows should not be detected as foreground but background. A typical example is given (see Fig. 1). It is a common transport scene. The foreground was detected by SuBSENSE [2] which is one of the top-class algorithms for change detection. Although this algorithm was able to suppress some soft shadows, it still cannot distinguish most shadows from foreground. The truck and the car merged because of shadows. Our motivation is to remove shadows from the foreground areas. The result detected by our method is given in the last image (see Fig. 1). It is obvious that the proposed method can distinguish shadows from the foreground mask and reduce merging while other five methods have difficulty in dealing with it.

Actually, there are some shadow properties that can be applied to distinguish objects, shadows and backgrounds. A significant property is that the shadow area has always lower brightness than that of corresponding background area. For instance, the RGB values of a pixel are lower than those values of corresponding background pixel. But it is different for Hue-Saturation-Value (HSV). In HSV color space, except value component, the components of hue and saturation change a little. This is called color constancy. Additionally, in despite of the brightness of cast shadow areas become lower, the shadow areas have similar texture features with those of corresponding background areas. This is called texture consistency.

Our method is based on the assumptions [1] for shadow detection: the area of shadows is darker than that of corresponding background, but keeps color constancy and texture consistency. So color information and texture features are taken into account in our approach. Firstly, candidate shadows are detected by RGB ratio method. Then, an improved local ternary pattern (ILTP) [8] is utilized to obtain partial accurate shadows that are used to evaluate the threshold parameters of all the detectors, which is aimed to establish an adaptive mechanism for crucial parameters adjustment. Moreover, after obtaining the accurate threshold parameters, three detectors: HSV detector, Gradient detector and ILTP detector, are utilized at pixel-level with neighboring information to detect shadows. Experimental results demonstrate that our approach outperforms other five methods.

2   Related Works

In recent years, many methods have been developed [9–17] and surveyed [1, 7, 18] for moving shadows detection. Prati et al. [1] organized shadow detection methods in a two-layer taxonomy early. The first layer classification depended on whether the decision process introduced uncertainty, which included two categories: deterministic approaches and statistical approaches. Deterministic approaches utilized an on/off decision process and can be further subdivided by model based knowledge or not, while statistical approaches utilized probabilistic functions and could be subdivided into parametric or non-parametric methods due to the importance of parameters selection. No matter which category the method belongs to, it could be identified by features utilized for shadow detection. There are three kinds of features: spectral, spatial and temporal. Among them, spectral and spatial features are the main features researched in the field. Spectral features include intensity, chromaticity and physical properties and spatial features are divided into two main categories: geometry and texture. The choices of features are placed more importance due to their greater impact on the detection results of moving shadows.

Therefore, according to the types of features, Sanin et al. [7] further divided the methods into four categories: chromaticity-based methods, physical-based methods, geometry-based methods and texture-based methods. Chromaticity-based (Chr) methods was based on the assumption that shadow regions are darker than the corresponding background regions but keep the chrominance. HSV color space-based was one of the most widely used methods [3]. It was easy to implement and run with a faster speed, but sensitive to noise and its parameters should be tuned in different scenes. Geometry-based (Geo) methods [4] did not depend on an accurate background frame and they worked straightforward on input frame with the prior knowledge such as illumination source, object shape, etc. But in most cases, the proper knowledge was not easy to be obtained, for instance, the orientations of multiple objects and their shadows limited these methods' applications. Physical (Phy) methods [5] tended to build non-linear attenuation models to predict the color change of shadow regions in various scenes by taking light sources and ambient illumination into account, but failed when the properties of objects are similar with those of background. Texture-based methods [6, 7] were mainly based on the assumption that shadows remained the most of their textures. These methods always extracted candidate shadows and then classified the candidate shadows into foreground or shadow through texture correlation between input frame and estimated background frame. The pixel-level operation was treated as small texture while region-level was treated as large texture. Small region (SR) texture-based method [6] were particular effective and robust for pixels when the texture information was rich, but might be computationally expensive. Although large region (LR) textures-based method [7] made use of color and texture information together, there were still significant difficulties in completely extracting the real shadows. Multiple features-based method [8]: chromaticity, gradient, and texture were utilized to detect moving shadows by a simple voting strategy. The fusion of multiple features can improve the adaptability to scenes, which is also carried out and verified in this paper.

For most of the mentioned methods, almost all the related threshold parameters needed to be tuned to achieve satisfied results when the applied scenarios changed. To tackle this issue, an adaptive threshold parameters should be obtained from the input frame. A practicable strategy is to utilize the real shadows in current frame to evaluate these parameters. In this paper, an adaptive and effective shadow detection method using texture and color information is proposed. We also compared our method with other five methods whose implementations were available in [19] which also includes the test dataset. The experiments on standard dataset showed that our method outperformed the five methods.

3   The Proposed Approach

Generally, the accurate shadow detection is with multiple steps [6,7]. The typical one is to extract candidate shadows and then classify them as foreground or shadow. The strategy treats foreground as the initial candidate shadows and narrows the range of candidate shadows continually. Inspired by this, our approach is designed through three steps (see Fig. 2). The first step is to detect candidate shadows. There are two functions for candidate shadows: one is used to extract partial accurate shadows and the other is used for partial accurate shadows detection. The second step is the threshold parameters estimation, which is aimed to produce relative accurate thresholds. The third step is the accurate shadow detection using the candidate shadows gained in first step and threshold parameters gained in second step. A vote and post-processing procedure is done to ensure the final integrity of objects.

3.1   The Proposed RGB Ratio for Candidate Shadows Detection

The typical assumption for is that shadow region has lower intensity than the corresponding background. Therefore, coarse shadows can be detected as candidate shadows that might contain some pixels of foreground. It is inevitable because some object regions are more likely to fit with the assumption. Here, we denote $I_i, F_i, B_i, (i \epsilon r, g, b)$ as input frame, foreground frame, and background frame respectively, and Ψ_r, Ψ_g and Ψ_b are three RGB ratios (see (1)).

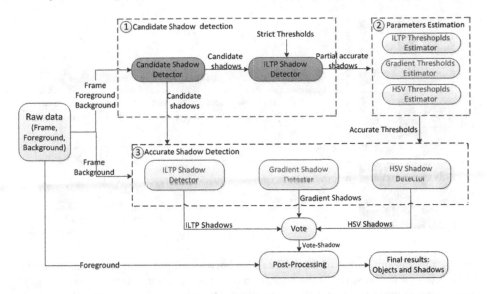

Fig. 2. The flowchart of the proposed approach for moving shadows detection. The two red dashed boxes are the contributions of this paper. The first red dashed box illustrates the process of candidate shadows and partial accurate shadows. The second red dashed box presents the parameters estimation by three estimators: ILTP, Gradient and HSV Thresholds Estimators. The third black dashed box is the process of accurate shadow detection by three detectors. (Color figure online)

$$\Psi_r = \frac{I_b/I_g}{B_b/B_g}, \Psi_g = \frac{I_b/I_r}{B_b/B_r}, \Psi_b = \frac{I_g/I_r}{B_g/B_r}. \tag{1}$$

$$Candidate\_Shadow = \begin{cases} 1, & if \left| \sum_{i\epsilon\{r,g,b\}} \Psi_i - \mu \right| < \lambda \ \ i \in \{b,g,r\} \\ 0, & otherwise \end{cases}. \tag{2}$$

We have observed that each RGB ratio is close to one for each pixel in shadow regions but not necessarily in object regions. The difference between the sum of RGB ratios $\sum_{i\epsilon\{r,g,b\}} \Psi_i$ of shadows and objects is obvious (see Fig. 3), and the blue, green and red areas represent Ψ_b, Ψ_g and Ψ_r respectively. The parameter μ in (2) is a reference value that revolves three around and λ is a small value (less than 0.2). By using RGB ratios, the candidate shadows could cover most of real shadows and possibly partial objects' pixels (see Table 1).

The shadow coverage rate (SCR) represents the proportion of detected shadow regions compared to the real shadow regions in ground truth, while the object coverage rate (OCR) represents the proportion of detected object regions compared to the real object regions in ground truth. Actually, the shadow coverage rate is equivalent to shadow detection rate (ξ) and the object coverage rate is equivalent to shadow discrimination rate (η), which will be presented in subsequent Sect. 4. Though the object coverage rates are relatively high, the shadow coverage rates for seven scenes are over 96%, with only "campus" achieving

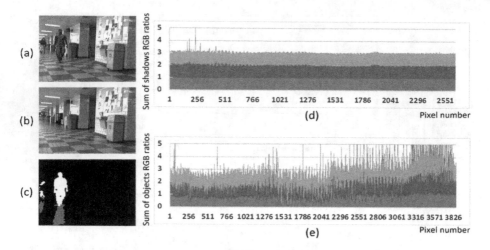

Fig. 3. The discrimination of the RGB ratios between objects and shadows. (a) is the input frame of Hallway, (b) is the background, (c) is the ground truth where the gray stands for moving shadows and the white stands for objects, (d) and (e) are the statistic sum of RGB ratios of shadows and objects, respectively. (Color figure online)

Table 1. The coverage rates of candidate shadows and misclassified objects.

	SCR	OCR		SCR	OCR
Campus	0.904	0.652	Highway1	0.962	0.682
Highway3	0.987	0.736	Hallway	0.978	0.339
Room	0.991	0.700	Lab	0.962	0.578
Caviar	0.997	0.770			

90.4%. This step is a coarse detection process for the sake of obtaining maximum shadow pixels and minimum object pixels.

Afterwards, ILTP detector is utilized to obtain partial accurate shadows under strict threshold parameters. The reason why we choose ILTP detector is that it can achieve relative high shadow coverage rate and relative low object coverage rate (see Table 2) in most sequences compared with other two detectors. A relatively fair and practicable way to evaluate the effectiveness and usefulness for partial accurate shadows is a high shadow coverage rate with a low object coverage rate. The ratio between shadow coverage rate and object coverage rate is a good choice. It can be seen from Fig. 4 that ILTP detector owns highest ratio values among three detectors. Despite the shadow coverage rate of "Lab scene" for ILTP detector is lower than that of HSV detector, the object coverage rate of ILTP detector (0.079) is much lower than that of HSV detector (0.322). Therefore, it can be concluded that ILTP shadow detector is more effective than other detectors and the partial accurate shadows are relatively reliable.

Table 2. The coverage rates of partial accurate shadows and misclassified objects detected by three detectors

	HSV detector		Gradient detector		ILTP detector	
	SCR	OCR	SCR	OCR	SCR	OCR
Campus	0.406	0.289	0.577	0.197	0.727	0.199
Highway1	0.811	0.427	0.386	0.144	0.466	0.111
Highway3	0.325	0.227	0.274	0.104	0.341	0.081
Hallway	0.865	0.124	0.811	0.035	0.937	0.039
Room	0.919	0.371	0.806	0.125	0.899	0.132
Lab	0.956	0.322	0.697	0.136	0.759	0.079
Caviar	0.882	0.354	0.733	0.256	0.901	0.243

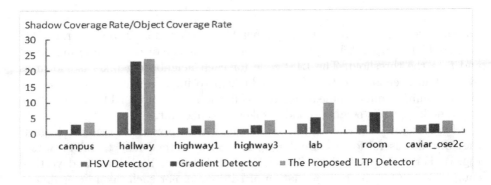

Fig. 4. The ratio between shadow coverage rate and object coverage rate by three detectors. The higher the ratio is, the more reliable the detector is.

3.2   The Combination of Three Shadow Detectors

The survey [7] had indicated that all feature-based shadow detection algorithms made different contributions and each had individual strength and weakness. The LR textures-based method not only utilized gradient texture feature but also chromaticity feature, and outperformed other four methods that used only single feature: chromaticity-based, physical-based, geometry-based and SR texture-based. The strategy of shadow pre-detection and accurate detection used by LR inspired us to take similar strategy with multiple processing steps (see Fig. 2). More importantly, this paper combines several shadow detectors for shadow detection and it has been proved to be effective. Three shadow detectors of ILTP, HSV, gradient are given in detail as follows. It should be noted that each pixel's neighboring pixels are considered to reduce the interference caused by noise and illumination changing.

The Proposed ILTP Shadow Detector. Local binary pattern (LBP) is one of excellent textures to describe local gray-level structure, but it is very sensitive to

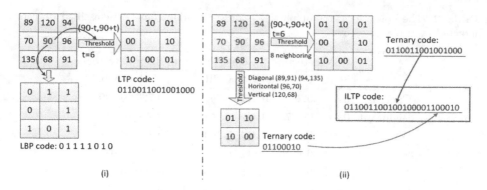

Fig. 5. The example code compárison of the proposed ILTP with LBP, LTP. (i) illustrates the codes of LBP and LTP. (ii) shows the code of ILTP.

random and quantization noise in the neighborhood region. Tan and Triggs [20] extended LBP to LTP. For pixel x, we denote i_c as one of its neighboring pixels and t as the threshold. The LTP code for each neighboring pixel is shown as (3). It had been proven that LTP could help to improve the accuracy rate of face recognition under difficult lighting conditions. As far as LTP and LBP are concerned, a pixel's neighborhood is taken into account.

However, the relations within the neighborhood pixels are ignored. In fact, they are also useful for texture description. So we extend LTP to ILTP [8] (see Fig. 5). For each pixel, the ILTP code adds diagonal, horizontal and vertical comparison results on the foundation of LTP code. For each pixel belonging to candidate shadow, both its code-values in input frame and background frame are computed and they are used to compare the similarity between input frame and background frame (see (4)). For pixel x, we denote $I_i(x)$ and $B_i(x)$ as the pixel values of ith neighboring pixel in input frame and background frame, respectively, and $I_i^c(x)$ and $B_i^c(x)$ represent the ILTP code-values, respectively. If $I_i^c(x)$ is equal to $B_i^c(x)$, the similarity $s_x(i)$ is set to one otherwise zero. The number n represents the number of neighboring pixels for given pixel x and the selected neighboring size is 3×3. The similarity threshold is defined as δ. The result of ILTP detector is as shown in (5).

$$LTP(x, i_c, t) = \begin{cases} 10, & i_c \geq x + t \\ 01, & |x - i_c| < t \\ 00, & i_c \leq x - t \end{cases} . \tag{3}$$

$$s_x(i) = \begin{cases} 1, & if\ I_i^c(x) = B_i^c(x) \\ 0, & otherwise \end{cases} . \tag{4}$$

$$ILTPShadow = \begin{cases} 1, & if\ \sum_{i=1}^{n} s_x(i)/n > \delta \\ 0, & otherwise \end{cases} . \tag{5}$$

HSV Shadow Detector. Since the HSV color space corresponds closely to the human perception of color, HSV color space-based [3] method has been given more attention. The chromaticity and brightness of HSV color space has revealed more accuracy in distinguishing shadows. This is because (1) the brightness in shadow regions is lower than that of background, and (2) the hue of a shadow cast on a background does not change significantly, which is not suitable for objects of foreground.

$$HSVShadow = \begin{cases} 1, & if \; \frac{1}{n}\sum_{i=1}^{n} \left| I_i^h(x) - B_i^h(x) \right| < \tau_h \wedge \\ & \frac{1}{n}\sum_{i=1}^{n} \left(I_i^s(x) - B_i^s(x) \right) < \tau_s \\ 0, & otherwise \end{cases} \tag{6}$$

Due to the detection of candidate shadows by the assumption that a shadow with lower brightness in comparison to the background pixels, there is no need to take value component of HSV into account for shadow detection again. So we mainly make use of hue and saturation components. For each pixel, we denote $I_i^h(x)$ and $B_i^h(x)$ as the hue values of the ith neighboring pixel in input frame and background frame, respectively. Similarly, we denote $I_i^s(x)$ and $B_i^s(x)$ as the saturation values. In the HSV shadow detector, τ_h and τ_s are two deciding thresholds (see (6)).

Gradient Shadow Detector. Since shadow tends to preserve the underlying textures, high texture correlation between shadow and background can be computed at region-level. Gradient information [7] consists of magnitude and direction. The shadows with rich textures can be detected in easier ways especially for indoor scenes whose textures are rich and simple. Gradient detector is operated in RGB color space. We denote ∇_i and θ_i as gradient magnitude and direction, respectively (see (7)). For a candidate shadow pixel, the $I\left(\nabla_i^j\right)$ represents the jth channel gradient magnitude of its ith neighboring pixel in the input frame, and $B\left(\nabla_i^j\right)$ is for the corresponding background pixel. The $I\left(\theta_i^j\right)$ represents the jth channel gradient direction of its ith neighboring pixel in the input frame, and $B\left(\theta_i^j\right)$ is for the corresponding background pixel. The deciding thresholds are ϕ_m and ϕ_d. The classification is as shown in (8).

$$\nabla_i = \sqrt{\nabla_x^2 + \nabla_y^2}, \; \theta_i = \arctan\left(\frac{\nabla_y}{\nabla_x}\right). \tag{7}$$

$$GradientShadow = \begin{cases} 1, & if \; \dfrac{\sum_{i=1}^{n}\sum_{j\in\{b,g,r\}} \left| I(\nabla_i^j) - B(\nabla_i^j) \right|}{n} < \phi_m \wedge \\ & \dfrac{\sum_{i=1}^{n}\sum_{j\in\{b,g,r\}} \left| I(\theta_i^j) - B(\theta_i^j) \right|}{n} < \phi_d \\ 0, & otherwise \end{cases} \tag{8}$$

Three detected results could be gained by these three detectors. Then the combination process would be operated. For each pixel, it will be classified as shadow pixel if there are more than 2 (including 2) detected shadow results. Subsequently the post-processing, which includes open/close morphology processing and filling for small holes, will be carried out to segment shadow properly. The experimental results that demonstrated the effectiveness of combination would be given in Sect. 4.

3.3   Adaptive Parameters Estimation

The process of parameters estimation is a crucial step for the framework of shadow detection. For the three detectors above, multiple thresholds are utilized to classify the candidate shadow pixel: $\delta, \tau_{h/s}, \phi_m, \phi_d$. They should be relatively reliable values for different scenes or changings. One effective way to tackle this problem is to make use of partial shadows for the evaluation of global shadows statistically. Each threshold parameter is set by an empirical value initially. The relative threshold parameters are evaluated by partial accurate real shadows generated by ILTP detector. The equations are given by (9) (10) (11). We denote $\delta^{ini}, \tau_{h/s}^{ini}, \phi_m^{ini}, \phi_d^{ini}$ as the default initialized values while $\delta^{cor}, \tau_{h/s}^{cor}, \phi_m^{cor}, \phi_d^{cor}$ as the correlation values computed by partial accurate shadows. Fixed values are given to $\alpha, \beta_1, \beta_2, \gamma, \kappa_1, \kappa_2$. These parameters are suitable to the soft shadows in indoor scenes and moderate hard shadows in outdoor scenes, which will be demonstrated by experimental results.

$$\delta = \delta^{ini} - \alpha * |\delta^{ini} - \delta^{cor}|. \tag{9}$$

$$\tau_{h/s} = \begin{cases} \tau_{h/s}^{cor} + \beta_1 * \tau_{h/s}^{ini}, & if \ \tau_{h/s}^{cor} < \tau_{h/s}^{ini} \\ \tau_{h/s}^{cor} + \beta_2 * \tau_{h/s}^{ini}, & otherwise \end{cases}. \tag{10}$$

$$\phi_m = \phi_m^{cor} - \gamma * \phi_m^{ini}, \phi_d = \begin{cases} \phi_d^{cor} + \kappa_1 * \phi_d^{ini}, & if \ \phi_d^{cor} < \phi_d^{ini} \\ \phi_d^{cor} + \kappa_2 * \phi_d^{ini}, & otherwise \end{cases}. \tag{11}$$

4   Experimental Results

In our experiments, seven sequences are used for shadow detection. They are typical scenes including three outdoor scenes (Campus, Highway1, and Highway3) and four indoor scenes (Room, Hallway, Lab, and Caviar_ose2c). The detailed description can be found in Table 3. All the ground truths are available online [19]. To test the performance of six methods, we used shadow detection rate η and shadow discrimination rate ξ [1] which was widely used to evaluate the performance of shadow detection methods. In order to analyze the comprehensive performance, F-measure was used as the cogent evaluate metric. These standard metrics are defined in (12).

$$\eta = \frac{TP_S}{TP_S + FN_S}, \ \xi = \frac{TP_F}{TP_F + FN_F}, F-measure = \frac{2\eta\xi}{\eta + \xi}. \tag{12}$$

where TP_S and FN_S are the true positives and false negative for shadows, TP_F and FN_F are the true positives and false negatives for objects. The proposed method compared with other five methods: chromaticity-based method

Table 3. Test sequences utilized for evaluation. They are described in terms of: scene type, sequence number, size, shadow size, shadow strength and object type.

	Campus	Highway1	Highway3	Hallway
Scene type	Outdoor	Outdoor	Outdoor	Indoor
Sequence number	53	8	7	13
Image size	352 × 288	320 × 240	320 × 240	320 × 240
Shadow size	Very large	Large	Small	Medium
Shadow strength	Weak	Strong	Very Strong	Weak
Object type	Vehicles/people	Vehicles	Vehicles	People
	Room	Lab	Caviar	
Scene type	Indoor	Indoor	Indoor	
Sequence number	22	14	164	
Image size	320 × 240	320 × 240	384 × 288	
Shadow size	Large	Medium	Medium	
Shadow strength	Weak	Medium	Medium	
Object type	People	People/other	People	

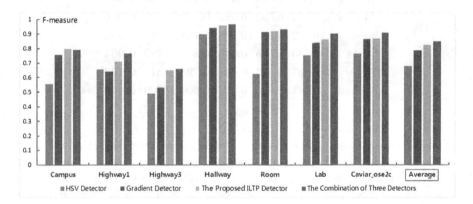

Fig. 6. Comparison of shadow detection by single detector and the combination of three detectors. The performance metric is F-measure between the shadow detection rates and shadow discrimination rates.

[3], geometry-based method [4], physical-based method [5], SR textures-based method [6] and LR textures-based method [7]. For fairness and the respect for original authors, each compared method made use of a suit of parameters [19]. The control parameters are $\alpha = 0.33, \beta_1 = 0.5, \beta_2 = 1, \gamma = 0.5, \kappa_1 = 0.5, \kappa_2 = 0.75$ and they were used to generate the adjusted threshold parameters. Figure 6 and Table 4 illustrate the quantitative results while Fig. 7 shows the qualitative comparison results.

Figure 6 gives the shadow detection results of three detectors singly and the combination of these detectors. It can be seen that ILTP detector owns the higher F-measure values than those of HSV detector and Gradient detector in all the seven sequences, but lower than those of the combination. On one hand, this demonstrates the effectiveness of ILTP texture. On the other hand, the combination of multiple features outperforms the single component detector.

Table 4 compares the F-measure comparison of seven methods. Our method shows better performance than the five methods. The average F-measure value of the proposed method is over 7% than that of LR textures-based method. The ranking third is chromaticity-based method with average F-measure value 71.1%. Moreover, it can be seen that the figures for indoor scenes are higher than those for outdoor scenes.

Table 4. The quantitative results of the proposed method with five methods.

F-m	Chr	Geo	Phy	SR	LR	Ours
Campus	0.532	0.589	0.553	0.674	0.668	**0.791**
Highway1	0.704	0.705	0.568	0.287	0.740	**0.766**
Highway3	0.534	0.546	0.488	0.109	0.536	**0.657**
Hallway	0.864	0.592	0.686	0.779	0.956	**0.966**
Room	0.782	0.613	0.716	0.751	0.886	**0.931**
Lab	0.843	0.565	0.545	0.847	0.780	**0.902**
Caviar	0.687	0.573	0.541	0.798	0.814	**0.906**
Average	0.711	0.600	0.563	0.606	0.769	**0.846**

The F-measure values of four indoor scenes are all over 90% (Hallway scene up to 96.6%), while the F-measure values of three outdoor scenes are no more than 80% (Highway3 scene only 65.7%). The difference between Hallway and Highway3 is over 30%. Indoor scenes are easy to be detected, for example, the Hallway's soft shadows with rich chromaticity and texture information. By contrast, outdoor scenes like Highway3, the hard shadows with less color and even no texture information constitute the majority of candidate shadow areas, which makes the features unobvious and those feature-based methods fail.

Figure 7 shows the qualitative results for all the six methods. In general, it can be noted that our method performs good performance with relatively accurate objects and outlines. The chromaticity-based method is easily influenced by noise and it fails when the object pixels have the similar colors with corresponding background as in the Highway1 scene. The geometry-based method shows its strength when its required conditions are met, for example, in Campus scene the orientation of object is different from that of shadow. The physical-based method makes use of color information like chromaticity-based method but is more likely to classify real shadows as objects. Specially, only minor shadows on Highway1 and Highway3 scenes are detected by SR. Actually, the results of all

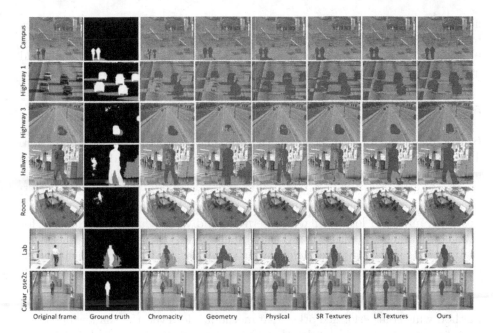

Fig. 7. The qualitative comparison results of our approach and other five methods: Chromaticity [3], Geometry [4], Physical [5], SR Textures [6], LR Textures [7]. The blue represents the object pixel and the green represents the shadow pixel. The first column and second column are original frame and corresponding ground truth [19], respectively. The remaining columns illustrate the detected results for all the sequences and all the methods. (Color figure online)

the methods on Highway1 and Highway3 are relatively worse in all the results of sequences. This is in accordance with the quantitative results in Table 4. It also indicates that hard shadow is a challenge because the chromaticity and texture information is not significant under high illumination condition. One possible way to solve this problem is to integrate multiple features. For LR textures-based method, the operation is at region-level. If candidate shadow region is not separated from object, the misclassification would happen (see Fig. 6, Campus Hallway results of LR textures-based method). Therefore, it is unsuited for the situation that the boundaries of objects and shadows are difficult to distinguish.

Actually, the running time is also of importance for shadow detection methods. The experiments were operated under the running environment of Visual Studio 2013, OpenCV2.4.9 and PC with Intel Core i7-3770 CPU. The average running time of above methods for each 352×288 resolution image is approximately: chromaticity (25.66 ms), Geometry (32.05 ms), Physical (40.11 ms), SR texture (432 ms), LR Texture (297 ms), and our proposed method (293 ms). It can be concluded that the time-consuming of our method and LR textures-based method is roughly equivalent but the average F-measure value of our method is 7.7% higher than that of LR textures-based method. In our method, what needs to be pointed out is that the ILTP detector constitutes nearly half of the total time-consuming because of the pixel-level computation referring to neighboring information. Overall, the results from both quantitative experiments and qualitative observations demonstrated our approach's effectiveness.

5 Conclusion

In this paper, we present an adaptive approach to moving shadows detection. The strategy of shadow detection is to detect candidate shadows initially as much as possible by RGB ratios and then for accurate shadows detection by the combination of three detectors: HSV detector, Gradient detector and ILTP detector. The main innovations of this paper include two parts: firstly, we proposed an improved local ternary pattern; secondly, an adaptive mechanism for threshold parameters estimation of three shadow detectors was given. Moreover, each pixel's local neighboring information is utilized to determine the classification of candidate shadow. Qualitative and quantitative results validated that our approach is effective in detecting moving shadows of indoor and outdoor scenes, with average F-measure value over 7.7% higher than that of LR textures-based method which outperformed other four methods. However, the shadow detection in outdoor scenes is still harder than indoor scenes. In this regard, more work to this should be invested in future.

References

1. Prati, A., Mikic, I., Trivedi, M.M., Cucchiara, R.: Detecting moving shadows: algorithms and evaluation. IEEE Trans. Pattern Anal. Mach. Intell. **25**(7), 918–923 (2003)
2. St-Charles, P.L., Bilodeau, G.A., Bergevin, R.: Subsense: a universal change detection method with local adaptive sensitivity. IEEE Trans. Image Process. **24**(1), 359–373 (2015)
3. Cucchiara, R., Grana, C., Piccardi, M., Prati, A.: Detecting moving objects, ghosts, and shadows in video streams. IEEE Trans. Pattern Anal. Mach. Intell. **25**(10), 1337–1342 (2003)
4. Hsieh, J.W., Hu, W.F., Chang, C.J., Chen, Y.S.: Shadow elimination for effective moving object detection by Gaussian shadow modeling. Image Vis. Comput. **21**(6), 505–516 (2003)
5. Huang, J.B., Chen, C.S.: Moving cast shadow detection using physics-based features. In: IEEE Conference on Computer Vision and Pattern Recognition, pp. 2310–2317 (2009)
6. Leone, A., Distante, C.: Shadow detection for moving objects based on texture analysis. Pattern Recogn. **40**(4), 1222–1233 (2007)
7. Sanin, A., Sanderson, C., Lovell, B.C.: Shadow detection: a survey and comparative evaluation of recent methods. Pattern Recogn. **45**(4), 1684–1695 (2012)
8. Wang, B., Zhu, W., Zhao, Y., Zhang, Y.: Moving cast shadow detection using joint color and texture features with neighboring information. In: Huang, F., Sugimoto, A. (eds.) PSIVT 2015. LNCS, vol. 9555, pp. 15–25. Springer, Heidelberg (2016). doi:10.1007/978-3-319-30285-0_2

9. Zhang, W., Fang, X.Z., Yang, X.K., Wu, Q.M.J.: Moving cast shadows detection using ratio edge. IEEE Trans. Multimed. **9**(6), 1202–1214 (2007)
10. Qin, R., Liao, S., Lei, Z., Li, S.Z.: Moving cast shadow removal based on local descriptors. In: 2010 International Conference on Pattern Recognition, pp. 1377–1380 (2010)
11. Russell, M., Zou, J.J., Fang, G.: Real-time vehicle shadow detection. Electron. Lett. **51**(16), 1253–1255 (2015)
12. Choi, J.M., Chang, H.J., Yoo, Y.J., Jin, Y.C.: Robust moving object detection against fast illumination change. Comput. Vis. Image Underst. **116**(2), 179–193 (2012)
13. Jiang, K., Li, A.H., Cui, Z.G., Wang, T., Su, Y.Z.: Adaptive shadow detection using global texture and sampling deduction. J. IET Comput. Vis. **7**(2), 115–122 (2013)
14. Wang, J., Wang, Y., Jiang, M., Yan, X., Song, M.: Moving cast shadow detection using online sub-scene shadow modeling and object inner-edges analysis. J. Vis. Commun. Image Represent. **25**(5), 978–993 (2014)
15. Dai, J., Han, D., Zhao, X.: Effective moving shadow detection using statistical discriminant model. Optik - Int. J. Light Electron Opt. **126**(24), 5398–5406 (2015)
16. Huerta, I., Holte, M.B., Moeslund, T.B., Gonzàlez, J.: Chromatic shadow detection and tracking for moving foreground segmentation. Image Vis. Comput. **41**(C), 42–53 (2015)
17. Kar, A., Deb, K.: Moving cast shadow detection and removal from video based on HSV color space. In: International Conference on Electrical Engineering and Information Communication Technology (2015)
18. Al-Najdawi, N., Bez, H.E., Singhai, J., Edirisinghe, E.A.: A survey of cast shadow detection algorithms. Pattern Recogn. Lett. **33**(6), 752–764 (2012)
19. The Test Sequences are from: https://sourceforge.net/projects/arma/files/
20. Tan, X., Triggs, B.: Enhanced local texture feature sets for face recognition under difficult lighting conditions. IEEE Trans. Image Process. **19**(6), 1635–1650 (2010)

Workshop on Mathematical and Computational Methods in Biomedical Imaging and Image Analysis

Cell Lineage Tree Reconstruction from Time Series of 3D Images of Zebrafish Embryogenesis

Robert Spir[1,2]([✉]), Karol Mikula[1,2], and Nadine Peyrieras[3]

[1] Department of Mathematics, Slovak University of Technology,
Radlinskeho 11, 810 05 Bratislava, Slovakia
spir@math.sk
[2] Algoritmy:SK s.r.o., Sulekova 6, 811 06 Bratislava, Slovakia
[3] Institut de Neurobiologie Alfred Fessard, CNRS UPR 3294,
Av. de la Terrasse, 91198 Gif-sur-Yvette, France

Abstract. The paper presents numerical algorithms, postprocessing and validation steps for an automated cell tracking and cell lineage tree reconstruction from large-scale 3D+time two-photon laser scanning microscopy images of early stages of zebrafish (Danio rerio) embryo development. The cell trajectories are extracted as centered paths inside segmented spatio-temporal tree structures representing cell movements and divisions. Such paths are found by using a suitably designed and computed constrained distance functions and by a backtracking in steepest descent direction of a potential field based on these distance functions combination. Since the calculations are performed on big data, parallelization is required to speed up the processing. By careful choice and tuning of algorithm parameters we can adapt the calculations to the microscope images of vertebrae species. Then we can compare the results with ground truth data obtained by manual checking of cell links by biologists and measure the accuracy of our algorithm. Using automatic validation process and visualisation tool that can display ground truth data and our result simultaneously, along with the original 3D data, we can easily verify the correctness of the tracking.

1 Introduction

The comprehensive image analysis for complex stages of embryogenesis is a difficult problem not yet solved satisfactory. Modern imaging technologies generate terabytes of image data, comprising up to tens of thousands of cells imaged for thousands of time points. However, existing manual or semi-automated approaches to reconstructing cell lineages do not scale to data sets of such complexity and size. Automated computational approaches have been developed to analyse such image data for small model organisms such as Caenorhabditis elegans [2] embryos and for early developmental stages of more complex organisms such as the early zebrafish blastula [11,20] and the Drosophila blastoderm [11,23]. However, a development of methods for accurate, automated cell lineaging in later stages of development is still a hot topic and open problem.

© Springer International Publishing AG 2017
C.-S. Chen et al. (Eds.): ACCV 2016 Workshops, Part II, LNCS 10117, pp. 539–554, 2017.
DOI: 10.1007/978-3-319-54427-4_39

In [1], methods based on sequential Bayesian approach with Gaussian mixture models were developed. First, a partition of the 3D image volume recorded at each time point into supervoxels is performed. A supervoxel is a connected set of voxels in space that all belong to a single nucleus, and each nucleus can be represented by multiple supervoxels. Second, an interconnection of supervoxels in space (segmentation) and time (tracking) is done to recover full cell lineages. The authors developed a sequential Bayesian approach with Gaussian mixture models (GMMs) to perform both tasks simultaneously using parametric contour evolution. The parametric model reduces the segmentation and tracking problem to finding ten parameters per nucleus: its 3D center, 3D covariance matrix (shape) and parent identity.

A recent work towards building the cell lineage tree for the complex stages of Zebrafish embryo development based on stochastic simulated annealing minimization of a heuristic energy functional has been presented in [9]. After the construction of the tree, the individual cells or cell populations are tracked producing lineage "forest" as an union of several disjoint trees representing cell lineage. First, the edges between detected cell centers at consecutive time steps are created using nearest-neighbour heuristic method. Next, the simulated annealing, variant of Metropolis algorithm, is used to progressively enforce a set of predefined constrains summarizing together a certain number of biological requirements, such as no cell should have more than two "daughters" and divisions should not occur too frequently.

The method presented in this paper is based on extraction of the cell trajectories as centered paths inside 4D spatio-temporal tree structures obtained by segmentation of 4D images. In addition to the approaches described by Mikula et al. in [18,19], in the presented approach the 4D segmentation is obtained by creating spatio-temporal tubes using cell nuclei diameters obtained from real nuclei segmentations [4,6,9,12,17,21] around the cell identifiers given as a result of suitable image filtering [4,8,9,14] followed by a cell detection algorithm [4,9,10]. Then a computation of constrained distance functions inside 4D segmentation is performed by solving numerically a spatially 4D eikonal equation. By a suitable combination of computed distance functions we build a potential field which is backtracked in steepest descent direction in order to get the cell trajectories. In contrast to [18,19], in this paper we significantly improve the tracking results by introducing a new parameter α in the construction of the potential field, weighting the distance functions influence. The cell lineage tree can be constructed by detecting merging trajectories when going backward in time indicating mitosis and thus a branching node of the cell lineage tree. In the paper we discuss the results of our improved method and perform its validation on real ground-truth data. This ground truth data contains 38797 manually checked, correct cell links. First we find the correspondence between trajectory points in our result and ground truth data and then we check for the exactly matching cell links. Using the presented method we are able to obtain 96.5% of correct cell links accuracy which is on the top of existing tracking methods.

The data we are dealing with are given by two-photon laser scanning microscopy and represent the first hours of zebrafish embryo development,

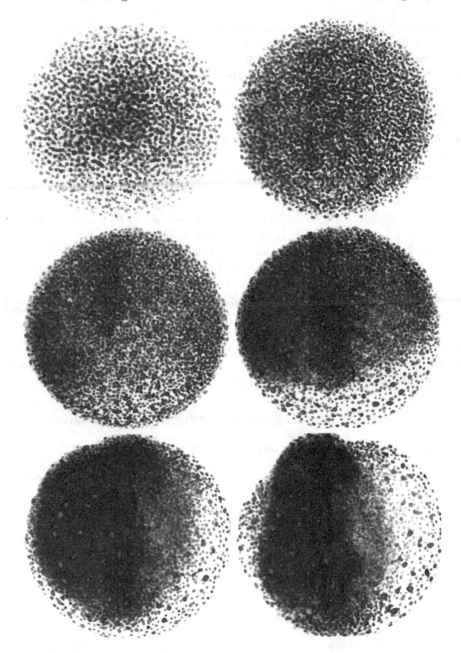

Fig. 1. Volume rendering of the cell nuclei data during the embryogenesis starting at 4 h after fertilization until 13 h after fertilization in time steps 1, 96, 192, 288, 384 and 480.

approximately from the 4th until 10th–20th h. The labeling of cell nuclei is obtained by expression of the fluorescence protein through its RNA injection performed at the one-cell stage. The 3D images are obtained by moving the focal plane from the top more deeply inside the embryo and their quality depends on the speed of scanning in one plane, we refer to a web page for various quality datasets (http://bioemergences.iscpif.fr/bioemergences/). The 3D image acquisition step ranges from 50 s to 5 min. A longer time step produces better image quality and such data is well suited for segmentation purposes e.g. for obtaining a shape of cells and other their characteristics during the embryogenesis [4,17]. On the other hand, such data is not suitable for tracking since the cells move too far between single 3D images and consequently mother-daughters cell correspondences can be lost. In Fig. 1 we plot an example of embryo development from the beginning to the end of the imaging. We see visualization of 3D cell nuclei which are tracked (Fig. 4) by our method. One can clearly observe how the zebrafish embryo grows from an unorganized set of cells to complex stages of development containing presumptive organs of the future zebrafish adult.

The paper is organized as follows. In the next section we present our approach to cell trajectories and lineage tree extraction, discuss numerical approaches used in the tracking method and parameters that can be used to tune and improve tracking results. Then we discuss numerical experiments devoted to processing of real 3D+time image sequences of the early zebrafish embryogenesis. We also present comparison of our results with ground truth data containing correct cell links in time verified manually.

2   The Cell Lineage Tree Reconstruction Algorithm

2.1   Filtering, Cell Nuclei Detection and Segmentation

These first three algorithmic steps are taken from [10,14,15] and for completeness are described below. First step in our approach is nonlinear diffusion filtering of the input data. The noise is intrinsically linked to the microscopy technique and its level increases with decreasing the time step $d\theta$ of the scanning. Here we use geodesic mean curvature flow (GMCF) [9,14] which is based on discretization of the following nonlinear diffusion equation

$$u_t - |\nabla u| \nabla \cdot \left(g\left(|\nabla G_\sigma * u|\right) \frac{\nabla u}{|\nabla u|} \right) = 0, \tag{1}$$

where $u(t, x)$, $t > 0$, represents the filtered image intensity function. We start from the initial condition $u(0, x) = u^0(x)$, where u^0 is an original 3D image and we consider few discrete time steps of the discretized model. We consider the zero Neumann boundary conditions on the boundary $\partial\Omega$ of the 3D image domain Ω. In this model, the mean curvature motion of the level sets of function u is determined by the edge indicator function

$$g(s) = \frac{1}{1 + Ks^2}, \qquad K \geq 0 \tag{2}$$

where K is the edge detection sensibility parameter, that is applied to the image gradient presmoothed by the Gaussian kernel G_σ with a small variance σ. The essential property of this function is that its negative gradient points towards the edges in the image and overall nonlinear diffusion process given by (1) causes accumulation of the level sets of u along the boundaries of objects in the image and therefore the filtering is also edge preserving. The filtered image u^F is obtained as the solution of (1) at a time $t = T_F$. The optimal choice of the model and discretization parameters was studied in [14] on the basis of the mean Hausdorff distance of the level sets of the filtered image and a gold standard. For discretization of (1) in time we use the semi-implicit approach that guarantees unconditional stability of the numerical scheme, for details see [14] and for spatial discretization we use the finite volume method.

The second step of our approach is the detection of cell nuclei centers, we call them also cell identifiers. The cell center detection method is based on a fact that objects visible in the image can be seen as humps of relatively higher image intensity. Any such hump is represented by certain image intensity level sets. The diameter of these level sets allow us to distinguish between significant objects, e.g. cell nuclei, and spurious inner structures which still remain after GMCF filtering. For cell nuclei, the diameter d is relatively large, $0 << c_1 \leq d \leq c_2$, while the diameter of the spurious inner structures is much smaller, $0 < d << c_1$. If the level sets are moving (advected) at a constant speed in the direction of the inner normal, the encompassed volume is decreasing and finally the hump disappears. Our model is based on the fact that the level sets with small diameter corresponding to spurious structures disappear quickly while level sets representing cell nuclei are observable in a much longer time scale. Since the motion of every level set is given by the normal velocity $V = \delta + \mu k$ where δ and μ are constants (model parameters) and k is the mean curvature, we formulate our level set center detection (LSCD) method in the form of the following nonlinear advection-diffusion equation [9, 10]

$$u_t + \delta \frac{\nabla u}{|\nabla u|} . \nabla u - \mu |\nabla u| \nabla . \left(\frac{\nabla u}{|\nabla u|} \right) = 0 \qquad (3)$$

which is applied to the initial condition given by u^F, the result of GMCF filtering. Again, we consider the zero Neumann boundary condition and the equation is solved in time interval $[0, T_C]$. Due to the shrinking and smoothing of all (real and spurious) structures in the evolutionary process represented by (3), we observe decrease of the number of local maxima M of the solution u as time proceeds. This decrease is fast in the beginning and much slower later. We stop this process when the slope of decrease is below a certain threshold, and then we use visual inspection in few 3D images of the whole sequence in order to choose an optimal evolutionary step of LSCD for finding cell identifiers. The time discretization of LSCD Eq. (3) is explicit in advective and semi-implicit in diffusion parts and it uses the finite volume method together with up-wind principle for space discretization [10].

The third step in our approach is segmentation of cell nuclei to obtain approximate cell diameters used in our new 4D segmentation approach. Here we use the generalized subjective surface method for 3D image segmentation [9,15,21]. Let $I^0 : \Omega \rightarrow R, \Omega \subset R^3$ represent the intensity function of an image, usually the image after GMCF filtering. If we want to segment an object, we need a segmentation seed - the starting point that determines the approximate position of the object in the image. Then we construct an initial segmentation function $u^0(x)$. For nuclei segmentation, all isosurfaces of the initial segmentation function were equal ellipsoids centered in the detected cell center. The radii were approximated from average nuclei radius given by biologists. The partial differential equation for generalized subjective surface (GSUBSURF) method is given by

$$u_t - w_a \nabla g \cdot \nabla u = w_c g \sqrt{\varepsilon^2 + |\nabla u|^2} \nabla \cdot \left(\frac{\nabla u}{\sqrt{\varepsilon^2 + |\nabla u|^2}} \right) \tag{4}$$

and comes from the level set formulation of the geodesic active contour model [6,7,12,13,21]. In the model (4), g is an edge detector function, for which we again use $g(s) = \frac{1}{1+Ks^2}$, where K is the edge detection sensibility parameter and $s = |\nabla I^0|$, where I^0 is the input image intensity function. Parameters w_a and w_c are weights for the advection and curvature terms of the model, respectively and ε is the regularization parameter, usually $\varepsilon << 1$. The same ε-regularization is used in Eqs. (1) and (3). We choose zero Dirichlet boundary condition for the Eq. (4). In order to discretize (4) in time, we apply the semi-implicit approach that guarantees unconditional stability with respect to the diffusion term. In order to discretize (4) in space, we apply the so called flux-based level set finite volume method for advective part and in curvature part we use approach similar to the ones used in GMCF and LSCD discretizations [15,17].

2.2   Tracking Algorithm Steps

Our final step, the method for cell trajectories extraction and cell lineage tree reconstruction is composed of the following steps:

- construction of a 4D segmentation yielding the 4D spatio-temporal tubular tree structure, Sect. 2.3,
- computation of the first constrained distance function D giving distance of any point of 4D segmentation to the most far (backwards in time) cell identifier to which it is continuously connected, Sect. 2.4,
- computation of the second constrained distance function D_B giving distance of any point of 4D segmentation to its boundary, Sect. 2.4,
- building a potential field V for tracking by using a suitable combination of two computed distance functions, Sect. 2.4,
- extraction of the steepest descent paths of the potential field inside all simply connected 4D segmentation regions, Sect. 2.5,
- centering the extracted paths inside the 4D spatio-temporal trees in order to get unique cell trajectories, Sect. 2.5,
- postprocessing and validation of the results, Sect. 2.6.

We note that after successful cell trajectories extraction the reconstruction of the cell lineage tree can be performed by detecting trajectories which merge together when going backward in time indicating mitosis and thus a branching node of the cell lineage tree.

2.3   Building the 4D Segmentation

From mathematical point of view, the 3D+time image sequence is understood as a function $u(x_1, x_2, x_3, \theta)$, $u : \Lambda \to [0, 1]$, where Λ is a bounded spatio-temporal (rectangular) subdomain of R^4, (x_1, x_2, x_3) is a spatial point and θ represents a time.

The 4D segmentation is a spatio-temporal structure which approximates the space-time movement of cell nuclei. Due to [4,15] the shape of cell nuclei during zebrafish embryogenesis is reasonably approximated by spheres or ellipsoids. Thus, in order to construct 4D segmentation we use cell identifiers detected in all time steps, s_m^l, $m = 1, \ldots, n_C^l$, $l = 1, \ldots, N_\theta$ (m denotes cell identifier index at time step l and N_θ is number of time steps) by method from [4,10], and create 4D ellipsoids around all these points. To determine halfaxes of the ellipsoids we use real cell nuclei segmentations obtained using GSUBSURF method [4,15] paired with cell coordinates from cell detection step. We calculate the volume of real segmented nucleus and compute the radius of a sphere with the same volume. This radius is then used as spatial halfaxes for constructed ellipsoids. Here, we also introduce a parameter S representing shrinking of the halfaxes (if $S < 0$) or expanding of them (if $S > 0$). A slight shrinking of real radius is used later in the tracking algorithm since it helps to have spatially non-overlapping tubular structure representing the cell movement. This parameter is tuned by comparison of tracking with ground truth data and its optimal choice improve the quality of tracking results. In temporal direction we are using halfaxis equal to $d\theta$ corresponding to the image acquisition interval. The nonzero temporal halfaxis is important due to the time overlap which we create and thus we improve connectivity of 4D spatio-temporal tree structures. Thanks to the time overlap we interconnect branches of the 4D spatio-temporal tree where a cell center was not detected in one frame but it was detected in two neighboring frames and thus we correct false negative errors of the cell center detection algorithm.

2.4   Building the Potential Field for Tracking

As noticed above, for building the potential field V we compute two types of distance functions, D and D_B, inside the 4D spatio-temporal segmented tree structures. The distance functions are computed by OpenMP implementation of 4D Rouy-Tourin scheme [5] or they can be determined by fast-marching or fast-sweeping methods [22,24]. These distance functions are called constrained because all the calculations are constrained by the boundaries of the 4D segmentation. Due to that fact, the computed distances between doxels of the 3D+time image sequence are not a standard Euclidean distances in R^4 but they approximate a minimal Euclidean paths between the points inside the 4D segmentation.

We represent the 4D segmentation by a 4D piecewise constant function, with some *BIG* value (value that is bigger than the biggest distance in dataset) outside of the segmentation and with zero value inside it. The 4D distance function $D(x_1, x_2, x_3, \theta)$ is calculated gradually inside all simply connected regions, starting from cell centers in lowest possible time step θ. After the calculation is completed in all regions reachable from these cell centers, we fix the computed values and continue the calculation from centers in next time step, but only in regions where the values are not yet fixed. Using this approach we calculate the distance function D inside whole 4D segmentation. At the end all doxels inside the 4D segmentation contain the value of distance to the most far (backwardly in time) cell identifier to which it is continuosly connected. In Fig. 2 left we show that inside the regions encompassed by the *BIG* values, the value of D is growing from zero, in cell identifier where the simply connected component "begins", up to a locally maximal value, where the simply connected component "ends".

We could think about D as a potential field and traverse it in the steepest descent direction from the local maxima at every simply connected component to the zero value. The paths obtained in such way would represent good approximation of the space-time cell trajectories. And, if the 4D segmentation would contain only perfectly separated 4D spatio-temporal tree structures, we would obtain correctly all (also partial) cell trajectories which can be extracted from the data. Unfortunately, in the real 4D data it is not always possible and we must deal with imperfections given mainly by a cells overlapping. In order to overcome this difficulty we have to keep the extracted paths in a certain distance from the spatio-temporal cell boundaries or, in other words, they should be more centered inside the 4D segmentation.

This can be achieved by using the constrained distance function $D_B(x_1, x_2, x_3, \theta)$ [3,16] values of which grow from boundaries to the center of the 4D spatio-temporal trees, see Fig. 2 right.

Finally, we build a potential field

$$V(x_1, x_2, x_3, \theta) = D(x_1, x_2, x_3, \theta) - \alpha D_B(x_1, x_2, x_3, \theta) \tag{5}$$

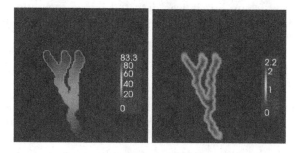

Fig. 2. On the left one can see the plot of constrained distance function D in one simply connected component of the 4D segmentation, on the right one can see the plot of the constrained distance function D_B in the same simply connected component.

which is used for the extraction of cell trajectories. Parameter $\alpha > 0$ is introduced here to adjust the weight of D_B function. It is used to tune and improve tracking results.

2.5   Extraction of the Cell Trajectories

The cell trajectory will be represented by a series of points in space-time (discrete spatio-temporal curve) for which we prescribe the condition that there is exactly one point in every time step $l = N_b, \ldots, N_e$, $1 \leq N_b < N_e \leq N_\theta$. The extraction of cell trajectories is realized in two steps

- first, we use backtracking in time by the steepest descent direction of the potential V built in (5) starting from all cell identifiers $s_m^l, m = 1, \ldots, n_C^l$ detected in all time steps $l = N_\theta, \ldots, 2$,
- then, we center all the extracted paths inside the 4D spatio-temporal trees by using constrained distance function D_B only, in order to eliminate duplicates and thus to obtain the unique cell trajectories.

The first step is realized as follows: Let s_m^l be one of the cell identifiers detected in the lth time step. Let us define a temporary point $P_T^l = s_m^l$. Then, we search recursively in the nearest vicinity of P_T^l, but only in the current time step l and previous time step $l - 1$, for a doxel with the minimal value of the potential V which is also strictly less than the value of potential at the temporary point. The extracted path point P^l for the time step l is defined as the (last in search) doxel from which we move to a point in the previous time step $l - 1$. The point where we moved becomes the temporary point P_T^{l-1} for time step $l - 1$ and we continue the descent as above. We end the process when we cannot move from a time step N_b to a previous time step $N_b - 1$ by decreasing value of the potential V. Then the last point of the search in the time step N_b must be some detected center $s_m^{N_b}$ and it becomes the first point of the extracted path starting in the time step N_b and ending in the time step N_e where we started the descent. As an output of this first step, we get as many extracted paths as is the number of cell identifiers in all time steps except the first one, which means that we have $\sum_{l=2}^{N_\theta} n_C^l$ steepest descent paths.

After the first step of trajectories extraction there exist many duplicated paths (representing the same cell space-time movement but for a shorter time). To illustrate the above fact, let us consider a long cell trajectory going from the first to the last 3D volume of the image sequence. Since we start the descent from centers detected at every $l = N_\theta, \ldots, 2$, and they all lay inside the branch corresponding to that cell, we obtain $N_\theta - 1$ extracted paths laying inside the same branch of the 4D spatio-temporal tree. These paths can slightly differ because the steepest descent search does not give necessarily the same set of points when starting at different time steps from different temporary points. Since all such paths lay in the same branch of the tree we center their points in 3D volumes by using the steepest growth direction of the constrained distance function D_B. After the centering step, the points of the shorter path become subset of points of the longer path and we can remove the shorter one just by comparing their points. The remaining (longest) path represents the cell trajectory. After this step we obtain a set of unique cell trajectories in the sense that a mother cell representative point is presented in as many trajectories as is the number of her descendant cells.

2.6 Lineage Tree Reconstruction

In this last step we build from unique cell trajectories the binary cell lineage tree by using biological coherence. First, we store the results in a tree-like data structure, where each point is stored only once with reference to mother cell and array of daughter cells. If the mother or daughters don't exist, we use null references.

Depending on the quality of input data, the resulting unique cell trajectories can be disjointed in a short time interval and sometimes more than two trajectories can merge going backward in time in single time step. To correct these situations we use the following postprocessing steps:

– We disconnect all trajectories in points where there more than two of them are merged. Then we allow only two trajectories having two nearest daughters to mother to merge, all others will remain disjointed.
– We allow reconnection of all disjointed trajectories. For each ending trajectory we are searching for nearest starting trajectory in the next two time steps and interconnect them by gradually increasing the search radius.

After such postprocessing steps we have built the cell lineage tree which is optimally stored and can be used for validation of the results and for visualization of cell dynamics.

3 Numerical Experiment on Real Zebrafish Embryogenesis Data

We performed experiments on two representative real zebrafish embryogenesis datasets. First dataset has acquisition step $d\theta = 67\,\text{s}$, $N_\theta = 480$ number of time steps and dimension of every 3D image is $512 \times 512 \times 104$ voxels. The real voxel size is $dx_1 = dx_2 = dx_3 = 1.37\,\mu\text{m}$ in every spatial direction. In the last time step $N_\theta = 480$ biologist selected manually cells forming seven presumptive organs in the brain (hypoblast, presumptive hypothalamus, ventral telencephalon, right eye, right optic stalk, left eye, left optic stalk). Using developed approach we can track those cell populations backwards (and then also forward) in time and thus follow their dynamics and clonal history. For this dataset we also have the ground truth data which contains about 39000 manually checked cell links, that can be used for validation of the tracking. We have to note that building such ground truth is extremely difficult and time consuming task for expert in biology. We highly appreciate this unique work, which can be used for tuning algorithm parameters and thus allowing to use the methods in practice when processing similar type zebrafish datasets. Second dataset has acquisition step $d\theta = 154\,\text{s}$, $N_\theta = 200$ and dimension of every 3D image is $512 \times 512 \times 120$ voxels.

Before tracking, all 3D images of the processed data were filtered by 10 steps of geodesic mean curvature flow (GMCF) model [4,9,14] and the cell nuclei identifiers were detected by 15 steps of level set center detection (LSCD) algorithm [4,9,10] for the first dataset and by 4 steps of LSCD for the second dataset. The cell nuclei were segmented using generalized subjective surface (GSUBSURF) method [4,9]. From several millions of cell identifiers we built the 4D segmentation and then the cell trajectories were extracted by the approach developed in Sect. 2. The correctness of mother-daughter cell links for the first dataset were validated using ground truth data and the results are presented in Table 1 and Fig. 3.

Table 1. Comparison of the tracking result with ground truth data depending on α, with $S = -0.5$.

α	Correct mother links	Correct daughter links	Wrong mother links	Wrong daughter links
0.4	34856	34719	3935	4072
0.8	35510	35371	3280	3419
1.2	36064	35946	2726	2844
1.6	36551	36493	2240	2298
2.0	37125	37126	1668	1667
2.4	37288	37309	1506	1485
2.8	37349	37386	1448	1411
3.2	37402	37437	1395	1360
3.6	37357	37415	1440	1382
4.0	37377	37434	1420	1363

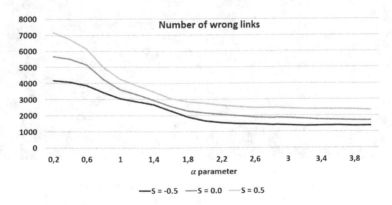

Fig. 3. Number of wrong links in tracking compared to ground truth data, depending on α and S. With α increasing towards 2, the number of wrong links is decreasing, then it stabilizes and the best result is obtained for $S = -0.5$, cf. also Table 1.

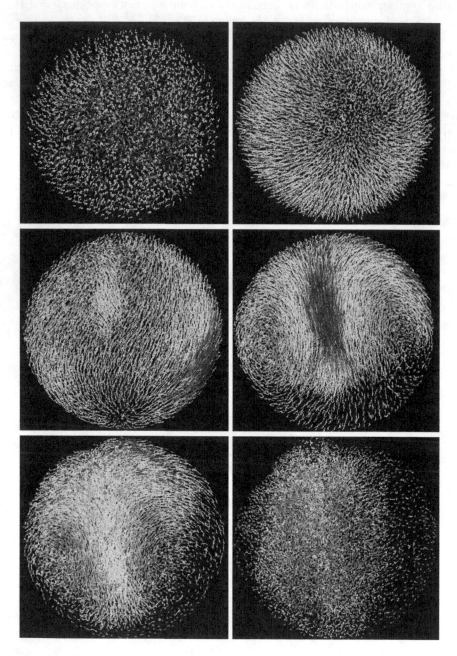

Fig. 4. Visualization of the cell movement speed during embryogenesis in time steps 1, 96, 192, 288, 384 and 480. Red trajectories are the fastest moving cells, while blue are the slowest. (Color figure online)

To tune the tracking results we adjust two parameters mentioned in Sect. 2, S and α. First, we can expand or shrink the nuclei radii used for building the 4D segmentation. By comparison with ground truth data we concluded that the best results for these two datasets were obtained when we shrink the radii by $S = -0.5$, cf. Fig. 3. The second parameter is α, used in the construction of the potential V. We tested tracking for $\alpha \in [0.5, 4]$ and obtained the best results around $\alpha = 3.2$, see Table 1.

We present here also Figs. 4 and 5 showing results of the tracking procedure on our two datasets. For the cell trajectories visualization we built software,

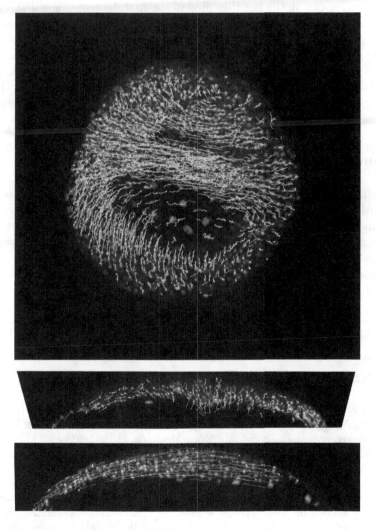

Fig. 5. Tracked cells with trajectories displayed along the slices of original data. Middle image is slightly tilted for better visibility since the trajectories are going outwards from the slice.

running on graphics card, where we can fluently zoom, rotate and animate in time a 3D scene even with very high number of trajectories. The trajectories are displayed as short lines (in various colors, according to speed, direction or manually defined color of cell population) connecting a few subsequent spatio-temporal points with a freely chosen starting time. In Fig. 4 we show trajectories of a second dataset with colors depending on the cell movement speed. Dark blue are the slowest, while red are the fastest moving cells. One can see the evolution of cells from the 1st still chaotic stage, through 96th, 192nd, 288th, 384th time steps where the cells are becoming more compactly localized up to 480th time step, cf. Fig. 1.

As noticed above, in the last time step of the first dataset, $N_\theta = 480$, biologist manually selected cells forming seven presumptive organs during the zebrafish brain early embryogenesis.

In Fig. 5 we show trajectories along with the slices of original data in time step $N_\theta = 260$. Here all trajectories have the same color and original data is displayed as black and white slices obtained from 3D volume. Only the cells that are near the slices are displayed along with the trajectories showing their movement in next and previous 20 time steps.

4   Conclusions

In this paper we presented algorithm for the cell tracking and lineage tree recon-struction from 3D+time microscopy data. We validate the tracking results by comparing them with manually verified ground truth data. The best results which we can achieve show more than 96% correctness of mother-daughter links. We applied the tracking method to complex stages of the zebrafish early embryo-genesis images and visualize extracted cell trajectories and present mean velocity of selected cell populations.

Acknowledgement. This work was supported by grants APVV-15-0522 and VEGA 1/0608/15.

References

1. Amat, F., Lemon, W., Mossing, D.P., McDole, K., Wan, Y., Branson, K., Myers, E.W., Keller, P.J.: Fast, accurate reconstruction of cell lineages from large-scale fluorescence microscopy data. Nat. Methods **11**, 951–958 (2014)
2. Bao, Z., et al.: Automated cell lineage tracing in Caenorhabditis elegans. Proc. Natl. Acad. Sci. USA **103**, 2707–2712 (2016)
3. Bellaïche, Y., Bosveld, F., Graner, F., Mikula, K., Remešková, M., Smíšek, M.: New robust algorithm for tracking cells in videos of Drosophila morphogenesis based on finding an ideal path in segmented spatio-temporal cellular structures. In: Proceeding of the 33rd Annual International IEEE EMBS Conference, Boston Marriott Copley Place, Boston, MA, USA, 30 August–3 September 2011. IEEE Press (2011)

4. Bourgine, P., Čunderlík, R., Drblíková, O., Mikula, K., Peyrieras, N., Remešíková, M., Rizzi, B., Sarti, A.: 4D embryogenesis image analysis using PDE methods of image processing. Kybernetika **46**, 226–259 (2010)

5. Bourgine, P., Frolkovič, P., Mikula, K., Peyriéras, N., Remešíková, M.: Extraction of the intercellular skeleton from 2D Images of embryogenesis using Eikonal Equation and advective subjective surface method. In: Tai, X.-C., Mørken, K., Lysaker, M., Lie, K.-A. (eds.) SSVM 2009. LNCS, vol. 5567, pp. 38–49. Springer, Heidelberg (2009). doi:10.1007/978-3-642-02256-2_4

6. Caselles, V., Kimmel, R., Sapiro, G.: Geodesic active contours. Int. J. Comput. Vis. **22**, 67–79 (1997)

7. Caselles, V., Kimmel, R., Sapiro, G.: Geodesic active contours. In: Proceedings International Conference on Computer Vision 1995, pp. 694–699, Boston (1995)

8. Chen, Y., Vemuri, B.C., Wang, L.: Image denoising and segmentation via nonlinear diffusion. Comput. Math. Appl. **39**, 131–149 (2000)

9. Faure, E., Savy, T., Rizzi, B., Melani, C., Remešíkova, M., Špir, R., Drblíková, O., Čunderlík, R., Recher, G., Lombardot, B., Hammons, M., Fabrèges, D., Duloquin, L., Colin, I., Kollár, J., Desnoulez, S., Affaticati, P., Maury, B., Boyreau, A., Nief, J.Y., Calvat, P., Vernier, P., Frain, M., Lutfalla, G., Kergosien, Y., Suret, P., Doursat, R., Sarti, A., Mikula, K., Peyriéras, N., Bourgine, P.: An algorithmic workflow for the automated processing of 3d+time microscopy images of developing organisms and the reconstruction of their cell lineage. Nat. Commun. **7** (2016). Article no: 8674

10. Frolkovič, P., Mikula, K., Peyriras, N., Sarti, A.: A counting number of cells and cell segmentation using advection-diffusion equations. Kybernetika **43**, 817–829 (2007)

11. Kausler, B.X., et al.: A discrete chain graph model for 3d+t cell tracking with high misdetection robustness. In: Fitzgibbon, A., Lazebnik, S., Perona, P., Sato, Y., Schmid, C. (eds.) ECCV 2012. LNCS, vol. 7574, pp. 144–157. Springer, Heidelberg (2012). doi:10.1007/978-3-642-33712-3_11

12. Kichenassamy, S., Kumar, A., Olver, P., Tannenbaum, A., Yezzi, A.: Conformal curvature flows: from phase transitions to active vision. Arch. Ration. Mech. Anal. **134**, 275–301 (1996)

13. Kichenassamy, S., Kumar, A., Olver, P., Tannenbaum, A., Yezzi, A.: Gradient flows and geometric active contours model. In: Proceedings of International Conference on Computer Vision 1995, Boston (1995)

14. Kriva, Z., Mikula, K., Peyriéras, N., Rizzi, B., Sarti, A., Stašová, O.: 3D early embryogenesis image filtering by nonlinear partial differential equations. Med. Image Anal. **14**, 510–526 (2010)

15. Mikula, K., Peyriéras, N., Remešíková, M., Sarti, A.: 3D embryogenesis image segmentation by the generalized subjective surface method using the finite volume technique. In: Eymard, R., Herard, M. (eds.) Finite Volumes for Complex Applications V, Problems and Perspectives, pp. 585–592. ISTE and Wiley, London (2008)

16. Mikula, K., Peyriéras, N., Remešíková,M., ,Smíšek, M.: 4D numerical schemes for cell image segmentation and tracking. In: Fořt, J. et al. (ed.) Finite Volumes in Complex Applications VI, Problems and Perspectives, Proceedings of the Sixth International Conference on Finite Volumes in Complex Applications, Prague, 6–10 June 2011, pp. 693–702. Springer, Heidelberg (2011)

17. Mikula, K., Peyriéras, N., Remešíková, M., Stašová, O.: Segmentation of 3D cell membrane images by PDE methods and its applications. Comput. Biol. Med. **41**, 326–339 (2011)

18. Mikula, K., Peyriéras, N., Špir, R.: Numerical algorithm for tracking cell dynamics in 4D biomedical images. Discrete Continuous Dyn. Syst. - Ser. S **8**, 953–967 (2015)
19. Mikula, K., Špir, R., Smíšek, M., Faure, E., Peyriéras, N.: Nonlinear PDE based numerical methods for cell tracking in zebrafish embryogenesis. Appl. Numer. Math. **95**, 250–266 (2015)
20. Olivier, N., Luengo-Oroz, M.A., Duloquin, L., Faure, E., Savy, T., Veilleux, I., Solinas, X., Dbarre, D., Bourgine, P., Santos, A., Peyriras, N., Beaurepaire, E.: Cell lineage reconstruction of early zebrafish embryos using label-free nonlinear microscopy. Science **329**, 967–971 (2010)
21. Sarti, A., Malladi, R., Sethian, J.A.: Subjective surfaces: a method for completing missing boundaries. Proc. Natl. Acad. Sci. U.S.A. **97**, 6258–6263 (2000)
22. Sethian, J.A.: A fast marching level set method for monotonically advancing fronts. Proc. Natl. Acad. Sci. **93**, 1591–1595 (1996)
23. Tomer, R., Khairy, K., Amat, F., Keller, P.J.: Quantitative high-speed imaging of entire developing embryos with simultaneous multiview light-sheet microscopy. Nat. Methods **9**, 755–763 (2012)
24. Zhao, H.: A fast sweeping method for Eikonal equations. Math. Comput. **74**, 603–627 (2005)

Binary Pattern Dictionary Learning for Gene Expression Representation in *Drosophila* Imaginal Discs

Jiří Borovec[✉] and Jan Kybic

Center for Machine Perception, Department of Cybernetics,
Faculty of Electrical Engineering, Czech Technical University in Prague,
Prague, Czech Republic
jiri.borovec@fel.cvut.cz

Abstract. We present an image processing pipeline which accepts a large number of images, containing spatial expression information for thousands of genes in Drosophila imaginal discs. We assume that the gene activations are binary and can be expressed as a union of a small set of non-overlapping spatial patterns, yielding a compact representation of the spatial activation of each gene. This lends itself well to further automatic analysis, with the hope of discovering new biological relationships. Traditionally, the images were labeled manually, which was very time consuming. The key part of our work is a binary pattern dictionary learning algorithm, that takes a set of binary images and determines a set of patterns, which can be used to represent the input images with a small error. We also describe the preprocessing phase, where input images are segmented to recover the activation images and spatially aligned to a common reference. We compare binary pattern dictionary learning to existing alternative methods on synthetic data and also show results of the algorithm on real microscopy images of the Drosophila imaginal discs.

1 Introduction

The fruit fly Drosophila is a frequently used valuable subject in modern experimental biology due to their short life cycle and genetic similarity to humans [1]. Large scale mapping of the gene expressions was performed in embryos [2,3] as well as in imaginal discs [4,5], which are essential for the initial development of the adult fly. The expressed gene is highlighted using molecular biology methods and microscopy images of many thousands samples are acquired.

The final goal is to understand the role of the different genes by comparing locations, where the genes are expressed, with the known information about the function of the different areas. To reduce the dimensionality of the problem and enable an efficient statistical analysis, the observed spatial expressions are described by a set of labels from a limited, application specific dictionary, called an 'atlas'. Example labels for the leg imaginal disc are 'dorsal' or 'ventral' but

© Springer International Publishing AG 2017
C.-S. Chen et al. (Eds.): ACCV 2016 Workshops, Part II, LNCS 10117, pp. 555–569, 2017.
DOI: 10.1007/978-3-319-54427-4_40

also 'stripes' or 'ubiquitous'. Given such sets of labels, correlations with gene ontologies can be then found using data mining methods [6–8].

The dictionary as well as the labels for individual images are usually determined manually or semi-manually [2,9,10], which is extremely intensive work. Some automatic methods exists, based on e.g. sparse Bayesian factor models [11] or non-negative matrix factorization [12]. In contrast to these methods, we assume in this work that the activation and the patterns are inherently binary. We further assume that the patterns, corresponding to anatomically defined zones, are compact and non-overlapping. These constraints should increase the robustness of the estimation and yield more biologically plausible results.

1.1 State-of-the-Art

Let a matrix $X \in \mathbb{R}^{|\Omega| \times N}$ be a rearranged set of pixels of N images with pixel coordinates Ω. In our case, the images are assumed to be aligned and the pixel intensities to correspond to the gene activations. A linear decomposition of X can be found by minimizing

$$\min_{Y,W} \|X - Y \cdot W\|^2 \tag{1}$$

where $Y \in \mathbb{R}^{|\Omega| \times L}$ corresponds to a dictionary (or 'atlas') with L patterns and $W \in \mathbb{R}^{L \times N}$ are image specific weights. We shall give a few examples of known methods, differing in additional assumptions and constraints. Built on the well-known PCA, sparse Principal Component Analysis [13] (sPCA) assumes the weights W to be sparse. Fast Independent Component Analysis [14] (FastICA) seeks for spatial independence of the patterns. Dictionary Learning [15] (DL) with Matching Pursuit is a greedy iterative approximation method with many variants, mainly in the field of sparse linear approximation of signals. Non-negative Matrix Factorization [16] (NMF) adds the non-negativity constraints, while sparse Bayesian models add a probabilistic prior on the weights, encouraging sparsity. Both methods were used for estimating gene expression patterns in Drosophila embryos [11,12] (see Berkeley Drosophila Genome Project[1]).

There is far less literature in the case of binary $X, Y,$ or W. If the requirement of spatial compactness of the patterns is dropped, then the problem is called binary matrix factorization [17,18] and is often used in data mining. Simplifying further to allow only one pattern per image leads to the problem of vector quantization [19].

2   Method

We shall now describe the complete pipeline consists of preprocessing (segmentation and registration) and atlas estimation via Binary Pattern Dictionary Learning (BPDL), as illustrated in Fig. 2.

---

[1] http://insitu.fruitfly.org/cgi-bin/ex/insitu.pl.

2.1   Preprocessing

Given a set of images of imaginal discs (see Fig. 1a) containing both anatomical and gene expression information, preprocessing is applied to obtain a set of segmented and aligned gene expression images, which serves as input for the subsequent binary dictionary learning (see Fig. 2).

Segmentation. We first segment the input images into three classes (background, imaginal disc, gene activation) by the following steps [20]:

(a) calculate SLIC superpixels [21];
(b) calculate superpixel colour features — mean, median, and variance;
(c) estimating a Gaussian Mixture Model (GMM) with one component per class,
(d) calculate superpixel-wise class probabilities based on the GMM;
(e) apply Graph Cut [22] to estimate a spatially regularized segmentation;
(f) post-process, e.g. suppress very small regions and identifying the imaginal disc component.

Registration. For each of the four disc types, a reference shape is calculated as the mean disc shape over all images. Then all other images are registered to the reference shape. We use a fast elastic registration algorithm [23], which works directly on the segmented images, transforming the disc shapes by aligning their contours, ignoring the activations (see Fig. 1b). The activations (Fig. 1c) are then aligned using the recovered transformation.

2.2   Binary Pattern Dictionary Learning — Problem Definition

Let us define the image pixels as $\Omega \subseteq \mathbb{Z}^d$, with $d = 2$, and the input binary image as $\mathbf{g} : \Omega \to \{0,1\}$. Our task is to find an atlas $\mathbf{y} : \Omega \to \mathbb{L}$, with labels $\mathbb{L} = [0, \ldots, K]$, assigning to each pixel either a background (label $l = 0$), or one of the labels (patterns, set of equal labels) $1, \ldots, K$. Each binary weight vector $\mathbf{w} : \mathbb{L} \to \{0,1\}$ yields an image $\hat{\mathbf{g}}$ as a union of the selected patterns in atlas $\mathbf{y}$

$$\hat{\mathbf{g}} = \sum_{l \in \mathbb{L}} \mathbf{w}_l \cdot [\![\mathbf{y} = l]\!] \tag{2}$$

where $[\![\cdot]\!]$ denotes the Iverson bracket. Note that this is a special case of (1), with all variables binary. Note also, that in this representation, the patterns cannot overlap. The approximation error on one image $\mathbf{g}$ and its representation by $\mathbf{y}$ and $\mathbf{w}$ is the Hamming distance

$$F(\mathbf{g}, \mathbf{y}, \mathbf{w}) = \sum_{i \in \Omega} [\![\mathbf{g}_i \neq \hat{\mathbf{g}}_i]\!] = \sum_{i \in \Omega} \left| \mathbf{g}_i - \sum_{l \in \mathbb{L}} \mathbf{w}_l \cdot [\![\mathbf{y} = l]\!] \right| \tag{3}$$

To encourage spatial compactness of the estimated atlas, we shall penalize differences between neighboring pixels i, j in the atlas

$$H(\mathbf{y}) = \sum_{\substack{i,j \in \Omega,\, i \neq j, \\ d(i,j)=1}} [\![\mathbf{y}_i \neq \mathbf{y}_j]\!] \tag{4}$$

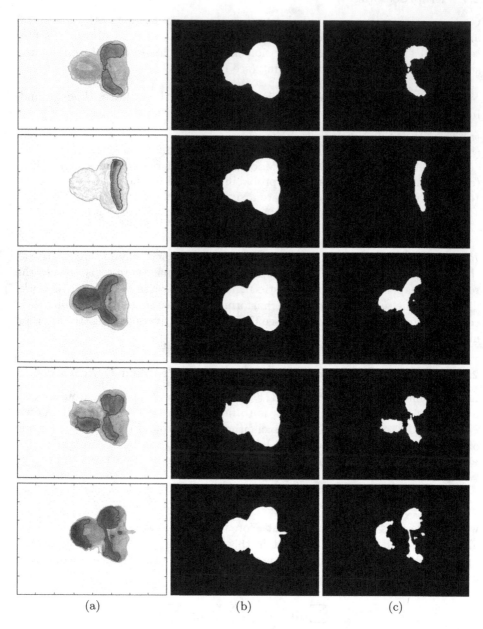

(a)   (b)   (c)

Fig. 1. Presenting samples of aligned Drosophila imaginal discs (eye antenna discs type) and their segmentation. First, we show the sensed images (a) with marked contour of segmented disc (orange) and gene expression (red) followed by visualisation of the segmented discs (b) and segmented gene expressions (c). (Color figure online)

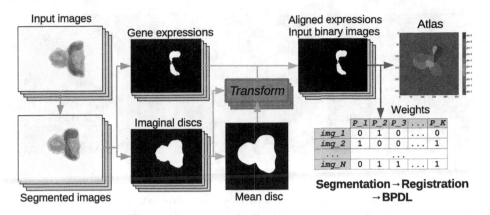

Fig. 2. A flowchart of the complete pipeline for processing images of Drosophila imaginal discs: (i) a pixel-wise segmentation into 3 classes (background, imaginal disk and gene expression); (ii) registration of binary segmented imaginal discs onto a reference shape (disc prototype); (iii) atlas estimation from aligned binary gene expressions.

where the Kronecker delta $[\![\mathbf{y}_i \neq \mathbf{y}_j]\!]$ for all combinations of $\mathbf{y}_{i,j} \in \mathbb{L}$ can be represented as a square matrix with zeros on the main diagonal and ones otherwise

The optimal atlas and the associated weights are found by optimizing the mean approximation error for all N images

$$\mathbf{y}^*, \mathbf{w}^* = \arg\min_{\mathbf{y}, \mathbf{W}} \frac{1}{N} \sum_n F(\mathbf{g}^n, \mathbf{y}, \mathbf{w}^n) + \beta \cdot H(\mathbf{y}) \qquad (5)$$

where the matrix $\mathbf{W}$ contains all weights $\mathbf{w}^n$ for $n \in [0, \ldots, N]$, and β is the spatial regularization coefficient. Sufficiently large β force the labelling to have all patterns to be connected.

2.3 BPDL — Alternating Minimization

The criterion (5) is minimized alternately with respect to atlas $\mathbf{y}$ and weights $\mathbf{w} \in \mathbf{W}$, (see Algorithm 1).

Algorithm 1. General schema of BPDL algorithm.

1: initialise atlas $\mathbf{y}$
2: **while** not converged **do**
3:     update weights $\mathbf{w} \in \mathbf{W}$
4:     reinitialise empty patterns in $\mathbf{y}^*$
5:     update atlas $\mathbf{y}^*$ via Graph Cut
6: **end while**

Initialization. We initialize the atlas with randomly labeled patches on a regular grid, with user-defined sizes; see Fig. 3 for examples.

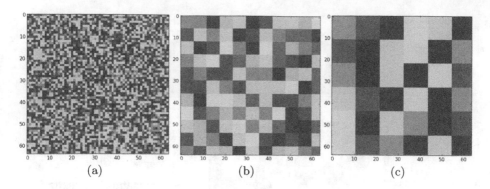

(a)   (b)   (c)

Fig. 3. Random atlas initialization with patch sizes 1 pixel (a), $m/2K$ pixels (b), and m/K pixels (c), for $K = 6$ and $m = 64$ pixels being the image size.

Update Weights **W**. With the atlas **y** fixed, we estimate the weights $\mathbf{w}^n$ for each image $\mathbf{g}^n$ independently. It turns out that $F(\mathbf{g}^n, \mathbf{y}, \mathbf{w})$ is minimized with respect to w_l, if the majority of pixels in the pattern $l \in \mathbf{y}$ agree with the image. We set

$$w_l = \llbracket P(\mathbf{g}, \mathbf{y}, l) \geq \sigma \rrbracket \qquad \text{where } \sigma = 1 \tag{6}$$

$$\text{and} \quad P(\mathbf{g}, \mathbf{y}, l) = \frac{\sum_{i \in \Omega, \mathbf{y}_i = l} \llbracket \mathbf{g}_i = 1 \rrbracket}{\sum_{i \in \Omega, \mathbf{y}_i = l} \llbracket \mathbf{g}_i \neq 1 \rrbracket} = \frac{\| \llbracket \mathbf{y} = l \rrbracket \|}{\sum_{i \in \Omega, \mathbf{y}_i = l} (1 - \mathbf{g}_i)} - 1 \tag{7}$$

We temporarily reduce σ in the initial stage of the algorithm, otherwise very few patterns might be selected.

Reinitialize Empty Patterns. During the weight calculation step, some pattern may not have been used for any image. This is wasteful, unless the reconstruction is already perfect, we can always improve it by adding another pattern. We iterate the following procedure until all K labels are used:

1. find an image $\mathbf{g}^n$ with the largest unexplained residual $\llbracket \mathbf{g}^n \wedge \neg \hat{\mathbf{g}}^n \rrbracket$:
2. find the largest connected component c of this residual and assign label $l \notin \mathbf{y}$;
3. calculate weights w_l^n for the new label l for all images $\mathbf{g}^n \in \mathbf{G}$ using (6).

Update of Atlas **y**. With the weight vectors **W** fixed, finding the atlas **y** is a discrete labeling problem. We can rewrite the criterion in (5) as

$$\frac{1}{N} \sum_{i \in \Omega} \underbrace{\sum_n \left| \mathbf{g}_i^s - \sum_{l \in \mathbb{L}} \mathbf{w}_l^s \cdot \llbracket \mathbf{y} = l \rrbracket \right|}_{U_i(y_i)} + \sum_{\substack{i,j \in \Omega,\, i \neq j, \\ d(i,j)=1}} \llbracket \mathbf{y}_i \neq \mathbf{y}_j \rrbracket \tag{8}$$

which can be solved for example with Graph Cut [22] and alpha expansion.

3   Experiments

We evaluate the performance (atlas similarity and descriptiveness and elapsed time) of the algorithm on both synthetic and real images.

3.1 Alternative Methods

We have compared our BPDL with the following methods: NMF [16], FastICA [14], SparsePCA [13] and Dictionary Learning [15] (DL). All methods were implemented in the scikit-learn[2] library.

Binarization of Continues Components. To obtain a binary atlas **y** from a continuous matrix $Y \in \mathbb{R}^{|\Omega| \times L}$, we select the component with a maximal value in each pixel position $i \in \Omega$, i.e.

$$\mathbf{y}_i = \arg \max_{l \in \mathbb{L}} Y_i^l$$

3.2 Synthetic Dataset

We generated three synthetic datasets (v0, v1, v2) representing already segmented and aligned binary images based on a random atlas and random pattern weights. The patterns are deformed ellipses. The datasets differ in image size and the true number of patterns in atlas (see Fig. 4).

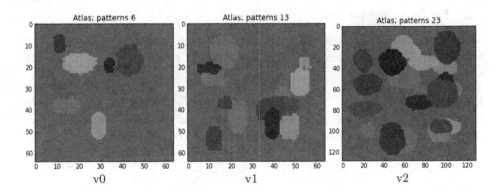

Fig. 4. Visualisation of the generated atlases for the three created synthetic datasets containing $K = 6$ (v0), $K = 13$ (v1) and $K = 23$ (v2) patterns, with image sizes 64×64 (v0, v1) and 128×128 (v2) pixels.

Each dataset is further divided into three sub-sets, each containing 1200 input images (Fig. 5):

1. **pure:** images generated from Eq. (2)
2. **deform:** pure images (1) independently transformed by a small elastic B-spline deformation with the maximum amplitude of $0.2\,m$, with $m = \sqrt{|\Omega|}$.
3. **deform & noise (D&N):** deformed images (2) with random binary noise (randomly flipping 10% pixels).

---

[2] http://scikit-learn.org/stable/.

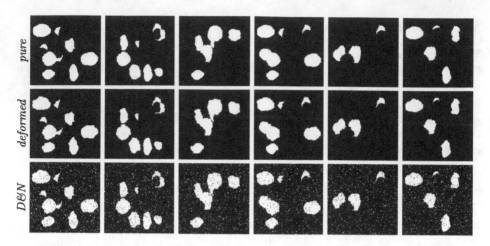

Fig. 5. We show sample images from the synthetic dataset v2. The tree rows represents the 3 sub-sets of input images: pure, deformed and deformed with random binary noise (denoted D&N).

3.3 Evaluation Metrics

Atlas Comparison. The difference between atlases is given by the Adjusted Rand Score[3] (ARS), which gives similarity value in the range $(0, 1)$, with 1 being a perfect match.

Reconstruction Difference. With the estimated atlas $\mathbf{y}$ and pattern weights $\mathbf{w}^n \in \mathbf{W}$ for each particular image $\mathbf{g}^n \in \mathbf{G}$ we reconstruct each input image $\hat{\mathbf{g}}^n$ (see Eq. (2), Fig. 6), the approximation error is averaged over all images:

$$R(\mathbf{G}, \mathbf{y}, \mathbf{W}) = \frac{1}{N \cdot |\Omega|} \sum_n F(\mathbf{g}^n, \mathbf{y}, \mathbf{w}^n) = \frac{1}{N \cdot |\Omega|} \sum_n \sum_i |\mathbf{g}_i^n - \hat{\mathbf{g}}_i^n| \quad (9)$$

In case of the synthetic datasets we always compare the reconstructed images to the pure input images.

3.4 Comparison on Synthetic Datasets

In Table 1 we show the accuracy of reconstructing the atlas (measured by ARS), the mean approximation error R, and elapsed time for all datasets and their modifications. The number of patterns was set to the true value K.

We can say that on the 'pure' images, all methods work well. In other cases, the accuracy of our method (as measured by ARS and R) is better. The fastest method is the NMF (on average twice faster than BPDL) but its results are poor. On the other hand FastICA gives the second best quality results after BPDL but is much slower (on average 40 times slower then BPDL).

---

[3] https://en.wikipedia.org/wiki/Rand_index.

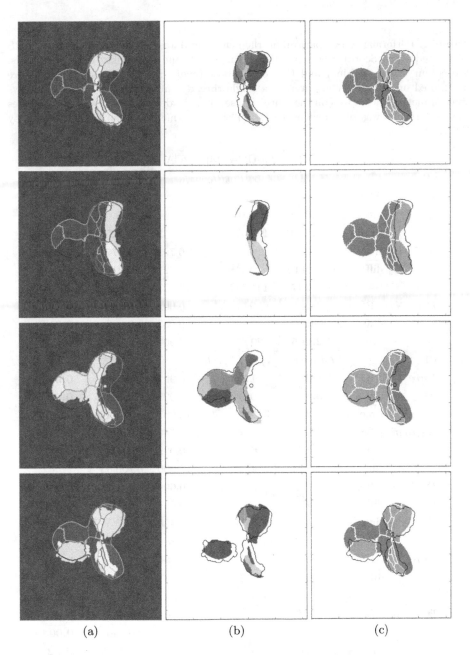

(a)   (b)   (c)

Fig. 6. Visualization of the reconstruction of real images (already showen in Fig. 1) in three different ways: (a) The input binary segmentation of gene expression overlapped by the atlas pattern contours. (b) The individual atlas patterns (in color) with the binary input gene expression segmentation overlaid (black contour). (c) Used (green) versus unused (red) atlas patterns with contour of the segmented input expression boundary (black). (Color figure online)

Table 1. Performance comparison on the synthetic datasets. We show the atlas ARS (Adjusted Rand Score), approximation error R (9), and processing time in seconds. We colour the best (blue) and the second best (cyan) result. All experiments were performed on the same computer, in a single thread configuration. The results shows that all methods work well on the 'pure' sub-set. For the deformed and also noise images the best results was obtained by BPDL. The fastest method was NMF, followed by BPDL.

Datasets		NMF [16]	FastICA [14]	sPCA [13]	DL [15]	BPDL *
v0		*(size 64 × 64 px, 6 patterns)*				
Pure	ARS	**1.0**	**1.0**	0.961	**1.0**	**0.999**
	diff.	**0.0**	**0.0**	0.002	**0.0**	**0.0**
	time	**2.780**	168.476	30.842	304.51	6.658
Deform	ARS	0.775	0.921	0.769	0.777	**0.993**
	diff.	0.014	0.004	0.0213	0.014	**0.0**
	time [s]	**1.697**	141.527	22.833	279.87	4.766
D & N	ARS	0.048	0.778	0.002	0.066	**0.999**
	diff.	0.033	0.014	0.033	0.033	**0.0**
	time [s]	**2.005**	229.47	24.907	598.83	6.774
v1		*(size 64 × 64 px, 13 patterns)*				
Pure	ARS	**1.0**	**1.0**	0.992	0.995	**0.999**
	diff.	**0.0**	**0.0**	0.0298	0.019	**0.0**
	time	**2.333**	340.32	18.291	737.47	6.029
Deform	ARS	0.785	0.948	0.780	0.779	**0.992**
	diff.	0.017	0.004	0.029	0.033	**0.005**
	time [s]	**4.001**	312.18	15.000	700.03	7.561
D & N	ARS	0.091	0.878	0.009	0.0727	**0.951**
	diff.	0.048	0.010	0.061	0.0499	**0.003**
	time [s]	**4.490**	439.04	11.420	697.599	9.562
v2		*(size 128 × 128 px, 23 patterns)*				
Pure	ARS	**1.0**	**1.0**	0.989	**1.0**	**0.999**
	diff.	**0.0**	**0.0**	0.037	**0.0**	0.005
	time [s]	**82.329**	5533.4	460.82	14786.	88.260
Deform	ARS	0.818	0.846	0.801	0.807	**0.970**
	diff.	0.019	0.015	0.056	0.046	**0.004**
	time [s]	**144.10**	5683.2	477.47	13619.	165.22
D & N	ARS	0.120	0.612	0.024	0.144	**0.877**
	diff.	0.036	0.036	0.092	0.039	**0.013**
	time [s]	**77.399**	6912.9	485.44	13729.	289.51

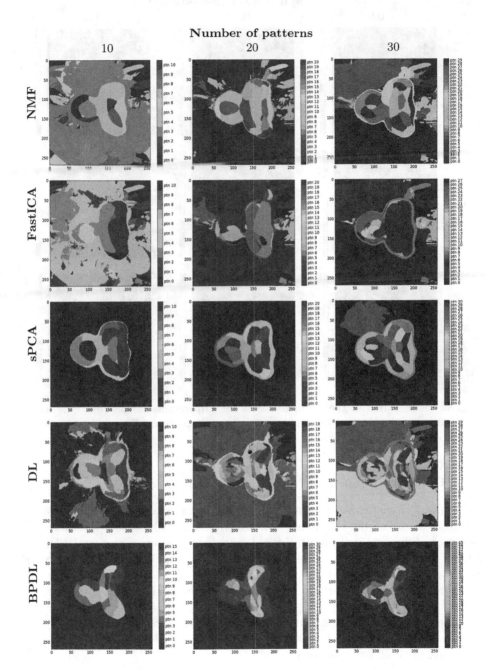

Fig. 7. Presenting estimated atlases by all methods with different number of estimated patterns K.

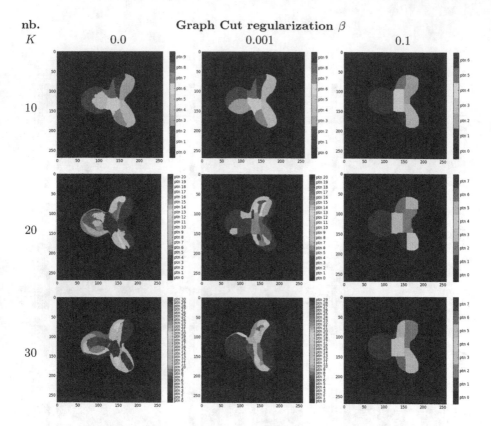

Fig. 8. Visualization of the estimated atlas for Drosophila images (eye type imaginal discs) as a function of the number of estimated patterns K and the Graph Cut regularization parameter β.

3.5   Comparison on Real Images

We applied all methods on segmented gene expressions images of the Drosophila imaginal discs varying the number of patterns $K \in \{10, 20, 30\}$ (see the reconstruction difference in Table 2). Several reconstruction examples for BFDL are shown in Fig. 6. Looking at the estimated atlases (Fig. 7) we found that NMF, FastICA and DL have difficulty to identify background and often produce very small regions. Example atlases by BFDL on all four considered disc types are shown in Fig. 9.

The effect of the Graph Cut regularization parameter β is shown in Fig. 8. A value of $\beta = 0.001$ was found to perform best by subjective evaluation and it was used in all other experiments.

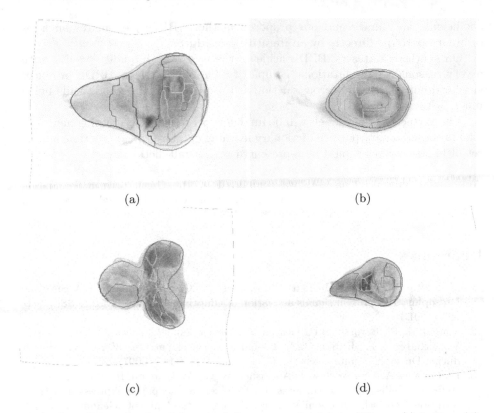

(a)   (b)

(c)   (d)

Fig. 9. Sample images of each imaginal disc types: wing (a), leg (b), eye (c), haltere (d) with the atlases estimated by BPDL shown as contour overlays for number of patterns $K = 20$.

Table 2. Reconstruction difference R on real images of imaginal disc (eye type) by all tested methods for three different assumed numbers of patterns K.

Method	Number of patterns K			Time [min]
	10	20	30	
NMF [16]	0.0939	0.0823	0.0723	**10**
FastICA [14]	0.1197	0.0779	0.0485	24
sPCA [13]	0.0476	0.0413	**0.0352**	477
DL [15]	0.0939	0.0648	0.0596	338
BPDL *	**0.0467**	**0.0395**	0.0361	20

4   Conclusion

This paper addresses automatic image analysis of Drosophila imaginal discs, focusing on the problem of finding an atlas of atomic gene expression from the images. Unlike alternative methods, we assume that the atlas and its coefficients are binary and our proposed method (BPDL) estimates an atlas of binary patterns directly by an iterative procedure.

On synthetic datasets, BPDL achieves the best overall quality results, with a very reasonable computational complexity. On real datasets, BPDL produces similar quality atlas and reconstruction as the SparsePCA method, while being much faster.

The extracted image labels will be further processed by data mining methods. The proposed binary pattern dictionary learning can be applied any time a large set of binary images should be represented by a small dictionary.

Acknowledgement. This work was supported by the Czech Science Foundation project 14-21421S and by the Grant Agency of the Czech Technical University in Prague under the grant SGS15/154/OHK3/2T/13.

References

1. Medzhitov, R., Preston-Hurlburt, P., Janeway Jr., C.A.: A human homologue of the Drosophila Toll protein signals activation of adaptive immunity. Nature **388**, 394–397 (1997)
2. Tomancak, P., Berman, B.P., Beaton, A., Weiszmann, R., Kwan, E., Hartenstein, V., Celniker, S.E., Rubin, G.M.: Global analysis of patterns of gene expression during Drosophila embryogenesis. Genome Biol. **8**, R145 (2007)
3. Hammonds, A.A.S., Bristow, C.A., Fisher, W.W., Weiszmann, R., Wu, S., Hartenstein, V., Kellis, M., Yu, B., Frise, E., Celniker, S.E.: Spatial expression of transcription factors in Drosophila embryonic organ development. Genome Biol. **14**, R140 (2013)
4. Brower, D.L.: Engrailed gene expression in Drosophila imaginal discs. EMBO J. **5**, 2649–2656 (1986)
5. Ahammad, P., Harmon, C.L., Hammonds, A., Sastry, S.S., Rubin, G.M.: Joint nonparametric alignment for analizing spatial gene expression patterns in Drosophila imaginal discs. In: Proceedings of CVPR (2005)
6. Jiang, D., Tang, C., Zhang, A.: Cluster analysis for gene expression data: a survey. IEEE Trans. Knowl. Data Eng. **16**, 1370–1386 (2004)
7. Kim, J., Kim, K., Kim, J.H.: Semantic signature: comparative interpretation of gene expression on a semantic space. Comput. Math. Methods Med. **2016**, 1–10 (2016)
8. Klema, J., Malinka, F., Zelezny, F.: Semantic biclustering: a new way to analyze and interpret gene expression data. In: Bourgeois, A., Skums, P., Wan, X., Zelikovsky, A. (eds.) Bioinformatics Research and Applications, pp. 332–333. Springer, Heidelberg (2016)
9. Tweedie, S., Ashburner, M., Falls, K., Leyland, P., McQuilton, P., et al.: FlyBase: enhancing Drosophila gene ontology annotations. Nucleic Acids Res. **37**, 555–559 (2009)
10. Tomancak, P., Beaton, A., Weiszmann, R., Kwan, E., Shu, S., Lewis, S.E., Richards, S., Ashburner, M., Hartenstein, V., Celniker, S.E., Rubin, G.M.: Systematic determination of patterns of gene expression during Drosophila embryogenesis. Genome Biol. **3** (2002). RESEARCH0088, https://genomebiology.biomedcentral.com/articles/10.1186/gb-2002-3-12-research0088

11. Pruteanu-Malinici, I., Mace, D.L., Ohler, U.: Automatic annotation of spatial expression patterns via sparse Bayesian factor models. PLOS Comput. Biol. **7**, e1002098 (2011)

12. Wu, S., Joseph, A., Hammonds, A.S., Celniker, S.E., Yu, B., Frise, E.: Stability-driven nonnegative matrix factorization to interpret spatial gene expression and build local gene networks. Proc. Natl. Acad. Sci. **113**, 201521171 (2016)

13. Zou, H., Hastie, T., Tibshirani, R., Johnstone, I., Lu, A.: Sparse principal component analysis. J. Comput. Graph. Stat. **15**, 1–29 (2006)

14. Hyvarinen, A.: Fast and robust fixed-point algorithm for independent component analysis. IEEE Trans. Neural Netw. **10**, 626–634 (1999)

15. Mairal, J., Bach, F., Ponce, J., Sapiro, G.: Online dictionary learning for sparse coding. In: Proceedings of the 26th Annual International Conference on Machine Learning, ICML 2009, pp. 1–8 (2009)

16. Lin, C.J.: Projected gradient methods for nonnegative matrix factorization. Neural Comput. **19**, 2756–2779 (2007)

17. Belohlavek, R., Vychodil, V.: Discovery of optimal factors in binary data via a novel method of matrix decomposition. J. Comput. Syst. Sci. **76**, 3–20 (2010)

18. Zhang, Z.Y., Li, T., Ding, C., Ren, X.W., Zhang, X.S.: Binary matrix factorization for analyzing gene expression data. Data Mining Knowl. Discov. **20**, 28–52 (2010)

19. Gersho, A., Gray, R.M.: Vector Quantization and Signal Compression, vol. 159. Kluwer Academic Press, Dordrecht (1992). 760

20. Borovec, J.: Fully automatic segmentation of stained histological cuts. In: Husník, L. (ed.) 17th International Student Conference on Electrical Engineering, pp. 1–7. CTU in Prague, Prague (2013)

21. Achanta, R., Shaji, A.: SLIC superpixels compared to state-of-the-art superpixel methods. IEEE Pattern Anal. Mach. Intell. **34**, 2274–2282 (2012)

22. Boykov, Y., Veksler, O.: Fast approximate energy minimization via graph cuts. Pattern Anal. Mach. Intell. **23**, 1222–1239 (2001)

23. Kybic, J., Dolejsi, M., Borovec, J.: Fast registration of segmented images by normal sampling. In: Bio Image Computing (BIC) Workshop at CVPR, pp. 11–19 (2015)

T-Test Based Adaptive Random Walk Segmentation Under Multiplicative Speckle Noise Model

Ang Bian and Xiaoyi Jiang[(✉)]

Department of Mathematics and Computer Science,
University of Münster, Münster, Germany
xjiang@uni-muenster.de

Abstract. Segmentation algorithms typically require some parameters and their optimal values are not easy to find. Training methods have been proposed to tune the optimal parameter values. In this work we follow an alternative goal of adaptive parameter setting. Considering the popular random walk segmentation algorithm it is demonstrated that the parameter used for the weighting function has a strong influence on the segmentation quality. We propose a hypothesis testing based adaptive approach to automatically setting this parameter, thus adapting the segmentation algorithm to the statistic properties of an image. Our data-driven weighting function is developed under the multiplicative speckle noise model. Since the additive Gaussian noise model is its special case, our method is applicable to a broad range of imaging modalities. Experimental results are presented to demonstrate the usefulness of the proposed approach.

1 Introduction

Image segmentation is one of the mostly studied and challenging problems in image analysis. Semi-automatic segmentation methods have been introduced to overcome the complexity of understanding image patterns by user interaction. This kind of algorithms are very helpful in providing targeted segments by roughly labeling scribbles by users. The popular random walk algorithm [3] belongs to this category of segmentation methods. In addition to image segmentation the random walk approach has meanwhile been applied to solve other tasks like ensemble segmentation [10], clustering [1], and semi-automatic 2D-to-3D conversion [6].

Random walk segmentation is a graph-based method that maximizes the image entropy satisfying the Markov property [15]. In [3] a Gaussian weighting function as proposed in [17] was introduced to measure the correspondence of connected nodes by normalizing the difference of intensities with a manual parameter β. For random walk based data classification it is already shown that only suitable parameter β can yield good results [16]. In this paper we propose a data-driven weighting function to solve the parameter setting problem for image segmentation.

C.-S. Chen et al. (Eds.): ACCV 2016 Workshops, Part II, LNCS 10117, pp. 570–582, 2017.
DOI: 10.1007/978-3-319-54427-4_41

Application fields like remote sensing and medical imaging pose additional challenges to image segmentation. In particular, many important imaging sensors deliver data contaminated by signal-dependent noise, i.e., the noise variance depends on the underlying signal intensity, in contrast to the additive Gaussian noise. One example is the so-called speckle noise characteristic for synthetic aperture radar and medical ultrasound imaging [7,9]. Multiple scales spectral clustering has been discussed in [14], in which the point-to-point distance is normalized by the estimated local scales. In our work the data-driven weighting function is developed under the speckle noise model and the pixel similarity can be directly calculated by the hypothesis testing statistics of their neighborhood patches. Since the additive Gaussian noise model is a special case of our speckle model, our method is applicable to a broad range of imaging modalities.

The remainder of this paper is organized as follows. In Sect. 2 we give a brief overview of the random walk segmentation method [3]. After formally defining the multiplicative speckle noise based image model in Sect. 3 we motivate our work by demonstrating the influence of parameter β on segmentation quality in Sect. 4. Section 5 presents our adaptive weighting function. The usefulness of our approach is shown in Sect. 6 on synthetic and real data. Finally, some discussions conclude this paper.

2   Random Walk Image Segmentation

The fundamental idea of random walk segmentation is to assign an unlabeled pixel to the label of the highest probability that a random walker starting from that pixel reaches the user provided scribbles (seeds).

An image is defined as a connected and undirected graph $G(V, E)$, where V is the set of all nodes v_i and E is the set of all edges e_{ij} connecting nodes v_i and v_j. Denoting by x_i^s the probability that a node v_i reaches seeds of label $s = 1, \ldots, l$, for each unseeded node it holds:

$$x_i^s = \frac{1}{d_i} \sum_{e_{ij}} \omega_{ij} x_j^s \tag{1}$$

where ω_{ij} is the weight of edge e_{ij} and $d_i = \sum_{e_{ij}} \omega_{ij}$ is the degree of node v_i. In [3] the weight is defined as Gaussian in a way typical to the dissimilarity measure as used in many other works:

$$\omega_{ij} = exp(-\beta(f_i - f_j)^2) \tag{2}$$

where f_i, f_j are the intensities of two connected nodes v_i, v_j, and β is a manual parameter.

The constrained probabilities (1) can be solved as a combinatorial Dirichlet problem with boundary conditions by its Laplace equation. Define the combinatorial Laplacian matrix L as:

$$L_{ij} = \begin{cases} d_i & \text{if } i = j \\ -\omega_{ij} & \text{if } v_i \text{ and } v_j \text{ are adjacent nodes} \\ 0 & \text{otherwise} \end{cases} \tag{3}$$

By partitioning the nodes into two sets V_M (marked seed nodes, regardless of their label) and V_U (unseeded nodes) and rearranging the probability vector of vertices (for a particular label s, which is skipped to simplify the notation) and matrix L, we can depose the Dirichlet integral into:

$$D[x_U] = \frac{1}{2} \begin{bmatrix} x_M^T & x_U^T \end{bmatrix} \begin{bmatrix} L_M & B \\ B^T & L_U \end{bmatrix} \begin{bmatrix} x_M \\ x_U \end{bmatrix} = \frac{1}{2}(x_M^T L_M x_M + 2x_U^T B^T x_M + x_U^T L_U x_U) \tag{4}$$

where x_M and x_U correspond to the probabilities of seeded and unseeded nodes, respectively, and B is the upper off-diagonal submatrix of the rearranged L with elements $B_{ij} = -\omega_{ij}$, $v_i \in V_M$, $v_j \in V_U$. Since L is positive semidefinite, the minimal solution of $D[x_U]$ would be the only solution of the following system of linear equations:

$$L_U x_U = -B^T x_M \tag{5}$$

This computation is repeated for all labels. Given the probabilities x_i^s of an unseeded node v_i reaching seeds of label $s = 1, \ldots, l$ determined this way, v_i is assigned the label with the highest probability.

3   Multiplicative Speckle Noise Based Image Model

One important characteristic of multiplicative speckle noise is that its variance is proportional to the mean. We can describe the related image formation procedure by the following model [9]:

$$f = u + u^{\gamma/2}\eta, \quad \eta \sim N(0, \sigma^2) \tag{6}$$

where f is the noisy image, u is the original image. In addition, η is random Gaussian variable of variance σ^2, and $\gamma \in \{1, 2\}$ is a parameter determining the signal-dependency of noise variance. Note that when $\gamma = 0$, the noise model described above is Gaussian noise. Another special case is $\gamma = 1$, which indicates the so-called Loupas noise model [4]. In this case the induced noise model is obviously signal-dependent local Gaussian and the perturbations on the image are amplified by the square root of image intensities.

Assuming $G = (V, E)$ as the image graph and f_i as the intensity of any node v_i, the image segmentation refers to a partition of the image graph into l disjoint image regions:

$$V = \cup_{s=1}^{l} V^s, \quad V^i \cap V^j = \emptyset, \quad \forall i \neq j \tag{7}$$

where V^s is the set of all the nodes in an image region with label s. Each region is supposed to be coherent so that the intensity f_i of any node $v_i \in V^s$ is an independent and identical realization of a Gaussian distribution with a regional mean u^s and variance $u^{s\gamma}\sigma^2$, where σ^2 is identical for all regions. The corresponding probability density function is:

$$p(f_i; u^s, u^{s\gamma}\sigma^2) = \frac{1}{\sqrt{2\pi u^{s\gamma}\sigma^2}} \exp \frac{-(f_i - u^s)^2}{2u^{s\gamma}\sigma^2} \tag{8}$$

An interactive image segmentation should achieve a partition of homogeneous regions by their *hidden* regional parameters $u^s, s = 1, \cdots, l$, identified by the user labeled seeds.

4   Influence of Global Parameter β on Segmentation Quality

The segmentation result critically relies on how the edge weights ω_{ij} in (2) are indirectly defined by the global parameter β. However, different settings of β can result in various segment results and the optimal one is not easy to choose. Also, for images of multiple foreground objects, one global β would be debatable as it normalizes the intensity differences of all the regions uniformly. In this section we use two groups of synthetic images to demonstrate the influence of global parameter β on segmentation quality.

The first group of images (256 × 256 pixels) contains four foreground circles with intensities $24, 48, 72, 96$ symmetrically arranged in the middle of the upper-left, upper-right, lower-left, lower-right subsquare regions of the image. The background intensity is set to 12. The contrast ratios of the foreground objects' intensities to background intensity are thus $2, 4, 6, 8$, respectively. Loupas noise of parameter σ ranging from 0.2 to 3 with step 0.2 is generated. For each of these 15 settings we randomly generate 30 images. β is set varying from 1 to 1001 with step 10. For each β and each image, we run six times the random walk method with different number of seeds. For each iteration, we add one foreground seed and one background seed in the upper-right square part of the image, and then symmetrically mark their corresponding points into the other three parts. In the last iteration we thus have 6 seeds for each of the foreground regions and 24 for the background region. We calculate the Dice index ($\frac{2|A \cap B|}{|A|+|B|}$ for set A and B) of the segmented foreground object according to the ground truth. Then, we pick up the β that achieves the best Dice index. Figure 1 shows the distribution of the optimal β values and the average result for each region with all different β values. As a concrete example, five results for one of the 450 test images (corresponding to $\sigma = 2.4$, $\beta = 51, 201, 401, 701, 1001$) are shown in Fig. 1 as well.

The second group of images (256 × 256 pixels) contains multiple foreground objects of various intensities while the background intensity is set to 12. The contrast ratio of foreground to background intensity r_{fb} depends on the foreground object (circle: 1.7–2.9, ellipse: 1.3–2.5, triangle: 1.5–2.7, irregular shape: 1.4 to 2.6, all with step 0.2). Loupas noise of σ from 0.2 to 2 with step 0.2 is added. In total 2100 test images are collected by randomly generating 30 images for each of these 70 settings. Four seeds are marked for each region (foreground and background regions). With β varying from 1 to 1001 with step 10 the test procedure results in the distribution of the optimal β values shown in Fig. 2. As a concrete example, five results for one of the 2100 test images (corresponding to $\sigma = 1$, $r_{fb} = 1.9, 1.5, 1.7, 1.6$ for circle, ellipse, triangle, and irregular shape respectively, $\beta = 51, 201, 401, 701, 1001$) are shown in Fig. 2 as well.

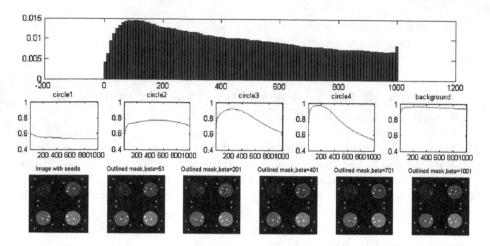

Fig. 1. Test image group 1: distribution of the optimal β (top). Average dice index of each β for all test images (middle). Segmentation results with five different β values for one particular image of this group (bottom).

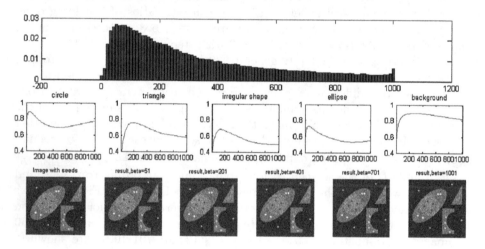

Fig. 2. Test image group 2: distribution of the optimal β (top). Average dice index of each β for all test images (middle). Segmentation results with five different β values for one particular image of this group (bottom).

It is easy to see that the optimal β substantially varies for different images. In Figs. 1 and 2 one can also see the average Dice index of each β for all test images. These results suggest that regions of various statistic properties require locally optimal β. Especially for regions of lower contrast or of higher noise level, the selection of optimal β would be strictly limited in a narrower range. Overall, it is thus not possible to fix the optimal β value. This motivates us to develop an approach to automatically adapting the segmentation algorithm to images contaminated by multiplicative noise.

Dealing with parameters is an important task of image segmentation in general. Traditionally, supervised training approaches have been applied [5]. The critical disadvantage there is the need of ground truth segmentation for the training. Recently, unsupervised parameter learning has been studied, e.g. by means of ensemble consensus learning [12]. Our work follows this unsupervised line and will develop a statistical modeling based approach for the task at hand.

5   T-Test Based Adaptive Random Walk Segmentation

The central task at hand is to design a weight function w_{ij} for each edge of an image graph with two connected nodes v_i and v_j. In [3] this function (2) is defined as Gaussian in terms of $f_i - f_j$ with a global parameter β. Instead of considering v_i and v_j (their intensity f_i and f_j respectively) individually, we use the optimal neighborhood of v_i and v_j that is supposed to build a homogeneous local smooth region. Then, an adaptive weighting is realized by means of statistical testing that indicates the similarity of the two local regions. To ease the reading Table 1 provides a list of the most important symbols.

Table 1. Definition of symbols.

V^s	Set of image nodes in an image region labeled s
u^s	Mean intensity of V^s
f_i	Intensity of node v_i
x_i^s	Probability that unseeded node v_i reaches seeds of label s
$\bar{f}_i$	Estimated mean of intensities in neighborhood f^i
s_i^2	Estimated variance of intensities in neighborhood f^i
σ^2	Global parameter according to our image model

5.1   T-Test Based Weight Function

Assuming that the pixels in an optimal neighborhood belong to one coherent region (smooth cues), we consider the intensities of the nodes in such a neighborhood as independent and identical random samples of a random variable. As image intensities contaminated by speckle noise are locally Gaussian, we can introduce hypothesis testing, concretely two-sample t-test, to measure the similarity of two disjoint neighborhoods.

Without loss of generality, let $N_i = \{v_i^1, \cdots, v_i^{n_i}\}$, $N_j = \{v_j^1, \cdots, v_j^{n_j}\}$, $N_i \cap N_j = \emptyset$, denote the sets of nodes in two disjoint neighborhoods of nodes v_i and v_j, where n_i, n_j are the number of elements. The corresponding intensities $F_i = \{f_i^1, \cdots, f_i^{n_i}\}$ are assumed to come from the same distribution, the same assumption is made for $F_j = \{f_j^1, \cdots, f_j^{n_j}\}$. Considering F_i, F_j as random samples of two random variables, we are interested in testing the following two hypotheses without assumption whether $\text{var}(F_i)$ equals to $\text{var}(F_j)$:

$$H_0 : \text{mean}(F_i) = \text{mean}(F_j), \quad \text{versus} \quad H_1 : \text{mean}(F_i) \neq \text{mean}(F_j) \quad (9)$$

By calculating

$$\bar{f}_i = \frac{1}{n_i} \sum_{k=1}^{n_i} f_i^k, \qquad \bar{f}_j = \frac{1}{n_j} \sum_{k=1}^{n_j} f_j^k$$

$$s_i^2 = \frac{1}{n_i - 1} \sum_{k=1}^{n_i} (f_i^k - \bar{f}_i)^2, \quad s_j^2 = \frac{1}{n_j} \sum_{k=1}^{n_j} (f_j^k - \bar{f}_j)^2 \quad (10)$$

Welch's t-test [11] is defined with the test statistic

$$T = \frac{\bar{f}_i - \bar{f}_j}{\sqrt{\frac{s_i^2}{n_i} + \frac{s_j^2}{n_j}}} \quad (11)$$

and T approximately follows a t-distribution of freedom m under hypothesis H_0 [2], where m is the closest integer of m^*:

$$m^* = \left(\frac{s_i^2}{n_i} + \frac{s_j^2}{n_j}\right)^2 \bigg/ \left[\frac{1}{n_i - 1}\left(\frac{s_i^2}{n_i}\right)^2 + \frac{1}{n_j - 1}\left(\frac{s_j^2}{n_j}\right)^2\right] \quad (12)$$

Thus, the similarity of two disjoint neighborhoods can be measured by the probability density function of the test statistic:

$$\omega_{ij} = \text{pdf}(T; m) = \frac{\Gamma\big((m+1)/2\big)}{\sqrt{m\pi}\,\Gamma(m/2)(1 + T^2/m)^{(m+1)/2}} \quad (13)$$

where Γ is a Gamma function $\Gamma(z) = \int_0^\infty \frac{t^{z-1}}{e^t} dt$. Clearly, this weighting function is positive and symmetrical, and achieves its maximum when $T = 0$, i.e., $\bar{f}_i = \bar{f}_j$, while the function is self-adapted to the local variances of the neighborhoods to be compared.

5.2   Determination of Optimal Neighborhoods

For each edge of an image graph with two connected nodes v_i and v_j we need to determine their respective optimal neighborhood. In addition, these two neighborhoods should be disjoint so that we can apply the t-test based weight function defined above.

The optimal neighborhood of any node v is supposed to build a homogeneous local smooth region around v. Generally, a neighborhood of size $(2k+1) \times (2k+1)$, where k is a positive integer, centered at v may not be a good choice, in particular in the vicinity of a discontinuity. We propose the following optimal neighborhood selection method, whose fundamental idea was already used very early, for instance for local surface fitting [13]. In fact, for each node v we have $(2k+1) \times (2k+1)$ neighborhood candidates $N^1, \cdots, N^{(2k+1)^2}$, each of size $(2k+1) \times (2k+1)$, that contain v. These are simply those neighborhoods centered at each of the $(2k+1) \times (2k+1)$ neighbors of v. Let $F^1, \cdots, F^{(2k+1)^2}$ denote their respective intensity values. As the image intensities are locally Gaussian distributed based on the speckle noise model, we can pick up the optimal neighborhood N^{opt} by the criterion that the intensity f of v most likely comes from the distribution of this neighborhood's intensities F^{opt}:

$$\mathrm{opt} = \mathrm{argmax}_p \quad \mathrm{pdf}_{\mathrm{Gaussian}}\left(f \mid F^p\right)$$

$$= \mathrm{argmax}_p \quad \mathrm{pdf}_{\mathrm{Gaussian}}\left(f; \mathrm{mean}(F^p), \mathrm{var}(F^p)\right) \tag{14}$$

This method is applied to both v_i and v_j to obtain their respective optimal neighborhood N_i^{opt} and N_j^{opt}.

In general, however, these two optimal neighborhoods might be overlapping. To guarantee the independence of the optimal sampling neighborhoods, we need to assign the pixels in the common part $N_i^{opt} \cap N_j^{opt}$ to a unique neighborhood. There are several options for this step and in our current implementation a simple rule is applied. The pixels in $N_i^{opt} \cap N_j^{opt}$ are sorted based on their Euclidean distance to v_i. The nearest ones remain in N_i^{opt} and removed from N_j^{opt} while the others remain in N_j^{opt} and are removed from N_i^{opt}. This is done in such a way that N_i^{opt} and N_j^{opt} finally have the same size.

Overall, we calculate the edge weights using the weighting function (13) and then apply the procedure as detailed in Sect. 2 to generate the final segmentation result. Taking the image in Figs. 1 and 2 as examples, Fig. 3 shows the computed probability maps and the final segmentation. For comparison purpose the original random walk algorithm is run with $\beta = 111, 51$ respectively, which are of the highest frequencies in Figs. 1 and 2.

6   Experimental Results

In this section we assess the performance of our adaptive weighting method on synthetic and real data. The window size parameter k is set to 3, thus 7×7.

Synthetic Data of Geometric Shapes. On the two groups of test images (450 and 2100, respectively) from Sect. 4 we run the random walk algorithm using β varying from 1 to 1001 with step 10. The performance comparison with our adaptive weighting is done in two different ways:

- For each β value the difference $D_{\mathrm{ours}} - D_\beta$ is averaged over all test images, where D_x denotes the Dice index of method x, see Fig. 4. No matter which β value is used, our adaptive weighting method consistently has superior performance.
- For each test image we study $D_{\mathrm{ours}} - D_{opt\_\beta}$, where D_{opt_β} is the Dice index achieved using the optimal β value for each individual image, which is in fact

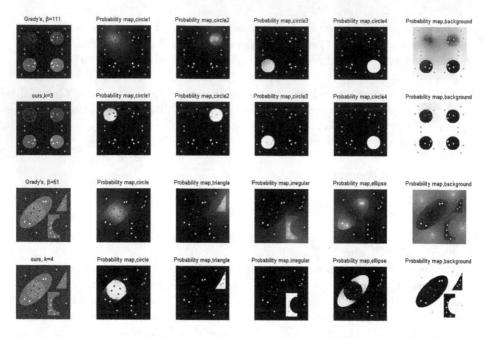

Fig. 3. Probability maps of original random walk algorithm (1st and 3rd row) vs. our method (2nd and 4th row).

unknown. The distributions over all images are shown in Fig. 5. Even when challenging the unknown optimal opt_β, our method is very favorable and achieves better Dice index for 91.01% on regional Dice Index and 98.30% for the average Dice index of the 5 foreground and background regions' Dice Index values of the test images from image group 1. For the test image group 2 the behavior is very similar: 85.95% better for regional results and 87.71% better for averaged results.

Using the second test image group we now study the algorithm behavior for the different shapes therein. Breaking the global statistics in Fig. 5 into that of individual shapes, Fig. 6 displays the distribution of $D_{ours} - D_{opt\_\beta}$ over all test images for each shape. The overall good performance of our adaptive approach is clearly reflected by each individual object in the scene.

Software Phantom Data. We take several images from [8] for testing. Figure 7 shows the segmentation results for a 2D ultrasound software phantom of an apical view of the human heart. Speckle noise was added by this rather simple simulation software with parameters $\gamma = 1.5$ and $\sigma \in [0.125, 0.75]$ as suggested by medical experts.

In [8] a realistic software phantom was also used to overcome the limitations of the simple 2D software phantom above. The extended cardiac-torso (XCAT) phantom provides data that has detailed anatomic structures. A realistic amount of three-dimensional speckle noise was added according to echocardiographic

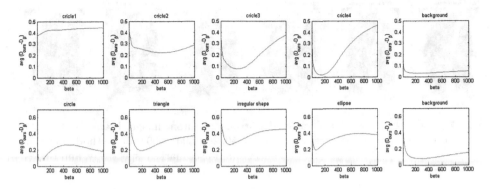

Fig. 4. Average of $D_{\text{ours}} - D_\beta$ over all test images for each β value. Top: test image group 1. Bottom: test image group 2.

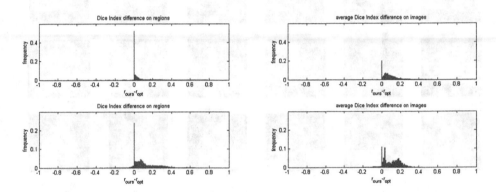

Fig. 5. Distribution of $D_{\text{ours}} - D_{opt\_\beta}$ over all test images. Top: test image group 1. Bottom: test image group 2.

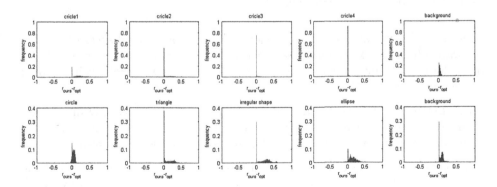

Fig. 6. Distribution of $D_{\text{ours}} - D_{opt\_\beta}$ over all test images (from test image group 2) for individual shapes.

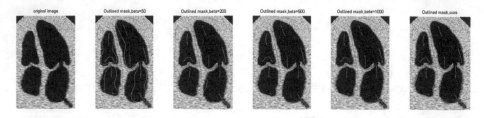

Fig. 7. Segmentation of 2D ultrasound software phantom image. From left to right: original image, random walk result with $\beta = 50, 200, 500, 1000$, our method. The blue and cyan scribbles are foreground and background user marks respectively, and the red curves demonstrate the region boundaries.

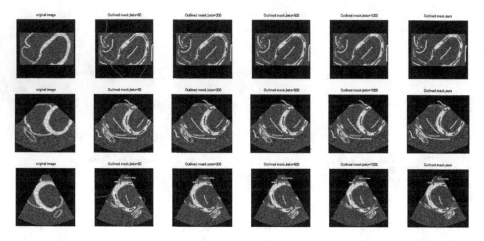

Fig. 8. Segmentation of XCAT phantom slices. From left to right: original image, random walk result with $\beta = 50, 200, 500, 1000$, our method. From top to bottom, the XY, XZ, YZ plane slices. The blue, yellow and cyan scribbles are two foreground objects and background user marks, respectively, and the red curves demonstrate the region boundaries. (Color figure online)

experts. Figure 8 shows the segmentation results for orthogonal slices of this 3D software phantom of a human heart with spatial resolution of $1\,mm^3$. While in all four cases our method achieves good results, none of the β values works consistently well.

Real Data. Figure 9 shows the segmentation results for a real 2D ultrasound B-mode slice of a patient's hypertrophic left ventricle in an apical four-chamber view. Results with $\beta = 50, 200, 500, 1000$ are contrasted with our results.

Discussion. The goal of this work is adaptive parameter setting to avoid the need of manually setting the parameter β. Our approach is not absolutely free of parameter; the window size parameter k has to be defined to specify the neighborhood for local sampling neighborhood selection. However, it is easy to fix this parameter in practice. A reasonable value of k should enable to catch sufficient

Fig. 9. Segmentation of real 2D ultrasound B-mode image. From left to right: original image, random walk result with $\beta = 50, 200, 500, 1000$, our method. The blue, cyan and yellow scribbles are foreground and background user marks, respectively, and the red curves demonstrate the region boundaries. (Color figure online)

statistics for local estimation on the one hand and avoid multiple distributions in the neighborhood on the other hand. In our experiments $k = 3$, i.e. 7×7 local neighborhood, turned out to be a good choice.

Our current way of optimal neighborhood generation is rather time-consuming. The average processing time of 20.65 s for an 256×256 image is mainly caused by 16.49 s for optimal neighborhood generation. We will work towards alternate approaches to improve this bottleneck. All experiments are implemented with Matlab, and run on a computer with Intel(R) Core(TM) i7-2600 CPU, 3.40 GHz, 20 GB.

7 Conclusion

In this work it is demonstrated that the parameter used for the weighting function has a strong influence on the segmentation quality of the popular random walk algorithm. This motivated us to propose a data-driven adaptive method. A statistical model based approach has been developed to automatically generate the edges' weights by their distribution similarity, thus adapting the segmentation algorithm to the statistic properties of an image. Our data-driven weighting function is developed under the multiplicative speckle noise model, which allows our method to be applicable to a broad range of imaging modalities. Experimental results on both synthetic and real data have been presented to demonstrate the usefulness of the proposed approach.

There are several issues for future research. The current implementation for solving the problems of overlapping neighborhoods as described in Sect. 5.2 is rather primitive and more sophisticated approaches may be more reliable. In the current work we have only studied the speckle noise model. In future we will include additional noise models into our consideration. In addition, the adaptive weighting function is investigated in the context of image segmentation. The question arises if this method (idea) can be used for other applications of random walk algorithms like ensemble segmentation [10], clustering [1], and semi-automatic 2D-to-3D conversion [6].

Acknowledgements. Ang Bian was supported by the China Scholarship Council (CSC). Xiaoyi Jiang was supported by the Deutsche Forschungsgemeinschaft (DFG): SFB656 MoBil (project B3) and EXC 1003 Cells in Motion – Cluster of Excellence.

References

1. Abdala, D.D., Wattuya, P., Jiang, X.: Ensemble clustering via random walker consensus strategy. In: International Conference on Pattern Recognition, pp. 1433–1436 (2010)
2. Fisher, R.A.: The fiducial argument in statistical inference. Ann. Eugen. **6**, 391–398 (1935)
3. Grady, L.: Random walks for image segmentation. IEEE Trans. Pattern Anal. Mach. Intell. **28**, 1768–1783 (2004, 2006)
4. Loupas, T., McDicken, W.N., Allan, P.L.: An adaptive weighted median filter for speckle suppression in medical ultrasonic images. IEEE Trans. Circ. Syst. **36**, 129–135 (1989)
5. Pignalberi, G., Cucchiara, R., Cinque, L., Levialdi, S.: Tuning range image segmentation by genetic algorithm. EURASIP J. Adv. Signal Process. **2003**, 780–790 (2003)
6. Phan, R., Androutsos, D.: Robust semi-automatic depth map generation in unconstrained images and video sequences for 2D to stereoscopic 3D conversion. IEEE Trans. Multimedia **16**, 122–136 (2014)
7. Sawatzky, A., Tenbrinck, D., Jiang, X., Burger, M.: A variational framework for region-based segmentation incorporating physical noise models. J. Math. Imaging Vis. **47**, 179–209 (2013)
8. Tenbrinck, D., Schmid, S., Jiang, X., Schäfers, K., Stypmann, J.: Histogram-based optical flow for motion estimation in ultrasound imaging. J. Math. Imaging Vis. **47**, 138–150 (2013)
9. Tenbrinck, D., Jiang, X.: Image segmentation with arbitrary noise models by solving minimal surface problems. Pattern Recogn. **48**, 3293–3309 (2015)
10. Wattuya, P., Rothaus, K., Praßni, J., Jiang, X.: A random walker based approach to combining multiple segmentations. In: Proceedings of International Conference on Pattern Recognition, pp. 1–4 (2008)
11. Welch, B.L.: The generalization of student's problem when several different population variances are involved. Biometrika **34**, 28–35 (1947)
12. Wu, Z., Jiang, X., Zheng, N., Liu, Y., Cheng, D.: Exact solution to median surface problem using 3D graph search and application to parameter space exploration. Pattern Recogn. **48**, 380–390 (2015)
13. Yokoya, N., Levine, M.D.: Range image segmentation based on differential geometry: a hybrid approach. IEEE Trans. Pattern Anal. Mach. Intell. **11**, 643–649 (1989)
14. Zelnik-Manor, L., Perona, P.: Self-tuning spectral clustering. In: NIPS, pp. 1601–1608 (2005)
15. Zhang, J.: The mean field theory in EM procedures for Markov random fields. IEEE Trans. Signal Process. **40**, 2570–2583 (1992)
16. Zhu, X., Ghahramani, Z., Lafferty, J.: Semi-supervised learning using Gaussian fields and harmonic functions. In: Proceedings of International Conference on Machine Learning, pp. 912–919 (2003)
17. Zhu, X., Lafferty, J., Ghahramani, Z.: Combining active learning and semi-supervised learning using Gaussian fields and harmonic functions. In: ICML Workshop on the Continuum from Labeled to Unlabeled Data in Machine Learning and Data Mining (2003)

Langerhans Islet Volume Estimation from 3D Optical Projection Tomography

Jan Švihlík[1,3][✉], Jan Kybic[1], David Habart[2], Hanna Hlushak[1], Jiří Dvořák[1,4], and Barbora Radochová[5]

[1] Biomedical Imaging Algorithms (BIA) Group, Faculty of Electrical Engineering, Department of Cybernetics, Center for Machine Perception, Czech Technical University in Prague, Technická 2, Prague, Czech Republic
svihlj1@fel.cvut.cz
[2] Institute for Clinical and Experimental Medicine, Vídeňská 1958/9, Prague, Czech Republic
[3] Faculty of Chemical Engineering, Department of Computing and Control Engineering, University of Chemistry and Technology, Technická 5, Prague, Czech Republic
[4] Faculty of Mathematics and Physics, Department of Probability and Mathematical Statistics, Charles University in Prague, Sokolovská 83, Prague 8, Czech Republic
[5] Department of Biomathematics, Institute of Physiology, Czech Academy of Sciences, Vídeňská 1083, Prague, Czech Republic

Abstract. This paper concerns the comparison of automatic volume estimation methods for isolated pancreatic islets. The estimated islet volumes are needed during the process of assessing the islet sample quality prior to the islet transplantation. We study several different methods for automatic volume estimation. For this purpose we acquired a set of projections using optical tomography for a sample of an islet population. Based on these projections we estimated the islet volumes using two stereological methods (the automatic Wulfsohn's method and the manual fakir method, considered to be the ground truth in this study), together with the filtered back projection followed by 3D segmentation. We have also employed two simple methods, currently used in medical practice, based on fitting a sphere or a prolate ellipsoid to a single binarized 2D islet projections.

1 Introduction

Transplantation of isolated pancreatic islets from cadaver donors is a promising therapy for patients with the type 1 diabetes [1]. To determine the quality of the isolated islets and their suitability for successful transplantation, microscopy images (2D) of islet graft samples are acquired and the volume of the islets is estimated.

The classical Ricordi approach for islet volume estimation is to use an optical microscope with a calibrated grid for manual islet diameter measurement [2].

© Springer International Publishing AG 2017
C.-S. Chen et al. (Eds.): ACCV 2016 Workshops, Part II, LNCS 10117, pp. 583–594, 2017.
DOI: 10.1007/978-3-319-54427-4_42

The islet volume, under the assumption of the spherical shape, is then estimated from the histogram of islet diameters (determined manually). This manual method is very time consuming. Image processing methods were employed during the last two decades in order to reduce the manual workload.

The currently used method consists of microscopy image acquisition (i.e. obtaining an islet projection), segmentation of the microscopy images to get binary images and fitting a circle (assuming a spherical shape) or an ellipse (assuming a prolate ellipsoidal shape of the islet) to the 2D segmented projection [3].

Several different approaches were used to segment the microscopy images of the Langerhans islets, e.g. Girman et al. [4] use multilevel thresholding of RGB microscopy images, where the thresholds are set manually. However, such an approach is very time consuming and medical expert dependent. In [5] authors proposed an automated algorithm for the analysis of microscopy images of Langerhans islets, where the segmentation method is based on a trained random forest classifier (HSV features) followed by a graph cut regularization.

The ultimate goal is to find a robust method for islet volume estimation from a single 2D projection (microscopy image) that could be quickly and easily used for large islet samples in the transplantation practice where the stringent time constraints prevent one from using slower and/or more expensive procedures. So far only the simple methods mentioned above (assuming the spherical or prolate ellipsoidal shape of the islet) are used.

To provide ground truth information in studies of volume estimation methods from 2D images of Langerhans islets one has to obtain also 3D information about the same islets, using e.g. optical tomography to obtain projections of the islets from different angles. From such data the islet volume can be estimated by the established stereological method called fakir [6]. This method is manual and time-consuming and we aim to investigate the possibility to use automatic methods for volume estimation from a set of projections in order to decide whether they can replace the fakir method in providing the ground truth information in future larger scale studies focusing on the volume estimation from a single projection.

The fakir method is based on applying perpendicular systems of linear probes and indicating the intersection of the probes with the object under study. Since the medical expert determines the boundary of the islets, this method is considered to be the ground truth in this study.

The fakir method is as reproducible and objective as the criteria for recognition of the object surface. When such criteria are more difficult to implement than to explain to human operator, i.e. it is difficult to segment the object, the interactive method is better. In fact, manual delineation is widely used for segmentation of 3D objects in medicine and biology. When the volume and/or surface area is of interest, our fakir approach requires fraction of time needed to manually delineate the object sections to obtain required precision, because only discrete points on the object surface have to be sampled. The information from neighboring pixels is highly correlated, so sampling rather sparse points may be sufficiently precise and more efficient. In practice, the fakir method is both precise and reproducible, and hence objective.

As for the automatic estimation methods based on a set of projections we employ a classical approach based on reconstruction of sections using filtered back projection (FBP) followed by segmentation of the reconstructed sections. However, the quality of reconstruction is poor due to insufficient transparency of the islets, see Fig. 3. To overcome the problem with islet transparency we employ the stereological Wulfsohn's method which allows to estimate the islet volume directly from the segmented projections [7]. It combines the Cavalieri principle with the known formula for the area in terms of the support function of a convex set. For comparison we also consider two simple methods, commonly used in practice, based on the spherical or ellipsoidal shape assumption and a single projection.

2   Method

2.1   Image Data and Segmentation Technique

We acquired a set of projections using optical tomography (OPT) for each of 60 islets from 6 donors. The volume of the individual islets was then estimated, based on this data, by a medical expert using the stereological fakir method [6].

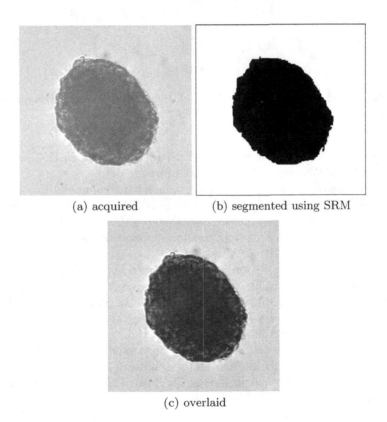

(a) acquired   (b) segmented using SRM

(c) overlaid

Fig. 1. Example of OPT projection and corresponding segmentation.

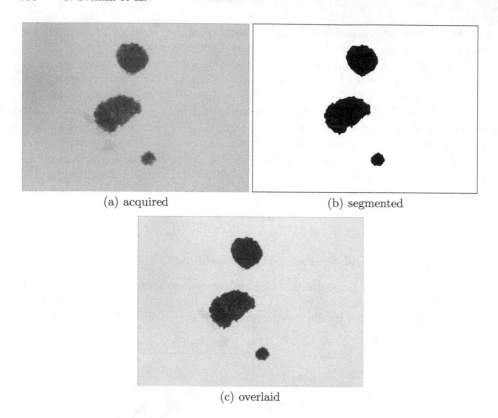

(a) acquired   (b) segmented

(c) overlaid

Fig. 2. Example of RGB microscopy images of islet projection and corresponding segmentation.

The methods for the islet volume estimation presented in this paper need binarized images. We work with two types of image data.

The first type of images—sets of projections for the volume estimation methods based on the 3D information, see Subsects. 2.2, 2.3 and 2.4—are 16 bpp grayscale optical tomography projections (1000×1000 pixels). In our experiment we use the optical tomography scanner Milano developed in cooperation with the Technical University in Milano.

For every islet there are 401 projections obtained by rotating the islet around an axis by a fixed angle. These projections are segmented by the statistical region merging (SRM) algorithm [8], see Fig. 1. SRM (implemented in Matlab) was chosen as a compromise between the segmentation quality and the time necessary for the segmentation of a single projection.

Images of the second type—single projections used for the methods of volume estimation from the 2D information, see Subsect. 2.5—contain 8 bpp RGB microscopy images (2048×1536 pixels), see Fig. 2(a). These images are segmented using method decribed in [5]. Firstly, the microscopy images are preprocessed to compensate for nonuniform illumination and to apply a color normalization.

We obtain a pixelwise probability map using a random forest classifier (using color HSV features). We consider two classes: islets vs. background + the so called exocrine tissue which wraps the islets and sometimes is not completely removed during the isolation process. The final classification (see Fig. 2(b)) is obtained using GraphCut. The algorithm was implemented in Fiji [9].

2.2   Fakir Method

The fakir method is a stereological method based on the so called fakir probes [6]. A fakir probe is a regular system of parallel lines intercepting the 3D object of interest. The volume estimation is based on manually determining the length of such intercepts between 3D object and the lines of the probe. In order to increase the efficiency of estimation, usually three perpendicular probes are used in practice. The volume estimate of a given islet is then

$$V_F = \frac{\lambda^2}{3}(L_1 + L_2 + L_3), \tag{1}$$

where λ is the grid constant, i.e. the distance between the neighboring parallel lines of the probe and $L_i, i = 1, 2, 3$, is the length of the intercepts between the object and the lines of the i-th probe.

2.3   Wulfsohn's Method

The Wulfsohn's method [7] is a stereological method for volume estimation from projections from different angles. The method assumes the object to be axially convex, i.e. we assume that there is a fixed axis L and that all the planar sections through the object, perpendicular to the axis L, are convex sets. The projections are obtained at angles φ_j with angular steps $\Delta\varphi_j$ in such a way that the axis L is always parallel to the projection plane.

The estimation is based on the Cavalieri principle and the known formula for the area $A(K)$ of a convex planar set K in terms of its support function h_K. In this case K or K_u will be a planar section through the object perpendicular to the axis L, at height u measured along L. It is rather surprising that it is possible to measure the value of the support function $h_{K_u}(\varphi)$ at the angle φ (measured in the section plane containing K_u) from the available projection data – it is enough to look at the projection at angle φ perpendicular to L. The value $h_{K_u}(\varphi)$ is then simply the distance from L (in the projection image) to the rightmost point of the object at height u in the direction perpendicular to L. For more technical details see [7]. In the following the symbol $h'_{K_u}(\varphi)$ will denote the derivative of $h_{K_u}(\varphi)$ with respect to φ.

The estimation then proceeds as follows:

1. Segmentation of OPT projections.
2. Small objects (artifacts) removal from the binary segmentation.
3. Determining the position of the axis L from the centroids computed from the projections.

4. Computing the values of the support function $h_{K_u}(\varphi)$.
5. The area of the planar section K_u at height u is given by the formula

$$A(K_u) = \frac{1}{2} \int_0^{2\pi} \left(h_{K_u}(\varphi)^2 - h'_{K_u}(\varphi)^2 \right) \mathrm{d}\varphi, \qquad (2)$$

which can be approximated numerically and estimated as

$$\widehat{A(K_u)} \approx \frac{1}{4} \sum_j (\varphi_{j+1} - \varphi_j) \left(H_{j+1}(u) + H_j(u) \right), \qquad (3)$$

where $H_j(u) = h_{K_u}^2(\varphi_j) - \widehat{h'_{K_u}(\varphi_j)}^2$ and $h_{K_u}(\varphi_j)$ denotes the support function computed for the projection at angle φ_j. The integral in (2) is approximated using the trapezoidal rule. The support function is filtered by a Gaussian filter due to noise contamination resulting from the segmentation procedure.

The derivative of the support function $h'_{K_u}(\varphi_j)$ was approximated by the difference $\widehat{h'_{K_u}(\varphi_j)} = \frac{1}{\Delta\varphi_j} [h_{K_u}(\varphi_{j+1}) - h_{K_u}(\varphi_j)]$, where $\Delta\varphi_j = \varphi_{j+1} - \varphi_j$.

6. The volume of the object $V = \int_{-\infty}^{\infty} A(K_u)\,\mathrm{d}u$ is then estimated by

$$V_W = \Delta u \sum_i A(K_{u_i}), \qquad (4)$$

where Δu is the discretization step along the axis L, i.e. the pixel size in the projection image, and u_i are the discrete values of u.

2.4   Filtered Back Projection

The filtered back projection is a well known method for the reconstruction of sections from the projections of a 3D object [10]. The reconstructed sections are of rather poor quality in our case because the islets are not sufficiently transparent. Hence, we segment the volume slice by slice using a level set algorithm [11]. Example of a reconstructed section and the corresponding segmentation can be seen in Fig. 3. The volume of a given islet is then directly calculated from the segmented slices, known pixel size and the distance between the neighboring slices.

2.5   Fitting an Ellipsoid and Sphere

In this subsection we deal with the methods of volume estimation from a single projection, as is the most commonly used method in clinical practice. A prolate ellipsoid and a sphere are the simple shape models currently used for islet volume estimation from a single projection [12]. In the case of the sphere model we measure the area A of the islet profile in a binarized image. The radius of the islet is estimated as $r = \sqrt{p^2 A/\pi}$, where p^2 denotes the pixel area. The islet volume is then estimated as $V_S = \frac{4}{3}\pi r^3$.

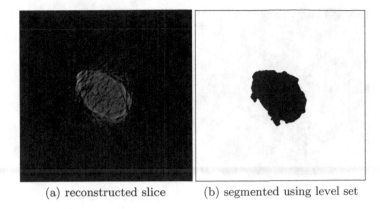

(a) reconstructed slice   (b) segmented using level set

Fig. 3. Example of FBP reconstruction from OPT projections and the corresponding segmentation.

In the case of the prolate ellipsoid model an ellipse is fitted to the binarized islet profile using the method based on the second moments [3]. The islet volume is then estimated as $V_P = \frac{4}{3}\pi a^2 b$, where a is the length of the minor semiaxis and b is the length of the major semiaxis of the fitted ellipse.

3 Results

By the different methods considered in this paper we have estimated the volumes of 60 Langerhans islets from 6 donors. We consider the volumes given by the manual fakir method to be the ground truth in this study [13].

The volumes of 30 Langerhans islets were evaluated by two independent experts using the fakir method. The relative error was computed for each islet as follows: $E_{FF}(i) = 2|V_{F1}(i) - V_{F2}(i)|/(V_{F1}(i)) + V_{F2}(i))$, where $V_{F1}(i)$ and $V_{F2}(i)$ are the volumes of the i-th islet determined by the fakir method by the first and the second expert. The mean relative error E_{FF} between the experts was 4.07%. The mean relative error of the volume estimates by fakir method evaluated by each (individual) expert was about 1% (the principle of the estimation of the relative error for one expert in the case of fakir method can be found in [14]).

For the islets, if a certain number of projections (in our case maximally 12 subsequent projections) turned out to be unusable (see Fig. 4), these projections were removed. Hence, $\Delta\varphi_j$ is in fact not constant through the given set in that case. However, the volume estimates (Wulfsohn's method is considered) can be computed also for these reduced sets of projections. Two examples of projections of the same islet are given in Fig. 4. Note that the left projection is overlaid by impurity and cannot be used for volume estimation – it cannot be segmented correctly.

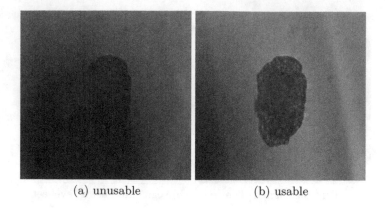

      (a) unusable   (b) usable

Fig. 4. Projections of Langerhans islets from OPT microscopy.

Table 1. The relative error of the islet volume estimates. Fakir method was used as the ground truth. Methods to compare: Wulfsohn's, FBP, Sphere, Ellipsoid.

Set	No. of islets	Relative error [%]							
		Wulfsohn's		FBP		Sphere		Ellipsoid	
		Mean	Std	Mean	Std	Mean	Std	Mean	Std
Donor 1	5	3.20	2.68	11.02	3.17	31.12	23.68	15.23	12.30
Donor 2	6	8.67	5.84	8.86	3.68	37.38	49.89	37.83	37.28
Donor 3	11	2.33	2.16	11.86	2.88	26.01	21.21	14.75	17.45
Donor 4	9	4.03	1.97	12.82	3.28	19.76	20.57	10.38	7.44
Donor 5	5	4.62	2.81	10.17	2.64	27.30	14.86	15.40	6.46
Donor 6	24	5.02	3.86	4.96	3.96	42.91	18.37	29.51	18.94
	Overall	4.56	3.47	8.73	3.49	33.50	24.39	22.40	19.04

Numerical comparison of the studied methods is shown in Tables 1 and 2. The relative error was computed for each islet as follows: $E_R(i) = |V_F(i) - V(i)|/V_F(i)$, where $V_F(i)$ is the volume of the i-th islet determined by the fakir method and $V(i)$ is the volume of the i-th islet estimated by tested method (i.e. Wulfsohn's, FBP, Sphere and Ellipsoid methods). The relative bias was computed for each islet as follows: $B_R(i) = (V_F(i) - V(i))/V_F(i)$. The relative error E_R and relative bias B_R were averaged for each individual donor. The number of islets used for the volume estimation for a given donor is listed in the column "No. of islets".

The tables show that the Wulfsohn's method and the FBP give better results and are in rather good agreement with the manual fakir method, considered to be the ground truth in this study. The numerical results also indicate that the Wulfsohn's method provides more precise estimates than FBP in terms of relative bias and relative error of the estimates.

Table 2. The relative bias of the islet volume estimates. Fakir method was used as the ground truth. Methods to compare: Wulfsohn's, FBP, Sphere, Ellipsoid.

Set	No. of islets	Relative bias [%]							
		Wulfsohn's		FBP		Sphere		Ellipsoid	
		Mean	Std	Mean	Std	Mean	Std	Mean	Std
Donor 1	5	−1.53	3.88	11.02	3.17	31.12	23.68	8.78	17.50
Donor 2	6	−8.47	5.98	8.06	5.99	5.90	67.08	−11.10	56.28
Donor 3	11	−0.70	−3.10	11.86	−2.88	21.94	−25.40	7.20	−21.68
Donor 4	9	−3.74	2.47	12.82	3.28	17.81	22.28	1.27	12.71
Donor 5	5	4.02	2.01	10.17	2.04	27.30	14.86	12.24	11.37
Donor 6	24	−2.50	5.81	2.65	5.77	31.41	34.52	16.82	30.76
	Overall	−3.05	4.66	7.73	4.63	24.72	34.41	8.88	29.02

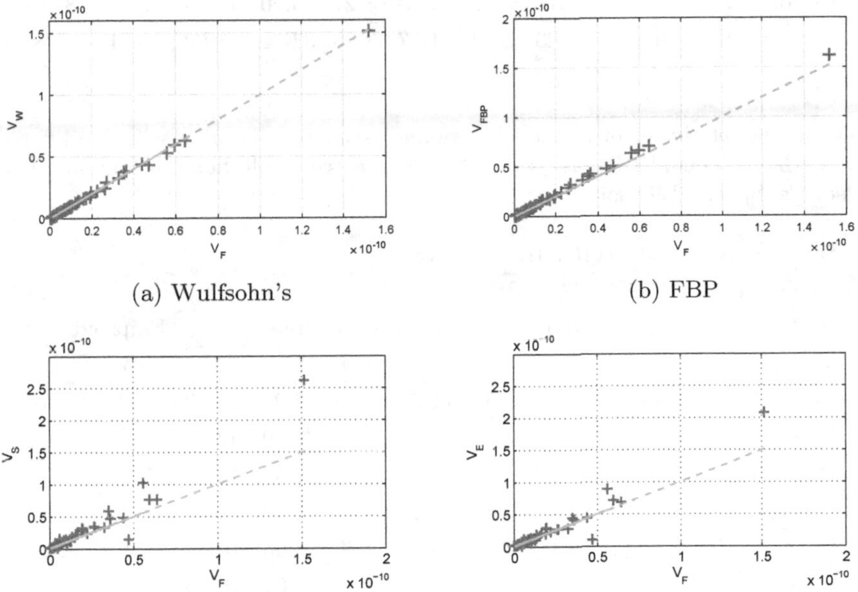

(a) Wulfsohn's

(b) FBP

(c) Sphere

(d) Ellipsoid

Fig. 5. Scatter plot of volumes [m³] estimated by tested methods in comparison with ground truth volumes (fakir method).

The two methods based on a single projection (assuming a spherical or ellipsoidal shape) provide estimates with large systematic bias and high variability, see also Fig. 5. Estimates with high positive bias are observed mainly in the case of large islets. This is caused by the fact that the larger islets are considerably more flat shaped and more irregular in comparison with small islets.

To assess the deviations between the methods we also calculated the numerical characteristics for the estimation methods also in two other situations where

Table 3. The relative error of the islet volume estimates. For assessing the difference between the methods, Wulfsohn's method was used as the ground truth. Methods to compare: Fakir, FBP, Sphere, Ellipsoid.

Set	No. of islets	Relative error [%]							
		GT: Wulfsohn's method							
		Fakir		FBP		Sphere		Ellipsoid	
		Mean	Std	Mean	Std	Mean	Std	Mean	Std
Donor 1	5	3.32	2.91	12.85	3.69	33.31	24.16	16.41	11.30
Donor 2	6	12.04	6.04	22.73	5.57	56.55	65.38	50.11	45.22
Donor 3	11	2.36	2.19	12.70	3.19	26.24	23.96	15.73	18.87
Donor 4	9	4.24	2.10	17.21	1.51	22.59	21.17	11.25	7.12
Donor 5	5	4.94	3.08	15.63	5.02	33.74	17.76	18.76	11.96
Donor 6	24	5.16	3.82	5.53	3.27	45.70	19.03	31.82	18.16
	Overall	5.03	3.54	11.77	3.58	37.72	28.63	25.24	20.78

Table 4. The relative error of the islet volume estimates. For assessing the difference between the methods, FBP was used as the ground truth. Methods to compare: Fakir, Wulfsohn's, Sphere, Ellipsoid.

Set	No. of islets	Relative error [%]							
		GT: FBP							
		Fakir		Wulfsohn's		Sphere		Ellipsoid	
		Mean	Std	Mean	Std	Mean	Std	Mean	Std
Donor 1	5	9.86	2.54	11.29	2.89	22.21	20.59	13.35	11.36
Donor 2	6	8.10	3.41	16.92	1.55	39.20	42.51	39.32	33.44
Donor 3	11	10.54	2.33	11.20	2.46	16.31	16.16	15.76	10.77
Donor 4	9	11.29	2.56	17.21	1.51	15.92	11.70	12.58	8.39
Donor 5	5	9.18	2.21	13.36	3.76	16.05	10.64	7.91	3.79
Donor 6	24	4.76	3.72	5.16	2.89	40.01	19.00	26.85	19.63
	Overall	7.93	3.10	10.44	2.63	28.49	20.94	21.22	17.60

we consider the Wolfsohn's method or the FBP, respectively, to be the ground truth. The results are given in Tables 3 and 4. Together with the results discussed above they provide a consistent picture of the relationship between the methods: the Wulfsohn's method and FBP are, loosely speaking, closer to the fakir method than to each other. Wulfsohn's method gives, on average, slightly smaller estimates than the fakir method while FBP gives larger values of the estimates than fakir.

4   Conclusion

We have estimated volumes of 60 islets from 6 donors based on the OPT projections using the fakir method, the Wulfsohn's method and the filtered back projection. We also estimated the islet volumes from a single projection using the simple spherical and ellipsoidal models.

The evaluated relative error and relative bias of the FBP and the Wulsohn's method are acceptable for medical experts. However, we recommend the use of the Wulfsohn's method as it estimates the volume directly from the segmented projections without the need of reconstruction of the whole islet and provide more precise estimates in terms of relative bias and relative error, as seen from Tables 1 and 2.

The numerical results also show that the ellipsoidal model gives relatively good results, taking into account its simplicity and the fact that it uses only a single projection. This suggests that at least some part of the islet population in fact has a rather ellipsoidal shape.

Acknowledgement. This work has been supported by the grant 14-10440S "Automatic analysis of microscopy images of Langerhans islets" of the Czech Science Foundation and by MEYS (LM2015062 Czech-BioImaging).

References

1. Alejandro, R., Barton, F.B., Hering, B.J., Wease, S.: Collaborative islet transplant registry investigators: update from the collaborative islet transplant registry. Transplantation **86**, 1783–1788 (2008)
2. Ricordi, C., Gray, D.W., Hering, B.J., et al.: Islet isolation assessment in man and large animals. Acta Diabetol. Lat. **27**, 185–195 (1990)
3. Mulchrone, K.F., Choudhury, K.R.: Fitting an ellipse to an arbitrary shape: implications for strain analysis. J. Struct. Geol. **26**, 143–153 (2004)
4. Girman, P., Kříž, J., Friedmanský, J., Saudek, F.: Digital imaging as a possible approach in evaluation of islet yield. Cell Transplant. **12**, 129–133 (2003)
5. Habart, D., Švihlík, J., Schier, J., et al.: Automated analysis of microscopic images of isolated pancreatic islets. Cell Transplant. **25**, 2145–2156 (2016)
6. Kubínová, L., Janáček, J., Guilak, F., Opatrný, Z.: Comparison of several digital and stereological methods for estimating surface area and volume of cells studied by confocal microscopy. Cytometry **36**, 85–95 (1999)
7. Wulfsohn, D., Gundersen, H.J.G., Jensen, E.B.V., Nyengaard, J.R.: Volume estimation from projections. J. Microsc. **215**, 111–120 (2004)
8. Nock, R., Nielsen, F.: Statistical region merging. IEEE Trans. Pattern Anal. Mach. Intell. **26**, 1452–1458 (2004)
9. Schindelin, J., Arganda-Carreras, I., Frise, E., Kaynig, V., Longair, M., Pietzsch, T., Preibisch, S., Rueden, C., Saalfeld, S., Schmid, B., Tinevez, J.Y., White, D.J., Hartenstein, V., Eliceiri, K., Tomancak, P., Cardona, A.: Fiji: an open-source platform for biological-image analysis. Nat. Methods **9**, 676–682 (2012)
10. Cheddad, A., Svensson, C., Sharpe, J., Georgsson, F., Ahlgren, U.: Image processing assisted algorithms for optical projection tomography. IEEE Trans. Med. Imaging **31**, 1–15 (2012)

11. Li, C., Xu, C., Gui, C., Fox, M.D.: Distance regularized level set evolution and its application to image segmentation. IEEE Trans. Image Process. **19**, 3243–3254 (2010)
12. Švihlík, J., Kybic, J., Habart, D., Berková, Z., Girman, P., Kříž, J., Zacharovová, K.: Classification of microscopy images of Langerhans islets. In: Sebastien Ourselin, M.A.S. (ed.) Medical Imaging 2014 Image Processing, pp. 1–8. SPIE, Bellingham (2014)
13. Hlushak, H.: Accuracy evaluation of langerhans islet volume estimation from microscopic images. Master's thesis, Czech Technical University in Prague (2016)
14. Janacek, J.: Variance of periodic measure of bounded set with random position. Comment. Math. Univ. Carol. **47**, 443–455 (2006)

Level Set Segmentation of Brain Matter Using a Trans-Roto-Scale Invariant High Dimensional Feature

Naveen Madiraju[1], Amarjot Singh[2(✉)], and S.N. Omkar[3]

[1] Department of ECE, National Institute of Technology, Warangal, India
[2] Faculty of Applied Sciences, Simon Fraser University, Burnaby, Canada
asa168@sfu.ca
[3] Department of Aerospace Engineering, Indian Institute of Science, Bangalore, India

Abstract. Brain matter extraction from MR images is an essential, but tedious process performed manually by skillful medical professionals. Automation can be a potential solution to this complicated task. However, it is an ambitious task due to the irregular boundaries between the grey and white matter regions. The intensity inhomogeneity in the MR images further adds to the complexity of the problem. In this paper, we propose a high dimensional translation, rotation, and scale-invariant feature, further used by a variational framework to perform the desired segmentation. The proposed model is able to accurately segment out the brain matter. The above argument is supported by extensive experimentation and comparison with the state-of-the-art methods performed on several MRI scans taken from the McGill Brain Web.

1 Introduction

Accurate brain matter segmentation is essential for the study of various ailments such as multiple sclerosis [1], Alzheimer [2], and Parkinson disease [2]. Hence, skilled medical professionals spend a considerable amount of their time manually labeling white and grey matter in the MR images. However, accurate labeling of brain matter is a tedious process due to which substantial difference in labeling is observed for different labelers. This difference can potentially be avoided by making use of automatic methods that are capable of automatically segmenting the brain matter.

Automated segmentation of the brain matter from the MR images is a particularly challenging task. The brain matter varies considerably in shape and size with irregular boundaries making gradient-based algorithms inefficient. Moreover, the corruption of images with a smoothly varying intensity inhomogeneity caused due to the image acquisition defects [3] further hinders the segmentation process. In addition, both tissues (grey and white) can have similar intensity levels in certain regions that further adds to the complexity of the problem. The within-tissue intensities for both tissues can also vary greatly, making it impossible to classify images based on pixel intensities.

© Springer International Publishing AG 2017
C.-S. Chen et al. (Eds.): ACCV 2016 Workshops, Part II, LNCS 10117, pp. 595–609, 2017.
DOI: 10.1007/978-3-319-54427-4_43

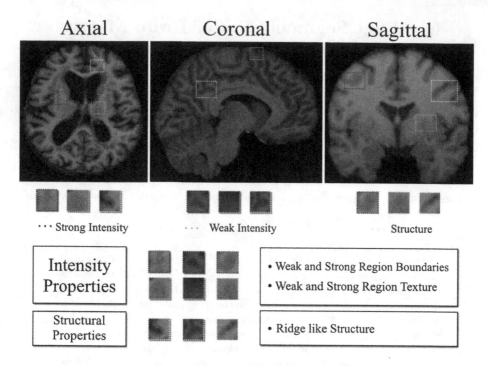

Fig. 1. Illustrates intensity and structural properties extracted from axial, coronal and sagittal view of brain MRI. It is shown that (weak, strong) region boundaries and texture are covered under intensity properties while ridge like structures are covered with structural properties.

In the recent past, numerous *two-stage* segmentation strategies have been recommended for brain matter segmentation [4–13]. Most of the strategies include a primary stage that extracts a high-dimensional feature, which combines the structural and intensity properties (as in Fig. 1) of brain matter in MR images. This high-dimensional feature is further used by a secondary supervised or a variational framework to achieve the desired brain matter segmentation. Supervised approaches use labeled data to automatically learn a model that is further used by a classifier such as Support vector machines [14], Markov random field [15], Neural networks [16] etc., for segmentation. However, statistical classification approaches are inefficient due to the presence of irregular boundaries and intensity inhomogeneity as mentioned above. On the contrary, variational methods have been relatively successful for medical image segmentation due to their ability to achieve sub-pixel accuracy and immunity towards topological changes [17, 18]. These methods can also effectively coordinate distinctive information in a principled manner (e.g., boundary information, region information, shape priors, texture for vector valued images) [2, 19, 20] to achieve optimal segmentation by minimizing the energy function defined in level sets framework.

Multiple methods have been introduced that incorporate local intensity information in an energy framework for image segmentation [21–25]. However, local

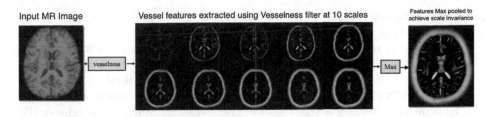

Fig. 2. Illustrates the vesselness feature extracted at 10 different scales. Next, max pooling is applied to achieve a scale invariant feature.

intensity itself may not be sufficient for accurate segmentation in the presence of heavy noise and intensity inhomogeneity [22]. Recently, Wang et al. [22] characterized an energy equation using local Gaussian distribution fitting of image intensities. This energy was minimized using level set to achieve the brain matter segmentation. He et al. [20] proposed a method that used local entropy feature derived from the grey level distribution of MR image instead of intensity, for segmentation. Further, Huang et al. [26] combined voxel probability, image gradient, and curvature information to achieve brain tissue segmentation. Finally, Popuria et al. [27] used various MRI modalities and their texture characteristics to construct a multi-dimensional feature set to achieve the same task.

The above-mentioned studies have proven that a combined feature set is well suited for brain matter segmentation as opposed to a single feature. However, no explanation is provided as to what kind of features best captures the properties of the brain matter in MR images. In this paper, a two-stage segmentation pipeline is used to achieve brain matter segmentation from MR images. The first stage of the pipeline uses a proposed translation, rotation and scale invariant high-dimensional feature derived by combining three sub-features that best represent the structural and intensity properties of the brain matter. The proposed high-dimensional feature is then used by a novel variational level sets method [28] to achieve the desired brain matter segmentation.

The main contributions of the first and second stage of the proposed segmentation pipeline are presented as below:

- *High-dimensional Feature*: The first stage of the architecture proposes a high dimensional feature that captures the intensity and structural properties of the brain matter. The intensity properties are represented in the brain matter by region boundaries and texture while the ridge-like structures correspond to the structural properties of the matter as shown in Fig. 1. Gabor [29] and Texture [30] filters were used to represent the intensity properties of the brain while the structural properties were extracted using a vesselness [31] filter. The intensity and structural properties can appear at any position, orientation and scale in the MR images. Hence, invariance to translation, rotation, and scale is introduced in each sub-feature. The invariant sub-features are weighted and combined to derive the high dimensional feature that precisely represents the brain matter properties.

- *Variational Framework*: The second stage of the pipeline uses an extended version of the variational framework proposed by Li et al. [28]. The variational framework is adapted to accommodate the proposed high dimensional feature as the objective function obtained by integration over the whole image domain. The objective function is minimized in the level-set framework to achieve the desired brain matter segmentation.

The remaining of this paper is organized as follows. The next section presents the proposed two-stage architecture for brain matter segmentation. The first subsection describes the formulation of the high dimensional feature and the reasons for selecting the specific sub-features. The second subsection illustrates the formulation of the variational framework and the minimization of the objective function with the incorporated high dimensional feature. Finally, Sect. 3 details the experimental results along with a qualitative and quantitative comparison with the state of the art methods. Section 4 presents the conclusions.

2   Proposed Model

This section details the proposed two-stage brain matter segmentation pipeline. In the first sub-section, the sub-features chosen to formulate the proposed high-dimensional feature are presented. Next, the mathematical formulation that presents the incorporation of the proposed high-dimensional feature within the Li et al.'s variational framework is introduced. Finally, the detailed derivation for the minimization of the objective function with the incorporated high-dimensional feature, that results in the desired brain matter segmentation, is presented.

2.1   Proposed High-Dimensional Feature

The proposed High-Dimensional feature is composed of three sub-features that attempt to capture the structural and intensity properties of the brain matter. Vesselness filter [31] is used to capture the ridge-like structural features of the brain matter while the intensity properties in the form of region boundaries and texture are captured using Gabor [32] and MR8 [33] texture filters respectively. The structural and intensity properties are shown in Fig. 1.

1. *Vesselness:* The brain matter contains ridge structures that can be exploited for superior segmentation. A few articles in the recent past have presented systems for vessel or ridge extraction using the eigen-decomposition of the Hessian calculated at every pixel [31, 34–38]. Further, Sofka et al. [31] developed an improvised filter to detect low contrast and narrow vessel structures. In this paper, the vessels are obtained by searching for tubular geometrical structures using second-order derivative information (principal curvatures of

Vesselness feature extracted at Scales = 1, 5 and 10

Pooled Scale Invariant Vesselness feature

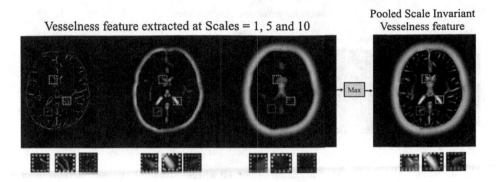

Fig. 3. Illustrates the max pooling of vesselness features at 3 different scales to achieve scale invariance. A 20 × 20 patch is selected at a spatial location from each feature map extracted at each scale. Max operation applied to the selected patches to achieve scale invariance.

image intensities). Vessels are distinguished by a locally prominent low curvature orientation (the vessel direction) and a plane of high-intensity curvature (the cross-sectional plane). The vesselness filter [31] uses the Hessian matrix to capture the above-mentioned geometric information. The local second-order structure of the image is decomposed into principal directions extracted using eigenvalue analysis of the Hessian. The eigenvalue decomposition extracts three orthonormal directions.

Invariance to scale is realized by extracting the vesselness using filters at different scales s, followed by the integration of their responses using a max operator to obtain a final estimate of vesselness as illustrated in Fig. 2. The process using which invariance is achieved is presented in detail in Fig. 3. A 10×10 patch is selected at a spatial location from each feature image extracted at every scale. A max operation is performed across patches selected from features extracted at each scales to achieve invariance as shown in Fig. 3. The process is detailed for 10 different scales due to lack of space.

2. *Edges*: The region boundaries separating the grey and white matter are important descriptors needed to distinguish between both matters. These descriptors can be extracted using edge detecting algorithms [32,39] developed in recent past. In this paper, one such method popularly known as Gabor filter is used to achieve rotation and scale invariance. Gabor [32] is a linear filter which is used to identify edges and line endings of white and grey matter over different scales and orientations. It's impulse response is determined by multiplication of harmonic function and gaussian function. Here, scale and rotational invariance are realized by extracting the Gabor features using filters at 3 scales and 8 orientation (as explained in the previous section), followed by the integration of their responses using a max operator to obtain rotation and scale invariance.

Feature Extraction and Concatenation

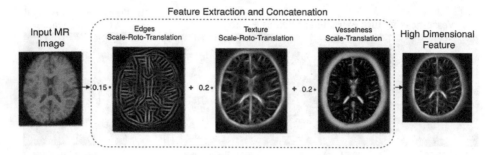

Fig. 4. Shows the weighted concatenation of all three invariant, gabor, texture and vessel feature to result into a trans-roto-scale high dimensional feature.

3. *Texture*: Texture is an another intensity property which can provide important information which can help to achieve optimal brain matter segmentation. It can be observed from the MR images that the grey matter has a more rough texture as compared to the white matter. The texture of an image is related to the spatial distribution of the intensity values of brain and white matter, and as such contains information regarding contrast, uniformity, rugosity, regularity, etc. This feature is efficient due to its sparse nature as it lies in lower dimensional subspaces. A few algorithms [33, 40, 41] were developed to extract this property. However, we make use of MR8 filter [33] to achieve roto-scale invariance. The texture information is captured using a MR8 filter bank [33] that consists of a Gaussian and a Laplacian of Gaussian (these filters have rotational symmetry) filter, an edge filter at 3 scales and a bar filter at the same 3 scales. The latter two filters are oriented and occur at 6 orientations at each scale. Measuring only the maximum response across orientations reduces the number of responses from 38 (6 orientations at 3 scales for 2 oriented filters, plus 2 isotropic) to 8 (3 scales for 2 filters, plus 2 isotropic) and achieves rotation invariance. A max operation is performed across the rotational invariant filters at different scales to achieve scale invariance in addition to rotation invariant to result into a roto-scale (rotation and scale) invariant feature (as explained above).

Features extracted at different scales and orientations using texture, edge, and vesselness filters are max pooled to achieve translational, rotational and scale invariance. The Fig. 2 shows vesselness features at 3 different scales, combined to achieve scale invariance. Similarly, Gabor based edges features and texture features at different scales and orientations are combined to achieve translational and rotational invariance. The invariant features are weighed and combined to give rise to a unified super feature that combines vesselness, edges and a texture filter features to derive a trans-roto-scale invariant high dimensional feature that is best suited for brain matter detection. The high dimensional feature is formulated by combining the above-mentioned sub features as presented below:

$$\mathcal{F} = \mathcal{V}_w \mathcal{V} + \mathcal{G}_w \mathcal{G} + \mathcal{T}_w \mathcal{T} \tag{1}$$

where $\mathcal{V}$ is vesselness feature, $\mathcal{G}$ is gabor feature and $\mathcal{T}$ is local texture feature while $\mathcal{V}_w$, $\mathcal{G}_w$ and $\mathcal{T}_w$ are the respective weights for each feature that decide the relative contribution of each feature. The similarity of the high dimensional feature with brain matter allows for better segmentation by the second stage of the architecture. The combined high dimensional feature is shown in Fig. 4.

2.2   Objective Function Formulation

In this section, the proposed high dimensional invariant feature is introduced in the variational segmentation framework proposed by [28] to achieve the brain matter segmentation. In this model, the MR image with the intensity inhomogeneity is formulated as:

$$I = bJ + n \tag{2}$$

Here I is the input image, b is the bias field which represents the amount of intensity inhomogeneity, n is the additive noise and J is the true image. In this model: (i) In a circular neighbourhood O_y centered at y the bias field is assumed to be constant (ii) Intensity belonging to the ith tissue in a small circular neighbourhood should take a specific value c_i as in [19]. Here Neighbourhood size of O_y is controlled by a truncated gaussian kernel $K(y - x)$ defined by:

$$K(z) = \begin{cases} \frac{1}{a}e^{-|z|^2/2\sigma^2} & |z| \leq r, \\ 0 & \text{otherwise} \end{cases}$$

Here a is the normalization constant such that $\int K(z) = 1$. σ represents the size of the neighbourhood. We choose an optimal σ based on the level of intensity inhomogeneity of image.

The aim is to segment the image domain $\Omega \subset \mathcal{R}^2$ into three distinct regions namely: grey matter (Ω_1), white matter (Ω_2) and background (Ω_3). In order to achieve this task, a n-dimensional feature $\mathcal{F} : \{F_i...\} \mid i \in \{1,2..n\}$ with $\mathcal{F} \in \Omega$ is extracted from the input image I. Next, each circular neighbourhood region O_y is initially segmented into N (N=3) clusters using a localised K-means clustering. All the segmented circular neighbourhoods are integrated over the image domain Ω to obtain the energy functional given by:

$$\mathcal{E}_y = \int_\Omega \left(\sum_{i=1}^3 \int_{\Omega_i} K(y - x)|\mathcal{F}(y)|(I(x) - b(y)c_i)|^2 dx \right) dy \tag{3}$$

The label assigned to each subregion within a circular neighbourhood may or may not be true. In order to determine the most probable label for each region, the energy functional is minimised as described in the next section.

2.3   Objective Function Minimization in Level-Set Framework

The energy equation is minimised by modifying the proposed energy functional equation as mentioned in the previous section into the level sets framework. In order to segment the image in N=3 disjoint regions, atleast two $(n > log(N))$ level set functions $\Phi = (\phi_1, \phi_2)$ are required [25]. The energy functional equation is modified using the membership function M_i for each disjoint region as below:

$$\mathcal{E}(\phi, c_i, b) = \sum_{i=1}^{3} \int_{\Omega} \left(\int_{\Omega} K(y - x)\mathcal{F}^2(y)|I(x) - b(y)c_i|^2 M_i(\phi(x))dy \right) dx \quad (4)$$

Here $ci = (c_1, c_2, c_3)$ and $M_i = (M_1, M_2, M_3)$. For simplicity, we use vector forms for $\mathbf{c} = (c_1, c_2, c_3)$ and $\mathbf{\Phi}(\mathbf{x}) = (\phi(x)_1, \phi(x)_2, \phi(x)_3)$. Equation 4 can be further simplified as below:

$$\mathcal{E}(\mathbf{\Phi}, \mathbf{c}, b) = \sum_{i=1}^{3} \int_{\Omega} (e_i(x)M_i(\mathbf{\Phi}(\mathbf{x}))dx) \quad (5)$$

where $e_i(x) = \int_{\Omega} K(y - x)\mathcal{F}^2(y)|I(x) - b(y)c_i|^2)dy$. For mathematical simplification we expand integral and rewrite e_i as

$$e_i(x) = I^2(x)T_1 - 2c_i I(x)T_2 + c_i^2 T_3 \quad (6)$$

where $T_1 = \int_{\Omega} K(y - x)\mathcal{F}^2(y)dy$, $T_2 = \int_{\Omega} K(y - x)\mathcal{F}^2(y)b(y)dy$ and $T_3 = \int_{\Omega} K(y - x)\mathcal{F}^2(y)b^2(y)dy$.

In order for the level set to effectively segment the image into the required disjoint regions, the energy functional is formulated using two regularization terms, $\mathcal{L}(\phi_j)$ and $\mathcal{R}(\phi_j)$, where j is the number of level set functions. The entire energy functional with regularization terms is mentioned below:

$$E(\mathbf{\Phi}, \mathbf{c}, b) = \mathcal{E}(\mathbf{\Phi}, \mathbf{c}, b) + \nu \left(\sum_{j=1}^{2} \mathcal{L}(\phi_j) \right) + \mu \left(\sum_{j=1}^{2} \mathcal{R}(\phi_j) \right) \quad (7)$$

where $\mathcal{L}(\phi)$ and $\mathcal{R}(\phi)$ are defined as in [42] as follows:

$$\mathcal{L}(\phi) = \int |\nabla H(\phi(x))|dx \quad (8)$$

$$\mathcal{R}(\phi) = \int \frac{1}{2} (|\nabla \phi(x)| - 1)^2 dx \quad (9)$$

where H is the heaviside function. The membership function can also be written using the heaviside function as: $M_1(\phi_1, \phi_2) = H(\phi_1)H(\phi_2)$, $M_2(\phi_1, \phi_2) = H(\phi_1)(1 - H(\phi_2))$ and $M_3(\phi_1, \phi_2) = 1 - H(\phi_1)$. The heaviside function H can be approximated by a smooth function H_ϵ defined by

$$H_\epsilon(u) = \frac{1}{2} \left[1 + \frac{2}{\pi} arctan \left(\frac{u}{\epsilon} \right) \right] \quad (10)$$

The derivative of H_ϵ is given by

$$\delta_\epsilon(u) = H'_\epsilon(u)$$
$$= \frac{1}{\pi} \frac{\epsilon}{\epsilon^2 + u^2} \quad (11)$$

The entire energy functional Eq. 7 is minimized with respect to Φ to achieve optimal image segmentation. The energy is minimized through the following gradient flow equations:

$$\frac{\partial \phi_1}{\partial t} = - \delta(\phi_1)\left(H(\phi_2)e_1 + (1 - H(\phi_2))e_2 - e_3\right)$$
$$+ \nu\delta_\epsilon(\phi_1)div\left(\frac{\nabla\phi_1}{|\nabla\phi_1|}\right) + \mu\left(\nabla^2\phi_1 - div\left(\frac{\nabla\phi_1}{|\nabla\phi_1|}\right)\right) \tag{12}$$

$$\frac{\partial \phi_2}{\partial t} = - \delta(\phi_2)\left(H(\phi_1)e_1 - H(\phi_1)e_2\right)$$
$$+ \nu\delta_\epsilon(\phi_2)div\left(\frac{\nabla\phi_2}{|\nabla\phi_2|}\right) + \mu\left(\nabla^2\phi_2 - div\left(\frac{\nabla\phi_2}{|\nabla\phi_2|}\right)\right) \tag{13}$$

To find optimal c, we keep Φ and b constant and minimize the energy equation with respect to c. The optimal c is given by:

$$c_i = \frac{\int_\Omega \left((\mathcal{F}^2 b) * K\right) I(x)M_i(\Phi(x))dx}{\int_\Omega \left((\mathcal{F}^2 b^2) * K\right) I(x)M_i(\Phi(x))dx} \tag{14}$$

Similarly by making Φ and c constant, we minimize the energy equation with respect to b. This can be written as:

$$b(y) = \frac{(S_1 I) * K}{(S_2 I) * K} \tag{15}$$

where $S_1 = \sum_{i=1}^3 c_i M_i(\Phi(x))$ and $S_2 = \sum_{i=1}^3 c_i^2 M_i(\Phi(x))$.

3   Experimentation

In this section, the parameter setting of the high-dimensional feature along with the experimentation details of the proposed algorithm is presented in detail. The parameters of the high dimensional feature are tuned on 60 brain MRI images chosen from the McGill Brain Web [43]. The performance of the proposed algorithm using the optimal high-dimensional feature is evaluated using the quantitative and qualitative results on 40 brain MR images. In addition, the qualitative and quantitative comparison are presented with 3 state-of-the-art brain segmentation methods. In addition, the quantitative comparison is further extended to 10 addition brain segmentation methods. The comparison is presented at different levels of intensity non-uniformity (INU) and noise in the subsequent sections.

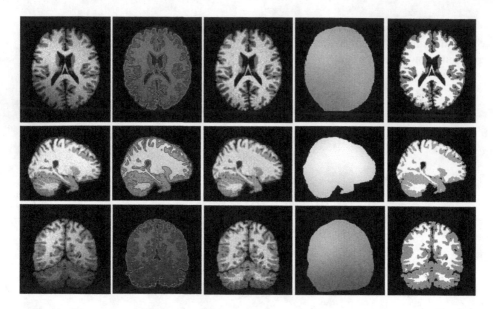

Fig. 5. Application of proposed method to axial, coronal and sagittal slices of brain MR images. Column 1: original images. Column 2: final contours. Column 3: corrected image. Column 4: bias field. Column 5: segmented image.

Parameters Setting: The parameters of the proposed method are set according to the attributes of the input images. In the subsequent discussion, the reasoning and values of the chosen parameters are provided. In the following experiments, all the parameters have values as follows:

1. We obtained the values of $\mathcal{V}_w$, $\mathcal{G}_w$ and $\mathcal{T}_w$ as 0.2, 0.15 and 0.2 respectively. The best feature weights ($\mathcal{V}_w$, $\mathcal{G}_w$, $\mathcal{T}_w$) for each feature are obtained heuristically using 60 training images. The process is discussed below: First, the weights for each feature (gabor, texture, and vessel) are randomly assigned using which the high dimensional feature is computed. This high dimensional feature is used in the objective function to obtain the cost value. The weights are modified after which the above-mentioned steps are repeated. The weights that result into the lowest cost value are finally decided that are used to form the final high dimensional feature.
2. As discussed by [28], a gaussian kernel of window size $64 * 64$ with a standard deviation $\sigma = 16$ is used to deal with high bias present in the Brain MRI.
3. The energy regularization coefficient μ and distance regularization coefficient ν are set as 1 and $0.001 * 255^2$ respectively as in [28]. The parameter ϵ in Heaviside function is set to 1 as discussed in chan and vese model [44].

Qualitative and Quantitative Results: Figure 5 shows the qualitative results obtained after applying the proposed algorithm to few images selected from the McGill Brain Web. The first column shows brain images of intensity non-uniformity(INU) = 100% and noise level of 3%. The second column shows final

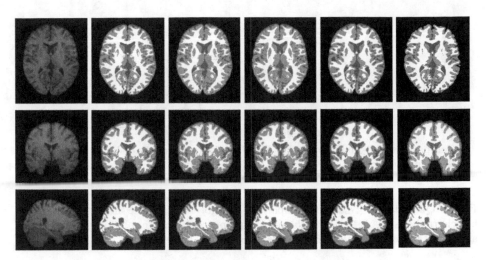

Fig. 6. Qualitative Comparison with Well et al.'s method [45], Leemput et al.'s method [1] and Wang et al.'s method [22] using simulated data obtained from McGill Brain Web. Column 1: original images. Column 2: Wells method. Column 3: Leemput's method. Column 4: Wang's method. Column 5: our proposed methodQualitative Comparison with Well et al.'s method [45], Leemput et al.'s method [1] and Wang et al.'s method [22] using simulated data obtained from McGill Brain Web. Column 1: original images. Column 2: Wells method. Column 3: Leemput's method. Column 4: Wang's method. Column 5: our proposed method

contour position while the third column shows the bias corrected images. The fourth column shows estimated bias field and the fifth column shows the segmented image. It can be observed that the intensities become sufficiently homogeneous within each tissue in the bias corrected images in column 3.

Next, Fig. 6 portrays the qualitative comparison of the proposed algorithm with the other state of the art methods. By careful observation of the segmentation results, as shown in Fig. 6, it can be inferred that our algorithm produces superior white matter segmentation results. For instance in Fig. 6, the lower left region of the sagittal image and in the lower region of the coronal image, the segmentation results of our algorithm looks most similar to ground truth. However, the segmentation results of other algorithms look similar to each other but are different from the ground truth. Further, we quantify the results and obtain that our algorithm yields more accurate results.

It is important to perform a quantitative evaluation in addition to the qualitative analysis to better understand the performance of the proposed method. In this paper, Jaccard similarity [3] measure is used to measure the similarity of the segmented regions with the ground truth. The measure is given by:

$$\mathcal{J}(S_1, S_2) = \frac{|S_1 \cap S_2|}{|S_1 \cup S_2|} \tag{16}$$

value of $\mathcal{J}$ lies between 0 and 1. Higher the value of $\mathcal{J}$ accurate the segmentation.

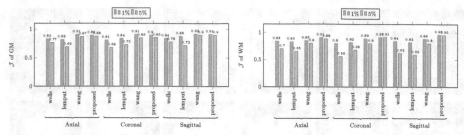

(a) Comparison of J similarity of Grey Mat-(b) Comparison of J similarity of White
ter segmentation results for various algo-Matter segmentation results for various al-
rithms   gorithms

Fig. 7. Quantitative (J) value comparison of algorithms for the 40 test images. The
bar chart shows results of [1, 22, 45] and the proposed method applied on axial, coronal
and sagittal slices of brain MR image at 1% and 5% noise levels.

A bar graph is shown in Fig. 7 portrays a Jaccard similarity J comparison of
segmentation results for both grey and white matter for the proposed and state
of the methods. From Fig. 7(a) shows grey matter segmentation results at 1%
noise levels. It can be observed that our algorithm produces slightly poor results
as compared to Wang et al.'s method [22] but yields superior results compared to
Well et al.'s method [45] and Leemput et al.'s method [1]. However, in presence
of 5% noise, our algorithm performs better than rest of the algorithms.

Table 1. The table shows comparison of average Jaccard (J) values of grey matter
and white matter of our algorithm with current state of the art algorithms.

Method	J of grey matter	J of white matter
Adaptive MAP [6]	0.564	0.567
Biased MAP [6]	0.558	0.562
Fuzzy c-means [6]	0.473	0.567
Maximum-a-posteriori (MAP) [6]	0.550	0.554
Maximum-likelihood [6]	0.535	0.551
Tree-structure k-means [6]	0.477	0.571
MPM-MAP [7]	0.662	0.683
BSE/BFC/PVC [8]	0.595	0.664
Constrained GMM (MAP) [9]	0.680	0.660
Spatial-varying GMM [10]	0.768	0.734
Coupled surface [11]	0.701	NA
FSL [12]	0.756^2	NA
SPM [13]	0.790^2	NA
MAP with histograms [6]	0.814	0.710
Decision forest classifier [12]	0.838	0.731
Our algorithm	**0.89**	**0.91**

Subsequently, Fig. 7(b) shows white matter segmentation results. It can be perceived that our algorithm performs better at both 1% and 5% noise setting than other methods. It can be concluded that our algorithm produces superior results in presence of high bias and high noise.

A more extensive comparison with other brain matter segmentation algorithms is presented in Table 1. The proposed algorithm outperforms the algorithms by a decent margin.

4   Conclusion

In this paper, we proposed a novel translation, rotation and scale invariant high dimensional feature that resembles the properties of brain matter. This proposed feature is used by level set based segmentation method to achieve sub-pixel segmentation of brain matter in the presence of noise and inhomogeneities. It is observed from the experiments that the high-dimensional feature gives superior segmentation performance over simple features. In addition, it can be concluded from the results that the similarity of the proposed feature to the structural and intensity properties of the brain matter further enhances the segmentation performance. Experiments demonstrate the superior performance of the proposed algorithm when compared to the state of the art methods.

References

1. Van Leemput, K., Maes, F., Vandermeulen, D., Suetens, P.: Automated model-based bias field correction of MR images of the brain. IEEE Trans. Med. Imaging 18, 885–896 (1999)
2. Wang, L., Chen, Y., Pan, X., Hong, X., Xia, D.: Level set segmentation of brain magnetic resonance images based on local Gaussian distribution fitting energy. IEEE Trans. Med. Imaging 188, 316–325 (2010)
3. Vovk, U., Pernus, F., Likar, B.: A review of methods for correction of intensity inhomogeneity in MRI. IEEE Trans. Med. Imaging 26, 405–421 (2007)
4. Angoth, V., Dwith, C., Singh, A.: A novel wavelet based image fusion for brain tumor detection. IEEE Trans. Med. Imaging 2, 1–7 (2013)
5. Dwith, C., Angoth, V., Singh, A.: Wavelet based image fusion for detection of brain tumor. IEEE Trans. Med. Imaging 5, 25 (2013)
6. Yi, Z., Criminisi, A., Shotton, J., Blake, A.: Discriminative, semantic segmentation of brain tissue in MR images. In: Yang, G.-Z., Hawkes, D., Rueckert, D., Noble, A., Taylor, C. (eds.) MICCAI 2009. LNCS, vol. 5762, pp. 558–565. Springer, Heidelberg (2009). doi:10.1007/978-3-642-04271-3_68
7. Marroquín, J.L., Vemuri, B.C., Botello, S., Calderón, F., Fernandez-Bouzas, A.: An accurate and efficient Bayesian method for automatic segmentation of brain MRI. IEEE Trans. Med. Imaging 21, 934–945 (2002)
8. Shattuck, D.W., Sandor-Leahy, S.R., Schaper, K.A., Rottenberg, D.A., Leahy, R.M.: Magnetic resonance image tissue classification using a partial volume model. IEEE Trans. Med. Imaging 13, 856–876 (2001)
9. Greenspan, H., Ruf, A., Goldberger, J.: Constrained Gaussian mixture model framework for automatic segmentation of MR brain images. IEEE Trans. Med. Imaging 25, 1233–1245 (2006)

10. Peng, Z., Wee, W., Lee, J.H.: Automatic segmentation of MR brain images using spatial-varying Gaussian mixture and Markov random field approach. In: Conference on Computer Vision and Pattern Recognition Workshop, CVPRW 2006, pp. 80–80. IEEE (2006)

11. Zeng, X., Staib, L.H., Schultz, R.T., Duncan, J.S.: Volumetric layer segmentation using coupled surfaces propagation. In: Proceedings of 1998 IEEE Computer Society Conference on Computer Vision and Pattern Recognition, pp. 708–715. IEEE (1998)

12. Zhang, Y., Brady, M., Smith, S.: Segmentation of brain MR images through a hidden Markov random field model and the expectation-maximization algorithm. IEEE Trans. Med. Imaging **20**, 45–57 (2001)

13. Ashburner, J., Friston, K.: Multimodal image coregistration and partitioning–a unified framework. Neuroimage **6**, 209–217 (1997)

14. Magnin, B., Mesrob, L., Kinkingnéhun, S., Pélégrini-Issac, M., Colliot, O., Sarazin, M., Dubois, B., Lehéricy, S., Benali, H.: Support vector machine-based classification of Alzheimers disease from whole-brain anatomical MRI. Neuroradiology **51**, 73–83 (2009)

15. Khayati, R., Vafadust, M., Towhidkhah, F., Nabavi, M.: Fully automatic segmentation of multiple sclerosis lesions in brain MR flair images using adaptive mixtures method and Markov random field model. IEEE Trans. Med. Imaging **38**, 379–390 (2008)

16. Shen, S., Sandham, W., Granat, M., Sterr, A.: MRI fuzzy segmentation of brain tissue using neighborhood attraction with neural-network optimization. IEEE Trans. Med. Imaging **9**, 459–467 (2005)

17. Cobzas, D., Birkbeck, N., Schmidt, M., Jagersand, M., Murtha, A.: 3D variational brain tumor segmentation using a high dimensional feature set. In: IEEE 11th International Conference on Computer Vision, ICCV 2007, pp. 1–8. IEEE (2007)

18. Singh, A., Karanam, S., Bajpai, S., Choubey, A., Raviteja, T.: Malignant brain tumor detection. Int. J. Comput. Theor. Eng. **4**, 1002–1006 (2011)

19. Cui, W., Wang, Y., Lei, T., Fan, Y., Feng, Y.: Level set segmentation of medical images based on local region statistics and maximum a posteriori probability. Comput. Math. Methods Med. **2013**, 1–12 (2013)

20. He, C., Wang, Y., Chen, Q.: Active contours driven by weighted region-scalable fitting energy based on local entropy. IEEE Trans. Med. Imaging **92**, 587–600 (2012)

21. Li, C., Kao, C.Y., Gore, J.C., Ding, Z.: Minimization of region-scalable fitting energy for image segmentation. IEEE Trans. Med. Imaging **17**, 1940–1949 (2008)

22. Wang, L., He, L., Mishra, A., Li, C.: Active contours driven by local Gaussian distribution fitting energy. IEEE Trans. Med. Imaging **89**, 2435–2447 (2009)

23. Wang, Y., Xiang, S., Pan, C., Wang, L., Meng, G.: Level set evolution with locally linear classification for image segmentation. IEEE Trans. Med. Imaging **46**, 1734–1746 (2013)

24. Hahn, J., Lee, C.O.: Geometric attraction-driven flow for image segmentation and boundary detection. IEEE Trans. Med. Imaging **21**, 56–66 (2010)

25. Vese, L.A., Chan, T.F.: A multiphase level set framework for image segmentation using the Mumford and Shah model. IEEE Trans. Med. Imaging **50**, 271–293 (2002)

26. Albert, H., Rafeef, A., Roger, T., Anthony, T.: Automatic MRI brain tissue segmentation using a hybrid statistical and geometric model. In: 3rd IEEE International Symposium on Biomedical Imaging: Nano to Macro, pp. 394–397 (2006)

27. Karteek, P., Dana, C., Martin, J., Sirish L., S., Albert, M.: 3D variational brain tumor segmentation on a clustered feature set. In: Medical Imaging (2009)
28. Li, C., Huang, R., Ding, Z., Gatenby, J., Metaxas, D.N., Gore, J.C.: A level set method for image segmentation in the presence of intensity inhomogeneities with application to MRI. IEEE Trans. Med. Imaging **20**, 2007–2016 (2011)
29. Zacharaki, E.I., Wang, S., Chawla, S., Soo Yoo, D., Wolf, R., Melhem, E.R., Davatzikos, C.: Classification of brain tumor type and grade using MRI texture and shape in a machine learning scheme. IEEE Trans. Med. Imaging **62**, 1609–1618 (2009)
30. Pitiot, A., Delingette, H., Thompson, P.M., Ayache, N.: Expert knowledge-guided segmentation system for brain MRI. IEEE Trans. Med. Imaging **23**, 985–996 (2004)
31. Frangi, A.F., Niessen, W.J., Vincken, K.L., Viergever, M.A.: Multiscale vessel enhancement filtering. In: Wells, W.M., Colchester, A., Delp, S. (eds.) MICCAI 1998. LNCS, vol. 1496, pp. 130–137. Springer, Heidelberg (1998). doi:10.1007/BFb0056195
32. Mehrotra, R., Namuduri, K.R., Ranganathan, N.: Gabor filter-based edge detection. IEEE Trans. Med. Imaging **25**, 1479–1494 (1992)
33. Geusebroek, J.M., Smeulders, A.W., Van de Weijer, J.: Fast anisotropic gauss filtering. IEEE Trans. Med. Imaging **12**, 938–943 (2003)
34. Aylward, S.R., Bullitt, E.: Initialization, noise, singularities, and scale in height ridge traversal for tubular object centerline extraction. IEEE Trans. Med. Imaging **21**, 61–75 (2002)
35. Lindeberg, T.: Edge detection and ridge detection with automatic scale selection. IEEE Trans. Med. Imaging **30**, 117–156 (1998)
36. Damon, J.: Properties of ridges and cores for two-dimensional images. IEEE Trans. Med. Imaging **10**, 163–174 (1999)
37. Lindeberg, T.: Feature detection with automatic scale selection. IEEE Trans. Med. Imaging **30**, 79–116 (1998)
38. Sato, Y., Nakajima, S., Atsumi, H., Koller, T., Gerig, G., Yoshida, S., Kikinis, R.: 3D multi-scale line filter for segmentation and visualization of curvilinear structures in medical images. In: Troccaz, J., Grimson, E., Mösges, R. (eds.) CVRMed/MRCAS -1997. LNCS, vol. 1205, pp. 213–222. Springer, Heidelberg (1997). doi:10.1007/BFb0029240
39. Canny, J.: A computational approach to edge detection. IEEE Trans. Pattern Anal. Mach. Intell. **6**, 679–698 (1986)
40. Leung, T., Malik, J.: Representing and recognizing the visual appearance of materials using three-dimensional textons. IEEE Trans. Med. Imaging **43**, 29–44 (2001)
41. Schmid, C.: Constructing models for content-based image retrieval. In: Proceedings of the 2001 IEEE Computer Society Conference on Computer Vision and Pattern Recognition, CVPR 2001, vol. 2, pp. II-39. IEEE (2001)
42. Li, C., Xu, C., Gui, C., Fox, M.D.: Distance regularized level set evolution and its application to image segmentation. IEEE Trans. Med. Imaging **19**, 3243–3254 (2010)
43. Web, B.: Simulated brain database. McConnell Brain Imaging Centre, Montreal Neurological Institute, McGill (2004). http://www.bic.mni.mcgill.ca/brainweb/
44. Chan, T.F., Vese, L., et al.: Active contours without edges. IEEE Trans. Med. Imaging **10**, 266–277 (2001)
45. Wells, W.M., Grimson, W.E.L., Kikinis, R., Jolesz, F.A.: Adaptive segmentation of MRI data. IEEE Trans. Med. Imaging **15**, 429–442 (1996)

Discriminative Subtree Selection for NBI Endoscopic Image Labeling

Tsubasa Hirakawa[1]([✉]), Toru Tamaki[1], Takio Kurita[1], Bisser Raytchev[1], Kazufumi Kaneda[1], Chaohui Wang[2], Laurent Najman[2], Tetsushi Koide[3], Shigeto Yoshida[4], Hiroshi Mieno[4], and Shinji Tanaka[5]

[1] Hiroshima University, Higashihiroshima, Japan
hirakawat@hiroshima-u.ac.jp
[2] Laboratoire d'Informatique Gaspard-Monge, Université Paris-Est, Champs-sur-Marne, France
[3] Research Institute for Nanodevice and Bio Systems (RNBS), Hiroshima University, Higashihiroshima, Japan
[4] Hiroshima General Hospital of West Japan Railway Company, Hiroshima, Japan
[5] Hiroshima University Hospital, Hiroshima, Japan

Abstract. In this paper, we propose a novel method for image labeling of colorectal Narrow Band Imaging (NBI) endoscopic images based on a tree of shapes. Labeling results could be obtained by simply classifying histogram features of all nodes in a tree of shapes, however, satisfactory results are difficult to obtain because histogram features of small nodes are not enough discriminative. To obtain discriminative subtrees, we propose a method that optimally selects discriminative subtrees. We model an objective function that includes the parameters of a classifier and a threshold to select subtrees. Then labeling is done by mapping the classification results of nodes of the subtrees to those corresponding image regions. Experimental results on a dataset of 63 NBI endoscopic images show that the proposed method performs qualitatively and quantitatively much better than existing methods.

1   Introduction

Colorectal cancer has been one of the major cause of cancer death in many advanced countries [1]. For early detection of colorectal cancer, colorectal endoscopy (colonoscopy) with Narrow-Band Imaging (NBI) system is widely used, where endoscopists observe the condition of tumor displayed on a screen. However, because of intra/inter-observer variability [2–4], the visual inspection of a tumor depends on the subjectivity and experience of endoscopists. Therefore, developing a computer-aided diagnosis system that provides objective measure of tumor to endoscopists would be greatly helpful [5]. To develop such a computer-aided system, Tamaki et al. [6] have proposed a recognition system to classify NBI endoscopic image patches into three-types (types A, B, and C3) based on the NBI magnification findings [4,7] (see Fig. 1). Furthermore, this

© Springer International Publishing AG 2017
C.-S. Chen et al. (Eds.): ACCV 2016 Workshops, Part II, LNCS 10117, pp. 610–624, 2017.
DOI: 10.1007/978-3-319-54427-4_44

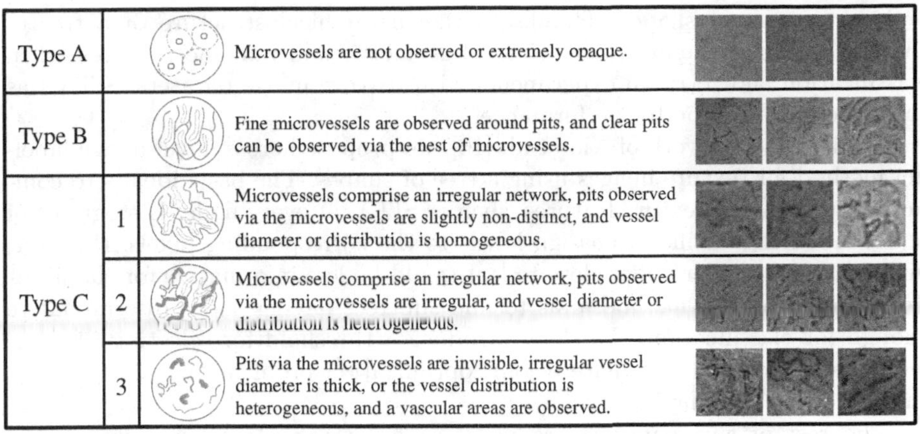

Type A		Microvessels are not observed or extremely opaque.	
Type B		Fine microvessels are observed around pits, and clear pits can be observed via the nest of microvessels.	
Type C	1	Microvessels comprise an irregular network, pits observed via the microvessels are slightly non-distinct, and vessel diameter or distribution is homogeneous.	
	2	Microvessels comprise an irregular network, pits observed via the microvessels are irregular, and vessel diameter or distribution is heterogeneous.	
	3	Pits via the microvessels are invisible, irregular vessel diameter is thick, or the vessel distribution is heterogeneous, and a vascular areas are observed.	

Fig. 1. NBI magnification findings [7].

system has been extended to a frame-wise recognition that classifies the center patch of each endoscopic video frame and shows classification results on a monitor by a frame-by-frame manner [8]. Although these systems have achieved high recognition rate and have been confirmed its medical significance, an important limitation lies in the fact that they can only a part of images of the frame. For instance, in case that a tumor is not in the center of the frame or multiple tumors exist in the frame, these systems cannot provide appropriate objective measures. Therefore, recognizing an entire endoscopic image would be a further assistance for endoscopists during examinations, and could be used to train inexperienced endoscopists.

In this paper, we aim to assign labels each pixel in an entire NBI endoscopic image. Also for the same purpose, previously we proposed an image labeling method for endoscopic images that uses a posterior probability obtained from an SVM classifier trained with a Markov Random Field (MRF) [9], but the obtained results were not satisfactory enough. One reason lies in the large variation of the texture caused by geometrical and illumination changes. Colorectal polyps and intestinal walls are not flat but undulating (wave-like or spherical shapes). Furthermore, endoscopic images have high contrast textures due to the lighting condition of the endoscope. In such a circumstance, recognition methods would fail because texture descriptors such as BoVW, Gabor, wavelet, and LBPs become unstable to be computed. Another reason is the lack of spatial consistency of MRF framework. In general, object shapes and boundaries are roughly modeled by the pairwise term of an MRF model with edges in the image. However NBI endoscopic images used in our task often do not have clear boundaries between categories and therefore it would be difficult to model the edge information by the MRF.

Towards a robust texture representation, Xia et al. [10] proposed a texture descriptor, shape-based invariant texture analysis (SITA), based on a tree of shapes [11]. SITA consists of histograms of texture features computed from all nodes in a tree of shapes. Thanks to the hierarchical structure of a tree of shapes, SITA has the invariance to local geometric and radiometric changes. In classification and retrieval experiments with texture image datasets, SITA was shown to achieve a better performance.

Inspired by the work of Xia et al. [10], we propose here a novel image labeling method for texture images using a tree of shapes. The basic idea is to compute histograms of texture features, such as SITA, at every node. Histograms of nodes are then classified to assign labels to the corresponding pixels. However, histograms of smaller nodes close to leaf would be less informative for classification. Therefore, our method aims to find subtrees having nodes discriminative enough for classification. We then introduce a threshold for selecting discriminative subtrees and formulate a joint optimization problem of estimating the threshold and training a classifier.

The rest of this paper is organized as follows. Section 2 reviews related medical and morphological work. Then, a tree of shapes and the SITA textures descriptor are briefly introduced in Sect. 3. We formulate the problem in Sect. 4. Section 5 shows some experimental results with an NBI endoscopic image dataset. Finally, we conclude this paper in Sect. 6.

2   Related Work

Polyp segmentation is a well studied task in endoscopic image analysis. Gross et al. [12] proposed a segmentation method of NBI colorectal polyps using the Canny edge detector and Non-Linear Diffusion Filtering (NLDF), which is the first attempt for polyp segmentation of NBI endoscopic images. Ganz et al. [13] proposed Shape-UCM, an extension of gPb-OWT-UCM [14] for segmentation of polyps in NBI endoscopic images. Shape-UCM solves a scale selection problem of gPb-OWT-UCM by introducing a prior about the shape of polyps. Collins et al. [15] proposed a method using Conditional Random Field (CRF) with Deformable Parts Model (DPM) and a response of positive Laplacian of Gaussian filter (negative responses clipped to zero). Some other methods using watershed and region merging [16] or GrabCut [17] have proposed.

A popular approach for polyp segmentation is the use of active contours. Breier et al. [18] proposed a method for localizing colorectal polyps in NBI endoscopic images. They assumed that polyps appear as convex objects in a image, and introduced active rays to obtain a smoother contour. Figueiredo et al. [19] proposed a segmentation method for assessing the aberrant crypt foci captured by an endoscope. They used variational level sets and active contours without the edge model of Chan and Vase [20].

All of the above existing methods try to find contours between polyps and non-polyp intestine walls. Instead, we aim to assign labels conditions of cancer to pixels. To avoid confusion, we mention the term *segmentation* for finding contours and *labeling* for assigning pixel labels like as our task. The most similar task to ours is one conducted by Nosrati et al. [21]. To increase a surgeon's visibility in an endoscopic view, they label visible objects (such as tumors, organs, and arteries) in an endoscopic video frame. Their approach is based on transferring 3D data into 2D images. The pose and deformation of objects estimated from preoperative 3D data are aligned into 2D images.

In the field of computer vision, image labeling is a well studied task and many methods have been proposed including ones using CNN features. Farabet et al. [22] proposed a labeling method for scene parsing. Their approach assigns estimated labels to pixels, and then refines the results using superpixels, CRF, and optimal-purity cover on a segmentation tree. Long et al. [23] used a fully convolutional network trained in a end-to-end manner, and some more methods have been proposed [24–26]. Although these methods have achieved high accuracies, these need a large amount of training samples, which is impractical to medical image analysis; in general, it is difficult to collect a large amount of medical data in a short period of time. If fact, we have only a dataset of 63 NBI endoscopic images (described in Sect. 5). In contrast, our method trains classifiers with histogram features from thousands of nodes in a tree of shapes, hence requires relatively few training images.

In the field of mathematical morphology, a hierarchical representation, so-called morphological tree, is a popular framework, and a number of hierarchical trees have been proposed such as min/max-trees [27,28], binary partition trees [29], minimum spanning forests [30], and tree of shapes [11]. Morphological trees have been applied to various images. One of the most popular application is biomedical imaging [31–33].

Meanwhile, Xia et al. [10] focused on the natural scale-space structure and invariance for contrast change of tree of shapes, they proposed a texture descriptor based on tree of shapes (details are described in Sect. 3.2). To the best of our knowledge, this is the first attempt to make texture descriptor from tree of shapes. Then, another texture descriptors based on tree of shapes have been proposed. Liu et al. [34] introduced a bag-of-words model of the branches in a tree of shapes and represented co-occurrence patterns of shapes. He et al. [35] adopted a basic idea of LBP, and proposed a texture descriptor. It divides a concentric circle of a shape into fan-based regions and computes the ratio of occlusion of a shape for each region. Histogram is computed from these ratios. However, these works handle only texture patch classification and retrieval and ignore multiple textures in a single image.

3   Tree of Shapes and Texture Feature

In this section, we briefly introduce the definition of the tree of shapes and the SITA histogram features.

3.1   Tree of Shapes

A tree of shapes [11] is an efficient image representation in a self-dual way. Given an image $u : \mathbb{R}^2 \to \mathbb{R}$, the upper and lower level sets of u are defined as $\chi_\lambda(u) = \{x \in \mathbb{R}^2 | u(x) \geq \lambda\}$ and $\chi^\lambda(u) = \{x \in \mathbb{R}^2 | u(x) < \lambda\}$ respectively, where $\lambda \in \mathbb{R}$.

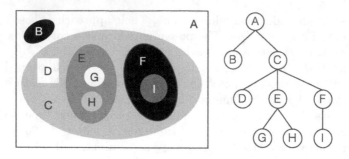

Fig. 2. An example of a synthetic image (left) and corresponding tree of shapes (right). Alphabets denote the correspondence between blobs and tree nodes.

From these level sets, we can obtain tree structures $\mathcal{T}_{\geq}(u)$ and $\mathcal{T}_{<}(u)$ that consist of connected components of upper and lower level sets: $\mathcal{T}_{\geq}(u) = \{\Gamma \in CC(\chi_\lambda(u))\}_\lambda$ and $\mathcal{T}_{<}(u) = \{\Gamma \in CC(\chi^\lambda(u))\}_\lambda$ where CC is an operator giving a set of connected components.

Furthermore, we define a set of upper shapes $\mathcal{S}_{\geq}(u)$ and lower shapes $\mathcal{S}_{<}(u)$. These sets are obtained by the cavity-filling (saturation) of components of $\mathcal{T}_{\geq}(u)$ and $\mathcal{T}_{<}(u)$. A *tree of shapes* of u is defined as the set of all shapes defined as $\mathcal{G}(u) = \mathcal{S}_{\geq}(u) \cup \mathcal{S}_{<}(u)$.

Thanks to the nesting property of level sets, the tree of shapes forms a hierarchical structure. Figure 2 shows an example of a tree of shapes. Given an image u whose image size is A. Let $T = \{V, E\}$ be a tree of shapes where V is a set of nodes, E a set of edges, $n_j \in V$ be nodes in the tree of shapes. We define parent and children nodes of n_j as $Pa(n_j) = \{n_k | (n_j, n_k) \in E, a_j < a_k\}$ and $Ch(n_j) = \{n_k | (n_j, n_k) \in E, a_j > a_k\}$ respectively, where a_j is area of n_j.

3.2   Shape-Based Invariant Texture Analysis

Xia et al. [10] proposed a texture descriptor based on the tree of shapes, *Shape-based Invariant Texture Analysis* (SITA). It consists of four shape features of the blob corresponding to a node.

Let s_j be a blob of n_j. The $(p + q)$-th order central moment μ_{qp} of s_j is defined by

$$\mu_{pq}(s_j) = \int\!\!\int_{s_j} (x_j - \bar{x}_j)^p (y_j - \bar{y}_j)^q dx_j dy_j, \tag{1}$$

where $(\bar{x}_j, \bar{y}_j)$ are the center of mass of s_j.

The normalized $(p + q)$-th order moments are

$$\eta_{pq}(s_j) = \frac{\mu_{pq}(s_j)}{\mu_{00}(s_j)^{(p+q+2)/2}}. \tag{2}$$

Then, two eigenvalues $\lambda_{1j}, \lambda_{2j}$ ($\lambda_{1j} \geq \lambda_{2j}$) of the normalized inertia matrix are computed as

$$\epsilon_j = \frac{\lambda_{2j}}{\lambda_{1j}} \tag{3}$$

and

$$\kappa_j = \frac{1}{4\pi\sqrt{\lambda_{1j}\lambda_{2j}}}, \tag{4}$$

where ϵ_j is elongation and κ_j is compactness. These are two shape features of a blob.

The third feature $\alpha(s_j)$ is computed from blob sizes and the parent-children relationship defined as

$$\alpha(s_j) = \frac{\mu_{00}(s_j)}{\sum_{s_k \in Pa^M(s_j)} \mu_{00}(s_k)/M}, \tag{5}$$

where $Pa^M(s_j) = \{s_m, \forall m \in (1, \ldots, M)\}$ is a set of M-th ancestor blobs. This feature is the ratio of blob sizes between s_j and the ancestor blobs, which is called a scale ratio. According to [10], we set $M = 3$ in our method.

The last feature is a normalized gray value computed for each pixel x in the image u as follows

$$\gamma(x) = \frac{u(x) - mean_{s(x)}(u)}{\sqrt{var_{s(x)}(u)}}, \tag{6}$$

where $s(x)$ is the smallest blob containing x. $mean_{s(x)}(u)$ and $var_{s(x)}(u)$ are mean and variance of pixel values over $s(x)$.

These are computed on every nodes (hence blobs) in the tree of shapes. The first three features are computed at all nodes, and the last feature is computed for all pixels in the image. Histograms of each feature are then constructed. These histograms are concatenated to form the SITA texture descriptor of the image, which is invariant to local geometric and radiometric changes because of the hierarchical structure of the tree of shapes.

4   Proposed Method

In this section, we develop a method for selecting discriminative subtrees in a tree of shapes by using SITA at each node. Figure 3 shows an overview of the proposed method. In the training phase, the histogram features of all nodes are used for train a classifier and estimate a size threshold. In the labeling phase, histogram features of only nodes whose sizes are larger than the estimated threshold are classified to assign labels to the nodes, then map to the corresponding blobs. Hereafter, we introduce the details of the proposed method.

We extend basic notions of the tree of shapes defined for a single image to that for a set of images. Let $\{u_i\}_{i=1}^N$ be a set of images and A_i be the image size of i-th image u_i. A tree of shapes of image u_i is defined as $T_i = \{V_i, E_i\}$, where V_i is a set of nodes and E_i a set of edges. Each node $n_{ij} \in V_i$ has the corresponding label $y_{ij} \in \{-1, 1\}$ and the area a_{ij} of the corresponding blob.

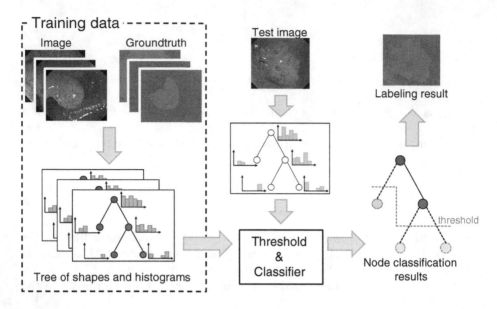

Fig. 3. Overview of the proposed method.

Herein, we explain the details of the histogram feature used in our method. In the work of Xia et al., a SITA descriptor is computed as a histogram feature for a given image. This means that the SITA is computed at the *root* node of the tree by aggregating feature from all descendant nodes. In contrast, our method constructs the histogram features at *all* nodes of the tree. Let $\boldsymbol{g}(n_{ij})$ be a histogram computed by the features from node n_{ij} only. Then the total histogram $\boldsymbol{h}(n_{ij})$ of node n_{ij} is computed recursively as

$$\boldsymbol{h}(n_{ij}) = \boldsymbol{g}(n_{ij}) + \sum_{n_{ik} \in Ch(n_{ij})} \boldsymbol{h}(n_{ik}), \tag{7}$$

then normalized to have a unit L1 norm. This means that $\boldsymbol{h}(n_{ij})$ is computed in a bottom-up manner, i.e., the computation is done from leaf nodes to the root node. For the sake of simplicity, we denote $\boldsymbol{h}(n_{ij})$ as $\boldsymbol{h}_{ij}$.

As we mentioned, discriminative subtrees useful for labeling are expected to exist in the tree of shapes. Figure 4 shows examples of labeling results with different subtrees. If we use smaller, less discriminative subtrees (such as leaf nodes), labeling fails; Using subtrees with large size nodes (e.g. the root node), labeling wouldn't be satisfactory because such nodes correspond to a large part of the image. Therefore, we introduce a threshold θ to node sizes for selecting discriminative subtrees, and define the objective function to estimate θ and classifier parameters as follows:

$$E(\theta, \boldsymbol{w}, b) = \|\boldsymbol{w}\|^2 + \frac{1}{NM_i} \sum_i^N \sum_j^{M_i} m_{ij}(\theta) \, \ell \left(y_{ij}(\boldsymbol{w}^T \boldsymbol{h}_{ij} + b) \right) + \lambda \frac{\theta}{\overline{A}}, \tag{8}$$

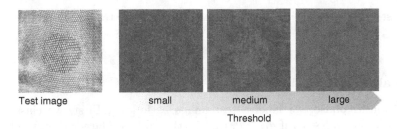

Test image   small   medium   large

Threshold

Fig. 4. Examples of labeling results using subtrees with different node sizes. Subtrees used for labeling are decided by an estimated threshold.

where $\boldsymbol{w}$ and b are the weight and bias of an SVM classifier, respectively. The first term denotes a regularizer for the weight. The third term $\lambda \frac{\theta}{\bar{A}}$ is a regularizer for the threshold θ, where λ is a scale parameter and $\bar{A} = \frac{1}{N} \sum_i^N A_i$ is the mean size of N training images. $\ell(\cdot)$ is the hinge loss function of the SVM classifier.

$m_{ij}(\theta)$ represents the sample weight for $\boldsymbol{h}_{ij}$. In our method, we need to select the threshold θ to find discriminative subtrees. In other words, we have to use histograms of nodes whose area is larger than θ, otherwise ignore. Therefore, we define $m_{ij}(\theta)$ as a step function;

$$m_{ij}(\theta) = \begin{cases} 0 & if \ a_{ij} < \theta \\ 1 & otherwise \end{cases}. \qquad (9)$$

4.1 Optimization

Given a training data, we need to estimate $\hat{\theta}, \hat{\boldsymbol{w}}, \hat{b}$ by minimizing the cost function;

$$\hat{\theta}, \hat{\boldsymbol{w}}, \hat{b} = \underset{\theta, \boldsymbol{w}, b}{\operatorname{argmin}} E\left(\theta, \boldsymbol{w}, b\right). \qquad (10)$$

However, it is difficult to estimate all parameters at once. Therefore, we use block-coordinate decent to solve this optimization; estimate the threshold θ and the classifier parameters $\boldsymbol{w}, b$ iteratively. For estimating θ, we solve

$$\theta_k = \underset{\theta}{\operatorname{argmin}} E(\theta, \boldsymbol{w}_{k-1}, b_{k-1}). \qquad (11)$$

This problem is non-convex because θ depends on histograms $\boldsymbol{h}_{ij}$. However, we experimentally confirmed that the cost function is rather smooth and have a single minimum (details are discussed in Sect. 5.1). For estimating $\boldsymbol{w}$ and b, we solve

$$\boldsymbol{w}_{k+1}, b_{k+1} = \underset{\boldsymbol{w}, b}{\operatorname{argmin}} E(\theta_k, \boldsymbol{w}, b). \qquad (12)$$

This is an SVM formulation with sample weights, which is convex. For a large number of training samples, it would be difficult to obtain a nonlinear SVM problem within a practical time, therefore we solve the SVM in a primal domain by using the primal solver of LIBLINEAR [36].

We stop the alternation when θ converges with the termination criterion of

$$|\theta_k - \theta_{k-1}| < \epsilon. \tag{13}$$

4.2 Labeling Procedure

After training phase, we label a test image as follows. First, we classify histograms h_{ij} of nodes n_{ij} if $a_{ij} \geq \hat{\theta}$, that is, the node are larger than the threshold, then assign the estimated labels to the nodes n_{ij}. For smaller nodes, we assign the label of their parent node. This procedure is done from the root node down to the leave nodes. Once labels are assigned to every nodes, labeling results are obtained by mapping labels of nodes into the corresponding blobs.

5   Experimental Results

We have prepared a dataset of 63 NBI endoscopic images. Example images in the dataset are shown in Fig. 5. Sizes of images are $1,000 \times 870$ pixels. There are two label categories (foreground and background) based on the NBI magnification findings (see Fig. 1). Foreground regions correspond to polyps of types B and C, and background regions are others (type A polyps, normal intestinal walls, and uninformative dark regions). Among 63 images, 20 images are negative samples which don't contain any foreground regions; the left-most image in Fig. 5 captures only a hyperplastic polyp (i.e. benign tumor and non-cancer, hence Type A) labeled as background. A tree of shapes created from an NBI endoscopic image contains a large number of nodes. The average number of nodes from images in the dataset is 24,070. We randomly divided the dataset into half for training and test. We set parameters λ as 1.0 and initial value of threshold θ_0 as 1000.

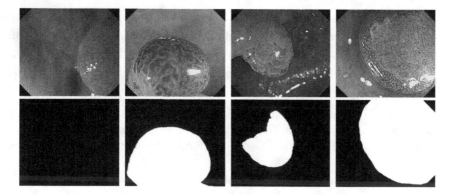

Fig. 5. Examples of images in the NBI endoscopic image dataset. Upper row shows NBI images and bottom row shows corresponding masks. White color of the mask represents foreground and black represents background. The left-most image is a negative sample which doesn't have any foreground region.

We used two methods for comparison. One is to simply classify histograms of nodes in a tree of shapes and assign labels to pixels, which is corresponding to $m_{ij}(\theta) = 1$ in Eq. (8). This is a simple application of SITA for every nodes and is an obvious extension, while our proposed method is not. In the following experiments, we refer this method as conventional method. The other is a patch based segmentation method using MRF and posterior probabilities obtained from a trained SVM classifier [9]. For training SVM, we used 1,608 NBI endoscopic image patches (type A: 484, types B and C3: 1,124) trimmed and labeled by endsocpists. In this method, densely sampled SIFT features are extracted from these patches and converted as BoVW histograms. BoVW histograms are then used for training an SVM classifier. Small square patches corresponding to each site of the MRF grid are classified to obtain posterior probabilities used as the MRF data term. The MRF energy is minimized by Graph Cut for obtaining labeling results.

5.1 Labeling Results of NBI Endoscopic Images

Figure 6 shows the cost function values over different threshold at each iteration in the training phase. We can see that the minimum of the cost function become smaller and threshold θ converges.

Figure 7 shows labeling results. As we mentioned above, we used the half of dataset (31 images) are used for training. The total number of nodes for training is 747,937 and the primal solver for SVM training is necessary. The numbers of nodes of each test images are also shown in Fig. 7. In SVM-MRF segmentation, labeling results are poor because the accuracy of MRF-based approaches highly depends on the data term. In other words, failures by the SVM classifier have the large impact on the poor accuracy. The conventional method provides cluttered labeling results because it classifies even small nodes. For instance, in the first two

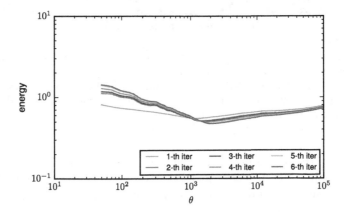

Fig. 6. Energies at each iteration. Horizontal axis shows θ value and vertical axis shows energy. Colorized curves are energies of each iteration. Initial value of $\theta_0 = 1000$ is used, $\lambda = 1.0$ is fixed.

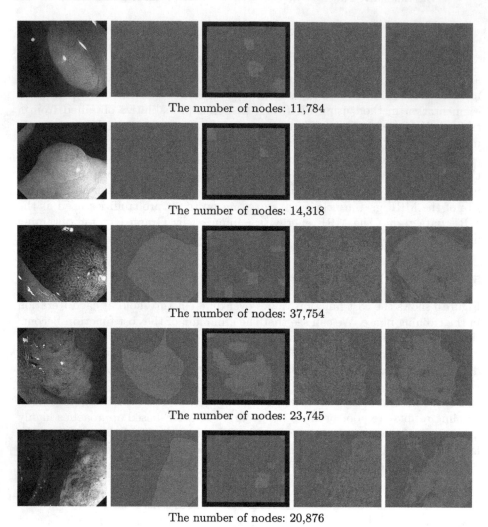

The number of nodes: 11,784

The number of nodes: 14,318

The number of nodes: 37,754

The number of nodes: 23,745

The number of nodes: 20,876

Fig. 7. Labeling results. From left to right: test image, ground truth, labeling result of SVM-MRF [9], conventional, and proposed. The number of nodes in the trees of shapes created from test images are shown below the images. Red color represents foreground and blue background. Black color of SVM-MRF results represents unlabeled region due to the boundary effect. (Color figure online)

rows shows that the results of the conventional method provide small foreground regions. Meanwhile the proposed method can suppress the cluttered labels by selecting discriminative subtrees. In the middle and last two rows, foreground shapes of the proposed results are similar to the ground truth.

For quantitative evaluation, we used the dice coefficient [37]. Table 1 shows dice coefficients of each method. For conventional and proposed methods, we

Table 1. Dice coefficients of labeling results.

Method	Dice coefficient
SVM-MRF [9]	0.555
Conventional	0.522 ± 0.056
Proposed	0.633 ± 0.041

tested the procedures mentioned above repeatedly ten times and for the SVM-MRF method we tested only once. Note that the dice coefficient is calculated only for samples containing foreground. We can see that the proposed method outperforms the other two methods because using discriminative subtrees suppresses cluttered labels.

The proposed method outperforms the others in both the qualitative and quantitative evaluations. However, we need to discuss failure labeling results. Some failure examples are shown in Fig. 8. In the case of top row, the image is almost labeled as foreground. A possible reason is that the used histogram is simply constructed from four low level features, which might be too few to be

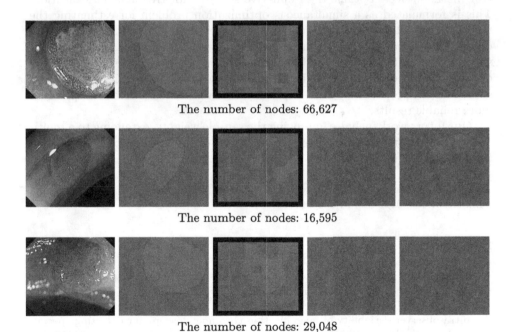

The number of nodes: 66,627

The number of nodes: 16,595

The number of nodes: 29,048

Fig. 8. Some failure examples. From left to right: test image, ground truth, labeling result of SVM-MRF [9], conventional, and proposed. The number of nodes in a tree of shapes created from the test image is shown in the bottom of images. Red color represents foreground and blue background. Black color of SVM-MRF results represents unlabeled region due to boundary effect. (Color figure online)

discriminative enough. Therefore, using richer texture features is included in our future work. The proposed method labels as background inside of the foreground region in the middle row. In our method, subtrees are selected by one threshold, but optimal thresholds may be different for different images, which is a limitation of the proposed method. Results of bottom row provide small foreground labels, which correspond to specular reflections (highlight) and the surrounding regions. Because highlights are large area nodes, texture features extracted from highlights may affect classification results, and dealing with highlights is also one of our future work.

About the computational time, our python implementation of the proposed method takes about 600 s for training and about 70 s for labeling an image. Although we handle ten thousand and more samples, our method can be trained within a practical time.

6   Conclusion

In this paper, we proposed an image labeling method for NBI endoscopic images using a tree of shapes and histogram features derived from the tree structure. Our method selects optimal discriminative subtrees for tree node classification. This is formulated as a simultaneous optimization problem for estimating the threshold and classifier parameters and is solved via iterative block-coordinate decent. Then, we label images using the estimated parameters and the tree of shapes, by classifying each node from the root node to leave nodes, and mapping classification results into pixels. Experimental results on NBI endoscopic images show that the proposed method outperforms conventional methods and provides more reliable results.

Our future work includes improving the sample weights, extending to a multi-class problem, and seeking a more effective way of the labeling procedure using the hierarchical structure.

Acknowledgement. This work was supported in part by JSPS KAKENHI grants numbers JP14J00223 and JP26280015.

References

1. Cancer Research, U.K.: Bowel cancer statistics (2015). http://www.cancer researchuk.org/cancer-info/cancerstats/types/bowel/. Accessed 7 Aug 2016
2. Meining, A., Rösch, T., Kiesslich, R., Muders, M., Sax, F., Heldwein, W.: Inter- and intra-observer variability of magnification chromoendoscopy for detecting specialized intestinal metaplasia at the gastroesophageal junction. Endoscopy **36**, 160–164 (2004)
3. Mayinger, B., Oezturk, Y., Stolte, M., Faller, G., Benninger, J., Schwab, D., Maiss, J., Hahn, E.G., Muehldorfer, S.: Evaluation of sensitivity and inter- and intra-observer variability in the detection of intestinal metaplasia and dysplasia in barrett's esophagus with enhanced magnification endoscopy. Scand. J. Gastroenterol. **41**, 349–356 (2006)

4. Oba, S., Tanaka, S., Oka, S., Kanao, H., Yoshida, S., Shimamoto, F., Chayama, K.: Characterization of colorectal tumors using narrow-band imaging magnification: combined diagnosis with both pit pattern and microvessel features. Scand. J. Gastroenterol. **45**, 1084–1092 (2010)
5. Takemura, Y., Yoshida, S., Tanaka, S., Kawase, R., Onji, K., Oka, S., Tamaki, T., Raytchev, B., Kaneda, K., Yoshihara, M., Chayama, K.: Computer-aided system for predicting the histology of colorectal tumors by using narrow-band imaging magnifying colonoscopy (with video). Gastrointest. Endosc. **75**, 179–185 (2012)
6. Tamaki, T., Yoshimuta, J., Kawakami, M., Raytchev, B., Kaneda, K., Yoshida, S., Takemura, Y., Onji, K., Miyaki, R., Tanaka, S.: Computer-aided colorectal tumor classification in NBI endoscopy using local features. Med. Image Anal. **17**, 78–100 (2013)
7. Kanao, H., Tanaka, S., Oka, S., Hirata, M., Yoshida, S., Chayama, K.: Narrow-band imaging magnification predicts the histology and invasion depth of colorectal tumors. Gastrointest. Endosc. **69**, 631–636 (2009)
8. Kominami, Y., Yoshida, S., Tanaka, S., Sanomura, Y., Hirakawa, T., Raytchev, B., Tamaki, T., Koide, T., Kaneda, K., Chayama, K.: Computer-aided diagnosis of colorectal polyp histology by using a real-time image recognition system and narrow-band imaging magnifying colonoscopy. Gastrointest. Endosc. **83**, 643–649 (2016)
9. Hirakawa, T., Tamaki, T., Raytchev, B., Kaneda, K., Koide, T., Kominami, Y., Yoshida, S., Tanaka, S.: SVM-MRF segmentation of colorectal NBI endoscopic images. In: 2014 36th Annual International Conference of the IEEE Engineering in Medicine and Biology Society, pp. 4739–4742 (2014)
10. Xia, G.S., Delon, J., Gousseau, Y.: Shape-based invariant texture indexing. Int. J. Comput. Vis. **88**, 382–403 (2010)
11. Monasse, P., Guichard, F.: Fast computation of a contrast-invariant image representation. IEEE Trans. Image Process. **9**, 860–872 (2000)
12. Gross, S., Kennel, M., Stehle, T., Wulff, J., Tischendorf, J., Trautwein, C., Aach, T.: Polyp segmentation in NBI colonoscopy. In: Meinzer, H.P., Deserno, T.M., Handels, H., Tolxdorff, T. (eds.) Bildverarbeitung für die Medizin 2009, pp. 252–256. Springer, Heidelberg (2009)
13. Ganz, M., Yang, X., Slabaugh, G.: Automatic segmentation of polyps in colonoscopic narrow-band imaging data. IEEE Trans. Biomed. Eng. **59**, 2144–2151 (2012)
14. Arbelaez, P., Maire, M., Fowlkes, C., Malik, J.: Contour detection and hierarchical image segmentation. IEEE Trans. Pattern Anal. Mach. Intell. **33**, 898–916 (2011)
15. Collins, T., Bartoli, A., Bourdel, N., Canis, M.: Segmenting the uterus in monocular laparoscopic images without manual input. In: Navab, N., Hornegger, J., Wells, W.M., Frangi, A.F. (eds.) MICCAI 2015. LNCS, vol. 9351, pp. 181–189. Springer, Heidelberg (2015). doi:10.1007/978-3-319-24574-4_22
16. Bernal, J., Sánchez, J., Vilariño, F.: A region segmentation method for colonoscopy images using a model of polyp appearance. In: Vitrià, J., Sanches, J.M., Hernández, M. (eds.) IbPRIA 2011. LNCS, vol. 6669, pp. 134–142. Springer, Heidelberg (2011). doi:10.1007/978-3-642-21257-4_17
17. Hegadi, R.S., Goudannavar, B.A.: Interactive segmentation of medical images using grabcut. Int. J. Mach. Intell. **3**, 168–171 (2011)
18. Breier, M., Gross, S., Behrens, A., Stehle, T., Aach, T.: Active contours for localizing polyps in colonoscopic NBI image data (2011)
19. Figueiredo, I.N., Figueiredo, P.N., Stadler, G., Ghattas, O., Araujo, A.: Variational image segmentation for endoscopic human colonic aberrant crypt foci. IEEE Trans. Med. Imaging **29**, 998–1011 (2010)

20. Chan, T.F., Vese, L.A.: Active contours without edges. IEEE Trans. Image Process. **10**, 266–277 (2001)
21. Nosrati, M.S., Amir-Khalili, A., Peyrat, J.M., Abinahed, J., Al-Alao, O., Al-Ansari, A., Abugharbieh, R., Hamarneh, G.: Endoscopic scene labelling and augmentation using intraoperative pulsatile motion and colour appearance cues with preoperative anatomical priors. Int. J. Comput. Assist. Radiol. Surg. **11**, 1409–1418 (2016)
22. Farabet, C., Couprie, C., Najman, L., LeCun, Y.: Learning hierarchical features for scene labeling. IEEE Trans. Pattern Anal. Mach. Intell. **35**, 1915–1929 (2013)
23. Long, J., Shelhamer, E., Darrell, T.: Fully convolutional networks for semantic segmentation. In: 2015 IEEE Conference on Computer Vision and Pattern Recognition (CVPR), pp. 3431–3440 (2015)
24. Girshick, R., Donahue, J., Darrell, T., Malik, J.: Rich feature hierarchies for accurate object detection and semantic segmentation. In: 2014 IEEE Conference on Computer Vision and Pattern Recognition, pp. 580–587 (2014)
25. Liu, F., Lin, G., Shen, C.: CRF learning with CNN features for image segmentation. Pattern Recogn. **48**, 2983–2992 (2015). Discriminative Feature Learning from Big Data for Visual Recognition
26. Noh, H., Hong, S., Han, B.: Learning deconvolution network for semantic segmentation. In: 2015 IEEE International Conference on Computer Vision (ICCV), pp. 1520–1528 (2015)
27. Jones, R.: Connected filtering and segmentation using component trees. Comput. Vis. Image Underst. **75**, 215–228 (1999)
28. Najman, L., Couprie, M.: Building the component tree in quasi-linear time. IEEE Trans. Image Process. **15**, 3531–3539 (2006)
29. Salembier, P., Garrido, L.: Binary partition tree as an efficient representation for image processing, segmentation, and information retrieval. IEEE Trans. Image Process. **9**, 561–576 (2000)
30. Cousty, J., Najman, L.: Incremental algorithm for hierarchical minimum spanning forests and saliency of watershed cuts. In: Soille, P., Pesaresi, M., Ouzounis, G.K. (eds.) ISMM 2011. LNCS, vol. 6671, pp. 272–283. Springer, Heidelberg (2011). doi:10.1007/978-3-642-21569-8_24
31. Xu, Y., Géraud, T., Najman, L.: Two applications of shape-based morphology: blood vessels segmentation and a generalization of constrained connectivity. In: Hendriks, C.L.L., Borgefors, G., Strand, R. (eds.) ISMM 2013. LNCS, vol. 7883, pp. 390–401. Springer, Heidelberg (2013). doi:10.1007/978-3-642-38294-9_33
32. Dufour, A., Tankyevych, O., Naegel, B., Talbot, H., Ronse, C., Baruthio, J., Dokládal, P., Passat, N.: Filtering and segmentation of 3D angiographic data: advances based on mathematical morphology. Med. Image Anal. **17**, 147–164 (2013)
33. Perret, B., Collet, C.: Connected image processing with multivariate attributes: an unsupervised Markovian classification approach. Comput. Vis. Image Underst. **133**, 1–14 (2015)
34. Liu, G., Xia, G.S., Yang, W., Zhang, L.: Texture analysis with shape co-occurrence patterns. In: Pattern Recognition (ICPR), pp. 1627–1632 (2014)
35. He, C., Zhuo, T., Su, X., Tu, F., Chen, D.: Local topographic shape patterns for texture description. IEEE Sig. Process. Lett. **22**, 871–875 (2015)
36. Fan, R.E., Chang, K.W., Hsieh, C.J., Wang, X.R., Lin, C.J.: LIBLINEAR: a library for large linear classification. J. Mach. Learn. Res. **9**, 1871–1874 (2008)
37. Dice, L.R.: Measures of the amount of ecologic association between species. Ecology **26**, 297–302 (1945)

Modelling Respiration Induced Torso Deformation Using a Mesh Fitting Algorithm

Haobo Yu[1], Harvey Ho[1(✉)], Adam Bartlett[2], and Peter Hunter[1]

[1] Auckland Bioengineering Institute, The University of Auckland,
Auckland, New Zealand
hyu754@aucklanduni.ac.nz, {harvey.ho,p.hunter}@auckland.ac.nz
[2] Department of Surgery, The University of Auckland,
Auckland, New Zealand
a.bartlett@auckland.ac.nz

Abstract. Precise positioning of an ablation probe in soft abdominal organs requires taking the respiration effects into account. Fast and reliable registration of a virtual abdominal organ with intra-operational imaging data remains a challenge in image-guided and Virtual Reality (VR) aided surgeries. In this paper we present a Host Mesh Fitting (HMF) algorithm to imitate the deformation of a torso due to aspiration effects. Displacements of the torso mesh are driven by virtual fiducial markers placed on the abdominal surface, which consequently deform abdominal organs in an implicit manner and with a small computational cost. In order to test the HMF algorithm a gelatine phantom was made with its internal channels detectable from ultrasonic imaging. Deformation of the channels due to a compression force was reproduced from the warping of the host mesh. After coupling with a fiducial marker tracking system the HMF algorithm can be used to model the torso deformation due to respiration effects.

Keywords: Virtual abdominal organ · Host Mesh Fitting · Respiration effects · Fiducial markers

1 Introduction

Percutaneous minimally invasive procedures have become alternatives to traditional open surgeries due to their lower complication rates, shorter hospital stays and less expenses [1]. These procedures are usually aided by pre-operational computed tomography (CT) or magnetic resonance imaging (MRI) scans and guided by intra-operational imaging data, e.g., from cone beam CT and/or ultrasonic images [1]. Since the respiration causes abdominal organ displacements, virtual models built from CT/MRI images need to be registered with intra-operational data by respiration gated or breathing holding techniques, so that an intervention procedure can be performed at the same phase as that of the CT/MRI scan [2]. An alternative solution is 4D CT where multiple images are acquired during the rotation of the CT gantry, and the patient's respiration is monitored

© Springer International Publishing AG 2017
C.-S. Chen et al. (Eds.): ACCV 2016 Workshops, Part II, LNCS 10117, pp. 625–634, 2017.
DOI: 10.1007/978-3-319-54427-4_45

by an exterior marker attached to the patient's abdomen [2,3]. As 4D CT introduces more radiation to patients and operators, computer algorithms have been proposed to interpolate image volume between the expiration and inspiration phases, thus the radiation dose can be reduced [2,4].

Among the computational interpolation schemes, a deformable, B-Spline based registration model is proposed in [2] where the contour of an organ is automatically mapped from one phase to another, and a mapping accuracy of 3 mm is achieved. In [4], the position of a tumour is estimated using a nonlinear registration algorithm and its new position is reproduced by image morphing.

In this paper we propose an approach based on a Host Mesh Fitting (HMF) algorithm. The core of the algorithm is two sets of finite element mesh, i.e. the *slave* and the *host* mesh, whereby the deformation of the host mesh drives the motion of the slave mesh. This algorithm has been described in literatures, e.g., in [5] for muscle modelling. In this paper we extend this algorithm to the context of image warping as analogous to CT image interpolation of [4]. This is achieved by incorporating image voxels into the host mesh, therefore a volume image can be morphed between difference breathing phases. The concept is illustrated in Fig. 1, where the image intensities of a stack of CT images are normalised into the texture space $(0, 1)$, and placed into a cube-shaped host mesh. By applying the HMF algorithm [5], the texture space is morphed after altering the direction vectors and positions of the nodes of the host mesh.

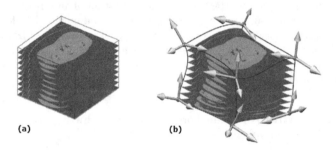

(a)   (b)

Fig. 1. The concept of 3D image morphing within a finite element mesh: (a) A 3D image volume is placed in a tricubic Hermite element; and (b) arbitrary morphing can be made by adjusting the orientation or location of mesh nodes.

In the following sections, we will outline the HMF algorithm and describe its application in a CT image containing a human torso. Since respiration-gated CT data are not available for validation, we will use a gelatine phantom to compare the displacements simulated from the HMF algorithm and that derived from ultrasound imaging. The main goal is to elaborate the fitting algorithm with fiducial markers. The visualisation tool used in the work is CMGUI (http://www.cmiss.org/cmgui), an open source visualisation and imaging software.

2   Methods

2.1   Host Mesh Fitting

The host-mesh-fitting algorithm is a subset of the free-form deformation technique [5]. The idea is to deform an object by enclosing it within a bounding object, and by deforming the bounding object the enclosed object will be deformed accordingly. The concept is illustrated in Fig. 2, where a surface mesh (the slave mesh) is completely enclosed within a 3D host mesh (thus the name). Since the local coordinates (η_1, η_2, η_3) of every node of the slave mesh can be written as a function of the coordinates of the host mesh (ξ_1, ξ_2, ξ_3) and the relative nodal positions of the slave mesh with respect to the host mesh remain intact, when the host mesh is deformed the slave mesh will be updated accordingly.

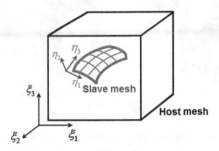

Fig. 2. Description of a host-mesh and a slave mesh and their respective coordinate systems. A surface mesh i.e. the slave mesh is completely enclosed within a 3D host mesh.

Parametric representation of the host and slave mesh has been described in many literatures. In particular, we refer the interested reader to [5] for mathematical details of the fitting algorithm. We should highlight that the host mesh can be viewed as a Finite Element mesh, and any internal points within the mesh can be expressed as a weighted sum of *basis functions*. It should also be stressed that although the slave mesh in Fig. 2 is a 2D surface patch, a more complex mesh made of tetrahedra can also be used. Indeed a tetrehedra mesh will be used to simulate the deformation of a phantom (described later in Sect. 2.3). Furthermore, the slave mesh can be multi-dimensional, i.e., containing a combination of 1D, 2D and 3D mesh of vasculature, surface and parenchyma of an organ as introduced in [6].

2.2   Meshing an Image Volume

In the image morphing example of Fig. 1, the host mesh is a tricubic Hermite element, i.e. any point X in it can be expressed as $X = \sum \phi_i \xi_j$ where ϕ is the basis function (in this case a Hermite function) and ξ the nodal coordinates. In order to model soft organs, a more sophisticated host mesh needs to be used to represent their respective locations in the texture space, so that their deformation can be simulated. For example, bones are rigid and therefore they cannot be morphed in the same way as elastic tissues/organs. This concept is illustrated in Fig. 3 where the vertebra (indicated by an arrow) are separated from the soft abdominal cavity.

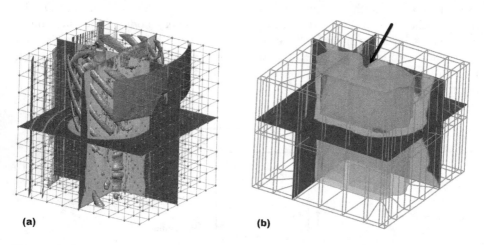

(a)   (b)

Fig. 3. A CT image containing a human torso and the Finite Element host mesh is constructed for the volume image: (a) a trilinear mesh ($8 \times 8 \times 8$) is used as the host mesh; and (b) a custom-made mesh arranged around the torso surface. The arrow indicates the location of vertebra which are not contained in the elastic torso mesh.

In Fig. 3(a) the host mesh is made of a trilinear mesh of 128 elements which are of the same size. The advantage of this method is that minimum efforts are required to create the host mesh, however the mesh does not define any specific organs, which have to be differentiated by an image segmentation algorithm. In Fig. 3(b) a custom-made host mesh was constructed around the torso surface so that its deformation can be simulated. This method has the advantage of being able to describe deformation of individual organs/tissues, but also bears the disadvantage of a high-cost in the mesh construction process.

In order to morph the host mesh, a set of *source* points within the mesh and their corresponding *target* points are fitted using a least square quasi Newton method [7] within OPT++, an open source library for nonlinear optimization algorithms.

Only four key landmark or fiducial markers are chosen as the source points, two on the diaphragm apexes and the other two on the surface of the chest (shown in Fig. 5 of the Results section). According to [1,8], the inferior-superior movement of the diaphragm apex is $27.3 \pm 10.2\,\text{mm}$, and the anterior-posterior translation of abdominal organs is around 8 to 10 mm. Thus a set of target points can be prescribed to guide the motion of the torso mesh. This feature is valuable when physical landmarks are not available on the torso, which is the case for this preliminary study.

2.3 Gelatine Phantom and Ultrasonic Imaging

Since respiratory gating CT images were not available for this study, a phantom was used to validate the HMF algorithm. The aim was to check the accuracy of simulated deformation of the phantom, in comparison with the ultrasound imaging data. The phantom (67 mm × 92 mm × 54 mm), shown in Fig. 4, was made of gelatine, a commonly used material for bio-tissues [9]. Softness of the phantom can be controlled by using different percentage of gelatine powder mixed with water. In this specific phantom 23% of gelatine power was used. Meanwhile two perpendicular tubes (diameter 10 mm) were placed in the water container. After the gelatine mixture was solidified the two tubes were removed forming two hollow channels, as shown in Fig. 4.

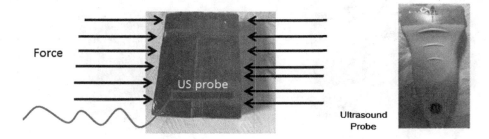

Fig. 4. Left: the phantom contains two perpendicularly arranged channels and is compressed by forces applied laternally; Right: the transducer of Voluson i BT14 was used for ultrasound scanning.

In the actual ultrasonic scan experiment, the hollow channels were filled with water so that there was no air inside the phantom to interfere with ultrasound signals. After a compression force was applied from lateral sides of the phantom, the channels were deformed and scanned.

We used a Voluson i BT14 (GE) scanner in its B-mode to collect real-time ultrasonic imaging data from the compressed phantom, as analogous to the respiration cycle experienced by a torso. We also created a virtual computer model for the phantom, shown in Fig. 8 of the Results section, which consists of 4,925 tetrahedra elements. By fitting the ultrasound image into the virtual model the HMF algorithm can be tested.

3   Results

3.1   Simulation of Respiration Effects on Torso

A simulation was made for the respiration effects on the torso using the method described above. As mentioned earlier the data for the abdominal organ movements was adopted from [3,8] and prescribed to the source points. In Fig. 5(a), the torso was assumed to be at the end of the expiration phase and the red markers, representing the *source* points, are at their most posterior positions. The golden arrows represent trajectory vectors from the source points to their corresponding target points. The additional two markers are placed on the apex of the diaphragm, and are hidden behind the iso-surface of the torso. In Fig. 5(b), the minimization and smoothing problem was solved in a desktop computer (Intel i7-4790 CPU @ 3.60 GHz, RAM 32 GB, GeForce 2 GB GTX 745 GPU). The computation took 0.7 s to complete.

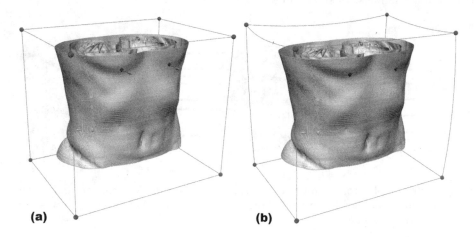

(a)   **(b)**

Fig. 5. A simulation of respiration effects on torso. (a) The torso at a quiet breath holding state after expiration; and (b) Simulated torso position at the start of expiration (or end of inspiration) after applying the HMF algorithm. Note the red points are the set of source points and the arrows are the projection vectors from the source points to target points. Also note the deformed host mesh, which drives the slave torso mesh. (Color figure online)

Figure 6 yields more information about the fitting results. The black arrows indicate the anterior-posterior displacement of the torso surface. Meanwhile the image volume warps between respiration phases. This is shown from the three cross sections of the image, where the blue colour stands for the air and the light blue colour represents tissue. Also note the host mesh, i.e. the cube deforms during the fitting process, which in turn drove the torso displacements.

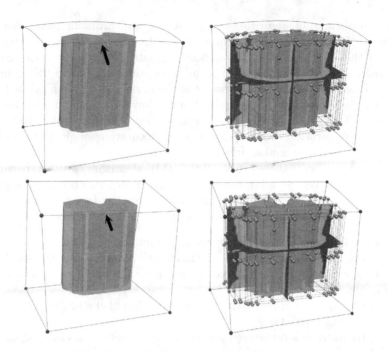

Fig. 6. Host mesh fitting algorithm applied to torso deformation due to respiration. The arrows indicate mesh displacements occurring at the anterior-posterior perspective. (Color figure online)

3.2   Experiments with the Phantom

The phantom shown in Fig. 4 was used to validate the HMF algorithm by comparing the deformed channels computed from the model with that derived from ultrasound imaging. In this case the eight corners of the phantom were used as the set of source points, and their new positions after applying a compression force as the target points. The computation took 0.4 s to complete in the same computer. The results are shown in Fig. 8.

In Fig. 8(a), the virtual phantom model and its host mesh are shown. The virtual phantom consists of a tetrahedra mesh and two channels. Profiles of the channels were shown via a cross-section of the phantom, which are compared with ultrasound images of Fig. 8(b). After superimposing the cross-section of the virtual model with the ultrasound image it was found that the profiles of deformed channel agreed with that shown in the ultrasonic images (Fig. 8c).

4   Discussion

Accurate prediction of tumour locations inside an organ *intra-operationally* is crucial for surgeons and interventional radiologists to achieve an optimal operational outcome. Breathing-induced organ displacement and deformation have been well studied, and algorithms to overcome the problem have also been described in literatures (see references [1–4]). In general, it is not trivial to account for the breathing effects in an image registration process because organ displacement and deformation occur in all directions. Adding to this challenge is the fact that only part of the torso or organ surface are visible, thus to register the whole torso/organ model with limited intra-operational data becomes questionable. The key issue here is to seek a solution which not only yields a prediction for respiration-induced deformation accurate enough, but also has a light computational cost and a sufficient robustness.

It is in this context that we proposed the HMF algorithm, which is essentially an optimisation process yielding the best fitting between two set of data, that of the source and of the target points. The workflow of the algorithm is illustrated in Fig. 7, where the 5 steps in the diagram are described as follows:

1. Choose landmark points in a *slave mesh* as land mark points;
2. Compute elemental coordinates of the landmark points in the *host mesh*. Also note the corresponding target data points have the same elemental coordinates in the *host mesh*;
3. Generate a transformation matrix by minimisation of the projection from the source and target points;
4. Use the transformation matrix to drive the deformation of the *host mesh*;
5. The deformation of the host mesh in turn drive the deformation of the *slave mesh*.

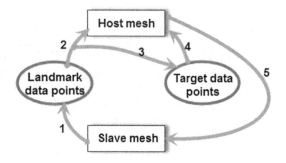

Fig. 7. Workflow of the HFM algorithm.

The advantage of this algorithm is that imaging warping comes with a minimal cost as it is implicitly done once the fitting algorithm is completed. Also the computation based on a small set of landmarks is fast (within one second), which is consistent with the surgical timeframe. The fast computation is achieved because solving elasticity equations is not required in the HMF algorithm. Therefore, even though fast finite element methods have been proposed (e.g., in [10]), applying these methods to nonlinear elasticity problems is non-trivial. Compared to the B-Spline interpolation model described in [2], where the displacement of one point is determined by an interpolation polynomial of its adjacent points, the HMF algorithm functions in a global scale rather than a local scale.

There are some limitations pertaining to the current method. For example, it is cumbersome to construct a slave mesh that encloses the target organ of interest. For instance, the liver organ was not distinguished from the torso mesh thus the hepatic motion secondary to respiration could not be effectively monitored. Since hepatic motion is a significant obstacle to precise needle placement [1], and also different abdominal organs have slightly different displacements [3], a new meshing strategy remains as our future work.

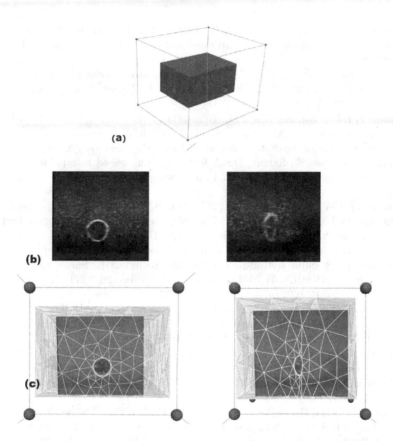

Fig. 8. Validation experiments for the HMF algorithm were done in a gelatine phantom: (a) A virtual model for the phantom (slave mesh) is contained within a cube (host mesh) so that the deformation of the phantom is driven by the deformation of the cube. Also visible are the two channels whose profiles are compared with ultrasound images; (b) Ultrasonic imaging reveals a cross-section of the phantom before and after a compression force was applied; and (c) Deformation of the channel yielded by the host-mesh algorithm agrees with the ultrasonic data.

5   Conclusion

In this paper we introduced a Host Mesh Fitting algorithm that uses fiducial markers to drive the deformation of a host mesh which in turn transforms a slave mesh. The algorithm is able to morph CT images between different respiration states with minimum cost, therefore could be useful to help surgeons to better place ablation probes.

References

1. Clifford, M.A., Banovac, F., Levy, E., Cleary, K.: Assessment of hepatic motion secondary to respiration for computer assisted interventions. Comput. Aided Surg. **7**, 291–299 (2002)
2. Schreibmann, E., Chen, G.T.Y., Xing, L.: Image interpolation in 4D CT using a BSpline deformable registration model. Int. J. Radiat. Oncol. Biol. Phys. **64**, 1537–1550 (2006)
3. Brandner, E.D., Wu, A., Chen, H., Heron, D., Kalnicki, S., Komanduri, K., Gerszten, K., Burton, S., Ahmed, I., Shou, Z.: Abdominal organ motion measured using 4D CT. Int. J. Radiat. Oncol. Biol. Phys. **65**, 554–560 (2006)
4. Atoui, H., Miguet, S., Sarrut, D.: A fast morphing-based interpolation for medical images: application to conformal radiotherapy. Image Anal. Stere. **25**, 95–103 (2011)
5. Fernandez, J.W., Mithraratne, P., Thrupp, S.F., Tawhai, M.H., Hunter, P.J.: Anatomically based geometric modelling of the musculo-skeletal system and other organs. Biomech. Model. Mechanobiol. **2**, 139–155 (2004)
6. Ho, H., Bartlett, A., Hunter, P.: Geometric modelling of patient-specific hepatic structures using cubic hermite elements. In: Yoshida, H., Sakas, G., Linguraru, M.G. (eds.) ABD-MICCAI 2011. LNCS, vol. 7029, pp. 264–271. Springer, Heidelberg (2012). doi:10.1007/978-3-642-28557-8_33
7. Haelterman, R., Degroote, J., Van Heule, D., Vierendeels, J.: The quasi-newton least squares method: a new and fast secant method analyzed for linear systems. SIAM J. Numer. Anal. **47**, 2347–2368 (2009)
8. Kolr, P., Neuwirth, J., Sanda, J., Suchnek, V., Svat, Z., Volejnk, J., Pivec, M.: Analysis of diaphragm movement during tidal breathing and during its activation while breath holding using MRI synchronized with spirometry. Physiol. Res. **58**, 383–392 (2009)
9. Farrer, A.I., Oden, H., de Bever, J., Coats, B., Parker, D.L., Payne, A., Christensen, D.A.: Characterization and evaluation of tissue-mimicking gelatin phantoms for use with MRgFUS. J. Ther. Ultrasound **3** (2015). doi:10.1186/s40349-015-0030-y
10. Bro-Nielsen, M.: Finite element modeling in surgery simulation. Proc. IEEE **86**, 490–503 (1998)

Author Index